Lecture Notes in Computer Science 10425

Commenced Publication in 1973
Founding and Former Series Editors:
Gerhard Goos, Juris Hartmanis, and Jan van Leeuwen

Editorial Board

More information about this series at http://www.springer.com/series/7412

Michael Felsberg · Anders Heyden
Norbert Krüger (Eds.)

Computer Analysis
of Images and Patterns

17th International Conference, CAIP 2017
Ystad, Sweden, August 22–24, 2017
Proceedings, Part II

 Springer

Editors
Michael Felsberg
Linköping University
Linköping
Sweden

Anders Heyden
Lund University
Lund
Sweden

Norbert Krüger
University of Southern Denmark
Odense
Denmark

ISSN 0302-9743   ISSN 1611-3349   (electronic)
Lecture Notes in Computer Science
ISBN 978-3-319-64697-8   ISBN 978-3-319-64698-5   (eBook)
DOI 10.1007/978-3-319-64698-5

Library of Congress Control Number: 2017947818

LNCS Sublibrary: SL6 – Image Processing, Computer Vision, Pattern Recognition, and Graphics

Printed on acid-free paper

This Springer imprint is published by Springer Nature
The registered company is Springer International Publishing AG
The registered company address is: Gewerbestrasse 11, 6330 Cham, Switzerland

Preface

We are very happy to present the contributions accepted for the 17th international Conference on Computer Analysis of Images and Patterns (CAIP 2017), which was held at Ystad Saltsjöbad, Ystad, Sweden, August 22–24.

CAIP 2017 was the 17th in the biennial series of conferences, which is devoted to all aspects of computer vision, image analysis and processing, pattern recognition, and related fields. Previous conferences were held, for instance, in Valletta, York, Seville, Münster, Vienna, Paris, etc. The contributions for CAIP 2017 were carefully selected based on a minimum of two but mostly three reviews. Among 144 submissions 72 were accepted, leading to an acceptance rate of 50%.

The conference included a tutorial on "Pose Estimation" by Anders G. Buch and a workshop on "Recognition and Action for Scene Understanding" (REACTS). Three keynote talks provided by world-renowned experts in the area of robotics (Markus Vincze), machine learning (Christian Igel), and image and video processing (Alan Bovik) were additional highlights.

The program covered high-quality scientific contributions in 2D-to-3D, 3D vision, biomedical image and pattern analysis, biometrics, brain-inspired methods, document analysis, face and gestures, feature extraction, graph-based methods, high-dimensional topology methods, human pose estimation, image/video indexing and retrieval, image restoration, keypoint detection, machine learning for image and pattern analysis, mobile multimedia, model-based vision, motion and tracking, object recognition, segmentation, shape representation and analysis, and vision for robotics.

CAIP has a reputation of providing a friendly and informal atmosphere, in addition to high-quality scientific contributions. We focused on maintaining this reputation, by designing a stimulating technical and social program that was hopefully inspiring for new research ideas and networking. We also hope that the venue at Ystad Saltsjöbad contributed to a fruitful conference.

We thank the authors for submitting their valuable work to CAIP. This is of course of prime importance for the success of the event. However, the organization of a conference also depends critically on a number of volunteers. We would like to thank the reviewers and the Program Committee members for their excellent work. We also thank the local Organizing Committee and all the other volunteers who helped us organize CAIP 2017.

We hope that all participants had a joyful and fruitful stay in Ystad.

August 2017

Michael Felsberg
Anders Heyden
Norbert Krüger

Organization

CAIP 2017 was organized by the Computer Vision Laboratory, Department of Electrical Engineering, Linköping University, Sweden.

Executive Committee

Conference Chair

Michael Felsberg   Linköping University, Sweden

Program Chairs

Anders Heyden   Lund University, Sweden
Norbert Krüger   University of Southern Denmark, Denmark

Program Committee

Muhammad Raza Ali, Pakistan
Furqan Aziz, Pakistan
Andrew Bagdanov, Spain
Donald Bailey, New Zealand
Antonio Bandera, Spain
Ardhendu Behera, UK
Michael Biehl, The Netherlands
Adrian Bors, UK
Kerstin Bunte, UK
Kenneth Camilleri, Malta
Kwok-Ping Chan, China
Rama Chellappa, USA
Dmitry Chetverikov, Hungary
Guillaume Damiand, France
Carl James Debono, Malta
Joachim Denzler, Germany
Mariella Dimiccoli, Spain
Francisco Escolano, Spain
Taner Eskil, Turkey
Giovanni Maria Farinella, Italy
Gernot Fink, Germany
Patrizio Frosini, Italy
Eduardo Garea, Cuba
Daniela Giorgi, Italy
Rocio Gonzalez-Diaz, Spain

Cosmin Grigorescu, The Netherlands
Miguel Gutierrez-Naranjo, Spain
Michal Haindl, Czech Republic
Anders Heyden, Sweden
Atsushi Imiya, Japan
Xiaoyi Jiang, Germany
Maria-Jose Jimenez, Spain
Dakai Jin, USA
Martin Kampel, Austria
Nahum Kiryati, Israel
Reinhard Klette, New Zealand
Gisela Klette, New Zealand
Andreas Koschan, USA
Ryszard Kozera, Australia
Walter Kropatsch, Austria
Guo-Shiang Lin, Taiwan
Agnieszka Lisowska, Poland
Josep Llados, Spain
Rebeca Marfil, Spain
Manuel Marin, Spain
Heydi Mendez, Cuba
Eckart Michaelsen, Germany
Majid Mirmehdi, UK
Matthew Montebello, Malta
Radu Nicolescu, New Zealand

Mark Nixon, UK
Darian Onchis, Austria
Mario Pattichis, Cyprus
Nicolai Petkov, The Netherlands
Gianni Poggi, Italy
Pedro Real, Spain
Paul Rosin, UK
Samuel Rota Bulo, Italy
Robert Sablatnig, Austria
Alessia Saggese, Italy
Hideo Saito, Japan
Albert Salah, Turkey
Lidia Sanchez, Spain
Angel Sanchez, Spain
Gabriella Sanniti di Baja, Italy

Sudeep Sarkar, USA
Klamer Schutte, The Netherlands
Giuseppe Serra, Italy
Francesc Serratosa, Spain
Antonio Jose Sanchez Salmeron, Spain
Akihiro Sugimoto, Japan
Bart ter Haar Romeny, The Netherlands
Bernie Tiddeman, UK
Klaus Toennies, Germany
Ernest Valveny, Spain
Thomas Villmann, Germany
Michael Wilkinson, The Netherlands
Richard Wilson, UK
Christian Wolf, France
Weiqi Yan, New Zealand

Invited Speakers

Markus Vincze   Technische Universität Wien, Austria
Christian Igel   University of Copenhagen, Denmark
Alan Bovik   The University of Texas at Austin, USA

Tutorial

Anders G. Buch   University of Southern Denmark, Denmark

Additional Reviewers

M. Al-Sarayreh   F. Hagelskjær   M.A. Pascali
F. Castro   J. Hilty   P. Radeva
F. Diaz-del-Rio   C. Istin   N. Saleem
J. Dong   T.B. Jørgensen   S. Sudholt
A. Eldesokey   F. Kahn   J. Toro
R. Grzeszick   S.F. Noor   M. Wallenberg
Q. Gu   D. Onchis   S. Zappala

Sponsoring Institutions

Computer Vision Laboratory, Department of Electrical Engineering, Linköping University
Swedish Society for Automated Image Analysis
The Swedish Research Council
Springer Lecture Notes in Computer Science
And all our industrial sponsors listed in the conference program booklet

Contents – Part II

Biometrics

Machine Learning

Image Restoration I

Poster Session III

Image Restoration II

Contents – Part I

Image/Video Indexing and Retrieval

Shape Representation and Analysis

Biomedical Image Analysis

Poster Session II

A New Scoring Method for Directional Dominance in Images

Bilge Suheyla Akkoca-Gazioglu$^{(\boxtimes)}$ and Mustafa Kamasak

Faculty of Computer and Informatics Engineering,
Istanbul Technical University, Istanbul, Turkey
{bakkoca,kamasak}@itu.edu.tr

Abstract. We aim to develop a scoring method for expressing directional dominance in the images. It is predicted that this score will give an information of how much improvement in system performance can be achieved when using a directional total variation (DTV)-based regularization instead of total variation (TV). For this purpose, a dataset consists of 85 images taken from the noise reduction datasets is used. The DTV values are calculated by using different sensitivities in the direction of the directional dominance of these images. The slope of these values is determined as the directional dominance score of the image. To verify this score, the noise reduction performances are examined by using direction invariant TV and DTV regulators of images. As a result, we observe that the directional dominance score and the improvement rate in noise reduction performance are correlated. Therefore, the resulting score can be used to estimate the performance of DTV method.

Keywords: Total Variation · Directed Total Variation · Directional Dominance Score

1 Introduction

Total variation regularization is most commonly used technique in digital image processing while reducing noises. Extreme and fake details in the images possess very high total variation. It means that the gradient of those details is very high. Hence, some regularizations on total variation should be handled in order to obtain the original details of the images.

Total Variation (TV) was first described by Rudin, Osher and Fatemi in 1992 as a regulatory criterion that could be used to solve inverse problems [1]. It was stated that this method is a very efficient method for rearranging the images without smoothing the boundaries of the objects. TV method is seen as one of the most successful image manipulation approaches that can effectively improve the edges of the image.

Some advantages can be acquired by denoising with Total Variation method especially compared to linear smoothing or median filtering techniques. Total variation denoising softens noises in flat regions, even at low signal-to-noise ratios, with remarkable effect while maintaining the edges at the same time [2].

© Springer International Publishing AG 2017
M. Felsberg et al. (Eds.): CAIP 2017, Part II, LNCS 10425, pp. 3–13, 2017.
DOI: 10.1007/978-3-319-64698-5_1

TV is commonly used in different image processing applications in the literature. We can give some examples of these applications areas: noise reduction [1–5] reconstruction from sparse samples [6], deblurring [7,8].

TV structure is an isotropic regulator, as described in Sect. 2.1. In image processing applications, it is generally chosen to reduce the TV value to the minimum within possible solutions. In some cases which there is no prior knowledge of the structure of the image, it is quite appropriate to use a direction invariant regulator. However, in some image types, objects in the image content may have a dominant direction. It seems more plausible to use an anisotropic regulator which is more sensitive to changes in certain directions instead of using a direction invariant regulator on such images. It seems that these kinds of directional regulators have begun to be used in existing studies [5,9].

In a study discussed in 2015, it is proposed a weighted difference of direction invariant and direction variant total variation values [9] It can be used as a regulator for and image processing studies based on the natural image statistics and TV model. The method developed in the study is tried on the noise reduction, deblurring and reconstruction of the magnetic resonance images and gives better performance than the TV model.

Yan and Lu generalize the power of TV to $0 \leq p \leq 1$ to TV to increase the gradient sparseness [5]. They also propose a generalized TV method that regulates the least squares method to remove noise from the images. In this study, it is aimed to approve solution by generalized TV method by weighting each item of TV which is discretized in horizontal and vertical directions.

Regulators based on Directional total variation (DTV) gain better performance compared to TV as the directional dominance in the images increases. However, when related studies are examined, there is no study on the directional dominance of an image. This study aims to develop a scoring method that shows the directional dominance in the images. This score is intended to be an indication of how much DTV will perform compared to TV. Therefore, the scores obtained by the scoring method developed within the scope of the study have been verified using the performance ratios obtained with the DTV and TV regulators for noise reduction application.

2   Methodology

First of all, the definitions of TV and DTV will be made in this section. It will then be explained how the direction dominance score will be obtained and how it will be validated.

2.1   Total Variation (TV)

The TV value of an image (f) is defined as follows.

$$TV(f) = \frac{1}{|f|} \sum_{i,j} \sqrt{(\Delta_1 f(i,j))^2 + (\Delta_2 f(i,j))^2} \tag{1}$$

In this equation, $|f|$ is the number of pixels in the (f) image; $f(i, j)$ is the pixel at the (i, j) location and Δ_1 and Δ_2 are the horizontal and vertical difference operators, respectively. These operators can be calculated as

$$\Delta_1 f(i, j) = f(i, j) - f(i - 1, j) \tag{2}$$

$$\Delta_2 f(i, j) = f(i, j) - f(i, j - 1). \tag{3}$$

If TV's Eq. (1) is examined, it is seen that the roots of the squares of horizontal and vertical differences of the image are summed. That is, no distinction or weighting between horizontal and vertical differences is made. For this reason, TV considers the changes in all directions to be equally independent, regardless of the direction. For this reason, TV is regarded as a direction invariant regulator.

2.2 Directional Total Variation (DTV)

Especially in the images we know to have a predominant direction, this method is used for noise reduction [1] and reconstruction from sparse samples [10] applications because it is thought that using a more sensitive regulator against these changes will give better performance. In these applications, it was observed that the DTV gives better results than the TV regulator when the images have the dominant direction [1,10]. In DTV Eqs. (2) and (3), Δf_1 and Δf_2 values are rotated at a certain angle, the rotated values are weighted, and the value per image element is calculated. Hence, DTV is calculated as

$$DTV(f, \alpha, \theta) = \frac{1}{|f|} \sum_{i,j} \sqrt{b_{1,\alpha,\theta}(i, j)^2 + b_{2,\alpha,\theta}(i, j)^2} \tag{4}$$

$b_{1,\alpha,\theta}(i, j)$ and $b_{2,\alpha,\theta}(i, j)$ values in the equation are calculated as the follows:

$$\begin{bmatrix} b_{1,\alpha,\theta}(i, j) \\ b_{2,\alpha,\theta}(i, j) \end{bmatrix} = \begin{bmatrix} \alpha \\ 1 \end{bmatrix}^T \begin{bmatrix} \cos(\theta) & -\sin(\theta) \\ \sin(\theta) & \cos(\theta) \end{bmatrix} \begin{bmatrix} \Delta_1 f(i, j) \\ \Delta_2 f(i, j) \end{bmatrix} \tag{5}$$

In DTV equation, θ defines the angle and α defines the precision of variation in the rotated direction. As, the α value increases, the contribution of the changes (edges) in the rotated direction increases. Therefore, (for $\alpha > 1$) in image processing applications, where DTV is used as a regulator, the solution that minimizes the changes in θ direction will be preferred.

In Fig. 1(a), there is a notebook image. In this image, a predominant directional dominance is observed in the horizontal direction. The DTV values of the notebook image are given for $\alpha = 5$ and different θ directions in Fig. 1(b). As it can be seen from this figure, the DTV value is lowest at 0°. In Fig. 1(c), the DTV values of the notebook image are given for $\theta = 0°$ directions and different α values. As expected, the value of DTV increases with the increase of α value. However, due to the dominance in the horizontal direction, the increase due to α value is lower than in other images. This is shown in the next section (Sect. 3).

Another subject to note here is that DTV and TV have the same result when α is 1. Therefore, it can be considered that TV is a special case of the DTV.

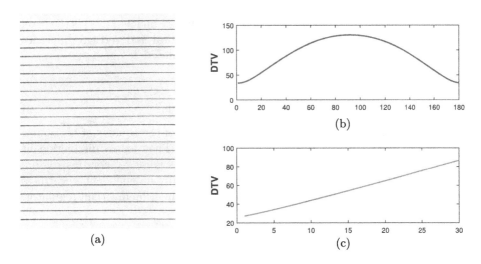

Fig. 1. (a) Notebook image. (b) The DTV values of the notebook image for $\alpha = 5$ at different θ directions. (c) The DTV values of the notebook image for $\theta = 0°$ directions at different α values.

2.3   Directional Dominance Score

It is predicted that an image with a high directional dominance will not be very sensitive to the increasing α coefficient of variation in the dominant direction. Thus, with the increase of α, it is expected that the DTV value will increase more rapidly for an image which does not have dominant direction (than image with dominant direction).

For this reason, by calculating the DTV values of an image for increasing α values, the rate of increase of DTV values is observed. In order to determine the rate of increase as a score, a slope of a straight line fitted to the DTV values obtained by different α values was calculated. Using this slope, a directional dominance score was obtained for the images.

$$m(f) = \arg\min_{m,b} \sum_i (DTV(f, \alpha_i, \hat{\theta}) - (m\alpha_i + b))^2 \tag{6}$$

$\hat{\theta}$ specifies the dominant direction in the image, namely, the direction with the minimum DTV value.

Nevertheless, $m(f, \theta)$ slope is decreasing with the dominant direction. On the other hand, it seems more appropriate to increase the directional dominance score with the directional dominance of the image. For this reason, the directional dominance score is used as the inverse of the slope instead of the direct slope. The direction dominance score is scaled and calculated as:

$$s(f) = \frac{1}{m(f)} \tag{7}$$

2.4   The Verification of Directional Dominance Score

We use 4 different images as in the Fig. 2 to verify the obtained score. It is observed subjectively that the directional dominances of these images are at different levels.

Fig. 2. The images at different directional dominance levels.

The purpose of developing this score is to estimate how much performance can be achieved with DTV, which is used as a regulator, compared to TV in image processing applications. For this reason, Gaussian noises are added to these images, and noise reduction is performed using TV and DTV regulators. It is assumed that the dominant direction of images is known as preliminary information. Noise reduction has been performed as shown below, as detailed in [3].

$$\hat{f}^{TV} = \arg\min_{f} \sum_{i,j} (f^g(i,j) - f(i,j))^2 + \lambda TV(f) \qquad (8)$$

$$\hat{f}^{DTV} = \arg\min_{f} \sum_{i,j} (f^g(i,j) - f(i,j))^2 + \lambda DTV(f, \alpha, \hat{\theta}) \qquad (9)$$

where, f^g shows the observed noisy image, $\hat{f}^{TV}$ and $\hat{f}^{DTV}$ are the noise-free images by using TV and DTV regulators respectively.

Root Mean Squared Error (RMSE) is used as the measure of performance of the noise reduction.

$$RMSE(f)^{TV/DTV} = \frac{1}{|f|} \sum_{i,j} \sqrt{(f(i,j) - \hat{f}^{TV/DTV}(i,j))^2} \qquad (10)$$

Improvement in performance achieved with DTV compared to TV is obtained as a percentage. This improvement is calculated as:

$$I(f) = \frac{RMSE^{DTV} - RMSE^{TV}}{RMSE^{TV}} \times 100 \tag{11}$$

where $RMSE^{DTV}$ and $RMSE^{TV}$ represents the root mean square error values which is obtained by the DTV and TV methods in Eq. (10) in the noise reduction application respectively. The correlation between the score obtained for the 4 images shown in Fig. 2 and the improvement ratios obtained with the Eq. (11) for verifying the directional dominance score was examined.

3   Experimental Studies

3.1   Data Set

We propose a new scoring method for the images in this study. We call this score as Directional Dominance Score. For this aim, we use the dataset which is used in TV denoising studies. It has 85 images with different directional dominances. Some images in the dataset can be seen in the Fig. 3.

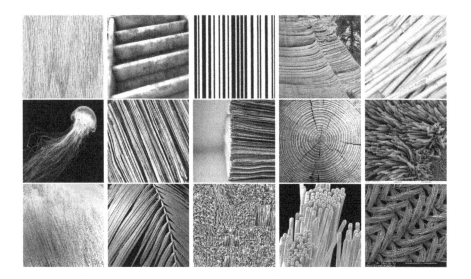

Fig. 3. Some selected images from dataset.

3.2   Experimental Results

In this study, determining a score for the image is discussed about the determination of the total variation in an image. For this purpose, this method has been applied to the images in the dataset discussed in the study which is used the total variance method for noise reduction [3].

The noise of Gaussian distribution with the average value ($\mu = 0$) and the variance value ($\sigma^2 = 0.04$) is added to the images with the lowest and highest score values. After that, the noise reduction method described in [3] is applied with different α values by using the cost function given by the DTV regulator as in the Eq. (9).

The images obtained after reducing the noises with application of TV and DTV methods to the notebook image. As can be seen in the Fig. 4, the noise reduction method with DTV gives very successful results.

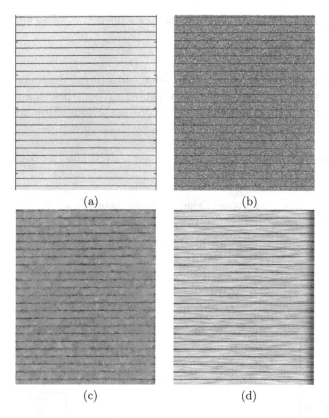

Fig. 4. (a): The original notebook image without any noise. (b): The noise with Gaussian distribution is added to notebook image. (c): The noise free image with TV Regulator. (d): The noise free image with DTV Regulator with the parameters of $\alpha = 5$ and $\theta = 88$.

This method has been applied to images with different patterns and the changes in DTV values are interpreted by the obtained graphics. These images and graphs of the images are shown in Figs. 2 and 5.

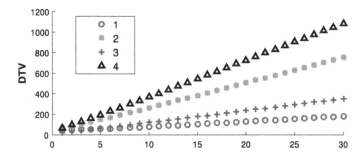

Fig. 5. The graph of DTV values corresponding to different α values for the images in Fig. 2.

The angle with the smallest DTV value, directional dominance score and the slope for the 4 images are given in Table 1. When these values are examined, the obtained directional dominance scores seem compatible with the visual evaluations made.

Table 1. Directional dominance angle, slope and score values of 4 images are shown in Fig. 2.

Image	Slope	Angle	Score
1	0,01892	89°	52,84
2	0,09435	13°	10,59
3	0,04472	40°	22,36
4	0,13886	70°	7,20

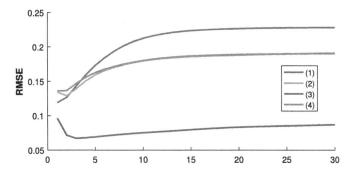

Fig. 6. The RMSE values of noise reduction results by using DTV regulators with different α values for all images. The RMSE value at $\alpha = 1$ is the result of using TV regulator.

RMSE performances of the results obtained with different values of α were calculated. The RMSE values obtained as a result of applying the same process to all the images are shown in Fig. 6. In this figure, the RMSE value of $\alpha = 1$ indicates the noise removal performance obtained using the TV regulator.

The average and the standard deviation of improvement ratio and the directional dominance score are shown in Table 2 when the noise reduction algorithm is applied to each image. As analyzed from the table, there is a correlation between the improvement ratios and the directional dominance scores of the images (Fig. 7).

Table 2. The directional dominance scores, average improvement ratios and standard deviations of 4 images when the algorithm is applied with different σ values.

Image	Score $m(f)$	$I(f) \pm$ std
1	52,84	$-35,67 \pm 9,16$
2	10,59	$19,46 \pm 11,72$
3	22,36	$30,50 \pm 12,34$
4	7,20	$21,15 \pm 10,89$

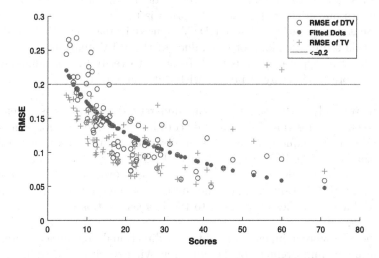

Fig. 7. The graphic shows the median of RMSE values of DTV and RMSE values of TV versus scope values of each 85 images. The purple line at 0.2 indicates that Gaussian noise with $\sigma = 0.2$ is added to each image and the blue dots show the fitted model of RMSE of DTV values. (Color figure online)

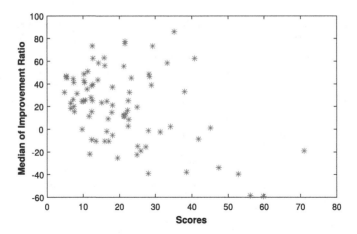

Fig. 8. The graphic shows the median of improvement ratios versus scope values of each 85 images.

4   Conclusion

In this study, a scoring system is proposed to define the directional dominance in the images numerically. The aim of scoring is to estimate how much better improvement can be achieved when DTV is used instead of TV. In order to calculate this score, the slope of the values of the DTV quantity of an image over the different α parameters is used. In order to verify the usefulness of the calculated score in this way, 4 images with different directional dominance are selected.

When we analyze the results we can understand that there is a relationship between the directional dominance score and the rate of improvement in the RMSE of noise reduction. Negative improvement values shown in the Fig. 8 indicate that when the DTV regulator is used (according to the result obtained by TV), the RMSE is decreasing, and the positive improvement values indicate the increasing RMSE values. When the contents of these images are examined, objects of different angles are caused to the increase of this error especially in the case of θ.

For the 85 different images, the obtained directional dominance scores and the performance improvement of the noise removing results of these images are compared. It is observed that this comparison is linearly correlated to the recovery rate with directional dominance score. Therefore, it can be resulted that developed scoring system can be used to estimate the performance of DTV.

As a future work, it is planing that the direction dominance score will be experimented on different datasets with more images. Furthermore, different image processing applications will be applied by getting more specific scores.

Acknowledgment. This study is partially supported by The Scientific and Technological Research Council of Turkey with the grant number 115R285.

References

1. Rudin, L.I., Osher, S., Fatemi, E.: Nonlinear total variation based noise removal algorithms. Physica D **60**, 259–268 (1992)
2. Strong, D., Chan, T.: Edge-preserving and scale-dependent properties of total variation regularization. Inverse Prob. **19**, S165–S187 (2003)
3. Bayram, İ., Kamasak, M.E.: Directional total variation. IEEE Signal Process. Lett. **19**(12), 781–784 (2012)
4. Beck, A., Teboulle, M.: Fast gradient-based algorithms for constrained total variation image denoising and deblurring problems. IEEE Trans. Image Process. **18**, 2419–2434 (2009)
5. Yan, J., Lu, W.S.: Wu-Sheng: Image denoising by generalized total variation regularization and least squares fidelity. Multidimension. Syst. Signal Process. **26**(1), 243–266 (2015)
6. Ritschl, L., Bergner, F., Fleischmann, C., Kachelrieß, M.: Improved total variation-based CT image reconstruction applied to clinical data. Phys. Med. Biol. **56**(6), 1545–1561 (2011)
7. Wang, Y., Yin, W., Zhang, Y.: A fast algorithm for image deblurring with total variation regularization (2007)
8. Liu, H., Gu, J., Huang, C.: Image deblurring by generalized total variation regularization and least squares fidelity. In: IEEE International Conference on Information and Automation (ICIA), pp. 1945–1949 (2016)
9. Lou, Y., Zeng, T., Osher, S., Xin, J.: A weighted difference of anisotropic and isotropic total variation model for image processing. SIAM J. Imaging Sci. **8**(3), 1798–1823 (2015)
10. Demircan-Tureyen, E., Kamasak, M.E., Bayram, I.: Image reconstruction from sparse samples using directional total variation minimization. In: Proceedings of the 24th IEEE Signal Processing and Applications Conference (2016)

A Multilayer Backpropagation Saliency Detection Algorithm Based on Depth Mining

Chunbiao Zhu[1], Ge Li[1(✉)], Xiaoqiang Guo[2], Wenmin Wang[1], and Ronggang Wang[1]

[1] School of Electronic and Computer Engineering, Shenzhen Graduate School,
Peking University, Shenzhen, China
zhuchunbiao@pku.edu.cn, geli@ece.pku.edu.cn
[2] Academy of Broadcasting Science, SAPPRFT, Beijing, China

Abstract. Saliency detection is an active topic in multimedia field. Several algorithms have been proposed in this field. Most previous works on saliency detection focus on 2D images. However, for some complex situations which contain multiple objects or complex background, they are not robust and their performances are not satisfied. Recently, 3D visual information supplies a powerful cue for saliency detection. In this paper, we propose a multilayer backpropagation saliency detection algorithm based on depth mining by which we exploit depth cue from four different layers of images. The evaluation of the proposed algorithm on two challenging datasets shows that our algorithm outperforms state-of-the-art.

Keywords: Saliency detection · Depth cue · Depth mining · Multilayer · Backpropagation

1 Introduction

Salient object detection is a process of getting a visual attention region precisely from an image. The attention is the behavioral and cognitive process of selectively concentrating on one aspect within the environment while ignoring other things.

Early work on computing saliency aims to locate the visual attention region. Recently the field has extended to locate and refine the salient regions and objects. Many saliency detection algorithms have been used as a useful tool in the pre-processing, such as image retrieval [1], object recognition [2], object segmentation [3], compression [4], image retargeting [5], etc.

In general, saliency detection algorithm mainly use top-down or bottom-up approaches. Top-down approaches are task-driven and need supervised learning. While bottom-up approaches usually use low-level cues, such as color features, distance features, depth features and heuristic saliency features. One of the most used heuristic saliency feature [6–10] is contrast, such as pixel-based or patch-based contrast, region-based contrast, multi-scale contrast, center-surround contrast, color spatial compactness, etc. Although those methods have their own advantages, they are not robust to specific situations which lead to inaccuracy of results on challenging salient object detection datasets.

© Springer International Publishing AG 2017
M. Felsberg et al. (Eds.): CAIP 2017, Part II, LNCS 10425, pp. 14–23, 2017.
DOI: 10.1007/978-3-319-64698-5_2

To deal with the challenging scenarios, some algorithms [11–15] adopt depth cue. In [11], Zhu et al. propose a framework based on cognitive neuroscience, and use depth cue to represent the depth of real field. In [12], Cheng et al. compute salient stimuli in both color and depth spaces. In [13], Peng et al. provide a simple fusion framework that combines existing RGB-produced saliency with new depth-induced saliency. In [14], Ju et al. propose a saliency method that works on depth images based on anisotropic center-surround difference. In [15], Guo et al. propose a salient object detection method for RGB-D images based on evolution strategy. Their results show that stereo saliency is a useful consideration compare to previous visual saliency analysis. All of them demonstrate the effectivity of depth cue in improvement of salient object detection.

Although, those approaches can enhance salient object region. It is very difficult to produce good results when a salient object has low depth contrast compared to the background. The behind reason is that only partial depth cue is applied. In this paper, we propose a multilayer backpropagation algorithm based on depth mining to improve the performance.

2   Proposed Algorithm

The proposed algorithm is a multilayer backpropagation method based on depth mining of an image. In the preprocessing layer, we obtain the center-bias saliency map and depth map. In the first layer, we use original depth cue and other cues to calculate preliminary saliency value. In the second layer, we apply processed depth cue and other cues to compute intermediate saliency value. In the third layer, we employ reprocessed depth cue and other cues to get final saliency value. The framework of the proposed algorithm is illustrated in Fig. 1.

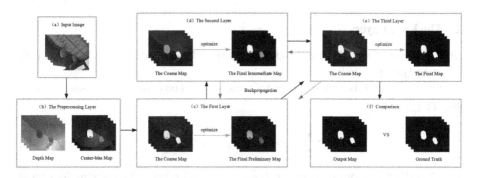

Fig. 1. The framework of the proposed algorithm.

2.1   The Preprocessing Layer

In the preprocessing layer, we imitate the human perception mechanism to obtain center-bias saliency map and depth map.

Center-bias Saliency Map. Inspired by cognitive neuroscience, human eyes use central fovea to locate object and make them clearly visible. Therefore, most of images taken by cameras always locate salient object around the center. Aiming to get centerbias saliency map, we use BSCA algorithm [6]. It constructs global color distinction and spatial distance matrix based on clustered boundary seeds and integrate them into a background-based map. Thus it can improve the center-bias, erasing the image edge effect. As shown in the preprocessing stage of Fig. 2 (c), the center-bias saliency map can remove the surroundings of the image and reserve most of salient regions. We denote this center-bias saliency map as C_b.

(a) I_o (b) I_d (c) C_b (d) $\neg I_d$ (e) I_e (f) I_{df} (g) I_{ef}

(h) S_{fc} (i) S_1 (j) S_{sc} (k) S_2 (l) S_{tc} (m) S (n) GT

Fig. 2. The visual process of the proposed algorithm.

Depth Map. Similarly, biology prompting shows that people perceive the distance and the depth of the object mainly relies on the clues provided by two eyes, and we call it binocular parallax. Therefore, the depth cue can imitate the depth of real field. The depth map used in the experimental datasets is taken by Kinect device. And we denote the depth map as I_d.

2.2   The First Layer

In the first layer, we extract color and depth features from the original image I_o and the depth map I_d, respectively.

First, the image I_o is segmented into K regions based on color via the K-means algorithm. Define:

$$S_c(r_k) = \sum_{i=1,i\neq k}^{K} P_i W_s(r_k) D_c(r_k, r_i), \tag{1}$$

where $S_c(r_k)$ is the color saliency of region k, $k \in [1, K]$, r_k and r_i represent regions k and i respectively, $D_c(r_k, r_i)$ is the Euclidean distance between region k and region i in L*a*b color space, P_i represents the area ratio of region r_i compared with the whole image, $W_s(r_k)$ is the spatial weighted term of the region k, set as:

$$W_s(r_k) = e^{-\dfrac{D_o(r_k, r_i)}{\sigma^2}}, \tag{2}$$

where $D_o(r_k, r_i)$ is the Euclidean distance between the centers of region k and i, σ is the parameter controlling the strength of $W_d(r_k)$.

Similar to color saliency, we define:

$$S_d(r_k) = \sum_{i=1, i \neq k}^{K} P_i W_s(r_k) D_d(r_k, r_i), \tag{3}$$

Where $S_d(r_k)$ is the depth saliency of I_d, $D_d(r_k, r_i)$ is the Euclidean distance between region k and region i in depth space.

In most cases, a salient object always locate at the center of an image or close to a camera. Therefore, we add the weight considering both center-bias and depth for both color and depth images. The weight of the region k is set as:

$$S_s(r_k) = \frac{G(\|P_k - P_0\|)}{N_k} W_d(d_k), \tag{4}$$

where G(•)represents the Gaussian normalization, $\|\cdot\|$ is Euclidean distance, P_k is the position of the region k, P_o is the center position of this map, N_k is the number of pixels in region k, W_d is the depth weight, which is set as:

$$W_d = (max\{d\} - d_k)^\mu, \tag{5}$$

where $max\{d\}$ represents the maximum depth of the image, and d_k is the depth value of region k, μ is a fixed value for a depth map, set as:

$$\mu = \frac{1}{max\{d\} - min\{d\}}, \tag{6}$$

where $min\{d\}$ represents the minimum depth of the image.

Second, the coarse saliency value of the region k is calculated as:

$$S_{fc}(r_k) = G(S_c(r_k) S_s(r_k) + S_d(r_k) S_s(r_k)), \tag{7}$$

Third, to refine the salient detection results, we optimize the coarse saliency map with the help of the center-bias and depth maps. The preliminary saliency map is calculated as following:

$$S_1(r_k) = S_{fc}(r_k) \neg I_d(r_k) C_b(r_k), \tag{8}$$

where $\neg$ is the negation operation which can enhance the saliency degree of front regions as shown in Fig. 2(d), because the foreground object has low depth value in depth map while the background object has high depth value.

2.3 The Second Layer

In the second layer, we exploit depth map further, we allocate the color values to the depth map according to different depth values, in this way, we can polarize the color attribute between foreground and background.

First, we set:

$$I_e \langle R|G|B \rangle = I_o \langle R|G|B \rangle \times I_d, \tag{9}$$

where I_e represents the extended color depth map. $\langle R|G|B \rangle$ represents processing of three RGB channels, respectively.

The extended color depth map is displayed in Fig. 2(e), from which the salient objects' edges are prominent.

Second, we use extended color depth map I_e to replace I_o. Then, we calculate intermediate coarse saliency value S_{sc} via the first stage' Eq. (1)–(6). We get:

$$S_{sc}(r_k) = G(S_c(r_k)S_s(r_k) + S_d(r_k)S_s(r_k)), \tag{10}$$

where $S_{sc}(r_k)$ is the intermediate saliency value.

Third, to refine coarse saliency value, we apply the backpropagation to enhance the intermediate saliency value by mixing the result of the first layer. And we define our intermediate saliency value as:

$$S_2(r_k) = S_1^2(r_k) + S_1(r_k)\left(1 - e^{-S_{sc}^2(rk) \, \neg I_d(r_k)}\right) \tag{11}$$

2.4 The Third Layer

In the third layer, we find that background noises can reduced by filtering the depth map, so, we exploit the depth map again.

First, we reprocess the depth cue by filtering the depth map via the following formula:

$$I_{df} = \begin{cases} I_d, & d \le \beta \times \max\{d\} \\ 0, & d > \beta \times \max\{d\}, \end{cases} \tag{12}$$

where I_{df} represents the filtered depth map. In general, salient objects always have the small depth value compared to background, thus, by Eq. (12), we can filter out the background noises. β is the parameter which controls the strength of I_{df}.

Second, we extend the filtered depth map to the color images via the Eq. (9). We denote the reprocessed depth map as I_{ef}.

We use filtered depth map I_{ef} to replace I_o. Then, we calculate third coarse saliency map S_{tc} via the first stage' Eq. (1)–(6), denoted as:

$$S_{tc}(r_k) = G(S_c(r_k)S_s(r_k) + S_d(r_k)S_s(r_k)), \tag{13}$$

Fourth, to refine $S_{tc}(r_k)$, we apply the backpropagation of $S_1(r_k)$ and $S_2(r_k)$ as following:

$$S(r_k) = S_2(r_k)\left(S_2(r_k) + S_{tc}(rk)\right)\left(S_{tc}(r_k) + 1 - e^{-S_{tc}^2(r_k)S_1(r_k)}\right). \tag{14}$$

From the Fig. 2, we can see the visual results of the proposed algorithm. The main steps of the proposed salient object detection algorithm are summarized in Algorithm 1.

Algorithm 1

Input: original map I_o, depth map I_d

Output: final saliency value $S(r_k)$

1: **for** each region $k = [1, K]$ **do**

2: compute color saliency value $S_c(r_k)$and depth saliency value $S_d(r_k)$

3: calculate the center-bias and depth weights $W_{cd}(r_k)$

4: get the preliminary saliency value $S_1(r_k)$

5: **end for**

6: **repeat** step 1-5 for the extended depth maps I_e , then calculate$S_2(r_k)$

7: **repeat** step 1-5 for the filtered depth maps I_{ef} , then calculate $S_{tc}(r_k)$

8: calculate $S(r_k)$ by applying the backpropagation of $S_1(r_k)$ and $S_2(r_k)$

9: **return** the final saliency value $S(r_k)$

---

3   Experimental Evaluation

3.1   Datasets and Evaluation Indicators

Datasets. We evaluate the proposed saliency detection algorithm on two RGBD standard datasets: RGBD1* [12] and RGBD2* [13]. RGBD1* has 135 indoor images taken by Kinect with the resolution of 640×480. This dataset has complex backgrounds and irregular shapes of salient objects. RGBD2* contains 1000 images with two different resolutions of both 640×480 and 480×640, respectively.

Evaluation Indicators.   Experimental evaluations are based on standard measurements including precision-recall curve, ROC curve, MAE (Mean Absolute Error), F-measure. The MAE is formulated as:

$$\text{MAE} = \frac{\sum_{i=1}^{N} |GT_i - S_i|}{N}. \tag{15}$$

And the F-measure is formulated as:

$$F-measure = \frac{2 \times \text{Precision} \times \text{Recall}}{\text{Precision} + \text{Recall}}. \qquad (16)$$

3.2   Ablation Study

We first validate the effectiveness of each layer in our method: the first layer result, the second layer result and the third layer result. Table 1 shows the MAE and F-measure validation results on two datasets. We can clearly see the accumulated processing gains after each layer, and the final saliency result shows a good performance. After all, it proves that each layer in our algorithm is effective for generating the final saliency map.

Table 1. Validation results on two datasets.

	RGBD1* Dataset			RGBD2* Dataset		
Layers	S_1	S_2	S	S_1	S_2	S
MAE values	0.1065	0.0880	0.0781	0.1043	0.0900	0.0852
F-measure values	0.5357	0.6881	0.7230	0.5452	0.7025	0.7190

3.3   Comparison

To illustrate the effectiveness of our algorithm, we compare our proposed methods with DES14 [12], RSOD14 [13], BSCA15 [6], LPS15 [10], ACSD15 [14], HS16 [9] and SE16 [11]. We use the codes provided by the authors to reproduce their experiments. For all the compared methods, we use the default settings suggested by the authors. And for the Eq. (2), we take $\sigma^2 = 0.4$ which has the best contribution to the results.

The MAE and F-measure evaluation results on both RGBD1* and RGBD2* datasets are shown in Figs. 3 and 4, respectively. From the comparison results, it can be observed that our saliency detection method is superior and can obtain more precise salient regions than that of other approaches. Besides, the proposed algorithm is the most robust.

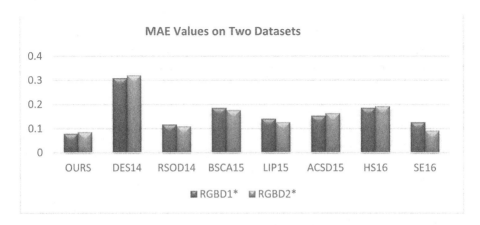

Fig. 3. The MAE results on two datasets. The lower value, the better performance.

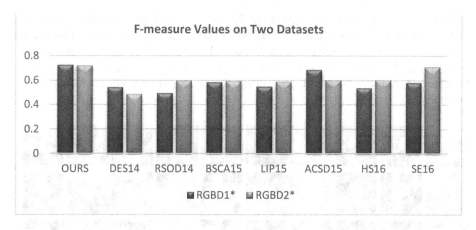

Fig. 4. The F-measure results on two datasets. The higher value, the better performance.

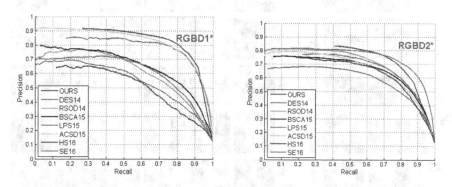

Fig. 5. From left to right: PR curve on RGBD1* dataset and PR curve on RGBD2* dataset.

The PR curve and ROC curve evaluation results are shown in Figs. 5 and 6, respectively. From the precision-recall curves and ROC curves, we can see that our saliency detection results can achieve better results on both RGBD1* and RGBD2* datasets.

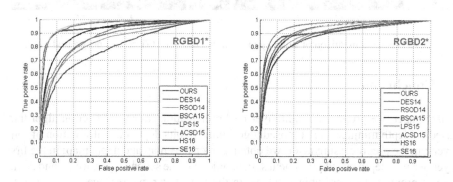

Fig. 6. From left to right: ROC curve on RGBD1* dataset and ROC curve on RGBD2* dataset.

The visual comparisons are given in Fig. 7, which clearly demonstrate the advantages of our method. We can see that our method can detect both single salient object and multiple salient objects more precisely. Besides, by intermediate results, it shows that by exploiting depth cue information of more layers, our proposed method can get more accurate and robust performance. In contrast, the compared methods may fail in some situations.

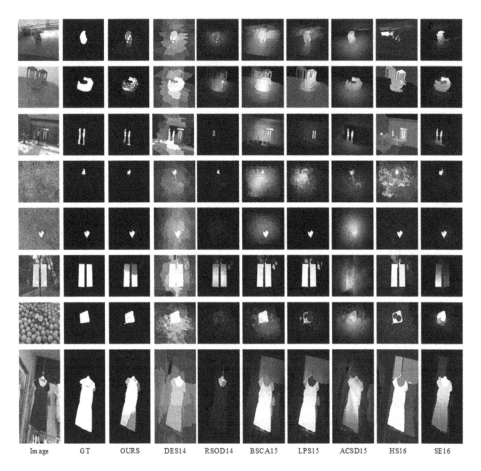

Image   GT   OURS   DES14   RSOD14   BSCA15   LPS15   ACSD15   HS16   SE16

Fig. 7. Visual comparison of saliency map on two datasets, GT represents ground truth.

4   Conclusion

In this paper, we proposed a multilayer backpropagation saliency detection algorithm based on depth mining. The proposed algorithm exploits depth cue information of four layers: in the preprocessing layer, we obtain center-bias map and depth map; in the first layer, we mix depth cue to prominent salient object; in the second layer, we extend depth map to prominent salient object' edges; in the third layer, we reprocess depth cue to eliminate background noises. And the experiments' results show that the proposed method outperforms the existing algorithms in both accuracy and robustness in different scenarios. To encourage future work, we make the source codes, experiment data and other related materials public. All these can be found on our project website: https://chunbiaozhu.github.io/CAIP2017/.

Acknowledgments. This work was supported by the grant of National Science Foundation of China (No.U1611461), the grant of Science and Technology Planning Project of Guangdong Province, China (No.2014B090910001), the grant of Guangdong Province Projects of 2014B010117007 and the grant of Shenzhen Peacock Plan (No.20130408-183003656).

References

1. Cheng, M.M., Mitra, N.J., Huang, X., et al.: SalientShape: group saliency in image collections. Vis. Comput. **30**(4), 443–453 (2014)
2. Alexe, B., Deselaers, T., Ferrari, V.: Measuring the objectness of image windows. IEEE Trans. Pattern Anal. Mach. Intell. **34**(11), 2189–2202 (2012)
3. Donoser, M., Urschler, M., Hirzer, M., et al.: Saliency driven total variation segmentation. In: IEEE International Conference on Computer Vision, pp. 817–824. IEEE (2009)
4. Sun, J., Ling, H.: Scale and object aware image retargeting for thumbnail browsing. In: IEEE International Conference on Computer Vision, pp. 1511–1518 (2011)
5. Itti, L.: Automatic foveation for video compression using a neurobiological model of visual attention. IEEE Trans. Image Process. **13**(10), 1304–1318 (2004)
6. Qin, Y., Lu, H., Xu, Y., et al.: Saliency detection via cellular automata. In: IEEE Conference on Computer Vision and Pattern Recognition, pp. 110–119. IEEE (2015)
7. Achanta, R., Hemami, S., Estrada, F., et al.: Frequency-tuned salient region detection. In: IEEE International Conference on Computer Vision and Pattern Recognition, pp. 1597–1604 (2009)
8. Murray, N., Vanrell, M., Otazu, X., et al.: Saliency estimation using a nonparametric low-level vision model. **42**(7), 433–440 (2011)
9. Shi, J., et al.: Hierarchical image saliency detection on extended CSSD. IEEE Trans. Patt. Anal. Mach. Intell. **38**(4), 717 (2016)
10. Li, H., Lu, H., Lin, Z., et al.: Inner and inter label propagation: salient object detection in the wild. IEEE Trans. Image Process. **24**(10), 3176–3186 (2015)
11. Zhu, C., Li, G., Wang, W., et al.: Salient object detection with complex scene based on cognitive. IEEE Third International Conference on Multimedia Big Data, pp. 33–37. IEEE (2017)
12. Cheng, Y., Fu, H., Wei, X., et al.: Depth enhanced saliency detection method. Eur. J. Histochem. Ejh **55**(1), 301–308 (2014)
13. Peng, H., Li, B., Xiong, W., Hu, W., Ji, R.: RGBD salient object detection: a benchmark and algorithms. In: Fleet, D., Pajdla, T., Schiele, B., Tuytelaars, T. (eds.) ECCV 2014. LNCS, vol. 8691, pp. 92–109. Springer, Cham (2014). doi:10.1007/978-3-319-10578-9_7
14. Ju, R., Liu, Y., Ren, T., et al.: Depth-aware salient object detection using anisotropic center-surround difference. Sig. Process. Image Commun. **38**(5), 115–126 (2015)
15. Guo, J., Ren, T., Bei, J.: Salient object detection for RGB-D image via saliency evolution. In: IEEE International Conference on Multimedia and Expo, pp. 1–6. IEEE (2016)

Laplacian Deformation with Symmetry Constraints for Reconstruction of Defective Skulls

Shudong Xie[1]([⊠]), Wee Kheng Leow[1], and Thiam Chye Lim[2,3]

[1] Department of Computer Science, National University of Singapore,
Singapore, Singapore
{xshudong,leowwk}@comp.nus.edu.sg

[2] Department of Surgery, National University of Singapore, Singapore, Singapore

[3] Division of Plastic, Reconstruction and Aesthetic Surgery,
National University Hospital, Singapore, Singapore
surlimtc@nuh.edu.sg

Abstract. Skull reconstruction is an important and challenging task in craniofacial surgery planning, forensic investigation and anthropological studies. Our previous method called FAIS (Flip-Avoiding Interpolating Surface) [17] is reported to produce more accurate reconstruction of skulls compared to several existing methods. FAIS iteratively applies Laplacian deformation to non-rigidly register a reference to fit the target. Both FAIS and Laplacian deformation have one major drawback. They can produce distorted results when they are applied on skulls with large amounts of defective parts. This paper introduces symmetric constraints to the original Laplacian deformation and FAIS. Comprehensive test results show that the Laplacian deformation and FAIS with symmetric constraints are more robust and accurate than their original counterparts in reconstructing defective skulls with large amounts of defects.

Keywords: Laplacian deformation · Non-rigid registration

1 Introduction

Skull reconstruction is an important and challenging task in craniofacial surgery planning, forensic investigation and anthropological studies. Existing skull reconstruction methods can be broadly categorized into three categories: symmetry-based, statistical, and geometric. Symmetry-based methods [4,5,16] rely on the approximate left-right symmetry of human skulls. They regard the reflection of the non-defective parts of a target skull about the lateral symmetry plane as the reconstruction of the defective parts. These methods are not applicable when both sides of a skull are defective. Statistical methods, in particular active shape models [11,18,20], map a target skull to a statistical skull model by computing the model parameters that best fit the non-defective parts of the target, and generate the reconstructed skull from the model parameters. Unlike human face

© Springer International Publishing AG 2017
M. Felsberg et al. (Eds.): CAIP 2017, Part II, LNCS 10425, pp. 24–35, 2017.
DOI: 10.1007/978-3-319-64698-5_3

images, it is very difficult to collect a wide variety of 3D models of human skulls to cover all normal skull variations across age, race, and gender. Thus, it is difficult to apply statistical methods to skull reconstruction. Geometric methods [6,10,13,19] perform non-rigid registration of a single reference skull model to fit the non-defective parts of the target skull, and regard the registered reference as the reconstruction. Due to its wide applicability, geometric methods are the most suitable methods for skull reconstruction. Among them, our method proposed in [17] called FAIS, which iteratively applies Laplacian deformation, is reported to produce the most accurate reconstruction results.

FAIS and Laplacian deformation have one major drawback. For defective skulls with large amounts of defects, FAIS and Laplacian deformation can produce distorted results (Sect. 3). This paper proposes an improved Laplacian deformation method (Sect. 3) and an improved FAIS method (Sect. 4) that incorporate symmetry constraints. Comprehensive test results show that the improved Laplacian deformation and the improved FAIS are more robust and accurate than their original counterparts in reconstructing defective skulls with large amounts of defects.

2   Related Work

Two broad categories of non-rigid registration methods have been used for geometric reconstruction, namely, approximation and interpolation. Approximating methods such as piecewise rigid registration [2] and non-rigid ICP [1,9] produce **approximating surfaces**. They register a reference surface non-rigidly to fit the target model by minimizing the average distance between corresponding reference and target surfaces. Their registered surface points have non-zero distance to the target surfaces because they regard positional correspondence as **soft constraints**. Therefore, shape reconstruction by approximating methods have non-zero errors even for the non-defective parts.

Interpolating methods such as thin-plate spline (TPS) [3] and Laplacian deformation [12,14] non-rigidly register the reference surface to pass through corresponding target surface points. Their registered surfaces have zero distance to the corresponding target surface points. Therefore, interpolating methods are more accurate than approximating methods for skull reconstruction. Among the interpolating methods, TPS is the most popular method for skull reconstruction [6,10,13,15,17,19]. It is robust against noise and can produce smooth surfaces by minimizing surface bending energy.

Laplacian deformation deforms a surface by preserving local surface shape. It is used in [7] for 3D segmentation of anatomical bodies and in [17] for skull reconstruction. In particular, our method developed in [17] called FAIS (Flip-Avoiding Interpolating Surface) iteratively moves the surface of a reference skull closer and closer to the target using Laplacian deformation. FAIS uses flip-avoidance technique to allow for very dense surface correspondence without causing surface flipping which is a consequence of surface self-intersection. We have reported that FAIS can handle a lot more corresponding points and achieve higher reconstruction accuracy than other methods such as [6,10,13,19]. The drawback of FAIS is its longer running time due to its multiple iterations of Laplacian deformation. In addition, Laplacian deformation, as well as FAIS, can produce distorted results given defective skulls with large amounts of defects (Sect. 3).

3   Laplacian Deformation with Symmetry Constraints

Consider the C-shape model shown in Fig. 1(a). We want to apply Laplacian deformation to deform it such that the bottom landmark is fixed and the left landmark is moved to the right slightly. The result is a deformed shape whose bottom part is fixed but the top part is shifted to the right (Fig. 1(b)).

Similar situation can occur when reconstructing a defective skull with a large amount of defective or missing parts (Fig. 1(c)). The reference model shown in Fig. 1(d) happens to be disconnected on the right like the C-shape model, and it is wider than the target. Many reference landmarks on the facial bones have no corresponding target landmarks due to the large amount of missing parts in the target. During non-rigid registration, Laplacian deformation reduces the width of the lower jaw of the reference to fit the target's lower jaw according

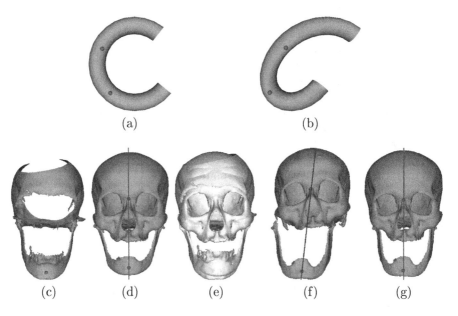

Fig. 1. Effect of Laplacian deformation. (a) C-shape model with 2 landmarks. (b) laplacian deformation of C-shape model with the bottom landmark fixed and the left landmark moved to the right. (c) defective skull model with large missing parts. (d) reference model that is disconnected on the right side like the C-shape model. Straight line denotes the symmetric mid-plane. (e) initial alignment. Reference is white and target is yellow. (f) registration of reference to target using ordinary Laplacian deformation produces distorted result. (g) laplacian deformation that preserves the vertical mid-plane produces undistorted result.

to the landmarks. This process moves the left and right landmarks of the reference inward onto the positions of the corresponding target landmarks. These movements, coupled with the C-shaped reference and lack of correspondence of reference landmarks, cause the craniofacial bones of the reference to shift to the right instead of reducing its width. Consequently, a distorted skull is produced (Fig. 1(f)).

To overcome this problem, we improve Laplacian deformation by imposing the **mid-plane constraint** as follow: Every reference skull model has some landmarks called the **mid-point landmarks** that fall on mid-line of the skull. These mid-point landmarks form a plane called the mid-plane (Fig. 1(d)). Before deformation, the mid-plane is a vertical, laterally symmetric plane. The mid-plane constraint states that after deformation, the mid-plane should still be a vertical, laterally symmetric plane. With this additional constraint, Laplacian deformation will produce an undistorted result (Fig. 1(g)).

Although the mid-plane stays vertical and laterally symmetric, the model after deformation can still be laterally distorted (Fig. 2(c)). This is due to the symmetrically unbalanced correspondence because some reference landmarks have no corresponding points on the target. To overcome this problem, we introduce the **symmetric constraint** which constrains every symmetric pair of landmarks to remain symmetric after deformation. With this constraint, Laplacian deformation will produce a symmetric and undistorted result (Fig. 2(d)).

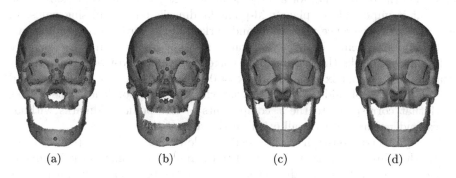

(a)	(b)	(c)	(d)

Fig. 2. Laplacian deformation with the symmetric constraint. (a) reference model. (b) target model with missing facial bone. (c) laplacian deformation with the mid-plane constraint produces a laterally distorted result (distorted right orbital bone), although the mid-plane remains vertical. (d) laplacian deformation with the additional symmetric constraint produces a symmetric result.

Laplacian deformation [12, 14] applies the discrete Laplacian operator $L(\mathbf{p}_i)$ to estimate the surface curvature and normal at vertex i:

$$L(\mathbf{p}_i) = \sum_{j \in \mathcal{N}_i} w_{ij}(\mathbf{p}_i - \mathbf{p}_j) \tag{1}$$

where $\mathcal{N}_i$ is the set of connected neighbours of vertex i. The weight w_{ij} can be cotangent weight [12] or equal weight $w_{ij} = 1/|\mathcal{N}_i|$. Equal weight leads to simpler optimization equations. With equal weight, Laplacian operator becomes

$$L(\mathbf{p}_i) = \mathbf{p}_i - \frac{1}{\mathcal{N}_i} \sum_{j \in \mathcal{N}_i} \mathbf{p}_j. \tag{2}$$

Laplacian deformation preserves the model's shape by minimizing the difference of Laplacian operators $L(\mathbf{p}_i^0)$ before and $L(\mathbf{p}_i)$ after deformation, which is $\|L(\mathbf{p}_i) - L(\mathbf{p}_i^0)\|^2$. These differences for all n mesh vertices are organized into a matrix equation: $\mathbf{L}\mathbf{x} = \mathbf{a}$, where $\mathbf{L}$ is a $3n \times 3n$ matrix that captures the Laplacian constraints, $\mathbf{x}$ is a $3n \times 1$ vector of unknown positions $\mathbf{x}_i$ of mesh vertices, $\mathbf{x} = [\mathbf{x}_1^\top \cdots \mathbf{x}_n^\top]^\top$, and $\mathbf{a}$ is a $3n \times 1$ vector that contains $L(\mathbf{p}_i^0)$ before deformation.

The additional mid-plane constraint is imposed as follows: Without loss of generality, let the skull model before deformation be oriented such that the mid-plane is located at $x = 0$ and its surface normal is parallel to the x-axis. Moreover, the landmark points coincide with some mesh vertices. Then, the mid-plane constraint requires that the x coordinates of the mid-point landmarks remain as 0 after deformation, which constrains the mid-plane to remain vertical and laterally symmetric after deformation. For a non-defective target skull, the mid-point landmarks of a reference skull always have corresponding landmarks on the target. On the other hand, for a target skull with large amount of missing facial bones, it is impossible to place mid-point landmarks on the missing parts. In this case, some reference mid-point landmarks will not have corresponding target landmarks. Then, mid-plane constraint has to be imposed on these reference mid-point landmarks that do not have correspondence. Mid-plane constraint is organized into a matrix equation: $\mathbf{M}\mathbf{x} = \mathbf{0}$, where $\mathbf{M}$ is a $k \times 3n$ matrix and k is the number of mid-point landmarks without correspondence. The entries in $\mathbf{M}$ that correspond to the x components of mid-point landmarks without correspondence are set to 1; all other entries are set to 0.

The symmetric constraint is imposed as follows: For every pair (l, r) of landmarks which are symmetric with respect to the mid-plane which is the YZ-plane, after deformation, their coordinates should have the relationships: $x_l + x_r = 0$, $y_l - y_r = 0$, and $z_l - z_r = 0$. The relationships can be organized into a matrix equation: $\mathbf{S}\mathbf{x} = \mathbf{0}$, where $\mathbf{S}$ is a $3s \times 3n$ matrix and s is the number of symmetric landmark pairs. The entries in $\mathbf{S}$ that correspond to x_l, x_r, y_l, and z_l are set to 1; those correspond to y_r and z_r are set to -1; all other entries are set to 0.

The Laplacian constraint, together with the mid-plane constraint and the symmetric constraint, are combined into the following objective function to be minimized:

$$\|\mathbf{A}\mathbf{x} - \mathbf{b}\|^2 = \left\| \begin{bmatrix} \mathbf{L} \\ \mathbf{M} \\ \mathbf{S} \end{bmatrix} \mathbf{x} - \begin{bmatrix} \mathbf{a} \\ \mathbf{0} \\ \mathbf{0} \end{bmatrix} \right\|^2. \tag{3}$$

The positional constraints of the corresponding points between the reference and the target are organized into a matrix equation of the form: $\mathbf{C}\mathbf{x} = \mathbf{d}$, where

$\mathbf{C}$ indicates the mesh vertices with positional constraints and $\mathbf{d}$ contains the desired vertex positions. Without loss of generality, we can arrange the mesh vertices with positional constraints as vertices 1 to $m < n$. Then, $\mathbf{C}$ is a $3m \times 3n$ matrix that contains a $3m \times 3m$ identity matrix and a $3m \times 3(n-m)$ zero matrix: $\mathbf{C} = [\mathbf{I}_{3m}\mathbf{0}]$. Correspondingly, the top $3m$ elements of $\mathbf{x}$ are the mesh vertices with positional constraints, the bottom $3(n-m)$ elements are those without positional constraints, and $\mathbf{d}$ is a $3m \times 1$ vector of the coordinates of the desired vertex positions. Then the Laplacian deformation with mid-plane constraint and symmetric constraint, namely $\mathbf{sLD}$, solves the following problem:

$$\min_{\mathbf{x}} \|\mathbf{Ax} - \mathbf{b}\|^2 \text{ subject to } \mathbf{Cx} = \mathbf{d}. \tag{4}$$

That is, the Laplacian, mid-plane constraint and symmetric constraint are soft constraints whereas the positional constraints are hard constraints.

This Laplacian deformation problem is an *equality-constrained least squares* problem, which can be solved using QR factorization [8,12] as follows: $\mathbf{C}^\top$ has QR factorization $\mathbf{C}^\top = \mathbf{QR}$, where $\mathbf{Q} = [\mathbf{Q}_1\mathbf{Q}_2]$ is orthogonal and $\mathbf{R} = [\mathbf{R}_1^\top \mathbf{0}^\top]^\top$ is upper triangular. Define vectors $\mathbf{u}$ and $\mathbf{v}$ such that

$$\mathbf{x} = \mathbf{Q}\begin{bmatrix} \mathbf{u} \\ \mathbf{v} \end{bmatrix} = \begin{bmatrix} \mathbf{Q}_1 & \mathbf{Q}_2 \end{bmatrix}\begin{bmatrix} \mathbf{u} \\ \mathbf{v} \end{bmatrix}. \tag{5}$$

Then, the objective function of (4) becomes

$$\|\mathbf{Ax} - \mathbf{b}\|^2 = \|\mathbf{AQ}_1\mathbf{u} + \mathbf{AQ}_2\mathbf{v} - \mathbf{b}\|^2. \tag{6}$$

Since $\mathbf{C} = [\mathbf{I}_{3m}\mathbf{0}]$, the QR factorization of $\mathbf{Q}^\top$ is

$$\mathbf{C}^\top = \begin{bmatrix} \mathbf{I}_{3m} & \mathbf{0} \\ \mathbf{0} & \mathbf{I}_{3(n-m)} \end{bmatrix}\begin{bmatrix} \mathbf{I}_{3m} \\ \mathbf{0} \end{bmatrix}. \tag{7}$$

That is,

$$\mathbf{Q}_1 = \begin{bmatrix} \mathbf{I}_{3m} \\ \mathbf{0} \end{bmatrix}, \quad \mathbf{Q}_2 = \begin{bmatrix} \mathbf{0} \\ \mathbf{I}_{3(n-m)} \end{bmatrix}, \quad \mathbf{R}_1 = \mathbf{I}_{3m}.$$

With QR factorization of $\mathbf{C}^\top$, the positional constraint equation $\mathbf{Cx} = \mathbf{d}$ becomes

$$\mathbf{Cx} = \mathbf{R}^\top\mathbf{Q}^\top\mathbf{x} = \mathbf{R}^\top\begin{bmatrix} \mathbf{u} \\ \mathbf{v} \end{bmatrix} = \mathbf{R}_1^\top\mathbf{u} = \mathbf{d}. \tag{8}$$

Right-hand-side of Eq. 8 yields $\mathbf{I}_{3m}\mathbf{u} = \mathbf{u} = \mathbf{d}$. So,

$$\mathbf{Q}_1\mathbf{u} = \begin{bmatrix} \mathbf{I}_{3m} \\ \mathbf{0} \end{bmatrix}\mathbf{u} = \begin{bmatrix} \mathbf{I}_{3m} \\ \mathbf{0} \end{bmatrix}\mathbf{d} = \begin{bmatrix} \mathbf{d} \\ \mathbf{0} \end{bmatrix}, \tag{9}$$

$$\mathbf{Q}_2\mathbf{v} = \begin{bmatrix} \mathbf{0} \\ \mathbf{I}_{3(n-m)} \end{bmatrix}\mathbf{v} = \begin{bmatrix} \mathbf{0} \\ \mathbf{v} \end{bmatrix}. \tag{10}$$

Organize the matrix $\mathbf{A}$ as $[\mathbf{A}_1 \mathbf{A}_2]$. Then,

$$\mathbf{AQ}_1\mathbf{u} = \mathbf{A}_1\mathbf{d}, \quad \mathbf{AQ}_2\mathbf{v} = \mathbf{A}_2\mathbf{v}. \tag{11}$$

Algorithm 1. FAIS with Symmetry Constraint (sFAIS)

Input: Reference F, target T, manually marked positional constraints of landmarks C^*.

1 Rigidly register F to T using reference landmarks with correspondence C^*.
2 Orientate F and T so that the mid-plane of F is located at $x = 0$ and its normal is parallel to x-axis.
3 Non-rigidly register F to T with positional constraints C^* using sLD; then set R as registered F.
4 **for** k *from* 1 *to* K **do**
5 | Find correspondence C from R to T using first correspondence search method.
6 | Choose a sparse subset C^+ from $C^* \cup C$.
7 | Non-rigidly register R to T with constraints C^+ using sLD.
8 **end**
9 Find correspondence C from R to T using second correspondence search method.
10 Remove crossings in $C^* \cup C$ giving C^+.
11 Non-rigidly register R to T with constraints C^+ using sLD.
Output: Reconstructed model R.

Then, the objective function (6) becomes

$$\|\mathbf{A}_2\mathbf{v} - (\mathbf{b} - \mathbf{A}_1\mathbf{d})\|^2. \tag{12}$$

Minimization of objective function (12) with linear least squares yields

$$\mathbf{v} = (\mathbf{A}_2^\top \mathbf{A}_2)^{-1}\mathbf{A}_2^\top(\mathbf{b} - \mathbf{A}_1\mathbf{d}). \tag{13}$$

Then, the positions of the mesh vertices after deformation can be computed as

$$\mathbf{x} = \mathbf{Q}_1\mathbf{u} + \mathbf{Q}_2\mathbf{v} = \begin{bmatrix} \mathbf{d} \\ \mathbf{v} \end{bmatrix}. \tag{14}$$

As $\mathbf{d}$ contains known desired positions, Laplacian deformation only needs to solve for $\mathbf{v}$, the coordinates of the mesh vertices without positional constraints after deformation. Therefore, sLD runs faster with increasing number of positional constraints.

4   FAIS with Symmetric Constraint

To use sLD, we improve the FAIS method proposed in [17] as follows:

– After rigidly registering the reference F to the target T, orientate both F and T so that the mid-plane of F is located at $x = 0$ and its normal is parallel to x-axis. This step prepares the skull models for imposing the mid-plane constraint.

– Apply the improved Laplacian deformation (sLD) with mid-plane constraint and symmetric constraint.

Our improved FAIS, named **sFAIS**, is summarized in Algorithm 1. The improvements of FAIS are marked in red. Details of correspondence search (Step 5, 9) and flip avoidance (Step 6, 10) are discussed in [17]. In the case that all the reference landmarks have correspondence, sFAIS reverts to the original FAIS.

5   Experiment

5.1   Data Preparation

3D mesh models of non-defective skulls were constructed from patient's CT images. One normal skull is used as the reference model (Fig. 2(a)) which has about 100,000 vertices. Five synthetic skulls with different amount of fractures were manually generated from the non-defective skulls (Fig. 3(1)) by moving and rotating the corresponding bones in a way similar to real fractured skulls for reconstruction test. In addition, 5 skull models of trauma patients with real fractures (Fig. 3(2)) were used for reconstruction test. Each skull model had 47 landmarks manually placed on them. The target skull models were initially aligned to the reference model by applying the best similarity transformation computed from their corresponding landmarks.

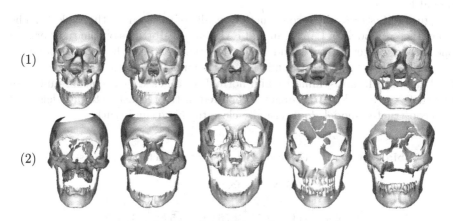

Fig. 3. Target skulls. (1) synthetic fractured skulls T1–5. (2) real fractured skulls.

5.2   PC Configuration and Running Time

The programs were implemented in Mathematical which used Intel® Math Kernel Library to solve linear systems. All tests were run on a PC with Intel i7-2600 CPU at 3.4 GHz and 8 GB RAM.

For the timing, sLD runs faster as the number of correspondence increases. It takes 6.4 s, 5.7 s, and 1.8 s, respectively, to register the reference with 1,000, 5,000 and 76,500 correspondence. Moreover, sFAIS takes 180 s to register the reference to the first target in Fig. 3(1) with $K = 20$ iterations.

5.3   Reconstruction of Synthetic Fractured Skulls

This experiment compares the reconstruction accuracy of FAIS and sFAIS on synthetic fractured skulls given the same reference model. Five synthetic fractured skulls were used, whose fractured parts were marked manually. The normal skulls used to generate the synthetic skulls served as the ground truth. Each testing skull was reconstructed by FAIS and sFAIS. The reconstruction results of FAIS and sFAIS were recorded. In addition, the first non-rigid registration results of the original Laplacian deformation and sLD in FAIS and sFAIS (Step 3), respectively, were also recorded. Reconstruction errors were measured between the reconstructed models and their ground truths. Reconstruction errors of the defective and non-defective parts of the testing skulls were measured separately.

Table 1 summarizes the reconstruction errors of synthetic fracture skulls. Both FAIS and sFAIS have much lower reconstruction errors on the non-defective parts than the original Laplacian deformation and sLD because they apply multiple iterations of Laplacian deformation with dense sets of automatically detected correspondence points. Their reconstruction errors for non-defective parts are very small but not exactly zero because some non-defective parts have no correspondence due to the application of flip-avoidance technique. For the defective parts, sFAIS has smaller mean reconstruction error than original FAIS. This is expected as sFAIS yield more symmetric and normal reconstruction than the original FAIS does.

Figure 4 illustrates the reconstruction results of selected synthetic skulls. The reconstructed models of Laplacian and FAIS have distortions on the right orbital bone for the target in row 1 and have distortions on the whole facial bone for the target in row 2, due to the missing correspondence on them, whereas those of sLD and sFAIS are symmetric and normal. In all cases, the discontinuities of bone surfaces caused by fractures are repaired in the reconstructed models, and the reconstructed models look visually close to the ground truth.

Table 1. Reconstruction errors (in mm) of synthetic fractured skulls.

(a) Non-defective parts

	T1	T2	T3	T4	T5	mean
Laplace	4.09	2.78	2.76	3.35	5.89	3.77
sLD	4.08	2.80	2.69	3.39	5.93	3.78
FAIS	0.30	0.27	0.19	0.18	0.58	0.30
sFAIS	0.15	0.18	0.21	0.21	0.82	0.31

(b) Defective parts

	T1	T2	T3	T4	T5	mean
Laplace	2.47	2.34	1.30	2.83	4.36	2.66
sLD	2.36	1.44	1.88	2.39	2.95	2.20
FAIS	1.92	2.74	2.32	3.04	2.20	2.44
sFAIS	1.66	0.66	2.19	2.44	1.93	1.78

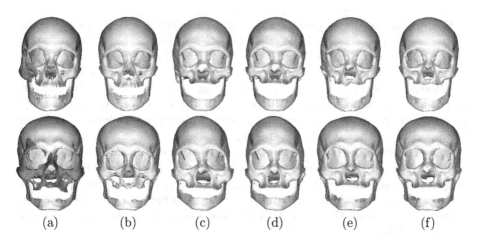

Fig. 4. Reconstruction results of synthetic fractured skulls. (a) synthetic fractured skulls with fractured parts denoted in pink. (b) ground truths. (c-f) reconstruction results of Laplacian deformation, sLD, FAIS, and sFAIS, respectively.

5.4 Reconstruction of Real Fractured Skulls

This experiment compares the reconstruction accuracy of the original FAIS and sFAIS on real fractured skulls given the same reference model. The test procedure was the same as that in Sect. 5.3. Ground truths were not available.

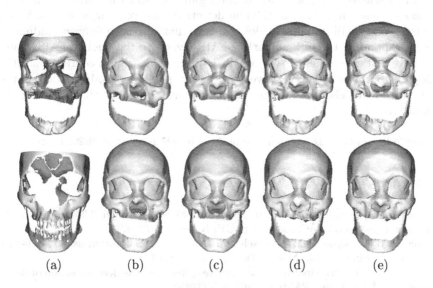

Fig. 5. Reconstruction results of real fractured skulls. (a) real fractured skulls with fractured parts (denoted in pink) and missing parts. Ground truths are not available. (b-e) reconstruction results of the original Laplacian deformation, sLD, orignal FAIS, and sFAIS, respectively.

Figure 5 shows the reconstruction results of selected real fractured skulls. The reconstruction results of the other targets are comparable between Laplacian and sLD as well as between FAIS and sFAIS. The targets in Fig. 5 have large amounts of defective or missing parts, and many mid-point landmarks on the reference models do not have corresponding target landmarks. As a result, the initial registration results of the original Laplacian deformation (Fig. 5(b)) have some lateral asymmetry and distortions. This can be obviously seen from the upper jaws. After iteratively applying Laplacian deformation, the original FAIS produces reconstruction results (Fig. 5(d)) with significant distortions and asymmetry, in particular, severe asymmetry of the upper jaws. On the other hand, the initial registration results of sLD (Fig. 5(c)) have less distortions and are more symmetric. Therefore, after iteratively applying sLD, sFAIS produces reconstruction results (Fig. 5(e)) that are much more symmetric than those of the original FAIS.

6   Conclusion

This paper proposes an improved version of Laplacian deformation that incorporates mid-plane constraint and symmetric constraint to ensure that the mid-plane of the registered reference model remains as a vertical, laterally symmetric plane and the registered reference model remains laterally symmetric. This improved Laplacian deformation, as known as sLD, is used to improve an existing skull reconstruction algorithm called FAIS. The improved FAIS, as known as sFAIS, iteratively applies sLD to non-rigidly register the surface of a reference model closer and closer to the non-defective parts of a target model. When the target has small amount of defects such that all the reference landmarks have corresponding target landmarks, sFAIS reverts to the original FAIS. Comprehensive test results show that sFAIS is more robust and accurate than the original FAIS in reconstructing severely defective skull models. In particular, the reconstructed models of sFAIS are laterally more symmetric than those of the original FAIS.

Acknowledgment. This research is supported by MOE grant MOE2014-T2-1-062.

References

1. Amberg, B., Romdhani, S., Vetter, T.: Optimal step nonrigid ICP algorithms for surface registration. In: Proceedings CVPR, pp. 1–8 (2007)
2. Bonarrigo, F., Signoroni, A., Botsch, M.: Deformable registration using patch-wise shape matching. Graph. Models **76**(5), 554–565 (2014)
3. Bookstein, F.L.: Principal warps: thin-plate splines and the decomposition of deformations. IEEE Trans. PAMI **11**, 567–585 (1989)
4. Cevidanes, L., Tucker, S., Styner, M., Kim, H., Reyes, M., Proffit, W., Turvey, T., Jaskolka, M.: 3D surgical simulation. Am. J. Orthod. Dentofac. Orthop. **138**(3), 361–371 (2010)

5. De Momi, E., Chapuis, J., Pappas, I., Ferrigno, G., Hallermann, W., Schramm, A., Caversaccio, M.: Automatic extraction of the mid-facial plane for cranio-maxillofacial surgery planning. Int. J. Oral Maxillofac. Surg. **35**(7), 636–642 (2006)
6. Deng, Q., Zhou, M., Shui, W., Wu, Z., Ji, Y., Bai, R.: A novel skull registration based on global and local deformations for craniofacial reconstruction. Forensic Sci. Int. **208**, 95–102 (2011)
7. Ding, F., Yang, W., Leow, W.K., Venkatesh, S.: 3D segmentation of soft organs by flipping-free mesh deformation. In: Proceedings WACV (2009)
8. Golub, G.H., Loan, C.F.V.: Matrix Computations, 3rd edn. Johns Hopkins, Baltimore (1996)
9. Hontani, H., Matsuno, T., Sawada, Y.: Robust nonrigid ICP using outlier-sparsity regularization. In: Proceedings CVPR, pp. 174–181 (2012)
10. Lapeer, R.J.A., Prager, R.W.: 3D shape recovery of a newborn skull using thin-plate splines. Comput. Med. Imaging Graph. **24**(3), 193–204 (2000)
11. Lüthi, M., Albrecht, T., Vetter, T.: Building shape models from lousy data. In: Yang, G.-Z., Hawkes, D., Rueckert, D., Noble, A., Taylor, C. (eds.) MICCAI 2009. LNCS, vol. 5762, pp. 1–8. Springer, Heidelberg (2009). doi:10.1007/978-3-642-04271-3_1
12. Masuda, H., Yoshioka, Y., Furukawa, Y.: Interactive mesh deformation using equality-constrained least squares. Comput. Graph. **30**(6), 936–946 (2006)
13. Rosas, A., Bastir, M.: Thin-plate spline analysis of allometry and sexual dimorphism in the human craniofacial complex. Am. J. Phys. Anthropol. **117**, 236–245 (2002)
14. Sorkine, O., Cohen-Or, D., Lipman, Y., Alexa, M., Rössl, C., Seidel, H.P.: Laplacian surface editing. In: Proceedings Eurographics/ACM SIGGRAPH Symposium Geometry Processing, pp. 175–184 (2004)
15. Turner, W.D., Brown, R.E., Kelliher, T.P., Tu, P.H., Taister, M.A., Miller, K.W.: A novel method of automated skull registration for forensic facial approximation. Forensic Sci. Int. **154**, 149–158 (2005)
16. Wei, L., Yu, W., Li, M., Li, X.: Skull assembly and completion using template-based surface matching. In: Proceedings International Conference 3D Imaging, Modeling, Processing, Visualization and Transmission (2011)
17. Xie, S., Leow, W.K.: Flip-avoiding interpolating surface registration for skull reconstruction. In: Proceedings ICPR (2016)
18. Zachow, S., Lamecker, H., Elsholtz, B., Stiller, M.: Reconstruction of mandibular dysplasia using a statistical 3D shape model. In: Proceedings Computer Assisted Radiology and Surgery, pp. 1238–1243 (2005)
19. Zhang, K., Cheng, Y., Leow, W.K.: Dense correspondence of skull models by automatic detection of anatomical landmarks. In: Wilson, R., Hancock, E., Bors, A., Smith, W. (eds.) CAIP 2013. LNCS, vol. 8047, pp. 229–236. Springer, Heidelberg (2013). doi:10.1007/978-3-642-40261-6_27
20. Zhang, K., Leow, W.K., Cheng, Y.: Performance analysis of active shape reconstruction of fractured, incomplete skulls. In: Azzopardi, G., Petkov, N. (eds.) CAIP 2015. LNCS, vol. 9256, pp. 312–324. Springer, Cham (2015). doi:10.1007/978-3-319-23192-1_26

A New Image Contrast Enhancement Algorithm Using Exposure Fusion Framework

Zhenqiang Ying[1], Ge Li[1($\boxtimes$)], Yurui Ren[1,2], Ronggang Wang[1], and Wenmin Wang[1]

[1] School of Electronic and Computer Engineering, Shenzhen Graduate School, Peking University, Shenzhen, China
geli@ece.pku.edu.cn
[2] Faculty of Electronic Information and Electrical Engineering, Dalian University of Technology, Dalian, China

Abstract. Low-light images are not conducive to human observation and computer vision algorithms due to their low visibility. Although many image enhancement techniques have been proposed to solve this problem, existing methods inevitably introduce contrast under- and over-enhancement. In this paper, we propose an image contrast enhancement algorithm to provide an accurate contrast enhancement. Specifically, we first design the weight matrix for image fusion using illumination estimation techniques. Then we introduce our camera response model to synthesize multi-exposure images. Next, we find the best exposure ratio so that the synthetic image is well-exposed in the regions where the original image under-exposed. Finally, the input image and the synthetic image are fused according to the weight matrix to obtain the enhancement result. Experiments show that our method can obtain results with less contrast and lightness distortion compared to that of several state-of-the-art methods.

Keywords: Image enhancement · Contrast enhancement · Exposure compensation · Exposure fusion

1 Introduction

Image enhancement techniques are widely used in image processing. In general, it can make the input images look better and be more suitable for specific algorithms [6,14]. Contrast enhancement, as one kind of enhancement techniques, can reveal the information of the under-exposed regions in an image. Many contrast enhancement techniques have been proposed including histogram-based [3,6,12,13], Retinex-based [7,9] and dehaze-based methods [4,5].

A color image can be represented as a three-dimensional array. The simplest scheme of contrast enhancement performs the same processing for each element. For example, the earliest image enhancement methods use a non-linear monotonic function (power-law [2], logarithm [11], etc.) for gray-level mapping.

© Springer International Publishing AG 2017
M. Felsberg et al. (Eds.): CAIP 2017, Part II, LNCS 10425, pp. 36–46, 2017.
DOI: 10.1007/978-3-319-64698-5_4

Fig. 1. Example of images that only differ in exposure.

Considering the uneven distribution of elements in different gray levels, histogram equalization (HE) is widely used for improving contrast. Many extensions to HE are proposed to take some restrictions into account such as brightness preservation [3,6,13] and contrast limitation [12]. However, HE-based methods always suffer from over-enhancement and lead to unrealistic results. Imitating the human visual system, Retinex theory is also widely used in image enhancement. By separating reflectance from the illumination, Retinex-based algorithms can enhance the details obviously [5,7,9]. However, these methods suffer from halo-like artifact in the high contrast region. In recent years, some methods borrowed the de-haze technique to contrast enhancement, and achieve good subjective visual effect [4]. But these work may cause color distortion because of contrast over-enhancement.

Although image contrast enhancement methods have been studied for decades, the definition of a good enhancement result is still not well-defined. In addition, no reference is provide for existing low-light enhancement algorithm to locate the over- and under-enhancement region. We noticed that the images that only differ in exposures can be used as a reference for enhancement algorithms, as shown in Fig. 1. As the exposure increases, some under-exposed regions become well-exposed. The enhancement result should keep the well-exposed regions unchanged and enhance the under-exposed region. Meanwhile, the contrast of enhanced regions should be consistent with the reference image that correct exposes the region.

In this paper, we proposed a new framework to help mitigate under- and over-enhancement problem. Our framework is based on exposure fusion among multi-exposure images synthesized from the input image by the camera response model. Based on our framework, we proposed an enhancement algorithm that can obtain results with less contrast and lightness distortion compared to several state-of-the-art methods.

2   Our Approach

2.1   Exposure Fusion Framework

In many outdoor scenes, cameras can not make all pixel well-exposed since its dynamic range is limited. As shown in the Fig. 1, although we can reveal some under-exposed regions by increasing exposure, the well-exposed regions may become over-exposed at the same time. To obtain an image with all pixel well-exposed, we can fuse these images:

$$\mathbf{R}^c = \sum_{i=1}^{N} \mathbf{W}_i \circ \mathbf{P}_i^c, \tag{1}$$

where N is the number of images, $\mathbf{P}_i$ is the i-th image in the exposure set, $\mathbf{W}_i$ is the weight map of the i-th image, c is the index of three color channels and $\mathbf{R}$ is the enhanced result. It is equal for three color components and nonuniform for all pixels: the well-exposed pixels are given a big weight while the poor-exposed pixels are given a small weight. The weight is normalized so that $\sum_{i=1}^{N} \mathbf{W}_i = 1$.

The problem is that images with another exposure setting is not available for image enhancement problem. Fortunately, pictures taken with different exposures are highly correlated. In our earlier work, we propose a camera response model to accurately describe the association between these images so that we can generate a series of images from the input image. The mapping function between two images that only differ in exposure is called Brightness Transform Function (BTF). Given exposure ratio k_i and BTF g, we can map the input image $\mathbf{P}$ to the i-th image in the exposure set as

$$\mathbf{P}_i = g(\mathbf{P}, k_i). \tag{2}$$

In this paper, we only fuse the input image itself with another exposure to reduce complexity, as shown in Fig. 2. The fused image is defined as

$$\mathbf{R}^c = \mathbf{W} \circ \mathbf{P}^c + (1 - \mathbf{W}) \circ g(\mathbf{P}^c, k) \tag{3}$$

The enhancement problem can be divided into three parts: the estimation of $\mathbf{W}$, g and k. In the following subsections, we solve them one by one.

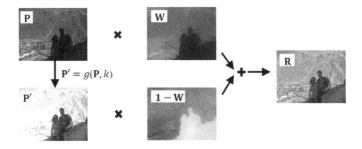

Fig. 2. Our framework.

2.2   Weight Matrix Estimation

The design of $\mathbf{W}$ is key to obtaining an enhancement algorithm that can enhance the low contrast of under-exposed regions while the contrast in well-exposed regions preserved. We need to assign big weight values to well-exposed pixels and small weight values to under-exposed pixels. Intuitively, the weight matrix is positively correlated with the scene illumination. Since highly illuminated regions have big possibility of being well-exposed, they should be assign with big weight values to preserve their contrast. We calculate the weight matrix as

$$\mathbf{W} = \mathbf{T}^{\mu} \tag{4}$$

where $\mathbf{T}$ is the scene illumination map and μ is a parameter controlling the enhance degree (See more details in Sect. 3.1). The estimation of scene illumination map $\mathbf{T}$ is solved by optimization.

Optimization Problem. The lightness component can be used as an estimation of scene illumination. We adopt the lightness component as the initial estimation of illumination:

$$\mathbf{L}(x) = \max_{c \in \{R,G,B\}} \mathbf{P}_c(x) \tag{5}$$

for each individual pixel x. Ideal illumination should has local consistency for the regions with similar structures. In other words, $\mathbf{T}$ should keep the meaningful structures of the image and remove the textural edges. As in [5], we refine $\mathbf{T}$ by solving the following optimization equation:

$$\min_{\mathbf{T}} ||\mathbf{T} - \mathbf{L}||_2^2 + \lambda ||\mathbf{M} \circ \nabla \mathbf{T}||_1, \tag{6}$$

where $|| * ||_2$ and $|| * ||_1$ are the ℓ_2 and ℓ_1 norm, respectively. The first order derivative filter ∇ contains $\nabla_h \mathbf{T}$ (horizontal) and $\nabla_v \mathbf{T}$ (vertical). $\mathbf{M}$ is the weight matrix and λ is the coefficient. The first term of this equation is to minimize the difference between the initial map $\mathbf{L}$ and the refined map $\mathbf{T}$, while the second term maintains the smoothness of $\mathbf{T}$.

The design of $\mathbf{M}$ is important for the illumination map refinement. A major edge in a local window contributes more similar-direction gradients than textures with complex patterns [15]. Therefore, the weight in a window that contains meaningful edges should be smaller than that in a window only containing textures. As a result, we design the weight matrix as

$$\mathbf{M}_d(x) = \frac{1}{|\sum_{y \in \omega(x)} \nabla_d \mathbf{L}(y)| + \epsilon}, \quad d \in \{h, v\}, \tag{7}$$

where $| * |$ is the absolute value operator, $\omega(x)$ is the local window centered at the pixel x and ϵ is a very small constant to avoid the zero denominator.

Closed-Form Solution. To reduce the complexity, we approximate Eq. 6 as in [5]:

$$\min_{\mathbf{T}} \sum_x \left((\mathbf{T}(x) - \mathbf{L}(x))^2 + \lambda \sum_{d \in \{h,v\}} \frac{\mathbf{M}_d(x)(\nabla_d \mathbf{T}(x))^2}{|\nabla_d \mathbf{L}(x)| + \epsilon} \right). \tag{8}$$

As can be seen, the problem now only involves quadratic terms. Let $\mathbf{m}_d$, $\mathbf{l}$, $\mathbf{t}$ and $\nabla_d \mathbf{l}$ denote the vectorized version of $\mathbf{M}_d$, $\mathbf{L}$, $\mathbf{T}$ and $\nabla_d \mathbf{L}$ respectively. Then the solution can be directly obtained by solving the following linear function.

$$(\mathbf{I} + \lambda \sum_{d \in \{h,v\}} (\mathbf{D}_\mathbf{d}^\intercal Diag(\mathbf{m}_d \oslash (|\nabla_d \mathbf{l}| + \epsilon)) \mathbf{D}_\mathbf{d}) \mathbf{t} = \mathbf{l} \tag{9}$$

where $\oslash$ is the element-wise division, $\mathbf{I}$ is the unit matrix, the operator $Diag(\mathbf{v})$ is to construct a diagonal matrix using vector $\mathbf{v}$, and $\mathbf{D}_\mathbf{d}$ are the Toeplitz matrices from the discrete gradient operators with forward difference. The main difference between our illumination map estimation method and that in [5] is the design of weight matrix $\mathbf{M}$. We adopted a simplified strategy which can yield similar results as in [5]. Other illumination decomposition techniques in Retinex-based methods can be borrowed here to find the weight matrix $\mathbf{W}$.

2.3   Camera Response Model

In our earlier work, we proposed a camera response model called Beta-Gamma Correction Model [16]. The BTF of our model is defined as

$$g(\mathbf{P}, k) = \beta \mathbf{P}^\gamma = e^{b(1-k^a)} \mathbf{P}^{(k^a)}. \tag{10}$$

where β and γ are two model parameters that can be calculated from camera parameters a, b and exposure ratio k. We assume that no information about the camera is provided and use a fixed camera parameters ($a = -0.3293, b = 1.1258$) that can fit most cameras.

2.4   Exposure Ratio Determination

In this subsection, we find the best exposure ratio so that the synthetic image is well-exposed in the regions where the original image under-exposed. First, we exclude the well-exposed pixels and obtain an image that is globally under-exposed. We simply extract the low illuminated pixels as

$$\mathbf{Q} = \{\mathbf{P}(x)|\mathbf{T}(x) < 0.5\}, \tag{11}$$

where $\mathbf{Q}$ contains only the under-exposed pixels.

The brightness of the images under different exposures changes significantly while the color is basically the same. Therefore, we only consider the brightness component when estimating k. The brightness component $\mathbf{B}$ is defined as the geometric mean of three channel:

$$\mathbf{B} := \sqrt[3]{\mathbf{Q}_r \circ \mathbf{Q}_g \circ \mathbf{Q}_b}, \tag{12}$$

where $\mathbf{Q}_r$, $\mathbf{Q}_g$ and $\mathbf{Q}_b$ are the red, green and blue channel of the input image $\mathbf{Q}$ respectively. We use the geometric mean instead of other definitions (*e.g.* arithmetic mean and weighted arithmetic mean) since it has the same BTF model parameters (β and γ) with all three color channels, as shown in Eq. 13.

$$\begin{aligned}
\mathbf{B}' &:= \sqrt[3]{\mathbf{Q}'_r \circ \mathbf{Q}'_g \circ \mathbf{Q}'_b} \\
&= \sqrt[3]{(\beta\mathbf{Q}_r^\gamma) \circ (\beta\mathbf{Q}_g^\gamma) \circ (\beta\mathbf{Q}_b^\gamma)} = \beta(\sqrt[3]{\mathbf{Q}_r \circ \mathbf{Q}_g \circ \mathbf{Q}_b})^\gamma \\
&= \beta\mathbf{B}^\gamma.
\end{aligned} \tag{13}$$

The visibility of a well-exposed image is higher than that of an under-exposed image and it can provide a richer information for human. Thus, the optimal k should provide the largest amount of information. To measure the amount of information, we employ the image entropy which is defined as

$$\mathcal{H}(\mathbf{B}) = -\sum_{i=1}^{N} p_i \cdot \log_2 p_i, \tag{14}$$

where p_i is the i-th bin of the histogram of $\mathbf{B}$ which counts the number of data valued in $[\frac{i}{N}, \frac{i+1}{N})$ and N is the number of bins (N is often set to be 256). Finally, the optimal k is calculated by maximizing the image entropy of the enhancement brightness as

$$\hat{k} = \underset{k}{\operatorname{argmax}} \ \mathcal{H}(g(\mathbf{B}, k)). \tag{15}$$

The optimized k can be solved by one-dimensional minimizer. To improve the calculation efficiency, we resize the input image to 50×50 when optimizing k.

3 Experiments

To evaluate the performance of our method, we compare it with several state-of-the-art methods (AMSR [9], LIME [5], Dong [4] and NPE [14]) on hundreds of low-light images from five public datasets: VV[1], LIME-data [5], NPE [14] (NPE-data, NPE-ex1, NPE-ex2 and NPE-ex3), MEF [10], and IUS [8]. MEF and IUS are multi-exposure datasets, we select a low-light image from each multi-exposure set for evaluation.

3.1 Implementation Details

In our algorithm, μ is a parameter that controls the overall degree of enhancement. When $\mu = 0$, the resulting $\mathbf{R}$ is equal to $\mathbf{P}$, *i.e.*, no enhancement is performed. When $\mu = 1$, both the under-exposed pixels and well-exposed pixels are enhanced. When $\mu > 1$, pixels may get saturated and the resulting $\mathbf{R}$ suffers from detail loss. In order to perform enhancement while preserve the well-exposed regions, we set μ to $1/2$.

---

[1] https://sites.google.com/site/vonikakis/datasets.

To maintain the fairness of the comparison, the parameters of our enhancement algorithm are fixed in all experiments: $\lambda = 1$, $\epsilon = 0.001$, $\mu = 1/2$, and the size of local window $\omega(x)$ is 5. The most time-consuming part of our algorithm is illumination map optimization. We employ the multi-resolution pre-conditioned conjugate gradient solver ($\mathcal{O}(N)$) to solve it efficiently. In order to further speedup our algorithm, we solve $\mathbf{T}$ using the down-sampled version of the input image and then up-sample the resulting $\mathbf{T}$ to the original size. No visual differences can be found in the enhanced results if we down-sample once, but the computational efficiency is greatly improved.

3.2 Contrast Distortion

As aforementioned, the image that only differ in exposures can be used as a reference for evaluating the accuracy of enhanced results. DRIM (Dynamic Range Independent Metric) [1] can measure the distortion of image contrast without the interference of change in image brightness. We use it to visualize the contrast difference between the enhancement result and the reference image.

As shown in Fig. 3, the proposed method obtains the most realistic results with the least distortion. The results of Dong suffer from severe contrast distortion. Although the details can be recovered by AMSR, the obvious loss of contrast makes the results look dim and unreal. By contrast, the results of LIME look a bit more vivid, but they suffer from amplification of invisible contrast. Figure 5 shows more examples for visual comparison.

Fig. 3. Comparison of contrast distortion. The loss of visible contrast is marked in green, the amplification of invisible contrast is marked in blue, and the reversal of visible contrast is marked in red. Different shades of color represent different degrees of distortion. (Color figure online)

3.3   Lightness Distortion

We use lightness order error (LOE) to objectively measure the lightness distortion of enhanced results. LOE is defined as

$$LOE = \frac{1}{m} \sum_{x=1}^{m} RD(x) \qquad (16)$$

where $RD(x)$ is the relative order difference of the lightness between the original image P and its enhanced version P' for pixel x, which is defined as follows:

$$RD(x) = \sum_{y=1}^{m} U\Big(\mathbf{L}(x), \mathbf{L}(y)\Big) \oplus U\Big(\mathbf{L}'(x), \mathbf{L}'(y)\Big), \qquad (17)$$

where m is the pixel number, $\oplus$ stands for the exclusive-or operator, $\mathbf{L}(x)$ and $\mathbf{L}'(x)$ are the lightness component at location x of the input images and the enhanced images, respectively. The function $U(p, q)$ returns 1 if $p >= q$, 0 otherwise.

As suggested in [5,14], down-sampling is used to reduce the complexity of computing LOE. We noticed that LOE may change significantly when an image is down-sampled to different sizes since RD will increase as the pixel number m increases. Therefore, we down-sample all images to a fixed size. Specifically, we collect 100 rows and columns evenly to form a 100×100 down-sampled image. As shown in Table 1, our algorithm outperforms the others in all datasets. This means that our algorithm can maintain the naturalness of images well. We also provide a visualization of lightness distortion on two cases in Fig. 4, from which,

LOE: 0 ~ 5000

Input   AMSR   LIME   Dong   NPE   Ours

Fig. 4. Comparison of lightness distortion. The odd rows show the original image and the results of various enhancement methods, and the even rows show the visualization of each method's lightness distortion (RD).

Table 1. Qantitative measurement results of lighness distortion (LOE).

Methods	VV	LIME-data	NPE-data	NPE-ex1	NPE-ex2	NPE-ex3	MEF	IUS
AMSR	3892	3300	3298	2969	3212	3006	3264	4658
Dong	853	1244	1012	1426	1096	1466	1065	758
NPE	821	1471	646	841	776	1130	1158	559
LIME	1275	1324	1120	1322	1215	1319	1079	1306
Ours	**287**	**479**	**308**	**320**	**324**	**379**	**326**	**311**

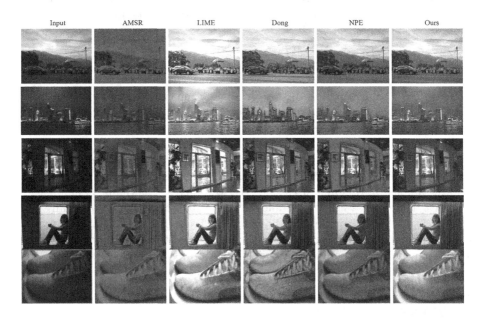

Fig. 5. Visual comparison among the competitors on different scenes.

we can find our results have the smallest lightness distortion. The results of AMSR lose the global lightness order and have the biggest lightness distortion. Although the results of LIME is visually pleasant, they also suffer from lightness distortion. The results of Dong and NPE can only retain the lightness order in the well-exposed regions.

4 Conclusion

In this paper, we propose an exposure fusion framework and an enhancement algorithm to provide an accurate contrast enhancement. Based on our framework, we solve three problems: (1) we borrow the illumination estimation techniques to obtain the weight matrix for image fusion. (2) we introduce our camera response model to synthesize multi-exposure images. (3) we find the best exposure ratio so that the synthetic image is well-exposed in the regions where the original image under-exposed. The final enhancement result is obtained by fusing the input image and the synthetic image according to the weight matrix. The experimental results have revealed the advance of our method compared with several state-of-the-art alternatives. To encourage future works and allow more experimental verification and comparisons, we make the source code open. More testing results can be found on our project website: https://baidut.github.io/OpenCE/caip2017.html.

Acknowledgments. This work was supported by the grant of National Science Foundation of China (No.U1611461), Shenzhen Peacock Plan (20130408-183003656), and Science and Technology Planning Project of Guangdong Province, China (No. 2014B090910001 and No. 2014B010117007).

References

1. Aydin, T.O., Mantiuk, R., Myszkowski, K., Seidel, H.P.: Dynamic range independent image quality assessment. ACM Trans. Graph. (TOG) **27**(3), 69 (2008)
2. Beghdadi, A., Le Negrate, A.: Contrast enhancement technique based on local detection of edges. Comput. Vis. Graph. Image Process. **46**(2), 162–174 (1989)
3. Chen, S.D., Ramli, A.R.: Minimum mean brightness error bi-histogram equalization in contrast enhancement. IEEE Trans. Consum. Electron. **49**(4), 1310–1319 (2003)
4. Dong, X., Wang, G., Pang, Y., Li, W., Wen, J., Meng, W., Lu, Y.: Fast efficient algorithm for enhancement of low lighting video. In: 2011 IEEE International Conference on Multimedia and Expo, pp. 1–6. IEEE (2011)
5. Guo, X.: Lime: a method for low-light image enhancement. In: Proceedings of the 2016 ACM on Multimedia Conference, pp. 87–91. ACM (2016)
6. Ibrahim, H., Kong, N.S.P.: Brightness preserving dynamic histogram equalization for image contrast enhancement. IEEE Trans. Consum. Electron. **53**(4), 1752–1758 (2007)
7. Jobson, D.J., Rahman, Z., Woodell, G.A.: A multiscale retinex for bridging the gap between color images and the human observation of scenes. IEEE Trans. Image Process. **6**(7), 965–976 (1997)
8. Karaduzovic-Hadziabdic, K., Telalovic, J.H., Mantiuk, R.: Subjective and objective evaluation of multi-exposure high dynamic range image deghosting methods (2016)
9. Lee, C.H., Shih, J.L., Lien, C.C., Han, C.C.: Adaptive multiscale retinex for image contrast enhancement. In: 2013 International Conference on Signal-Image Technology & Internet-Based Systems (SITIS), pp. 43–50. IEEE (2013)
10. Ma, K., Zeng, K., Wang, Z.: Perceptual quality assessment for multi-exposure image fusion. IEEE Trans. Image Process. **24**(11), 3345–3356 (2015)
11. Peli, E.: Contrast in complex images. JOSA A **7**(10), 2032–2040 (1990)
12. Reza, A.M.: Realization of the contrast limited adaptive histogram equalization (clahe) for real-time image enhancement. J. VLSI Signal Process. Syst. Signal Image Video Technol. **38**(1), 35–44 (2004)
13. Wang, C., Ye, Z.: Brightness preserving histogram equalization with maximum entropy: a variational perspective. IEEE Trans. Consum. Electron. **51**(4), 1326–1334 (2005)

14. Wang, S., Zheng, J., Hu, H.M., Li, B.: Naturalness preserved enhancement algorithm for non-uniform illumination images. IEEE Trans. Image Process. **22**(9), 3538–3548 (2013)
15. Xu, L., Yan, Q., Xia, Y., Jia, J.: Structure extraction from texture via relative total variation. ACM Trans. Graph. (TOG) **31**(6), 139 (2012)
16. Ying, Z., Li, G., Ren, Y., Wang, R., Wang, W.: A new low-light image enhancement algorithm using camera response model, manuscript submitted for publication (2017)

Quaternionic Flower Pollination Algorithm

Gustavo H. Rosa[1], Luis C.S. Afonso[2], Alexandre Baldassin[3], João P. Papa[1(✉)], and Xin-She Yang[4]

[1] School of Sciences, UNESP - São Paulo State University, Bauru, Brazil
gth.rosa@uol.com.br, papa@fc.unesp.br
[2] Department of Computing, UFSCar - Federal University of São Carlos,
São Carlos, Brazil
sugi.luis@dc.ufscar.br
[3] Institute of Natural Sciences and Technology,
UNESP - São Paulo State University, Rio Claro, Brazil
alex@rc.unesp.br
[4] School of Science and Technology, Middlesex University, London, UK
x.yang@mdx.ac.uk

Abstract. Metaheuristic-based optimization techniques offer an elegant and easy-to-follow framework to optimize different types of problems, ranging from aerodynamics to machine learning. Though such techniques are suitable for global optimization, they can still be get trapped locally under certain conditions, thus leading to reduced performance. In this work, we propose a quaternionic-based Flower Pollination Algorithm (FPA), which extends standard FPA to possibly smoother search spaces based on hypercomplex representations. We show the proposed approach is more accurate than five other metaheuristic techniques in four benchmarking functions. We also present a parallel version of the proposed approach that runs much faster.

Keywords: Flower Pollination Algorithm · Quaternions · Metaheuristics

1 Introduction

Metaheuristic-based techniques have been extensively employed to solve a number of problems in the literature, as well as in the real world. Applications that range from Computer Science, Engineering and Medicine are among the most pursued ones, with clear benefits in others. Usually, such techniques are nature-inspired, which means their mechanism is based on the way the nature finds a way to solve problems, such as the shortest path between ants and some source of food, echolocation in bats looking for preys, and a flock of birds trying to minimize the energy when flying.

Particle Swarm Optimization (PSO) [1], for instance, emulates three main principles in order to solve optimization problems: (i) mimicking, (ii) simulation, (iii) evaluation. Roughly speaking, the idea is to model the collective behavior of a flock of birds when trying to reach some common goal (e.g., source of food). The idea is to place a number of birds at random positions in the search space

M. Felsberg et al. (Eds.): CAIP 2017, Part II, LNCS 10425, pp. 47–58, 2017.
DOI: 10.1007/978-3-319-64698-5_5

(a given position means a possible solution to the problem), which evolves according to some dynamic equations and some organized behavior may emerge.

Later on, other metaheuristic techniques have been proposed as well. An interesting example concerns the Bat Algorithm (BA) [2], which aims at modeling the mechanism of social behaviors of bats, which use echolocation to send and hunt for preys. Also, Artificial Bee Colony (ABC) [3] mimics the collective and social mechanism of communication of bees. Another technique that has attracted attention in the last years concerns the so-called Flower Pollination Algorithm (FPA) [4], which is inspired by the flower pollination process that is usually carried out by plants to ensure their reproduction in nature.

Despite their outstanding results in a number of applications, they can still potentially get trapped at local optima under certain conditions, which somehow foster the research on different approaches to handle the problem of optimizing complex fitness functions. A promising direction is towards embedding the search space into different representations of those that are normally used. Fister et al. [5], for instance, presented a Firefly Algorithm embedded in a quaternion-based space. The idea is to map each possible solution encoded as an n-dimensional firefly vector as an $n \times 4$ tensor, where each decision variable is now encoded as a hypercomplex number, which contains four parts: one real-valued part and the remaining three parts as imaginary numbers. Later on, Fister et al. [6] proposed a quaternion-based Bat Algorithm, and Papa et al. [7] introduced Harmony Search in the context of Deep Belief Network fine-tuning using quaternion representations.

Normalized quaternions, also known as versors, are widely used to represent the orientation of objects in three-dimensional spaces, and thus can be efficient to perform rotations in such spaces. The idea behind using quaternionic search spaces is based on the possibility of having smoother fitness landscapes, although it has not been mathematically demonstrated. However, the results obtained previously support such assumption [5–7].

In this work, we proposed a hypercomplex-based variant of the Flower Pollination Algorithm called Quaternionic Flower Pollination Algorithm (QFPA), which is evaluated under four benchmarking functions under different numbers of dimensions, as well as it has been compared against five other metaheuristic techniques, including standard FPA. We show the proposed approach can achieve more accurate results, mainly at higher dimensional problems, as well as it can converge faster. Another contribution of this work is to propose a parallel version of Flower Pollination Algorithm, which is considerably faster than the original version.

The remainder of this paper is organized as follows. Sections 2 and 3 present the theoretical background related to quaternions, Flower Pollination Algorithm, and its quaternionic version, and the methodology employed in this work, respectively. Section 4 discusses the experiments adopted in this work, and Sect. 5 states conclusions and future works.

2   Theoretical Background

In this section, we present the theoretical background related to quaternions, and Flower Pollination Algorithm, as well as the proposed Quaternionic Flower Pollination Algorithm.

2.1   Quaternions

A quaternion q is composed of real and complex numbers, i.e., $q = z_0 + z_1 i + z_2 j + z_3 k$, where $z_0, z_1, z_2, z_3 \in \Re$ and i, j, k are imaginary numbers following the next set of equations:

$$ij = k, \tag{1}$$
$$jk = i, \tag{2}$$
$$ki = j, \tag{3}$$
$$ji = -k, \tag{4}$$
$$kj = -i, \tag{5}$$
$$ik = -j, \tag{6}$$

and

$$i^2 = j^2 = k^2 = -1. \tag{7}$$

Roughly speaking, a quaternion q is represented in a 4-dimensional space over the real numbers, i.e., $\Re^4$. Actually, we can consider the real numbers only, since most applications do not consider the imaginary part, as the one addressed in this work.

Given two quaternions $q_1 = z_0 + z_1 i + z_2 j + z_3 k$ and $q_2 = y_0 + y_1 i + y_2 j + y_3 k$, and a real-valued number θ, denoted as a scalar, the quaternion algebra defines a set of main operations. The addition, for instance, can be defined by:

$$\begin{aligned} q_1 + q_2 &= (z_0 + z_1 i + z_2 j + z_3 k) + (y_0 + y_1 i + y_2 j + y_3 k) \\ &= (z_0 + y_0) + (z_1 + y_1)i + (z_2 + y_2)j + (z_3 + y_3)k, \end{aligned} \tag{8}$$

while the subtraction is defined as follows:

$$\begin{aligned} q_1 - q_2 &= (z_0 + z_1 i + z_2 j + z_3 k) - (y_0 + y_1 i + y_2 j + y_3 k) \\ &= (z_0 - y_0) + (z_1 - y_1)i + (z_2 - y_2)j + (z_3 - y_3)k. \end{aligned} \tag{9}$$

Furthermore, we can define a scalar multiplication as follows:

$$\theta q_1 = \theta(z_0 + z_1 i + z_2 j + z_3 k) = \theta z_0 + \theta z_1 i + \theta z_2 j + \theta z_3 k, \tag{10}$$

where $\theta \in \Re$.

Another important operation is the norm, which maps a given quaternion to a real-valued number, as follows:

$$N(q_1) = N(z_0 + z_1 i + z_2 j + z_3 k)$$
$$= \sqrt{z_0^2 + z_1^2 + z_2^2 + z_3^2}. \tag{11}$$

Finally, Fister et al. [5,6] introduced two other operations, $qrand$ and $qzero$. The former initializes a given quaternion with values drawn from a Gaussian distribution, and it can be defined as follows:

$$qrand() = \{z_i = \mathcal{N}(0,1)|i \in \{0,1,2,3\}\}. \tag{12}$$

The latter function initializes a quaternion with zero values, as follows:

$$qzero() = \{z_i = 0|i \in \{0,1,2,3\}\}. \tag{13}$$

Although there are other operations, we define only the ones employed in this work.

2.2 Flower Pollination Algorithm

The Flower Pollination Algorithm proposed by [4] is inspired by the flow pollination process of flowering plants. The FPA is governed by four basic rules:

1. Biotic cross-pollination can be considered as a process of global pollination, and pollen-carrying pollinators move in a way that obeys Lévy flights;
2. For local pollination, abiotic pollination and self-pollination are used;
3. Pollinators such as insects can develop flower constancy, which is equivalent to a reproduction probability that is proportional to the similarity of two flowers involved; and
4. The interaction or switching of local pollination and global pollination can be controlled by a switch probability $p \in [0,1]$, slightly biased towards local pollination.

In order to model the updating formulas, the above rules have to be converted into proper updating equations. For example, in the global pollination step, flower pollen gametes are carried by pollinators such as insects, and pollen can travel over a long distance because insects can often fly and move over a much longer range. Therefore, Rules 1 and 3 can be represented mathematically as follows:

$$\mathbf{x}_i^{(t+1)} = \mathbf{x}_i^t + \alpha L(\lambda)(\mathbf{g}^* - \mathbf{x}_i^t), \tag{14}$$

where

$$L(\lambda) \sim \frac{\lambda \cdot \Gamma(\lambda) \cdot \sin(\lambda)}{\pi} \cdot \frac{1}{s^{1+\lambda}}, \quad s \gg s_0 > 0 \tag{15}$$

where $\mathbf{x}_i^t$ is the pollen i (solution vector) at iteration t, g^* is the current best solution among all solutions at the current generation, and α is a scaling factor to control the step size. $L(\lambda)$ is the Lévy-flights step size, that corresponds to the strength of the pollination, $\Gamma(\lambda)$ stands for the gamma function and s is the step size. Since insects may move over a long distance with various distance steps, a Lévy flight can be used to mimic this characteristic efficiently.

For local pollination, both Rules 2 and 3 can be represented as:

$$\mathbf{x}_i^{t+1} = \mathbf{x}_i^t + \epsilon(\mathbf{x}_j^t - \mathbf{x}_k^t), \tag{16}$$

where $\mathbf{x}_j^t$ and $\mathbf{x}_k^t$ are pollen from different flowers j and k of the same plant species at time step t. This mimics flower constancy in a limited neighborhood. Mathematically, if $\mathbf{x}_j^t$ and $\mathbf{x}_k^t$ come from the same species or are selected from the same population, it equivalently becomes a local random walk if ϵ is drawn from a uniform distribution in $[0, 1]$. In order to mimic the local and global flower pollination, a switch probability (Rule 4) or proximity probability p is used.

2.3   Quaternionic Flower Pollination Algorithm

The Quaternionic Flower Pollination Algorithm is a natural extension of standard FPA, where the arithmetic operations in the equations that model the interaction among pollinators are replaced by quaternionic algebra.

As aforementioned in Sect. 2.2, each pollinator $\mathbf{x}_i^t$ encodes a possible solution of the problem at time step t. Considering one has a problem with n decision variables, i.e., $\mathbf{x}_i^t \in \Re^n$, each jth decision variable to be optimized is now encoded as a quaternion, i.e. $\hat{x}_{ij}^t \in \Re^4$, where $\hat{x}_{ij}$ stands for the quaternionic representation of x_{ij}^t. Therefore, instead of modeling each possible solution as an n-dimensional array in the search space, one now has a $4 \times n$ tensor $\hat{\mathbf{x}}_i$, which encodes the set of n quaternions that are used to model each pollinator i.

Based on such assumption, we just need to adapt the quaternionic algebra described in Sect. 2.1 into Eqs. 14 and 16. Since Eq. 15 outputs an array with a real-valued distribution, there is no need to change its formulation. Roughly speaking, after initializing the new tensors, we need to evaluate them to asses the quality of each pollinator. Basically, for each $\hat{x}_{ij}^t$, we can compute its norm $N(\hat{x}_{ij}^t)$ according to Eq. 11, thus going back to the original space x_{ij}^t, since one needs a real-valued value to evaluate the fitness function. Essentially, we are looking for the best value of each quaternion $\hat{x}_{ij}^t$ that maximizes a real-valued fitness function.

3   Methodology

We propose to model the problem of optimizing benchmarking functions through meta-heuristic techniques and their hypercomplex representation, i.e. quaternions. As aforementioned in Sect. 2.1, quaternions employ a broader search space, where each parameter is encoded by an $\mathbb{R}^4$-space, and then mapped to an $\mathbb{R}$-space within the chosen limits. Roughly speaking, the proposed approach aims at selecting the set of parameters that minimizes the value of the evaluated functions.

The main problem when working with hypercomplex representations concerns the boundaries of $N(q_i^j)$, which is the norm of the quaternion associated to each decision variable $\mathbf{x}_i^j$. Therefore, we have devised a function to tackle this issue, mapping $\mathbf{x}_i^j$ to plausible boundaries, as follows:

$$span(x_i^j) = (U_j - L_j)\left(\frac{N(q_i^j)}{\sqrt{M}}\right) + L_j, \tag{17}$$

where U_j and L_j stand for the upper and lower bounds of decision variable $\mathbf{x}_i^j$, respectively, and $N(q_i^j)$ denotes the norm over quaternion q_i^j, and M corresponds to the number of dimensions of the hypercomplex representation (i.e., $M = 4$ concerning quaternions).

3.1 Benchmarking Functions

An interesting approach used to validate and compare the performance of optimization techniques is to apply them over benchmarking functions. However, it is important to include a wide variety of test functions, such as unimodal, multimodal, separable, non-separable, regular, irregular or even multi-dimensional functions.

Therefore, we have selected 4 different benchmarking functions in order to cross-check our proposed optimization methods. Table 1 depicts the employed functions, where the identifier column stands for their names, the function column stands for their mathematical formulations, the bounds column stands for their variables lower and upper bounds, and the $f(x^*)$ column stands for their optimum values.

Additionally, in order to provide a more comprehensive view of each function, we denote below some of their main characteristics:

- Lévy (f_1) - continuous, differentiable and multimodal;
- Rosenbrock (f_2) - continuous, differentiable, non-separable, scalable and unimodal;
- Schewefel (f_3) - continuous, differentiable, partially-separable, scalable and unimodal;
- Xin-She Yang #1 (f_4) - separable.

Table 1. Benchmarking functions.

Identifier	Function	Bounds	$f(x^*)$
Lévy	$f_1(x) = \sin^2(\pi w_1) + \sum_{i=1}^{D-1}(w_i - 1)^2[1 + 10\sin^2(\pi w_i + 1)]$ $+ (w_d - 1)^2[1 + \sin^2(2\pi w_d)]$, where $w_i = 1 + \frac{x_i - 1}{4}$	$-10 \le x_i \le 10$	0
Rosenbrock	$f_2(x) = \sum_{i=1}^{D-1}[100(x_{i+1} - x_i^2)^2 + (x_i - 1)^2]$	$-30 \le x_i \le 30$	0
Schewefel	$f_3(x) = (\sum_{i=1}^{D} x_i^2)^{\sqrt{\pi}}$	$-100 \le x_i \le 100$	0
Xin-She Yang #1	$f_4(x) = \sum_{i=1}^{D} \epsilon_i |x_i|^i$	$-5 \le x_i \le 5$	0

Table 2. Parameter configuration.

Technique	Parameters
ABC	Trials limit = 1000
BA	$f_{min} = 0, f_{max} = 2, A = 0.5, r = 0.5$
FPA	$\beta = 1.5, p = 0.8$
PSO	$c_1 = 1.7, c_2 = 1.7, w = 0.7$

3.2 Experimental Setup

In this work, we compared five meta-heuristic approaches (empirically chosen) against the proposed one:

- Artificial Bee Colony (ABC) [3];
- Bat Algorithm (BA) [2] and Quaternionic ABC [6];
- Flower Pollination Algorithm (FPA) [4] and Quaternionic FPA (QFPA); and
- Particle Swarm Optimization [1].

We also evaluated the robustness of parameter fine-tuning in three distinct dimensions D: 10, 50 and 100.

In order to provide a deeper analysis, we conducted experiments with 25 runs and the following metrics: best fitness value, mean of best fitness value and standard deviation of best fitness value. We also employed 100 agents over $2,000D$ iterations for convergence considering all techniques. Therefore, this means we have $20,000$ iterations for a $D = 10$ space, $100,000$ iterations for a $D = 50$ space and $200,000$ iterations for a $D = 100$ space. Table 2 presents the parameter configuration for each optimization technique[1]. In regard to the source-code, we used library LibOPT[2].

Regarding ABC, the number of trials limit stands for the amount of trials that a solution can be improved by an employee bee. Considering BA, we have minimum and maximum frequency ranges, f_{min} and f_{max}, respectively, as well as the loudness parameter A and pulse rate r. In regard to FPA, β is used to compute the Lévy distribution, while p is the probability of local pollination. Finally, PSO defines w as the inertia weight, and c_1 and c_2 as the control parameters.

4 Experiments

The results are presented for each number of dimensions considered in this work ($D = 10$, 50 and 100) within the following format: (BF, MBF, SDBF), where BF stands for "best fitness value", MBF for "mean of best fitness value", and SDBF for "standard deviation of best fitness value". Tables 3, 4, 5, and 6 present the results concerning functions f_1 to f_4, respectively, being the best values in

---

[1] Notice that all parameter values used in this work have been empirically set.
[2] https://github.com/jppbsi/LibOPT.

Table 3. Results concerning Lévy's (f_1) function.

Technique	$D = 10$	$D = 50$	$D = 100$
ABC	(7.77e-03, 1.12e+00, 0.00e+00)	6.12e+00, 6.12e+00, 0.00e+00)	(1.24e+01, 1.24e+01, 0.00e+00)
BA	(0.00e+00, 2.72e-01, 3.85e-01)	(7.45e+00, 1.26e+01, 1.79e+01)	(1.33e+01, 1.14e+01, 1.61e+01)
QBA	(2.69e-01, 5.44e-01, 7.69e-01)	(3.24e+00, 3.72e+00, 5.26e+00)	(8.59e+00, 7.39e+00, 1.05e+01)
FPA	**(0.00e+00, 0.00e+00, 0.00e+00)**	(8.95e-02, 1.49e-01, 2.58e-01)	(1.07e+00, 1.04e+00, 1.81e+00)
QFPA	**(0.00e+00, 0.00e+00, 0.00e+00)**	**(0.00e+00, 2.98e-02, 5.17e-02)**	**(8.95e-02, 2.98e-02, 5.17e-02)**
PSO	(0.00e+00, 6.71e-02, 1.34e-01)	(4.40e+00, 7.13e+00, 1.43e+01)	(8.56e+00, 2.49e+00, 4.98e+00)

Table 4. Results concerning Rosenbrock's (f_2) function.

Technique	$D = 10$	$D = 50$	$D = 100$
ABC	(4.91e-04, 4.91e-04, 0.00e+00)	(3.40e-03, 4.95e+00, 0.00e+00)	(8.27e+01, 9.28e+01, 0.00e+00)
BA	(1.10e-05, 3.63e-04, 5.13e-04)	(1.32e-02, 5.03e+00, 7.11e+00)	(5.62e+01, 4.79e+01, 6.77e+01)
QBA	(1.21e-01, 2.13e+00, 3.01e+00)	(4.05e+01, 2.43e+01, 3.43e+01)	(8.10e+01, 4.72e+01, 6.68e+01)
FPA	**(0.00e+00, 0.00e+00, 0.00e+00)**	**(0.00e+00, 0.00e+00, 0.00e+00)**	(0.00e+00, 1.35e+00, 2.35e+00)
QFPA	(0.00e+00, 7.00e-06, 1.20e-05)	**(0.00e+00, 0.00e+00, 0.00e+00)**	**(0.00e+00, 1.61e-02, 2.79e-02)**
PSO	(8.97e-01, 2.07e+00, 4.14e+00)	(1.83e+02, 3.28e+02, 6.57e+02)	(1.35e+02, 1.29e+03, 2.58e+03)

bold according to the Wilcoxon signed-rank statistical test with significance of 0.05 [8]. Notice the statistical test evaluated the MBF measure only.

The experimental results lead us to some interesting observations: (i) both FPA and QFPA obtained the best results concerning all functions so far; (ii) QFPA achieved better results when compared to the standard approaches, mainly at higher dimensions, where the fitness landscapes are more prone to present complex representations, i.e., we may find a more difficult optimization situation; and (iii) QFPA obtained more accurate results than its naïve version FPA. Actually, concerning two out of four functions, i.e., f_2 and f_4, FPA obtained slightly better results than QFPA at lower dimensions. Such results may imply the assumption that hypercomplex-based search spaces seem to be more suitable at higher-dimensional problems.

In order to evaluate the convergence process, we also considered an analysis that takes into account the value of the fitness function among the iterations. Figure 1 depicts the convergence plot considering Lévy's function. Clearly, one can observe QFPA converged faster than all compared techniques considering

Table 5. Results concerning Schewefel's (f_3) function.

Technique	$D = 10$	$D = 50$	$D = 100$
ABC	(0.00e+00, 0.00e+00, 0.00e+00)	(0.00e+00, 0.00e+00, 0.00e+00)	(0.00e+00, 0.00e+00, 0.00e+00)
BA	(0.00e+00, 0.00e+00, 0.00e+00)	(0.00e+00, 0.00e+00, 0.00e+00)	(0.00e+00, 0.00e+00, 0.00e+00)
QBA	(2.00e-06, 1.00e-06, 1.00e-06)	(8.14e-03, 4.81e-03, 6.80e-03)	(1.00e-01, 6.75e-02, 9.55e-02)
FPA	(0.00e+00, 0.00e+00, 0.00e+00)	(0.00e+00, 0.00e+00, 0.00e+00)	(0.00e+00, 0.00e+00, 0.00e+00)
QFPA	(0.00e+00, 0.00e+00, 0.00e+00)	(0.00e+00, 0.00e+00, 0.00e+00)	(0.00e+00, 0.00e+00, 0.00e+00)
PSO	(0.00e+00, 0.00e+00, 0.00e+00)	(1.44e-01, 1.02e+00, 2.05e+00)	(5.00e-06, 2.09e-03, 4.18e-03)

Table 6. Results concerning Xin-She Yang's #1 (f_4) function.

Technique	$D = 10$	$D = 50$	$D = 100$
ABC	(0.00e+00, 0.00e+00, 0.00e+00)	(0.00e+00, 0.00e+00, 0.00e+00)	(0.00e+00, 0.00e+00, 0.00e+00)
BA	(7.26e-03, 2.68e-02, 3.79e-02)	(7.80e+02, 2.34e+09, 3.31e+09)	(2.28e+13, 3.67e+23, 5.19e+23)
QBA	(2.36e-03, 4.81e-02, 6.81e-02)	(4.64e+00, 3.56e+04, 5.04e+04)	(1.78e+11, 1.26e+16, 1.78e+16)
FPA	(0.00e+00, 0.00e+00, 0.00e+00)	(0.00e+00, 0.00e+00, 0.00e+00)	(0.00e+00, 1.00e-06, 2.00e-06)
QFPA	(0.00e+00, 4.00e-06, 6.00e-06)	(0.00e+00, 0.00e+00, 0.00e+00)	(0.00e+00, 0.00e+00, 0.00e+00)
PSO	(3.60e-05, 1.10e-05, 2.10e-05)	(9.96e-02, 1.53e-01, 3.07e-01)	(2.77e+03, 7.01e+09, 1.40e+10)

all dimensions. Particularly, FPA and QFPA were the best ones in this example, with results considerably better than others.

We also considered another function, but now where QFPA did not outperform FPA at lower dimensions for the very same analysis. Figure 2 displays the convergence analysis considering Xin-She Yang's #1 function for different dimensions. A closer look at the bottom-left part of Fig. 2a reveals that QFPA converged faster than all compared techniques, although the results were not the best ones in the end of the convergence process. Concerning higher dimensions, QFPA took a little longer than others, but achieving the best results in the end.

Roughly speaking, we observed the proposed approach can achieve considerably better results in almost all functions and dimensions evaluated in this work, as well as hypercomplex spaces seem to provide smoother fitness landscapes, which can be stressed by the results obtained in the experiments. However, it is worth mentioning hypercomplex spaces require a higher computational load, since each decision variable is now represented by more variables.

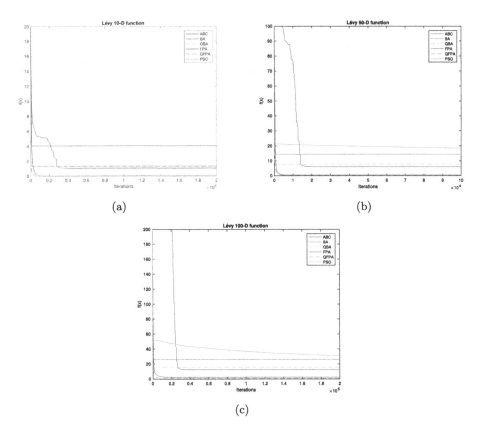

Fig. 1. Convergence plot considering Lévy's function. (a) $D = 10$, (b) $D = 50$ and (c) $D = 100$.

4.1   Parallel Quaternionic Flower Pollination Algorithm

In this section, we present the results related to the Parallel Quaternionic Flower Pollination Algorithm (PQFPA). Roughly speaking, QFPA has three main loops: (i) the outer one related to the iterations concerning the convergence process, (ii) the inner one that runs over all pollinators, and (iii) one more loop that is related to the quaternion coefficients for each decision variable. The outer loop can not be parallelized, since the iterations are time-dependent, and the loop related to quaternions has four iterations only, which does not worth parallelizing. Therefore, we decided to make the inner loop, i.e., the one related to the pollinators, to be parallelized by means of the OpenMP library [9]. Thus, each thread will be in charge of a set of pollinators.

The experiments were performed as follows: we considered $D = 10$ dimensions during $2,000 \times 10 = 20,000$ iterations (similarly to the previous section), as well as we also assessed the speed up considering 10, 100, 1,000 and 10,000 pollinators. This procedure was applied for both QFPA and PQFPA concerning the four benchmarking functions, and the results stand for the average over

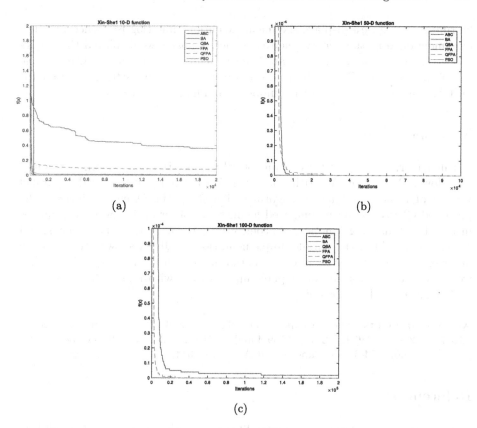

Fig. 2. Convergence plot considering Xin-She Yang's #1 function. (a) $D = 10$, (b) $D = 50$ and (c) $D = 100$.

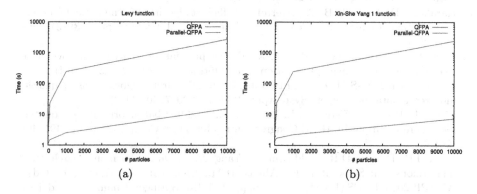

Fig. 3. Mean execution time concerning: (a) Lévy's and (b) Xin-She Yang's #1 benchmarking functions.

25 runnings. Figure 3 displays the mean execution time (log [s]) concerning the aforementioned methodology. Due to the lack of space, we displayed the results concerning Lévy's and Xin-She Yang's #1 benchmarking functions. Clearly, one can observe that QFPA has been in the order of 10^2 times faster than standard QFPA, being quite useful as the number of pollinators increases.

5   Conclusions

In this paper, we proposed two main contributions: (i) a quaternionic version of the Flower Pollination Algorithm for optimization purposes, and (ii) a parallel version of Flower Pollination Algorithm based on the OpenMP library. The proposed QFPA has been compared against five other optimization techniques in four benchmarking functions, outperforming them all in higher dimensional search spaces. Also, the parallel implementation of QFPA showed to be in the order of 10^2 times faster than its sequential implementation. In regard to future works, we aim at working on hypercomplex spaces with higher dimensions, as well as on parallel versions of them.

Acknowledgments. The authors would like to thank FAPESP grants #2014/16250-9, #2014/12236-1, #2015/25739-4 and #2016/21243-7, as well as Capes, CNPq grant #306166/2014-3, and Capes PROCAD 2966/2014.

References

1. Kennedy, J., Eberhart, R.C.: Swarm Intelligence. Morgan Kaufmann Publishers Inc., San Francisco (2001)
2. Yang, X.-S., Gandomi, A.H.: Bat algorithm: a novel approach for global engineering optimization. Eng. Comput. **29**(5), 464–483 (2012)
3. Karaboga, D., Basturk, B.: A powerful and efficient algorithm for numerical function optimization: Artificial Bee Colony (ABC) algorithm. J. Global Optim. **39**(3), 459–471 (2007)
4. Yang, S.-S., Karamanoglu, M., He, X.: Flower pollination algorithm: a novel approach for multiobjective optimization. Eng. Optim. **46**(9), 1222–1237 (2014)
5. Fister, I., Yang, X.-S., Brest, J., Fister Jr., I.: Modified firefly algorithm using quaternion representation. Expert Syst. Appl. **40**(18), 7220–7230 (2013)
6. Fister, I., Brest, J., Fister Jr., I., Yang, X.-S.: Modified bat algorithm with quaternion representation. In: IEEE Congress on Evolutionary Computation, pp. 491–498 (2015)
7. Papa, J.P., Pereira, D.R., Baldassin, A., Yang, X.-S.: On the harmony search using quaternions. In: Schwenker, F., Abbas, H.M., El Gayar, N., Trentin, E. (eds.) ANNPR 2016. LNCS (LNAI), vol. 9896, pp. 126–137. Springer, Cham (2016). doi:10.1007/978-3-319-46182-3_11
8. Wilcoxon, F.: Individual comparisons by ranking methods. Biom. Bull. **1**(6), 80–83 (1945)
9. Dagum, L., Menon, R.: Openmp: an industry-standard api for shared-memory programming. IEEE Comput. Sci. Eng. **5**(1), 46–55 (1998)

Automated Cell Nuclei Segmentation in Pleural Effusion Cytology Using Active Appearance Model

Elif Baykal[1]([✉]), Hulya Dogan[1], Murat Ekinci[1], Mustafa Emre Ercin[2], and Safak Ersoz[2]

[1] Department of Computer Engineering, Karadeniz Technical University, Trabzon, Turkey
{ebaykal,hulya,ekinci}@ktu.edu.tr
[2] Department of Pathology, Karadeniz Technical University, Trabzon, Turkey
{drmustafaemreercin,sersoz}@ktu.edu.tr

Abstract. Pleural effusion is common in clinical practice and it is a frequently encountered specimen type in cytopathological assessment. In addition to being time-consuming and subjective, this assessment also causes inter-observer and intra-observer variability, and therefore an automated system is needed. In visual examination of cytopathological images, cell nuclei present significant diagnostic value for early cancer detection and prevention. So, efficient and accurate segmentation of cell nuclei is one of the prerequisite steps for automated analysis of cytopathological images. Nuclei segmentation also yields the following automated microscopy applications, such as cell counting and classification. In this paper, we present an automated technique based on active appearance model (AAM) for cell nuclei segmentation in pleural effusion cytology images. The AAM utilizes from both the shape and texture features of the nuclei. Experimental results indicate that the proposed method separates the nuclei from background effectively. In addition, comparisons are made with the segmentation methods of thresholding-based, clustering-based and graph-based, which show that the results obtained with the AAM method are actually more closer to the ground truth.

Keywords: Pleural effusion · Active appearance model · Automated microscopy applications · Cytopathological images · Nuclei segmentation

1 Introduction

There are serous cavities in human body, including pleura, peritoneum and pericardium surrounded by visceral and parietal surfaces. In a healthy person, a small amount of fluid is secreted by the parietal surface in the serous space to facilitate easy movement of the visceral and parietal surfaces on top of each other. This imbalance in the reabsorption of the fluid by the visceral surface

© Springer International Publishing AG 2017
M. Felsberg et al. (Eds.): CAIP 2017, Part II, LNCS 10425, pp. 59–69, 2017.
DOI: 10.1007/978-3-319-64698-5_6

leads to accumulation of fluid in the serous spaces and thus formation of serous effusion [1]. A pleural effusion, a type of serous effusion, is an abnormal collection of fluid within the pleural space and it is a frequently encountered specimen type in cytopathological analysis [2]. Identification of the cancer cells in pleural effusion cytology allows for early diagnosis of cancer and also for staging, prognosis and monitoring these cells.

The cytopathological examination carried out by pathologists requires screening of the nuclei and cytoplasm structures contained in all the cells in the specimen by hand-eye coordination. In addition to being time-consuming and subjective, this examination also causes inter-observer and intra-observer variability, so an automated system is required [3]. The automated segmentation of cell nuclei in this examination is regarded as a corner stone for cancer diagnosis because the nuclei is the most salient structure in the cell as well as the fact that it offers important morphological features and changes [4]. Nuclei segmentation is an important step for the following automated microscopy applications, such as cell counting and classification.

There are many studies related to segmentation of cell nuclei in the literature that include thresholding-based, watershed transform, morphology-based, deformable models, clustering-based, graph-based, supervised classification, and so on [5]. Zhou et al. [6] used adaptive thresholding followed by watershed algorithm to segment nuclei. Although adaptive thresholding can effectively distinguish the cell nuclei from the background, it fails to distinguish the touching/overlapping nuclei. To overcome this problem, watershed transform is also applied to seperate cell clusters. Watershed transform generally results in oversegmentation so it requires region merging approach. Otherwise, Schmitt et al. [7] developed a new morphological technique generated by iterative erosion for segmentation of cell clumps. This technique may fail in dense cell clumps and may also result in undersegmentation results. Mathematical morphology operation is usually used as preprocessing step to enhance subsequent segmentation. Kothari et al. [8] addressed the issue of cell cluster segmentation. They have proposed a three-step approach, k-means clustering, morphological reconstruction and edge detection in pathology images. Zhou et al. [9] also proposed fuzzy c-means based multi threshold cell segmentation in RNAi images. Jung et al. [10] presented Expectation Maximization (EM) approach for nucleus segmentation in carcinoma and cervical images. Clustering-based methods might not result in the final segmentation, but give support for estimating the object boundaries. Al-Kohafi et al. [11] have proposed a graph-cut based method consisting of two stage for nuclei segmentation in histopathology images.

Although the thresholding-based, morphology-based and watershed algorithm can be implemented easily, they are not successful in touching/overlapping cell nuclei segmentation. To obtain good segmentation performance, thresholding-based methods are usually implemented along with other image processing techniques such as contrast enhancement, morphological operations. On the other hand, clustering-based and graph-based methods might have computational complexity and also require priori knowledge about the distribution so they might not be consistent for the arbitrary shapes of objects. Graph-based methods usually requires the user to interactively specify some pixels as a part of the foreground and background scene.

Cells in cytopathological images show very different appearances due to morphological structures and different staining as well as being highly variable, deformable objects. Deformable models are widely used because they are suitable for biomedical image segmentation. These models are also able to discriminate overlapping cells and nuclei [12–14]. In this paper, we propose to use AAM, which can be an alternative deformable model for nuclei segmentation. The AAM uses the shape and texture information in the cell nuclei together. Experimental results on pleural effusion cytology images demonstrate that the AAM is skillful for the cell nuclei segmentation and also it is consistent for the touching or overlapping nuclei. Unlike commonly used segmentation methods in the literature, the AAM provides a closed form of the boundary of the nuclei. Therefore, no other post processing steps are required, such as edge enhancement and border tracking.

The rest of the paper is organized as follows. The detailed analysis of proposed nuclei segmentation method AAM is presented in Sect. 2. The experimental results and quantitative comparisons with known methods are presented in Sect. 3. Conclusion is given in Sect. 4.

2   Theoretical Background

2.1   Active Appearance Model

The AAM algorithm was proposed in 1998 by Cootes et al. [15] and generalized for image segmentation purposes. The AAM algorithm, which combines the shape and the texture information in an image, can be divided into two stages: training a model and fitting a model. While training a model, both shape and texture model are established using the training samples. The model parameters are then computed using the Principal Component Analysis (PCA) method. Finally, considering the correlation of shape and texture of an object, the combined appearance model is built. While fitting this model, object contours are updated iteratively so that the difference between the combined model and the target image is reduced to the minimum [16].

Training a Model. Firstly, a shape model is constructed using the manually labeled landmark points of an object boundary. They are denoted as a vector x with $2n$ elements where (x_i, y_i) is the position of landmark points.

$$x = (x_1, y_1, x_2, y_2, ..., x_n, y_n)^T \tag{1}$$

Using Procrustes Analysis, all shapes are aligned to remove pose differences (i.e. rotation, translation and scale) and shape model is built using PCA:

$$x = \bar{x} + P_s b_s \tag{2}$$

In the above equation, $\bar{x}$ is the mean shape, P_s is a set of shape variances and b_s is a set of shape parameters.

Secondly, the AAM combines the gray level information and PCA is applied to establish the texture model:

$$g = \bar{g} + P_g b_g \tag{3}$$

where $\bar{g}$ is the mean texture, P_g is a set of texture variances and b_g is a set of texture parameters.

After that, both shape and texture model parameters are concatenated to create a combined appearance model:

$$b = \begin{pmatrix} W b_s \\ b_g \end{pmatrix} = \begin{pmatrix} W P_s^T (x - \bar{x}) \\ P_g^T (g - \bar{g}) \end{pmatrix} \tag{4}$$

where b is a concatenated parameter, W is a weight factor that specifies different units of shape and texture. Then, PCA is applied on these vectors and obtained:

$$b = Qc \tag{5}$$

where Q is the eigenvectors of b, c is the appearance model parameters that control both shape and texture of the object at the same time.

Finally, the combined appearance model is built according to the correlation of shape and texture:

$$x = \bar{x} + Q_s c \tag{6}$$
$$g = \bar{g} + Q_g c \tag{7}$$

where Q_s, Q_g are matrices that define the patterns of change derived from the training samples.

Fitting a Model. If we have appearance model parameters p, we can synthesize a model appearance vector g_m and normalized target image g_s. We can define a difference vector r which is a measure of correctness of the current contour and the error E as in (8) and (9) respectively.

$$r(p) = (g_m - g_s) \tag{8}$$
$$E = r^2 \tag{9}$$

During the segmentation stage the difference r is minimized by updating the parameters p. If the error cannot converge, the value of predicted displacements reduced and error is calculated again; otherwise the model and the parameters are updated until the error cannot be reduced further. The AAM search is implemented as below.

1. Normalize the target image g_s and project into the texture model.
2. Compute the residual image $r(p)$ and error E respectively.

3. Determine the displacement δ_p in model parameters.
4. Update the model parameters using $p = p + k\delta_p$ formula (k is initially set to 1) and compute the current position and texture of the model.
5. Compute new residual image $r(p)$ and error E'.
6. If $E' < E$, save p as new parameters, otherwise use different k values (ie $= 0.5$, 1.5) and go to 4.

3   Results and Discussion

3.1   Study Group

We used the AAM algorithm described in Sect. 2 to segment the cell nuclei in pleural effusion cytology images. The method is implemented in Matlab 2015a on a Laptop equipped with Intel Core i7 CPU with 8GB memory. Our data set is composed by 90 pleural effusion cytology cell images of size 400×400 pixels that are chosen from the Google search [17] (Fig. 1). Only 10 images of cell were used to train the cell nuclei AAM model and other 80 images were used to test the AAM model. The training set was chosen so as to best representation of the variations on the shape and texture of the nuclei images.

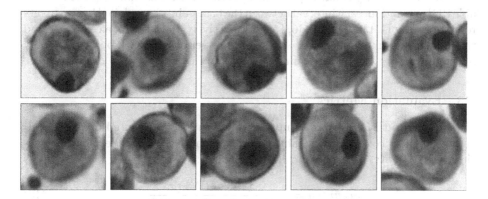

Fig. 1. Example of pleural effusion cytology cell images. Notice that the nuclei are at different positions in cell with various shapes.

3.2   Cell Nuclei Model

In model based segmentation methods, it is very important to determine the landmark points that will construct the object model. The goal of the cell nuclei modelling is to ensure that the nuclei region, called the region of interest (ROI), is chosen correctly in all cell images in the training set.

In the cell nuclei modelling, totally 12 landmark points at nearly even intervals around the cell nuclei boundary are determined by us. 4 landmark points represented by B1, B2, B3, B4 define the base points of the nuclei model. These points

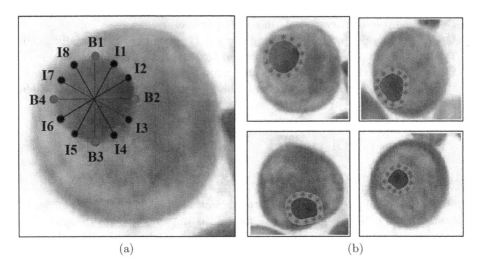

(a)   (b)

Fig. 2. (a) The landmark points used to construct the cell nuclei model (b) a sample set of landmark points is given on the training images.

are determined nearly equally spaced around the nuclei centroid. To provide a reliable ROI extraction, each interval between the base points is modelled by two points represented by I1, I2, I3, I4, I5, I6, I7, I8 respectively. In Fig. 2(a), a cell image and the landmark points are shown. Also, a sample set of landmark points manually selected by us is given on the training samples in Fig. 2(b).

3.3   Evaluation Metrics

The ground-truth images are compared to automated segmented images to calculate the most common segmentation evaluation metrics include accuracy, sensitivity, specificity, precision and Dice coefficient:

$$accuracy = \frac{(TP + TN)}{(TP + FP + FN + TN)} \tag{10}$$

$$sensitivity = \frac{TP}{(TP + FN)} \tag{11}$$

$$specificity = \frac{TN}{(TN + FP)} \tag{12}$$

$$precision = \frac{TP}{(TP + FP)} \tag{13}$$

where TP is the number of correctly segmented, TN is the number of correctly non-segmented, FP is the number of incorrectly segmented and FN is the number of missed segmented pixels. Dice coefficient is a similarity index used to measure the overlapping of two sets and calculated as:

$$DC = \frac{2\,|R_s \cap R_g|}{(|R_s| + |R_g|)} \tag{14}$$

where R_s is the segmented result and R_g is the ground truth.

3.4   Performance Evaluation

After the training phase is over, the resulting AAM model is used for nuclei segmentation on test set. All test images have been manually segmented by experts, which gives the ground-truth image for each sample. The performance of the proposed method in our experiments is measured by comparing the segmented region with the ground truth region. Figure 3 shows an example of cell image, its ground truth nuclei region, its nuclei segmentation result and its segmented nuclei region respectively.

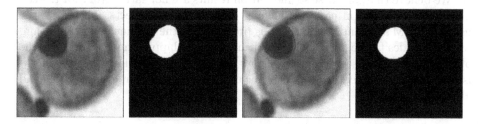

Fig. 3. An example of cell image, ground-truth region, final segmentation result (red line denotes nuclei boundary), segmented region. (Color figure online)

Various well known segmentation methods which have been used in the literature are applied on our pleural effusion cell database to evaluate the performance of the segmentation methods. These methods are thresholding-based Otsu method [18], clustering-based k-means method [19] and graph-based Grabcut method [20]. Compared to the final segmentation results given in Fig. 4, the proposed method produces better segmentation performance than other known methods.

The threshold-based Otsu method is not successful in case of touching or overlapping nuclei. The complexity between the nuclei and the background regions makes it difficult to determine a reliable threshold value. For instance, in Fig. 4(b), there exists part of another cell and it leads to erroneous segmentation.

Clustering-based k-means method does not give a definite segmentation result as seen in Fig. 4(c), it requires the extra process of object boundary extraction.

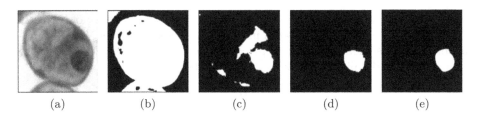

(a)   (b)   (c)   (d)   (e)

Fig. 4. Example results of segmentation using known methods and the proposed method (a) A sample cell image (b) Otsu method (c) k-means method (d) Grab-cut method (e) Proposed method.

Also, k-means clustering method is sensitive to initialization, and popular strategy for tackling this problem is to run the algorithm multiple times and select the one with the smallest distortion. Graph-based Grab-cut method may produce unacceptable results if there is low contrast between foreground and background colors. Furthermore, the most important disadvantage of this method is, the user need to specify some pixels for each image as a part of the foreground and background scene. This affects the result of segmentation as in Fig. 4(d) and it is also time-consuming.

We believe that, all these segmentation results are not satisfactory for the cell nuclei segmentation in case of touching or overlapping objects. Therefore, the usage of AAM segmentation method is proposed for the cell nuclei segmentation, which provides the segmentation of touching or overlapping nuclei. In Fig. 5, the obtained cell nuclei segmentation results are given. It is obvious that the AAM is powerful for segmentation of cell nuclei with different shapes and scales and also touching cells.

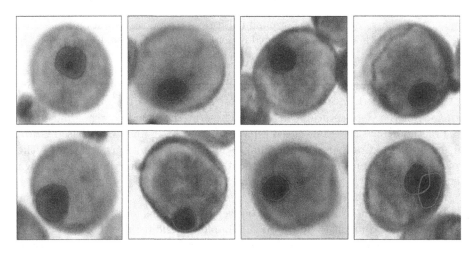

Fig. 5. Final segmentation results (red line denotes nuclei boundary). (Color figure online)

Table 1 summarizes the average performances obtained by these methods and the proposed method. Based on the quantitative results, AAM is found to outperform other methods with 98.77% accuracy, which demonstrates that model-based approach is very effective for nuclei segmentation. When the sensitivity results are examined, it is seen that the threshold-based Otsu method gives 100% success. However, when the segmentation results are analyzed, it is seen in Fig. 4(b) that this success is due to the Otsu method selecting the whole cell, and accordingly the precision value is reduced to the unacceptable value of 10.41%. The best precision value is obtained by the proposed method with a performance improvement of about 15% over the Grab-cut method. Finally, according to the results of the dice coefficient, it is seen that the highest similarity is obtained by the proposed method.

Table 1. Segmentation performances (%).

	Otsu	k-means	Grab-cut	Proposed method
Accuracy	43.84	85.09	92.31	**98.77**
Sensitivity	**100.0**	59.81	88.51	**88.54**
Specificity	40.09	87.05	92.60	**99.49**
Precision	**10.41**	27.42	76.25	**91.91**
Dice coefficient	18.53	35.84	80.62	**89.80**

4 Conclusion

In visual examination of cytopathological images of pleural effusion, cell nuclei present significant diagnostic value. Therefore, correct segmentation of cell nuclei in cytopathological images is a prerequisite step for the determination of accurate diagnosis. In this paper, we recommend that the AAM method be used to segment nuclei in cytopathological images. Estimating the cell nuclei boundary by the proposed method is highly accurate and satisfactory, and it also robust to different shapes and scales. Compared with the commonly used segmentation methods, the AAM method is also successful in the segmentation of touching nuclei. Five metrics including Accuracy, Sensitivity, Specificity, Precision and Dice coefficient were determined to evaluate the performance of the methods. AAM method achieves 98.77% accuracy using only 10 training example (10% of dataset). Although the proposed method is easy to implement, it supposed that there is only one cell in the image. For this reason, it is necessary to design a robust nuclei segmentation technique that can detect the nuclei positions in the whole-slide image when it contains millions of cells. At a next step, we intend to investigate the application of machine learning techniques for detection of cell nuclei in the cytopathological images.

References

1. Shidham, V.B., Atkinson, B.F.: Cytopathologic Diagnosis of Serous Fluids. Elsevier Health Sciences, London (2007)
2. Davidson, B., Firat, P., Michael, C.W.: Serous Effusions: Etiology, Diagnosis, Prognosis and Therapy. Springer Science and Business Media, New York (2011)
3. Schneider, T.E., Bell, A.A., Meyer-Ebrecht, D., Bcking, A., Aach, T.: Computer-aided cytological cancer diagnosis: cell type classification as a step towards fully automatic cancer diagnostics on cytopathological specimens of serous effusions. In: SPIE Medical Imaging 2007, Computer-Aided Diagnosis, p. 65140G (2007)
4. Phansalkar, N., More, S., Sabale, A., Joshi, M.: Adaptive local thresholding for detection of nuclei in diversity stained cytology images. In: 2011 International Conference on Communications and Signal Processing (ICCSP), pp. 218–220. IEEE (2011)
5. Xing, F., Yang, L.: Robust nucleus/cell detection and segmentation in digital pathology and microscopy images: a comprehensive review. IEEE Rev. Biomed. Eng. **9**, 234–263 (2016)
6. Zhou, X., Li, F., Yan, J., Wong, S.T.: A novel cell segmentation method and cell phase identification using Markov model. IEEE Trans. Inf. Technol. Biomed. **13**(2), 152–157 (2009)
7. Schmitt, O., Hasse, M.: Morphological multiscale decomposition of connected regions with emphasis on cell clusters. Comput. Vis. Image Underst. **113**(2), 188–201 (2009)
8. Kothari, S., Chaudry, Q., Wang, M.D.: Automated cell counting and cluster segmentation using concavity detection and ellipse fitting techniques. In: IEEE International Symposium on Biomedical Imaging: From Nano to Macro, ISBI 2009, pp. 795–798. IEEE (2009)
9. Zhou, X., Liu, K.-Y., Bradley, P., Perrimon, N., Wong, S.T.C.: Towards automated cellular image segmentation for RNAi genome-wide screening. In: Duncan, J.S., Gerig, G. (eds.) MICCAI 2005. LNCS, vol. 3749, pp. 885–892. Springer, Heidelberg (2005). doi:10.1007/11566465_109
10. Jung, C., Kim, C., Chae, S.W., Oh, S.: Unsupervised segmentation of overlapped nuclei using Bayesian classification. IEEE Trans. Biomed. Eng. **57**(12), 2825–2832 (2010)
11. Al-Kofahi, Y., Lassoued, W., Lee, W., Roysam, B.: Improved automatic detection and segmentation of cell nuclei in histopathology images. IEEE Trans. Biomed. Eng. **57**(4), 841–852 (2010)
12. Delgado-Gonzalo, R., Uhlmann, V., Schmitter, D., Unser, M.: Snakes on a Plane: a perfect snap for bioimage analysis. IEEE Sig. Process. Mag. **32**(1), 41–48 (2015)
13. Zimmer, C., Olivo-Marin, J.C.: Coupled parametric active contours. IEEE Trans. Pattern Anal. Mach. Intell. **27**(11), 1838–1842 (2005)
14. Plissiti, M.E., Nikou, C.: Overlapping cell nuclei segmentation using a spatially adaptive active physical model. IEEE Trans. Image Process. **21**(11), 4568–4580 (2012)
15. Cootes, T.F., Edwards, G.J., Taylor, C.J.: Active appearance models. In: Burkhardt, H., Neumann, B. (eds.) ECCV 1998. LNCS, vol. 1407, pp. 484–498. Springer, Heidelberg (1998). doi:10.1007/BFb0054760
16. Cootes, T.F., Edwards, G.J., Taylor, C.J.: Active appearance models. IEEE Trans. Pattern Anal. Mach. Intell. **23**(6), 681–685 (2001)
17. Google Image Search. http://images.google.com/

18. Otsu, N.: A threshold selection method from gray-level histograms. IEEE Trans. Syst. Man Cybern. **9**(1), 62–66 (1979)
19. MacQueen, J.: Some methods for classification and analysis of multivariate observations. In: Proceedings of 5th Berkeley Symposium on Mathematical Statistics and Probability, vol. 1, pp. 281–297 (1967)
20. Rother, C., Kolmogorov, V., Blake, A.: Grabcut: Interactive foreground extraction using iterated graph cuts. ACM Trans. Graph. (TOG) **23**(3), 309–314 (2004)

Parkinson's Disease Identification Using Restricted Boltzmann Machines

Clayton R. Pereira[1], Leandro A. Passos[1], Ricardo R. Lopes[2],
Silke A.T. Weber[3], Christian Hook[4], and João Paulo Papa[5(✉)]

[1] Department of Computing, UFSCAR - Federal University of São Carlos,
São Carlos, Brazil
{clayton.pereira,leandro.passosjr}@dc.ufscar.br
[2] Eldorado Research Institute, Campinas, Brazil
ricardoriccilopes@gmail.com
[3] Botucatu Medical School, UNESP - São Paulo State University, Botucatu, Brazil
silke@fmb.unesp.br
[4] Ostbayerische Technische Hochschule, Regensburg, Germany
christian.hook@oth-regensburg.de
[5] School of Sciences, UNESP - São Paulo State University, Bauru, Brazil
papa@fc.unesp.br

Abstract. Currently, Parkinson's Disease (PD) has no cure or accurate diagnosis, reaching approximately $60,000$ new cases yearly and worldwide, being more often in the elderly population. Its main symptoms can not be easily uncorrelated with other illness, being way more difficult to be identified at the early stages. As such, computer-aided tools have been recently used to assist in this task, but the challenge in the automatic identification of Parkinson's Disease still persists. In order to cope with this problem, we propose to employ Restricted Boltzmann Machines (RBMs) to learn features in an unsupervised fashion by analyzing images from handwriting exams, which aim at assessing the writing skills of potential individuals. These are one of the main symptoms of PD-prone people, since such kind of ability ends up being severely affected. We show that RBMs can learn proper features that help supervised classifiers in the task of automatic identification of PD patients, as well as one can obtain a more compact representation of the exam for the sake of storage and computational load purposes.

Keywords: Parkinson's Disease · Restricted Boltzmann Machines · Machine learning

1 Introduction

In the last decades, the number of people with Parkinson's disease (PD) has increased significantly worldwide. Also, research points out that approximately $60,000$ Americans are diagnosed with PD [1]. The disease was firstly described

---

C.R. Pereira and L.A. Passos—Both authors contributed equally.

© Springer International Publishing AG 2017
M. Felsberg et al. (Eds.): CAIP 2017, Part II, LNCS 10425, pp. 70–80, 2017.
DOI: 10.1007/978-3-319-64698-5_7

by the English physician James Parkinson [2], being more common in the elderly population and occurs when nerve cells that produce dopamine are destroyed [3]. Additionally, PD is a chronic, progressive and neuron-degenerative illness, which causes many symptoms, such as slowness of movement, freeze of gait, tremors and muscle stiffness.

One of the main problems related to PD concerns the dopamine, which is a neurotransmitter released by the brain and in charge of various tasks performed by our body, such as the movement, memory, sleep, learning and others. The absence of such substance in PD-affected individuals may trigger a number of symptoms, such as depression, sleep disturbances, memory impairment, and autonomic nervous system disorders. Also, the patient may have its speech and writing skills affected over time [4]. Furthermore, despite of not being considered lethal, people with PD have a shorter life expectancy than the general population.

Although some research indicates that PD may be trigged by hereditary factors [4], the cause of the disease is still unknown. Also, the difficulty to distinguish the difference from other neurological illness as well as to detect PD in its early stages are the main barriers related to automatic identification of Parkinson's Disease. In order to assist the computer-aided PD diagnosis, systems based on machine learning techniques have been employed, showing promising results [5]. Spadotto et al. [6], for instance, introduced the Optimum-Path Forest [7,8] classifier to aid the automatic identification of Parkinson's Disease. The same group of authors proposed an evolutionary-based approach to select the most discriminative set of features that help improving PD recognition rates [9]. Recently, Pereira et al. [10] proposed to extract features from writing exams using image processing techniques, and later on Pereira et al. [11] introduced Convolutional Neural Networks to learn features from handwriting dynamics in the context of automatic PD identification.

Another interesting approach that has been extensively used in a number of applications, mainly in the context of deep learning-oriented applications, concerns the so-called Restricted Boltzmann Machines (RBMs), which are an undirected generative model that use a layer of hidden variables to model a distribution over visible units [12]. Therefore, given an input set of images, an RBM basically learns how to effectively reconstruct them based on a learning process that aims at learning the weights that connect the input data (image) to a hidden layer. The values of the neurons' activation at that layer can be used as the features to describe the input data.

In this paper, we introduce Restricted Boltzmann Machines in the context of feature learning for the automatic identification of Parkinson's Disease by means of images acquired from handwritten exams. Further, the features learned are then used as inputs to supervised classifiers. In this work, we considered three state-of-the-art classifiers, say that Optimum-Path Forest (OPF) [7,8], Naïve Bayes (NB) [13] and Support Vector Machines [14]. The experiments are conducted over the "HandPD" dataset, which is publicly available at the internet.

The reminder of this paper is organized as follows: Sect. 2 describes the theoretical background related to RBMs, and Sect. 3 presents the methodology employed in this work. The experimental results are presented in Sect. 4, conclusions and final remarks are stated in Sect. 5.

2 Restricted Boltzmann Machines

Restricted Boltzmann Machines are energy-based stochastic neural networks composed of two layers of neurons (visible and hidden), in which the learning phase is conducted by means of an unsupervised fashion. A naïve architecture of a Restricted Boltzmann Machine comprises a visible layer $\mathbf{v}$ with m units and a hidden layer $\mathbf{h}$ with n units. Additionally, a real-valued matrix $\mathbf{W}_{m \times n}$ models the weights between the visible and hidden neurons, where w_{ij} stands for the weight between the visible unit v_i and the hidden unit h_j.

Let us assume both $\mathbf{v}$ and $\mathbf{h}$ as being binary-valued units. In other words, $\mathbf{v} \in \{0,1\}^m$ e $\mathbf{h} \in \{0,1\}^n$. The energy function of a Restricted Boltzmann Machine is given by:

$$E(\mathbf{v}, \mathbf{h}) = -\sum_{i=1}^{m} a_i v_i - \sum_{j=1}^{n} b_j h_j - \sum_{i=1}^{m} \sum_{j=1}^{n} v_i h_j w_{ij}, \tag{1}$$

where $\mathbf{a}$ e $\mathbf{b}$ stand for the biases of visible and hidden units, respectively.

The probability of a joint configuration $(\mathbf{v}, \mathbf{h})$ is computed as follows:

$$P(\mathbf{v}, \mathbf{h}) = \frac{1}{Z} e^{-E(\mathbf{v}, \mathbf{h})}, \tag{2}$$

where Z stands for the so-called partition function, which is basically a normalization factor computed over all possible configurations involving the visible and hidden units. Similarly, the marginal probability of a visible (input) vector is given by:

$$P(\mathbf{v}) = \frac{1}{Z} \sum_{\mathbf{h}} e^{-E(\mathbf{v}, \mathbf{h})}. \tag{3}$$

Since the RBM is a bipartite graph, the activations of both visible and hidden units are mutually independent, thus leading to the following conditional probabilities:

$$P(\mathbf{v}|\mathbf{h}) = \prod_{i=1}^{m} P(v_i|\mathbf{h}), \tag{4}$$

and

$$P(\mathbf{h}|\mathbf{v}) = \prod_{j=1}^{n} P(h_j|\mathbf{v}), \tag{5}$$

where

$$P(v_i = 1|\mathbf{h}) = \phi\left(\sum_{j=1}^{n} w_{ij} h_j + a_i\right), \tag{6}$$

and

$$P(h_j = 1|\mathbf{v}) = \phi\left(\sum_{i=1}^{m} w_{ij} v_i + b_j\right). \tag{7}$$

Note that $\phi(\cdot)$ stands for the logistic-sigmoid function.

Let $\theta = (W, a, b)$ be the set of parameters of an RBM, which can be learned through a training algorithm that aims at maximizing the product of probabilities given all the available training data $\mathcal{V}$, as follows:

$$\arg\max_{\Theta} \prod_{v \in \mathcal{V}} P(\mathbf{v}). \tag{8}$$

One can solve the aforementioned equation using the following derivatives over the matrix of weights $\mathbf{W}$, and biases $\mathbf{a}$ and $\mathbf{b}$ at iteration t as follows:

$$\mathbf{W}^{t+1} = \mathbf{W}^t + \underbrace{\eta(P(\mathbf{h}|\mathbf{v})\mathbf{v}^T - P(\tilde{\mathbf{h}}|\tilde{\mathbf{v}})\tilde{\mathbf{v}}^T) + \Phi}_{=\Delta\mathbf{W}^t}, \tag{9}$$

$$\mathbf{a}^{t+1} = \mathbf{a}^t + \underbrace{\eta(\mathbf{v} - \tilde{\mathbf{v}}) + \alpha\Delta\mathbf{a}^{t-1}}_{=\Delta\mathbf{a}^t} \tag{10}$$

and

$$\mathbf{b}^{t+1} = \mathbf{b}^t + \underbrace{\eta(P(\mathbf{h}|\mathbf{v}) - P(\tilde{\mathbf{h}}|\tilde{\mathbf{v}})) + \alpha\Delta\mathbf{b}^{t-1}}_{=\Delta\mathbf{b}^t}, \tag{11}$$

where η stands for the learning rate, and λ and α denote the weight decay and the momentum, respectively. Notice the terms $P(\tilde{\mathbf{h}}|\tilde{\mathbf{v}})$ and $\tilde{\mathbf{v}}$ can be obtained by means of the Contrastive Divergence [15] technique, which basically ends up performing Gibbs sampling using the training data as the visible units. Roughly speaking, Eqs. 9, 10 and 11 employ the well-known Gradient Descent as the optimization algorithm. The additional term Φ in Eq. 9 is used to control the values of matrix $\mathbf{W}$ during the convergence process, and it is formulated as follows:

$$\Phi = -\lambda\mathbf{W}^t + \alpha\Delta\mathbf{W}^{t-1}. \tag{12}$$

3 Methodology

In this section, we present the methodology employed to evaluate the proposed approach, as well as the datasets and the experimental setup.

3.1   Dataset

The dataset employed in this work is called "HandPD"[1], being firstly presented by Pereira et al. [16]. Roughly speaking, the dataset comprises images drawn in a form with a template for guideline purposes (Fig. 1), depicting exercises specifically designed to expose unique characteristics from PD patients. To create the images, the patient is aided with a digital pen $BiSP^{®}$ [17], which is equipped with six sensors capable of recording the hand movements during the exercises. However, differently from Pereira et al. [11] that used information from the sensors, we are considering only the visual features that can be learned from the drawing by means of RBMs.

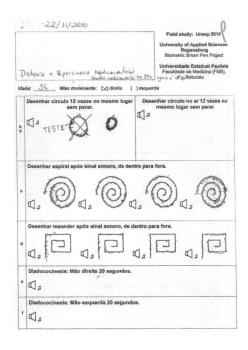

Fig. 1. Form used to assess the handwritten skills. Extracted from [16].

This work makes use of two kind of images extracted from the forms: Spiral and Meander. Also, the images are classified into two classes: patients (124 images) and healthy group (56 images). Furthermore, both datasets are resized in two different ways to evaluate the robustness of RBM for learning features in this context: 64×64 and 128×128 pixels each.

Since we are considering RBMs with binary-valued inputs, one needs to threshold the gray-scale images originally available from the dataset. In order to fulfill this task, we applied the well-known Otsu threshold [18] to all images,

---

[1] Available in http://wwwp.fc.unesp.br/~papa/pub/datasets/Handpd/.

which are finally used to feed the RBMs. After training, the images are used once more as the inputs, and the activation values at the hidden layer are used as the features for each image concerning the supervised learning methods employed in this work, i.e. OPF, SVM and Bayesian classifier.

3.2   Experimental Setup

The work employs an RBM architecture where the visible layer stands for the number of pixels from the input image, i.e. 4,096 (64×64 images) or 16,384 (128×128 images), as well as the hidden layer stands for the desirable number of features, which assume the values within the range $J \in \{10, 100, 500, 1,000, 2,000, 4,000, 7,000\}$[2]. The remaining parameters used during the RBM learning step were chosen empirically and fixed as follows: $\eta = 0.1$ (learning rate), number of epochs $= 1,000$, and mini-batches of size 20. Moreover, the RBM was trained with the Contrastive Divergence (CD) [15] algorithm.

Once the RBM learning process is finished, the new datasets composed of features extracted from the hidden layer units will feed three supervised learning algorithms: SVM, Bayes and the OPF. In regard to SVM, we used the Radial Basis Function kernel with parameters optimized by means of a grid search [19]. Finally, to evaluate the techniques considered in this work, a classification accuracy proposed by Papa et al. [7] that considers unbalanced datasets has been adopted.

4   Experiments

This section presents the experimental results concerning OPF, SVM and the Bayes classifier to the task of Parkinson's Disease identification by means of features learned from handwritten forms. The evaluation is taken upon a training set with size of 50%, as well as the remaining 50% is employed for testing purposes. Tables 1 and 2 present the accuracy results concerning Meander dataset with 64×64 and 128×128, respectively. The most accurate results according to the Wilcoxon signed-rank test for each number of features are in bold, and the best global results are underlined. Notice the experiments consider an average accuracy over 20 runnings.

Considering the smaller-sized Meander dataset (Table 1), one can observe RBM can provide reasonable results with SVM classifier using only 500 features. Since the dataset is composed of 180 images only, being 90 used for training, it is expected that higher dimensional spaces (i.e., 7,000 features) will not allow good results, since the mapped space will be shrank to a lower dimensional one. This same behavior can be observed for SVM considering 128×128 images (Table 2), but with OPF and Bayes obtaining the best results with 7,000 and 4,000 features, respectively.

In regard to the Spiral drawing, Table 3 presents the results with 64×64 images, where the best results were obtained by SVM with 1,000 features. Also,

---

[2] Notice the range values were empirically chosen.

Table 1. Mean accuracy results considering Meander 64 × 64 dataset.

Number of features	OPF	Bayes	SVM
10	**53.72 ± 0.78**	53.82 ± 0.18	**55.75 ± 1.37**
100	58.00 ± 1.42	57.20 ± 0.0	**71.94 ± 1.42**
500	62.23 ± 4.18	63.21 ± 2.32	**74.38 ± 3.02**
1,000	**67.59 ± 3.26**	**67.94 ± 1.74**	66.26 ± 5.66
2,000	**64.10 ± 1.67**	**64.07 ± 1.30**	62.72 ± 8.14
4,000	64.91 ± 4.53	64.58 ± 5.50	**69.99 ± 2.67**
7,000	62.94 ± 3.05	**64.62 ± 3.51**	**65.34 ± 3.57**

Table 2. Mean accuracy results considering Meander 128 × 128 dataset.

Number of features	OPF	Bayes	SVM
10	**55.26 ± 4.90**	50.0 ± 0.0	50.0 ± 0.00
100	64.89 ± 3.69	**65.69 ± 3.21**	63.57 ± 0.46
500	**60.63 ± 2.24**	**61.36 ± 2.69**	**62.94 ± 7.13**
1,000	60.17 ± 4.18	59.19 ± 4.18	**63.83 ± 3.95**
2,000	64.23 ± 1.38	66.22 ± 0.70	**68.18 ± 1.50**
4,000	66.33 ± 3.15	**68.18 ± 4.51**	65.05 ± 5.23
7,000	**71.05 ± 3.35**	69.55 ± 3.86	66.43 ± 8.12

Table 3. Mean accuracy results considering Spiral 64 × 64 dataset.

Number of features	OPF	Bayes	SVM
10	53.19 ± 1.39	58.26 ± 0.60	**58.85 ± 1.69**
100	67.75 ± 1.60	67.39 ± 5.15	**73.82 ± 2.26**
500	67.26 ± 3.58	68.00 ± 3.49	**79.93 ± 0.25**
1,000	74.59 ± 2.99	74.94 ± 2.77	**83.12 ± 2.55**
2,000	69.74 ± 2.68	71.58 ± 3.47	**80.71 ± 0.14**
4,000	68.76 ± 3.61	71.31 ± 4.55	**79.49 ± 3.81**
7,000	71.76 ± 3.13	**73.30 ± 3.59**	**75.31 ± 4.04**

one can observe the recognition rates were quite better than using Meanders. We believe that following the "circular" pattern of Spirals is way more difficult for those with reduced hand mobility, which turns out to be more discriminative to distinguish patients from the control group. However, the results using 128 × 128 images were slightly worse, but with SVM on top of the results once more. One can observe OPF has its accuracy degraded when the number of features increases, since it uses the distance among samples for classification purposes, and it can be influenced by higher dimensional spaces (Table 4).

Table 4. Mean accuracy results considering Spiral 128 × 128 dataset

Number of features	OPF	Bayes	SVM
10	56.00 ± 6.26	64.25 ± 2.74	**71.972 ± 2.74**
100	62.47 ± 2.33	63.68 ± 3.71	**73.87 ± 1.29**
500	67.22 ± 1.92	70.43 ± 0.81	**80.29 ± 3.19**
1,000	**70.17 ± 4.25**	**69.58 ± 2.27**	**72.69 ± 2.49**
2,000	74.44 ± 3.16	75.02 ± 4.41	**77.92 ± 3.61**
4,000	71.98 ± 3.20	73.32 ± 1.80	**79.73 ± 1.19**
7,000	67.49 ± 4.03	69.70 ± 3.80	**74.91 ± 4.64**

In order to keep track of the convergence process, we analyzed the mean squared error (MSE) of the RBM during the learning step. Figure 2 depicts the MSE obtained during the training considering 1, 000 epochs. Notice the error falls dramatically before epoch 100, and then traces a smooth curve that approaches zero as the epochs approximate 1, 000 iterations for models with

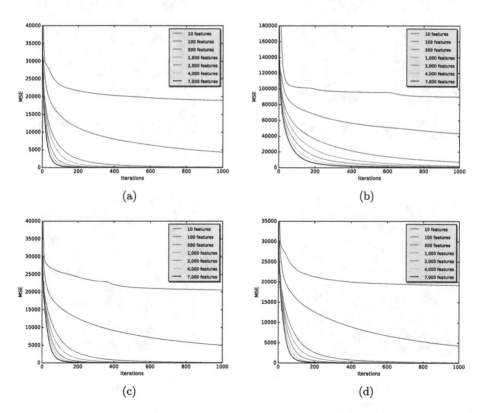

(a)   (b)

(c)   (d)

Fig. 2. Evolution of the Mean-Squared Error considering: (a) Meander 64 × 64, (b) Meander 128 × 128, (c) Spiral 64 × 64, and (d) Spiral 128 × 128.

500 or more hidden units. Although all approaches (i.e., different number of features) achieved similar results concerning MSE in the final of the learning step, one can observe the faster convergence during the first 200 iterations using 7,000 features. Since we have more latent variables in the hidden units, it is expected they can learn a more complex and detailed information about the input data.

In order to understand what is going on during the learning procedure of the RBM, we displayed what the network "sees" when is presented to an input image. Since each visible unit is connected to all hidden neurons, we can take all connection weights, normalize them and build an image, that is usually employed to understand what the network is learning. Figure 3 depicts some images from the weight matrices when Meanders and Spirals are presented to the network using 1,000 hidden units. Since all images from each dataset are somehow similar, most of the neurons are learning the shape of the drawing (i.e., spiral or meander), although some of them did not get excited by any shape in particular.

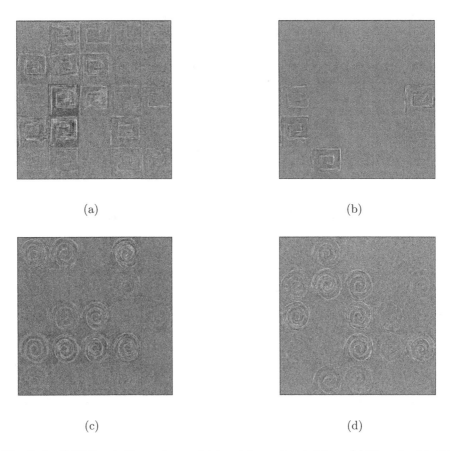

(a)   (b)

(c)   (d)

Fig. 3. An "RBM's mind": random weight matrices extracted from (a) Meander 64×64, (b) Meander 128 × 128, (c) Spiral 64 × 64 and (d) Spiral 128 × 128.

5   Conclusions

In this work, we dealt with the problem of automatic Parkinson's Disease identification by means of features learned from Restricted Boltzmann Machines. We considered two types of drawings in two different resolutions, which were then used to feed RBM models for further supervised classification purposes.

We observed the best results were obtained with spirals at lower-resolution images, but similar results can also be obtained with 128×128 images. Additionally, we observed that RBMs with more hidden layers allow a faster convergence during learning, but it does not necessarily imply in the better results over the test set.

With respect to future works, we aim at considering DBNs and DBMs for classification purposes, as well as to use Discriminative RBMs as the classification techniques as well.

Acknowledgments. The authors would like to thank FAPESP grants #2014/16250-9, #2014/12236-1, #2015/25739-4 and #2016/21243-7, as well as Capes, and CNPq grant #306166/2014-3.

References

1. Parkinson's Disease Fundation: Statistics on Parkinson's: Who has Parkinson's? (2016). http://www.pdf.org/en/parkinson_statistics. Accessed 15 July 2016
2. Parkinson, J.: An essay on the shaking palsy. J. Neuropsychiatry Clin. Neurosci. **14**(2), 223–236 (1817)
3. Lees, A.J., Hardy, J., Revesz, T.: Parkinson's disease. Lancet **373**(9680), 2055–2066 (2009)
4. Burke, R.E.: Evaluation of the braak staging scheme for Parkinson's disease: Introduction to a panel presentation. Mov. Disord. **25**(S1), S76–S77 (2010)
5. Sakar, B.E., Isenkul, M.E., Sakar, C.O., Sertbas, A., Gurgen, F., Delil, S., Apaydin, H., Kursun, O.: Collection and analysis of a Parkinson speech dataset with multiple types of sound recordings. IEEE J. Biomed. Health Inform. **17**, 828–834 (2013)
6. Spadotto, A.A., Guido, R.C., Papa, J.P., Falcão, A.X.: Parkinson's disease identification through optimum-path forest. In: IEEE International Conference of the Engineering in Medicine and Biology Society, pp. 6087–6090 (2010)
7. Papa, J.P., Falcão, A.X., Suzuki, C.T.N.: Supervised pattern classification based on optimum-path forest. Int. J. Imaging Syst. Technol. **19**(2), 120–131 (2009)
8. Papa, J.P., Falcão, A.X., Albuquerque, V.H.C., Tavares, J.M.R.S.: Efficient supervised optimum-path forest classification for large datasets. Pattern Recogn. **45**(1), 512–520 (2012)
9. Spadotto, A.A., Guido, R.C., Carnevali, F.L., Pagnin, A.F., Falcão, A.X., Papa, J.P.: Improving Parkinson's disease identification through evolutionary-based feature selection. In: IEEE International Conference of the Engineering in Medicine and Biology Society, pp. 7857–7860 (2011)
10. Pereira, C.R., Pereira, D.R., Silva, F.A., Masieiro, J.P., Weber, S.A.T., Hook, C., Papa, J.P.: A new computer vision-based approach to aid the diagnosis of Parkinson's disease. Comput. Methods Programs Biomed. **136**, 79–88 (2016)

11. Pereira, C.R., Weber, S.A.T., Hook, C., Rosa, G.H., Papa, J.P.: Deep learning-aided Parkinson's disease diagnosis from handwritten dynamics. In: Proceedings of the 29th SIBGRAPI Conference on Graphics, Patterns and Images, pp. 340–346 (2016)
12. Larochelle, H., Bengio, Y.: Classification using discriminative restricted Boltzmann machines. In: Proceedings of the 25th International Conference on Machine Learning, pp. 536–543. ACM (2008)
13. Duda, R.O., Hart, P.E., Stork, D.G.: Pattern Classification, 2nd edn. Wiley-Interscience, New York (2000)
14. Cortes, C., Vapnik, V.: Support vector networks. Mach. Learn. **20**, 273–297 (1995)
15. Hinton, G.E.: Training products of experts by minimizing contrastive divergence. Neural Comput. **14**(8), 1771–1800 (2002)
16. Pereira, C.R., Pereira, D.R., da Silva, F.A., Hook, C., Weber, S.A.T., Pereira, L.A.M., Papa, J.P.: A step towards the automated diagnosis of Parkinson's disease: analyzing handwriting movements. In: IEEE 28th International Symposium on Computer-Based Medical Systems, pp. 171–176 (2015)
17. Germany University of Applied Sciences Team, Regensburg. A novel multisensoric system recording and analyzing human biometric features for biometric and biomedical applications (2002)
18. Otsu, N.: A threshold selection method from gray-level histograms. IEEE Trans. Syst. Man Cybern. **9**(1), 62–66 (1979)
19. Pedregosa, F., Varoquaux, G., Gramfort, A., Michel, V., Thirion, B., Grisel, O., Blondel, M., Prettenhofer, P., Weiss, R., Dubourg, V., Vanderplas, J., Passos, A., Cournapeau, D., Brucher, M., Perrot, M., Duchesnay, E.: Scikit-learn: machine learning in Python. J. Mach. Learn. Res. **12**, 2825–2830 (2011)

Attention-Based Two-Phase Model for Video Action Detection

Xiongtao Chen, Wenmin Wang[(✉)], Weimian Li, and Jinzhuo Wang

School of Electronic and Computer Engineering, Shenzhen Graduate School,
Peking University, Shenzhen, China
{cxt,leewm,jzwang}@pku.edu.cn, wangwm@ece.pku.edu.cn

Abstract. This paper considers the task of action detection in long untrimmed video. Existing methods tend to process every single frame or fragment through the whole video to make detection decisions, which can not only be time-consuming but also burden the computational models. Instead, we present an attention-based model to perform action detection by watching only a few fragments, which is independent with the video length and can be applied to real-world videos consequently. Our motivation is inspired by the observation that human usually focus their attention sequentially on different frames of a video to quickly narrow down the extent where an action occurs. Our model is a two-phase architecture, where a temporal proposal network is designed to predict temporal proposals for multi-category actions in the first phase. The temporal proposal network observes a fixed number of locations in a video to predict action bounds and learn a location transfer policy. In the second phase, a well-trained classifier is prepared to extract visual information from proposals, to classify the action and decide whether to adopt the proposals. We evaluate our model on ActivityNet dataset and show it can significantly outperform the baseline.

1 Introduction

Compared with video classification, action detection in long, untrimmed video is a more challenging task, including temporal bound detection and action recognition. Most current approaches [21,36] tend to traverse the whole video using sliding window method to make localization and classification decision. The runner-up [26] method in ActivityNet Detection Challenge 2016 also needs to score every frame to assist localization and classification. These approaches are not efficient since the computational complexity inevitably depends on the video size.

In visual cognition literature, human focus their attention sequentially on different parts of the scene rather than the whole scene at a time. Inspired by this mechanism, attention model was proposed and recently has been applied to several visual tasks including image classification [19] and caption generation [41]. Processing only a few fixations of the scene, attention model can also achieve promising results [2,29].

This work was supported by Shenzhen Peacock Plan (20130408-183003656).

M. Felsberg et al. (Eds.): CAIP 2017, Part II, LNCS 10425, pp. 81–93, 2017.
DOI: 10.1007/978-3-319-64698-5_8

Ground Truth Segment

Fig. 1. Example of attention mechanism. Observing 4 frames (numbered from 1 to 4) in this video sequence, we can figure roughly the extent where the surfing action occurs (frames marked by red border). (Color figure online)

In this work, we introduce an attention-based model to automatically localize and recognize actions in long, untrimmed videos. Our model mainly consists of two parts, i.e. a temporal proposal network and a classifier. The first part is based on attention mechanism and aims to generate high recall temporal proposals while the classifier classifies the proposals in the second phase. Our intuition is that observing frames at a few positions in a video can gradually narrow down the extent where an action might occur (Fig. 1). Moreover, visual representation extracted only from the generated proposal is sufficient for a classifier to further distinguish the foreground and background.

Our temporal proposal network is implemented around a recurrent neural network (RNN), which takes a location and the corresponding video fragment as input and outputs a possible action bound at each timestep. This model is an end-to-end framework. It can be trained using back-propagation and reinforcement method [27,39]. After the temporal proposals are generated, the classifier just encodes a few frame-level features from each proposal and outputs scores for all classes. We experiment several classifiers with different training strategies and show the clear difference, which indicates the importance of classifier designed to improve the performance of action detection.

The rest of this paper is organized as follows. Section 2 reviews related works. Section 3 presents our approach with detail implementations of each component. Experiments and comparison are provided in Sect. 4. Finally, we conclude our paper in Sect. 5.

2   Related Works

There is a long history of work in video analysis and action detection [6,11,12, 14,24]. For more comprehensive studies, we refer to surveys [1,22,37]. Here, we first review action detection works from the view of methods. Also, we cover works of attention model, which is the main inspiration of this paper.

Hand-Crafted Features Based Methods. Early approaches interpret an action as a set of space-time trajectories of two-dimensional or three-dimensional points of human joints [7,20]. A series of local spatio-temporal features are designed to represent actions and videos [16,18,38]. A comprehensive evaluation in [34] compared different spatio-temporal interest point (STIP) detectors and descriptors. The authors drew conclusions that the performance of STIPs is dataset dependent. Later, remarkably, Wang et al. made use of point trajectories [32] to extract and align 3D volumes, and resorted to more rich low level descriptors for constructing effective video representations, including HOG/HOF and motion boundary histogram (MBH). An improved version of dense trajectory is updated in [33] to estimate camera motion, and obtained state-of-the-art results on a variety of benchmarks.

Deep Networks Based Methods. In contrast to the hand-crafted features, there is a growing trend of learning features directly from raw data using deep learning techniques, which has achieved great success in image-based tasks [10,17,28]. A number of attempts have developed deep architectures especially convolutional neural network (CNN) for video action detection [13,15,25,30]. In addition, a few works attempted to apply the CNN representations with RNN models to capture temporal information in videos and perform classification within the same network. Donahue et al. leveraged RNN model with LSTM units for action recognition [9] and Venugopalan et al. proposed to translate videos directly to sentences with the LSTM model by transferring knowledge from image description tasks [31]. Wang et al.[35] designed a deep alternative architecture by stacking convolutional layer and recurrent layer iteratively and showed its advantages over standard 3d convolutional networks (C3D) approaches.

However, both parts of the above methods can be viewed as leveraging visual information of the entire video, which have two main drawbacks, especially in real-world videos. The first is that scanning every frame is very time-consuming. The second is that huge feature space burdens the computation model.

Attention Model. Our work draws inspiration from recent popular attention model, which enjoys the advantage of lower computation compared with CNNs as it does not need to examine all the locations of input signals. In general, attention models can be classified into soft attention and hard attention models, while our model belongs to the latter. Soft attention models are deterministic and can be trained using back-propagation, whereas hard attention models are stochastic and can be trained by the REINFORCE algorithm [19,39] or by maximizing a variational lower bound or using importance sampling [3,4]. A very recent work [23] proposed a soft attention based model for the task of action recognition in videos, using multi-layered RNNs with LSTM units which are deep both spatially and temporally. Similarly, Bazzani et al. [5] presented a soft attentional model that learns "where to look" directly from the human scanpath data. Wu et al. [40] constructed a joint attention model on mulit-level features, which implicitly takes advantage of visual tracking and shares the robustness of both deep CNNs and RNNs.

The most related work to ours is [42], which used the hidden state of RNN to perform classification and localization simulaneously. However, we argue that the extra information from previous RNN steps may mislead the classification in current step since distinct actions sometimes take place in similar background. In contrast, we perform localization and classification separately in our work to escape from this limitation. Our model is simple, efficient and also effective in action detection. We have improved the mAP by more than 100% compared with the baseline provided in the original ActivityNet paper [6].

3   Approach

In this section, we show how we formulate and train our model for action detection. The main framework of our model is shown in Fig. 2. In the first phase, we build a temporal proposal network to process video glimpses and output temporal proposal at each timestep. In the second phase, we use a well-trained classifier to classify the proposals to assist final predictions.

3.1   Temporal Proposal Network

Unlike popular sliding window approach, the temporal proposal network is designed to predict high recall temporal proposals directly as well as learn a better location transfer policy. As shown in Fig. 2, the network consists of four sub-networks: glimpse network, recurrent network, locator network, and bounding regressor. We provide descriptions of these sub-networks as follows.

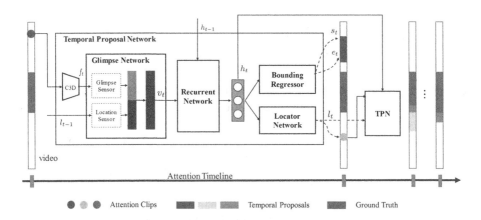

Fig. 2. Main structure of the Temporal Proposal Network (TPN). The network consists of 4 subnetworks: glimpse network, recurrent network, bounding regressor and locator network. At each time step, the temporal proposal network observes one clips at l_{t-1} in the video and predicts a temporal proposal (s_t, e_t) as well as outputs the next location l_t. The figure shows this procedure for 4 time steps alone the attention timeline.

Glimpse Network. The glimpse network combines visual information from a location $l_{t-1} \in [0,1]$ and the corresponding short fragment f_t to produces a glimpse representation v_t, which will be fed into the recurrent neural network next. To be specific, the glimpse network is structured by a glimpse sensor and a location sensor for extracting features from l_{t-1} and f_t respectively. In practice, we use the fc7 feature extracted from a pre-trained C3D network [30] as f_t input.

Recurrent Neural Network. The recurrent network is the core module of our model. The network encodes v_t and hidden state h_{t-1} from last timestep to produce h_t. The internal representation h_t accumulates information from current and previous RNN steps, which is prepared to be fed into the locator network, bounding regressor and the next state of RNN.

Bounding Regressor. The bounding regressor is a network simply containing several fully connected layers. The regressor propagates h_t and directly predicts a tuple $p_t(s_t, e_t) \in [0,1]^2$ as a temporal proposal, where s_t and e_t are start and end location respectively.

Locator Network. The locator network plays a role as a director, which utilizes h_t to choose the next location to focus on. The locator is composed of several fully connected layers along with a Gaussian location sampler. The Gaussian sampler uses information from h_t as mean and a constant *stdev* as standard deviation to sample the next location (in this paper we set $stdev = 0.11$).

Figure 2 shows how the temporal proposal network combines these sub-networks to interact with the video over time to generate proposals. Note that at the first timestep, we input a zero-vector to the locator to get the start location l_0.

3.2 Classifier

A softmax classifier is used in our work to classify temporal proposals. The classifier simply consists of 2 fully connected layers, a ReLU layer and a softmax layer. The classifier receives C3D features from a short video fragment and outputs a probability distribution for all action classes.

We experiment several training strategies on our classifier to explore good practices for video action recognition (see Sect. 3.3 for details). In addition to the softmax classifier, we also train a one-vs-all linear SVM for each action class. Experiments part shows the results and comparison of all these settings.

3.3 Training

Although the internal comprehension of RNN can be utilized for multi-kind of tasks for an attention-based model, however, in our perspective, extra information from previous RNN steps is not necessary to classify the current proposal. Therefore, to classify the proposals more accurately, we train the temporal proposal network and classifier separately in our model.

Training the Temporal Proposal Network. The temporal proposal network produces an action bound prediction $p_t(s_t, e_t)$ and a location l_t at each timestep. The bounding regressor is a fully connected network and thus can be trained using standard back-propagation. As for the locator network, the procedure for generating location sequences is not differentiable, which can be trained with reinforcement learning algorithm. A reasonable reward function needs to be designed to assess the location policy and direct the sensors where to focus on next.

When training the bounding regressor, a ground truth action segment needs to be selected to refine the prediction at each timestep. Given a ground truth set $G = \{g_i | i = 1, \cdots, N\}$ in a video, for each p_t, if there exists at least one g_i overlaping with p_t, the one with a maximum overlap is chosen as the ground truth $g_c(s_c, e_c)$. Otherwise, the one with a minimum distance with p_t should be selected. Besides, for each pair of p_t and g_i at timestep t, a matching indicator m_{ti} is set 1 if i equals to c, while $m_{ti} = 0$ for other situations. The overlap and distance function can be expressed as

$$\text{Overlap}(p_t, g_i) = \frac{\min(e_t, e_i) - \max(s_t, s_i)}{\max(e_t, e_i) - \min(s_t, s_i)} \tag{1}$$

$$\text{Distance}(p_t, g_i) = \min(|s_t - e_i|, |e_t - s_i|) \tag{2}$$

Once g_c is chosen, the loss function for the bounding regressor can be formulated using a smooth L_1 loss:

$$\text{L}_{loc}(p_t, g_c) = \text{smooth}_{L_1}(s_t - s_c) + \text{smooth}_{L_1}(e_t - e_c) \tag{3}$$

where

$$\text{smooth}_{L_1}(x) = \begin{cases} 0.5x^2 & \text{if } |x| < 1, \\ |x| - 0.5 & \text{otherwise} \end{cases} \tag{4}$$

At each single step, a scalar reward r_t is evaluated. If $\text{Overalp}(p_t, g_c) >= \alpha$ (threshold which is set 0.5 in our work), 1 will be assigned to r_t, otherwise r_t is set -0.1. The cumulative reward R_t after t steps is formulated as $R_t = \sum_{j=1,\cdots,t} r_j$. The goal of the training procedure is to minimize the difference between the predictions and the ground truth as well as maximize the final cumulative reward R_T. We define the total loss function over the prediction set P as following

$$\text{L}(P) = \sum_t \sum_i [m_{ti} = 1] \text{L}_{loc}(p_t, g_i) - \lambda R_T \tag{5}$$

where the hyper-parameter λ is set 1 in all experiments. We use back-propagation and reinforcement learning to optimize this loss function.

Training the Classifers. For a softmax classifier, we trim the videos of training set into foregrounds and backgrounds first, which are used for positive and negative samples respectively. Next we experiment three kinds of training strategies in our work:

Type-1. For each positive or negative trimmed video, we randomly sample n C3D features and join them into a feature vector. The classifier then propagates this feature vector and outputs scores for K classes and the background. Typically, n is set 5 in our experiments.

Type-2. Based on Type-1, on the left and right to each trimmed video s, we randomly generate 2 more video segments as training samples. These two segments have a high overlap (we use 0.6~0.9) with s, thus they are labeled the same as s naturally. Usually proposal predictions tend to contain both foreground and background information. Trained with these generated segments including similar information, classifier may achieve a more precise classification performance.

Type-3. Ignoring the negative samples, we randomly sample one C3D feature from each positive trimmed video to train a K-classes classifier. At test-time, the C3D feature extracted from the middle location of the proposal is used to perform classification.

For SVM classifier, we train a one-vs-all SVM classifier for each action class. When training a class-specific SVM classifier, we utilize all trimmed videos of other categories as negative samples, including foregrounds and backgrounds.

4   Experiments

4.1   Dataset and Implementation Details

We evaluate our model on the ActivityNet dataset [6]. The latest version of ActivityNet dataset consists of 19994 videos: 10,024 for training, 4,926 for validation and 5,044 for test. There are 1.54 activity instances per video on average and these instances belong to 200 activity categories. Since the ground truth of the testing set is not provided, we train our model on training set and evaluate it on validation set.

Our model is implemented using Torch7 framework [8]. We use C3D features provided by ActivityNet website http://activity-net.org/. Each C3D feature is a 500-dimentional vector extracted from every 16 frames.

For the temporal proposal network, we set $T = 7$ for the RNN, which means the agent will generate 7 temporal proposals for each video. A LSTM network with 256 units is used to implement RNN. The regressor is simply achieved with two fully connected layers ($256 \rightarrow 100, 100 \rightarrow 2$). Parameters of the network are initialized with uniform distribution between -0.1 and 0.1. At training time, we use 20 videos as a mini-batch and an initial learning rate of 0.01, which will decays by 0.0001 per 200 mini-batch iterations. Momentum is set 0.9 in our work.

4.2   Results

In this part, we first evaluate key configurations of our model in a sub-dataset and apply the optimal settings to the full ActivityNet dataset. Finally, we report the overall performance along with comparison, and provide analysis and discussions (Fig. 3).

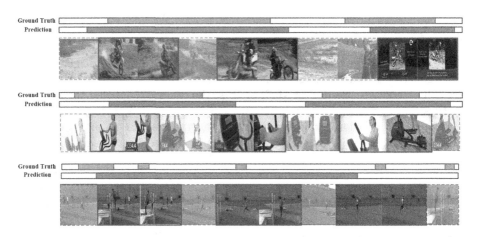

Fig. 3. Three illustrative examples of our temporal proposal predictions. The top 2 examples correctly cover the ground truth after observing several fragments of the video. The third one is a failure case of our model.

Selecting a Classifier. Using the same temporal proposal network, we test the performance of classifiers mentioned in Sect. 3 on the sub-dataset "Playing sports" of ActivityNet, which consists of 26 action classes. The detection results is shown in Table 1, which indicates that the Type-3 softmax classifier outperforms the Type-1 and Type-2. Without involving background information at training time, the Type-3 can learn a better representation of positive samples. When using linear one-vs-all SVM classifiers, the model can yield a better performance as shown in Table 1.

Training Manner. We test two kinds of strategies to train our agent. The first one is using an end-to-end manner, i.e. all the components are trained in a unified procedure like [23,42], denoted as "unified-sinlge/one-vs-all" which trains temporal regressor and classifier simultaneously. Specifically, "unified-single" trains a single model while "unified-one-vs-all" does for each class. On the other hand, similarly, "separate-single/one-vs-all" trains temporal regressor and classifier separately. In particular, "separate-single" trains a single model for all classes while "separate-one-vs-all" does for each class. The detailed comparisons are

Table 1. Performance comparison using different classifiers on "Playing Sports" subset of ActivityNet dataset at IoU of $\alpha = 0.5$.

Sports	Type-1	Type-2	Type-3	SVM
High jump	13.26	13.59	17.26	19.62
Long jump	4.05	5.21	6.68	1.86
Cricket	19.20	19.10	16.07	12.67
Discus throw	1.22	6.65	5.98	1.11
Rollerblading	19.46	26.98	23.27	24.74
Powerbomb	33.51	48.79	45.59	61.98
Javelin throw	3.03	6.71	6.45	7.08
Longboarding	28.44	25.80	21.69	37.00
Hurling	29.90	21.94	24.25	30.54
Shot put	2.06	2.40	0.44	1.38
Paintball	14.22	16.48	20.29	22.18
Bungee jumping	4.83	8.74	4.27	6.08
Triple jump	7.50	4.04	4.70	7.70
Pole vault	11.73	7.23	6.46	5.06
Powerbocking	37.26	36.26	43.39	43.09
Croquet	23.72	23.76	27.24	25.64
Hammer throw	10.13	16.44	15.29	5.33
Skateboarding	21.61	17.00	18.26	20.59
Dodgeball	41.26	49.71	55.63	63.72
Doing motocross	48.21	46.47	54.73	59.21
Starting campfire	45.44	44.48	49.50	62.50
Archery	13.28	12.44	21.71	14.38
Camel ride	78.96	69.38	70.99	83.61
Playing kickball	23.30	37.15	34.94	39.58
Baton twirling	72.30	71.27	71.15	77.64
Curling	14.50	7.10	11.94	14.37
mAP	23.94	24.81	26.08	**28.79**

shown in Table 2. From the table we can see that the separate models strongly outperform the unified models, which indicates that training temporal regressor and classifiers separately is more reasonable for action detection. We adopt separate model in our following experiments.

Overall Performance. A baseline on all 200 action classes of ActivityNet dataset is provided by [6]. The baseline utilizes Motion Features(MF), Deep Features(DF), Static Features(SF) and MF+DF+SF features respectively to

Table 2. Performance comparison using different training manners on "Playing Sports" subset of ActivityNet dataset with IOU threshold $\alpha = 0.5$.

	U-single	U-ova	S-single	S-ova
High jump	7.44	7.55	19.62	13.99
Long jump	0.47	6.38	1.86	6.37
Cricket	4.43	12.15	12.67	15.26
Discus throw	4.53	11.23	1.11	10.08
Rollerblading	5.33	10.87	24.74	22.62
Powerbomb	25.19	48.81	61.98	61.05
Javelin throw	5.04	2.07	7.08	14.64
Longboarding	18.22	32.52	37.00	38.56
Hurling	15.19	28.11	30.54	24.16
Shot put	1.32	0.29	1.38	0.57
Paintball	11.30	18.75	22.18	29.03
Bungee jump	0.34	5.09	6.08	15.53
Triple jump	1.19	11.58	7.70	16.89
Pole vault	1.39	12.72	5.06	6.68
Powerbocking	31.53	47.40	43.09	44.48
Croquet	15.11	15.64	25.64	28.96
Hammer throw	0.50	9.75	5.33	14.21
Skateboarding	1.89	12.91	20.59	20.45
Dodgeball	47.50	59.76	63.72	69.01
Doing moto	40.69	51.82	59.21	61.47
Starting camp	32.00	36.23	62.50	62.38
Archery	2.79	15.27	14.38	9.03
Camel ride	45.54	36.01	83.61	87.91
Playing ball	22.29	27.19	39.58	32.67
Baton twirling	53.58	77.03	77.64	75.19
Curling	4.94	0.58	14.37	16.08
mAP	15.37	22.99	28.79	**30.43**

conduct action detection. Since training 200 temporal proposal networks is troublesome, here we use the "separate-single" model with Type-3 softmax and SVM classifiers to report our results. As shown in Table 3, our model significantly outperforms the baseline for multiple overlap threshold, achieving a mAP of 22.0% for IoU of 0.5. The results show that our two-phase model is effective, and moreover, our model just needs to focus on a few fixations for a prediction decision.

Table 3. Performance comparison on ActivityNet dataset.

	IoU threshold(α)				
	0.1	0.2	0.3	0.4	0.5
MF [6]	11.7	11.4	10.6	9.7	8.9
DF [6]	7.2	6.8	4.9	4.1	3.7
SF [6]	4.2	3.9	3.1	2.1	1.9
MF+DF+SF [6]	12.5	11.9	11.1	10.4	9.7
S-single+Softmax(Ours)	27.3	25.8	23.7	21.8	19.5
S-single+SVM(Ours)	**30.2**	**27.7**	**24.5**	**23.1**	**22.0**

5   Conclusion

In this paper, we present an attention-based two-phase model for video action detection. As demonstrated on ActivityNet dataset, the model proves to be effective in detecting action instances while maintaining a low computational cost, ascribed to the attention mechanism and the classification strategy design. Besides, training the temporal proposal network and classifier separately offers more flexibility and potential for multi-task detection.

References

1. Aggarwal, J.K., Ryoo, M.S.: Human activity analysis: a review. ACM Comput. Surv. **43**(3), 16 (2011)
2. Ba, J., Mnih, V., Kavukcuoglu, K.: Multiple object recognition with visual attention. arXiv preprint arXiv:1412.7755 (2014)
3. Ba, J., Mnih, V., Kavukcuoglu, K.: Multiple object recognition with visual attention. In: ICLR (2015)
4. Ba, J., Salakhutdinov, R.R., Grosse, R.B., Frey, B.J.: Learning wake-sleep recurrent attention models. In: NIPS, pp. 2593–2601 (2015)
5. Bazzani, L., Larochelle, H., Torresani, L.: Recurrent mixture density network for spatiotemporal visual attention. arXiv preprint arXiv:1603.08199 (2016)
6. Caba Heilbron, F., Escorcia, V., Ghanem, B., Carlos Niebles, J.: Activitynet: a large-scale video benchmark for human activity understanding. In: CVPR, pp. 961–970 (2015)
7. Campbell, L.W., Bobick, A.F.: Recognition of human body motion using phase space constraints. In: ICCV, pp. 624–630 (1995)
8. Collobert, R., Kavukcuoglu, K., Farabet, C.: Torch7: A matlab-like environment for machine learning. In: BigLearn, NIPS Workshop. No. EPFL-CONF-192376 (2011)
9. Donahue, J., Anne Hendricks, L., Guadarrama, S., Rohrbach, M., Venugopalan, S., Saenko, K., Darrell, T.: Long-term recurrent convolutional networks for visual recognition and description. In: ICCV, pp. 2625–2634 (2015)
10. Girshick, R., Donahue, J., Darrell, T., Malik, J.: Rich feature hierarchies for accurate object detection and semantic segmentation. In: CVPR, pp. 580–587 (2014)

11. Gupta, A., Srinivasan, P., Shi, J., Davis, L.S.: Understanding videos, constructing plots learning a visually grounded storyline model from annotated videos. In: CVPR, pp. 2012–2019 (2009)
12. Jhuang, H., Gall, J., Zuffi, S., Schmid, C., Black, M.J.: Towards understanding action recognition. In: ICCV, pp. 3192–3199 (2013)
13. Ji, S., Xu, W., Yang, M., Yu, K.: 3d convolutional neural networks for human action recognition. TPAMI **35**(1), 221–231 (2013)
14. Kantorov, V., Laptev, I.: Efficient feature extraction, encoding and classification for action recognition. In: CVPR, pp. 2593–2600 (2014)
15. Karpathy, A., Toderici, G., Shetty, S., Leung, T., Sukthankar, R., Fei-Fei, L.: Large-scale video classification with convolutional neural networks. In: CVPR, pp. 1725–1732 (2014)
16. Klaser, A., Marszałek, M., Schmid, C.: A spatio-temporal descriptor based on 3d-gradients. In: BMVC, pp. 275:1–275:10 (2008)
17. Krizhevsky, A., Sutskever, I., Hinton, G.E.: Imagenet classification with deep convolutional neural networks. In: NIPS, pp. 1097–1105 (2012)
18. Laptev, I.: On space-time interest points. IJCV **64**(2–3), 107–123 (2005)
19. Mnih, V., Heess, N., Graves, A., et al.: Recurrent models of visual attention. In: NIPS, pp. 2204–2212 (2014)
20. Niyogi, S.A., Adelson, E.H.: Analyzing and recognizing walking figures in xyt. In: CVPR, pp. 469–474 (1994)
21. Oneata, D., Verbeek, J., Schmid, C.: The lear submission at thumos 2014 (2014)
22. Poppe, R.: A survey on vision-based human action recognition. Image Vis. Comput. **28**(6), 976–990 (2010)
23. Sharma, S., Kiros, R., Salakhutdinov, R.: Action recognition using visual attention. In: ICLR (2016)
24. Shi, Y., Bobick, A., Essa, I.: Learning temporal sequence model from partially labeled data. In: CVPR, vol. 2, pp. 1631–1638 (2006)
25. Simonyan, K., Zisserman, A.: Two-stream convolutional networks for action recognition in videos. In: NIPS, pp. 568–576 (2014)
26. Singh, G., Cuzzolin, F.: Untrimmed video classification for activity detection: submission to activitynet challenge. arXiv preprint arXiv:1607.01979 (2016)
27. Sutton, R.S., McAllester, D.A., Singh, S.P., Mansour, Y., et al.: Policy gradient methods for reinforcement learning with function approximation. In: NIPS, vol. 99, pp. 1057–1063 (1999)
28. Szegedy, C., Liu, W., Jia, Y., Sermanet, P., Reed, S., Anguelov, D., Erhan, D., Vanhoucke, V., Rabinovich, A.: Going deeper with convolutions. In: CVPR, pp. 1–9 (2015)
29. Tang, Y., Srivastava, N., Salakhutdinov, R.R.: Learning generative models with visual attention. In: NIPS, pp. 1808–1816 (2014)
30. Tran, D., Bourdev, L., Fergus, R., Torresani, L., Paluri, M.: Learning spatiotemporal features with 3d convolutional networks. In: ICCV, pp. 4489–4497 (2015)
31. Venugopalan, S., Xu, H., Donahue, J., Rohrbach, M., Mooney, R., Saenko, K.: Translating videos to natural language using deep recurrent neural networks. arXiv preprint arXiv:1412.4729 (2014)
32. Wang, H., Kläser, A., Schmid, C., Liu, C.L.: Dense trajectories and motion boundary descriptors for action recognition. IJCV **103**(1), 60–79 (2013)
33. Wang, H., Schmid, C.: Action recognition with improved trajectories. In: ICCV, pp. 3551–3558 (2013)
34. Wang, H., Ullah, M.M., Klaser, A., Laptev, I., Schmid, C.: Evaluation of local spatio-temporal features for action recognition. In: BMVC, pp. 124–1 (2009)

35. Wang, J., Wang, W., Wang, R., Gao, W., et al.: Deep alternative neural network: Exploring contexts as early as possible for action recognition. In: NIPS, pp. 811–819 (2016)
36. Wang, L., Qiao, Y., Tang, X.: Action recognition and detection by combining motion and appearance features. In: THUMOS14 Action Recognition Challenge, vol. 1, p. 2 (2014)
37. Weinland, D., Ronfard, R., Boyer, E.: A survey of vision-based methods for action representation, segmentation and recognition. Comput. Vis. Image Underst. **115**(2), 224–241 (2011)
38. Willems, G., Tuytelaars, T., Gool, L.: An efficient dense and scale-invariant spatio-temporal interest point detector. In: Forsyth, D., Torr, P., Zisserman, A. (eds.) ECCV 2008. LNCS, vol. 5303, pp. 650–663. Springer, Heidelberg (2008). doi:10.1007/978-3-540-88688-4_48
39. Williams, R.J.: Simple statistical gradient-following algorithms for connectionist reinforcement learning. Mach. Learn. **8**(3–4), 229–256 (1992)
40. Wu, J., Wang, G., Yang, W., Ji, X.: Action recognition with joint attention on multi-level deep features. arXiv preprint arXiv:1607.02556 (2016)
41. Xu, K., Ba, J., Kiros, R., Cho, K., Courville, A., Salakhudinov, R., Zemel, R., Bengio, Y.: Show, attend and tell: neural image caption generation with visual attention. In: ICML, pp. 2048–2057 (2015)
42. Yeung, S., Russakovsky, O., Mori, G., Fei-Fei, L.: End-to-end learning of action detection from frame glimpses in videos. arXiv preprint arXiv:1511.06984 (2015)

Learning Discriminative Representation for Skeletal Action Recognition Using LSTM Networks

Lizhang Hu and Jinhua Xu[(✉)]

Shanghai Key Laboratory of Multidimensional Information Processing,
Department of Computer Science and Technology, East China Normal University,
3663 North Zhongshan Road, Shanghai, China
51151201077@stu.ecnu.edu.cn, jhxu@cs.ecnu.edu.cn

Abstract. Human action recognition based on 3D skeleton data is a rapidly growing research area in computer vision due to their robustness to variations of viewpoint, human body scale and motion speed. Recent studies suggest that recurrent neural networks (RNNs) or convolutional neural networks (CNNs) are very effective to learn discriminative features of temporal sequences for classification. However, in prior models, the RNN-based method has a complicated multi-layer hierarchical architecture, and the CNN-based methods learn the contextual feature on fixed temporal scales. In this paper, we propose a framework which is simple and able to select temporal scales automatically with a single layer LSTM for skeleton based action recognition. Experimental results on three benchmark datasets show that our approach achieves the state-of-the-art performance compared to recent models.

Keywords: Skeleton based action recognition · Recurrent Neural Networks · Long short-term memory

1 Introduction

Action recognition has been an important research area of computer vision for the past several decades due to its applications in video surveillance, human-computer interaction, health care and so on. Generally, studies mainly focus on action recognition from video sequences of 2D frames with RGB channels [1], which may lose some information in the 3D space. Recently, with the developments of the low-cost RGB-D sensors such as Microsoft Kinect [13], the information based on depth has been employed in action recognition [5,28]. Since the seminal work by Shotton, et al. [20] for estimating the joint locations of a human body from depth map, the research in action recognition based on the locations of body joints has attracted an increasing attention due to their robustness to variations of viewpoint, human body scale and motion speed.

Much work has been done to extract discriminative features from the multidimensional skeleton sequences. The hand-crafted features are used in early work

© Springer International Publishing AG 2017
M. Felsberg et al. (Eds.): CAIP 2017, Part II, LNCS 10425, pp. 94–104, 2017.
DOI: 10.1007/978-3-319-64698-5_9

such as HOJ3D [26], EigenJoints [27] and covariance descriptor [14,21]. In recent years, some representation learning methods have been proposed, which can be broadly categorized into three groups: dictionary learning, unsupervised feature learning, and deep learning [12].

Deep learning models, including Convolutional Neural Networks (CNNs) and Recurrent Neural Networks (RNNs), have been used for action recognition of RGB video sequences. In [15,16], 2D or 3D convolutional neural networks were used for human action recognition. Grushin *et al.* [11] applied long short-term memory (LSTM) neural networks to action recognition. In [6], recurrent convolutional models were proposed for large-scale visual understanding tasks, which integrate CNN and RNNs and are doubly deep in that they learn compositional representations in space and time. Deep learning has also been applied to 3D skeletal based action recognition. In [7], Du *et al.* proposed an end-to-end hierarchical RNN for skeleton based action recognition. In [22], a Moving Poselets model was proposed for each body part, which includes one Convolutional layer and one pooling layer. These two models both achieved good performance, but the RNN model [7] has a complicated hierarchical architecture, and it needs to crop the variable-length sequence data to same length as input. The CNN model [22] is simple, and does not require the input sequence of fixed size, but it models the contextual information on fixed temporal scales with the fixed kernel size in the convolutional layer.

In this paper, we propose a two-layer network for skeleton based action recognition. The first layer is LSTM layer and we use variable-length skeleton joint sequence data as input, each time step receiving a frame. In this way, each time step's output can be regarded as the mid-level representation about all previous frames. The second layer is a temporal pooling layer, through which we obtain the fixed representation for each sequence. Our proposed model combines the advantages of the [7] and [22], which can not only learn discriminative feature representation through simple architecture but also model the contextual information on variable temporal scales. We conduct experiments on three benchmark datasets, and the results demonstrate the effectiveness of our approach. Our model achieves the state-of-the-art performance compared to previous models.

The rest of the paper is organized as follows. Section 2 gives a brief overview of related work which applys deep learning to skeleton based action recognition. In Sect. 3, we introduce the details of the proposed method. The Sect. 4 shows our experimental results on three datasets and do some analysis of our methods. Finally we conclude the paper in Sect. 5.

2   Related Work

Deep learning has been applied to action recognition for 3D skeletal joints in recent years. In this section, we limit our review to the most related models.

In [7], a novel hierarchical bidirectional RNN (HBRNN) was proposed for skeleton based action recognition. It used five body parts as input based on human physical structure, and each part was fed into a separated bidirectional

RNN. As the number of layers increases, the parts' representations are hierarchically fused and fed to a fully connected layer followed by a softmax layer for classification.

Tao et al. [22] developed a body-part motion based model called Moving Poselets. It was composed of one convolutional layer and one max pooling layer. Temporal pyramid pooling was performed over the longer sequences. After obtaining response maps of each body parts, all parts' features were concatenated to form the final global representation for action recognition.

A recurrent neural network structure called Part-aware LSTM was introduced for action classification in [2]. Instead of using entire body's motion in the cell for keeping a long-term memory, it was composed of part-based cells. Each part's cell has its individual input, forget, and modulation gates, but all body parts have a common output gate.

Liu et al. [18] designed a spatio-temporal LSTM network for skeleton based action recognition. In the spatial direction, body joints in a frame were fed in a sequence so that the spatial dependence among joints was considered. In the temporal direction, the locations of the corresponding joints were fed over time. And they designed a new gating function called trust gate to deal with noisy input.

Our model is most related to Moving Poselets [22] and HBRNN [7]. Instead of applying a CNN-based framework for each body parts, we learn the mid-level representation for entire body using a LSTM-based structure. The convolutional layer in [22] can only learn the temporal representation over a fixed temporal scale due to the fixed size of the convolutional kernel. But with the LSTM, we can learn the representation over variable temporal scales automatically. In HBRNN [7], there are three BRNN layers and one LSTM layer, and the body was divided into five parts, much more complicated than our full body model with a single LSTM layer. We doubt that the RNNs in the lower layers of HBRNN [7] are less efficient to learn the temporal representation than the LSTM in our model, which will be verified in our experiments.

3   Our LSTM Architecture

The architecture of the proposed network is shown in Fig. 1. We use the LSTM unit as described in [9]. In addition to a hidden unit $\mathbf{h}_t \in R^N$, the LSTM includes an input gate $\mathbf{i}_t \in R^N$, forget gate $\mathbf{f}_t \in R^N$, output gate $\mathbf{o}_t \in R^N$, and memory cell $\mathbf{c}_t \in R^N$. Given an input sequence $\mathbf{x} = [\mathbf{x}_1, \mathbf{x}_2, ..., \mathbf{x}_T]$, where $\mathbf{x}_t$ is the body joints data at time step t, and T is the total number of frames in the sequence, the LSTM updates for time step t given inputs $\mathbf{x}_t$, $\mathbf{h}_{t-1}$, and $\mathbf{c}_{t-1}$ are:

$$\mathbf{i}_t = \sigma \left(\mathbf{W}_{xi}\mathbf{x}_t + \mathbf{W}_{hi}\mathbf{h}_{t-1} + \mathbf{W}_{ci}\mathbf{c}_{t-1} + \mathbf{b}_i \right), \tag{1}$$

$$\mathbf{f}_t = \sigma \left(\mathbf{W}_{xf}\mathbf{x}_t + \mathbf{W}_{hf}\mathbf{h}_{t-1} + \mathbf{W}_{cf}\mathbf{c}_{t-1} + \mathbf{b}_f \right), \tag{2}$$

$$\mathbf{c}_t = \mathbf{f}_t \odot \mathbf{c}_{t-1} + \mathbf{i}_t \odot \tanh \left(\mathbf{W}_{xc}\mathbf{x}_t + \mathbf{W}_{hc}\mathbf{h}_{t-1} + \mathbf{b}_c \right), \tag{3}$$

$$\mathbf{o}_t = \sigma \left(\mathbf{W}_{xo}\mathbf{x}_t + \mathbf{W}_{ho}\mathbf{h}_{t-1} + \mathbf{W}_{co}\mathbf{c}_{t-1} + \mathbf{b}_o \right), \tag{4}$$

$$\mathbf{h}_t = \mathbf{o}_t \odot \tanh \left(\mathbf{c}_t \right). \tag{5}$$

where σ and tanh are the sigmoid function and the hyperbolic tangent respectively, $\odot$ denotes the element-wise product of two vectors. $\mathbf{W}$ and $\mathbf{b}$ are the weight matrixes and bias vectors respectively. By concatenating all time steps' output, we generate a response map:

$$\mathbf{H} = [\mathbf{h}_1, \mathbf{h}_2, ..., \mathbf{h}_T]. \tag{6}$$

Notice that the number of frames T may be different for different sequences. Therefore, a temporal max pooling is performed to compute its final representation of a sequence:

$$q_j = \max[h_{j,1}, h_{j,2}, ..., h_{j,T}], \quad 1 \le j \le N. \tag{7}$$

Then $\mathbf{Q} = [q_1, q_2, ..., q_N]^T$ is first passed through a rectified linear unit (ReLU) and then fed to a fully connected layer followed by a LogSoftMax layer. The model is trained using back-propagation through time (BPTT) algorithm.

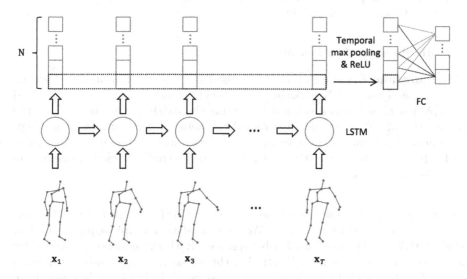

Fig. 1. The architecture of the proposed model. Notice that the max pooling is over the temporal dimension.

4   Experiments

In this section, we evaluated our method on three datasets: MSR Action3D dataset [17], MSR Daily Activity3D dataset [25] and UT-Kinect dataset [26]. We also compare our results to other models.

4.1   Datasets

MSR Action3D Dataset. The MSR Action3D dataset consists of 20 action types, such as high arm wave and draw circle. The 3D skeleton joint data are extracted from RGB-D videos, and the 3D coordinates of 20 joints are provided in this dataset. Each action is performed 2–3 times by 10 subjects. The action sequences are relatively short, with 30–50 frames, and the frame rate is 15 frames per second. In total there are 567 sequences.

MSR DailyActivity3D Dataset. The MSR DailyActivity3D dataset covers 16 daily activities performed by 10 subjects. Each subject performed an activity twice: once seated on a sofa and once standing. The sequences are longer, with 100–300 frames. There are 320 sequences in total. The body is also represented by 20 joints. It is more challenging than MSR Action3D, since the actions are more complex and contain human-object interactions.

UT-Kinect Dataset. The UT-Kinect dataset was captured by a stationary Kinect. It contains 200 sequences of 10 action classes performed by 10 subjects, and each action performed 2 times by each subject. The skeleton data also contains 3D positions of 20 joints.

4.2   Implementation Details

Configuration. In our experiments, we used Torch toolbox as the deep learning platform, and used RNN package [19] to build the network structure. We used variable-length sequences as input by setting the batch size as one, and set the output size of cell N as 500. We trained the network using stochastic gradient descent, and set learning rate, learning rate decay, momentum and weight decay as 1×10^{-1}, 1×10^{-4}, 1×10^{-3}, and 1×10^{-3} respectively. These parameters are same for all three datasets.

Data Segment. Data segment is only needed for MSR DailyActivity3D Dataset because the sequences are too long. We segmented each sample sequence in MSR DailyActivity3D Dataset into 7 sub-sequences. Each subsequence accounted for $3/5$ of the original sequence's length. For the ith subsequence, the start index is $round((i-1)/6 * 2L/5) + 1$, the subsequence length is $3L/5 - 1$. We uses mean prediction of the 7 subsequences as the original sequence's predicted class.

4.3   Results

MSR Action3D Dataset. About 10 skeleton sequences were not used in [25] because of missing data or highly erroneous joint positions. We followed this for fair comparisons. We used two setup protocols to conduct the experiments, as in [22,25]. In Setup 1, all sequences from subjects 1, 3, 5, 7 and 9 are used for

Table 1. Experimental results on MSR Action3D Dataset (setup 1).

Method	Accuracy (%)
Actionlet ensemble [25]	88.20
Lie group [23]	89.48
Hierarchical LDS [4]	90.00
Pose base approach [24]	90.22
Moving pose [29]	91.70
Moving poselets [22]	93.60
Ours	93.04

training and the remaining ones for testing. In Setup 2, the dataset is divided into three action sets, AS1, AS2 and AS3, and the same algorithm is tested on each of the three sets. The results are shown in Tables 1 and 2. In setup 1, our result was comparable to that of Moving Poselets [22]. In setup 2, our model achieved the state-of-the-art result and outperformed Moving Poselets [22], Deep unidirectional RNN(DURNN) [7] and deep bidirectional RNN (DBRNN) [7], but slightly worse than the hierarchical bidirectional RNN (HBRNN). DURNN and DURNN are directly stacked with several RNN layers with the whole human skeleton as the input, and only the last recurrent layer consists of LSTM neurons. From the aspect of structure, our model only uses the last LSTM layer in DURNN. It seems that the bidirectional connection, the part-based feature extraction and hierarchical fusion in [7] may not be critical, and only the LSTM layer plays an important role.

MSR DailyActivity3D Dataset. Due to the noisy estimated joint positions and skeleton variations across different subjects in this dataset, we normalized the skeleton data according to Algorithm 1 described in [29]. And we segmented data as previously described. We also used the sequences from subject 1, 3, 5, 7

Table 2. Experimental results on MSR Action3D Dataset (setup 2).

Method	AS1	AS2	AS3	Avg
Bag of 3D points [17]	72.90	71.90	79.20	74.70
HOD [10]	92.39	90.18	91.43	91.26
Lie group [23]	95.29	83.87	98.22	92.46
Moving poselets [22]	89.81	93.57	97.03	93.50
DURNN [7]	87.62	91.96	90.01	89.86
DBRNN [7]	88.57	93.75	95.50	92.61
HBRNN [7]	93.33	94.64	95.50	94.49
Ours	91.43	93.75	97.30	94.16

and 9 for training, and remaining ones for testing. We compared our result with previous models in Table 3. It can be seen that our result (74.4%) is comparable to that of Moving Poselets (74.4%) in [22]. HBRNN model was not tested on this dataset in [7] and we did not re-implement it, therefore no comparison was made with HBRNN.

Table 3. Results on the MSR DailyActivity3D Dataset.

Method	Accuracy (%)
Actionlet ensemble [25]	68.0
Efficient pose-based [8]	73.1
Moving pose [29]	73.8
Moving poselets [22]	74.5
Ours	74.4

UT-Kinect Dataset. One skeleton sequences were not used in our experiment due to the invalidation. We used the experimental protocol in [30], in which half of the subjects were used for training and the remaining for testing. As shown in Table 4, our model achieved 98.0% accuracy and outperformed previous models.

Table 4. Results of our method and other methods on the UT-Kinect Dataset.

Method	Accuracy (%)
Skeleton joint features [30]	87.9
Lie group [23]	93.6
Elastic functional coding [3]	94.9
ST-LSTM (Tree) + Trust Gate [18]	95.0
Ours	98.0

4.4   Analysis

Dimension of the LSTM Cell. To test the effects of the LSTM cell dimension on the recognition results, we did experiments with different values of N on the MSR Action3D Dataset (setup 1). As shown in Fig. 2, the results did not change much when the cell dimension was larger than 200. Therefore we set it as 500 in all experiments.

Comparison with RNNs Models. To show the efficiency of LSTM to learn the temporal representation, we run experiments with a RNN model, which has the same structure as our proposed model, but the LSTM layer is replaced by RNN. The performance of this RNN model on MSR Action3D is 86.81%, significantly worse than 93.04% of our LSTM method. This suggests that temporal dependency can be learned more efficiently using LSTM than RNN.

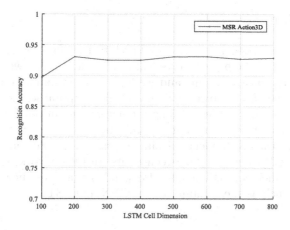

Fig. 2. Recognition accuracy under different LSTM cell dimension on the MSR Action3D Dataset (setup 1).

ReLU Activation. In order to demonstrate the role of the ReLU nonlinearity, we performed contrast experiments on the three datasets without ReLU. The results are shown in Table 5. It can be seen that ReLU activation increased the accuracy in all cases.

Table 5. Effects of the ReLU activation on the three datasets.

Dataset	w/o ReLU	with ReLU
MSR Action3D (setup 1)	94.09	94.16
MSR Action3D (setup 2)	91.94	93.04
MSR DailyActivity3D	70.62	74.38
UT-Kinect	94.94	97.98

Normalizing the Skeleton Data. To evaluate the contribution of the data normalization on MSR DailyActivity3D dataset, we performed the same experiment using the same network, but with raw data. The accuracy was 68.75% versus 74.38%. This suggests that for complex and noisy datasets like MSR DailyActivity3D, normalizing the skeleton data leads to better performance.

Data Segmentation. We also did the experiment on MSR Action3D Dataset which cropped each sequence to 3 sub-sequences. The accuracy reduced about 1%~2% from 93.04%. On MSR DailyActivity3D dataset, the performance improved about 4%(74.38%) compared to 70.63% without data segmentation. We conclude that segmentation can help in long sequences but deteriorates the performance for short sequences.

5 Conclusion

In this paper, we proposed a LSTM-based architecture for skeletal action recognition. Different from the prior feature learning models [7,22], we used variable-length skeletal sequence as input, and we learned discriminative feature representation from contextual information on variable temporal scales automatically. Our model had also be applied to long sequence recognition. From the experimental results, it can be seen that our results were comparable to or better than the state-of-the-art models. In our current model we did not take into account the spatial dependency between joints. In future work, we will explore an architecture which can model spatial dependency and temporal dependency simultaneously, such as the combination of CNN and LSTM.

References

1. Aggarwal, J.K., Ryoo, M.S.: Human activity analysis: a review. ACM Comput. Surv. **43**(3), 16 (2011)
2. Shahroudy, A., Liu, J., Ng, T.T., Wang, G.: NTU RGB+D: a large scale dataset for 3D human activity analysis. In: IEEE Conference on Computer Vision and Pattern Recognition (2016)
3. Anirudh, R., Turaga, P., Su, J., Srivastava, A.: Elastic functional coding of human actions: from vector-fields to latent variables. In: IEEE Conference on Computer Vision and Pattern Recognition, pp. 3147–3155 (2015)
4. Chaudhry, R., Ofli, F., Kurillo, G., Bajcsy, R., Vidal, R.: Bio-inspired dynamic 3D discriminative skeletal features for human action recognition. In: IEEE Conference on Computer Vision and Pattern Recognition (2013)
5. Chen, L., Wei, H., Ferryman, J.M.: A survey of human motion analysis using depth imagery. Pattern Recogn. Lett. **34**, 1995–2006 (2013)
6. Donahue, J., Hendricks, L.A., Rohrbach, M., Venugopalan, S., Guadarrama, S., Saenko, K., Darrell, T.: Long-term recurrent convolutional networks for visual recognition and description. IEEE Trans. Pattern Anal. Mach. Intell. (2016)
7. Du, Y., Wang, W., Wang, L.: Hierarchical recurrent neural network for skeleton based action recognition. In: IEEE Conference on Computer Vision and Pattern Recognition, pp. 1110–1118 (2015)
8. Eweiwi, A., Cheema, M.S., Bauckhage, C., Gall, J.: Efficient pose-based action recognition. In: Asian Conference on Computer Vision, pp. 428–443 (2014)
9. Gers, F.A., Schraudolph, N.N., Schmidhuber, J.: Learning precise timing with lstm recurrent networks. J. Mach. Learn. Res. **3**(1), 115–143 (2003)
10. Gowayyed, M.A., Torki, M., Hussein, M.E., El-Saban, M.: Histogram of oriented displacements (HOD): describing trajectories of human joints for action recognition. In: International Joint Conference on Artificial Intelligence, pp. 1351–1357 (2013)
11. Grushin, A., Monner, D.D., Reggia, J.A., Mishra, A.: Robust human action recognition via long short-term memory. In: International Joint Conference on Neural Networks, pp. 1–8 (2013)

12. Han, F., Reily, B., Hoff, W., Zhang, H.: Space-time representation of people based on 3D skeletal data: a review. arXiv preprint (2016). arXiv:1601.01006
13. Han, J.: Enhanced computer vision with microsoft kinect sensor: a review. IEEE Trans. Cybern. **43**(5), 1318–1334 (2013)
14. Hussein, M.E., Torki, M., Gowayyed, M.A., El-Saban, M.: Human action recognition using a temporal hierarchy of covariance descriptors on 3D joint locations. In: Proceedings of the 23rd International Joint Conference on Artificial Intelligence (2013)
15. Ji, S., Xu, W., Yang, M., Yu, K.: 3D convolutional neural networks for human action recognition. IEEE Trans. Pattern Anal. Mach. Intell. **35**(1), 221–231 (2013)
16. Karpathy, A., Toderici, G., Shetty, S., Leung, T., Sukthankar, R., Fei-Fei, L.: Large-scale video classification with convolutional neural networks. In: IEEE Conference on Computer Vision and Pattern Recognition (2014)
17. Li, W., Zhang, Z., Liu, Z.: Action recognition based on a bag of 3D points. In: Workshop on Human Activity Understanding from 3D Data, pp. 9–14 (2010)
18. Liu, J., Shahroudy, A., Xu, D., Wang, G.: Spatio-temporal lstm with trust gates for 3D human action recognition. In: European Conference on Computer Vision, pp. 816–833 (2016)
19. Lonard, N., Waghmare, S., Wang, Y., Kim, J.H.: rnn: Recurrent library for torch (2015). arXiv preprint arXiv:1511.07889
20. Shotton, J., Fitzgibbon, A., Cook, M., Sharp, T., Finocchio, M., Moore, R., Kipman, A., Blake, A.: Real-time human pose recognition in parts from single depth images. In: IEEE Conference on Computer Vision and Pattern Recognition, vol. 411(1), pp. 1297–1304 (2011)
21. Sivalingam, R., Somasundaram, G., Bhatawadekar, V., Morellas, V., Papanikolopoulos, N.: Sparse representation of point trajectories for action classification. In: 2012 IEEE International Conference on Robotics and Automation (ICRA), pp. 3601–3606 (2012)
22. Tao, L., Vidal, R.: Moving poselets: a discriminative and interpretable skeletal motion representation for action recognition. In: IEEE Conference on Computer Vision Workshop (2015)
23. Vemulapalli, R., Arrate, F., Chellappa, R.: Human action recognition by representing 3D skeletons as points in a lie group. In: IEEE Conference on Computer Vision and Pattern Recognition (2014)
24. Wang, C., Wang, Y., Yuille, A.: An approach to pose-based action recognition. In: IEEE Conference on Computer Vision and Pattern Recognition (2013)
25. Wang, J., Liu, Z., Wu, Y., Yuan, J.: Mining actionlet ensemble for action recognition with depth cameras. In: IEEE Conference on Computer Vision and Pattern Recognition, pp. 1290–1297 (2012)
26. Xia, L., Chen, C.C., Aggarwal, J.K.: View invariant human action recognition using histograms of 3D joints. In: IEEE Computer Society Conference on Computer Vision and Pattern Recognition Workshops (CVPRW), pp. 20–27 (2012)
27. Yang, X., Tian, Y.L.: Eigenjoints-based action recognition using na07ve-bayes-nearest-neighbor. In: IEEE Computer Society Conference on Computer Vision and Pattern Recognition Workshops, pp. 14–19 (2012)
28. Ye, M., Zhang, Q., Wang, L., Zhu, J., Yang, R., Gall, J.: A survey on human motion analysis from depth data. In: Grzegorzek, M., Theobalt, C., Koch, R., Kolb, A. (eds.) Time-of-Flight and Depth Imaging. Sensors, Algorithms, and Applications. LNCS, vol. 8200, pp. 149–187. Springer, Heidelberg (2013). doi:10.1007/978-3-642-44964-2_8

29. Zanfir, M., Leordeanu, M., Sminchisescu, C.: The moving pose: an efficient 3D kinematics descriptor for low-latency action recognition and detection. In: IEEE International Conference on Computer Vision, pp. 2752–2759 (2013)
30. Zhu, Y., Chen, W., Guo, G.: Fusing spatiotemporal features and joints for 3D action recognition. In: IEEE Conference on Computer Vision and Pattern Recognition Workshops, pp. 486–491 (2013)

Progressive Probabilistic Graph Matching with Local Consistency Regularization

Min Tang and Wenmin Wang[✉]

School of Electronic and Computer Engineering,
Shenzhen Graduate School, Peking University, Shenzhen, China
tangm@pku.edu.cn, wangwm@ece.pku.edu.cn

Abstract. Graph matching has attracted extensive attention in computer vision due to its powerful representation and robustness. However, its combinatorial nature and computational complexity limit the size of input graphs. Most graph matching methods initially reconstruct the graphs, while the preprocessing often results in poor performance. In this paper, a novel progressive probabilistic model is proposed in order to handle the outliers and boost the performance. This model takes advantage of the cooperation between process of correspondence enrichment and graph matching. Candidate matches are propagated with local consistency regularization in a probabilistic manner, and unreliable ones are rejected by graph matching. Experiments on two challenging datasets demonstrate that the proposed model outperforms the state-of-the-art progressive method in challenging real-world matching tasks.

Keywords: Graph matching · Progressive probabilistic model · Geometric consistency · Local regularization

1 Introduction

Graph matching has been widely used to deal with various problems in computer vision such as feature correspondence, object localization, image retrieval and 3-D construction [1–3].

Graph provides a flexible and powerful representation by modeling feature appearances and their relationships as nodes and edges respectively. Typically, graph matching is formulated as an integer quadratic programming (IQP) that is known to be NP-hard. Thus, many methods [1,4–9] have been proposed to find an approximated solution. However, there are still some challenges needed to be tackled in this field. For instance, to input complete graph structure of raw data is impracticable, due to the combinatorial nature and high computation complexity. Hence, by exploiting sparse discriminative features, most graph matching methods initially select some candidate correspondences. Disappointingly, those candidates might only include small-percentage true matches and the final performance suffers from the mismatches especially for real-world image matching.

This project was supported by Shenzhen Peacock Plan.

M. Felsberg et al. (Eds.): CAIP 2017, Part II, LNCS 10425, pp. 105–115, 2017.
DOI: 10.1007/978-3-319-64698-5_10

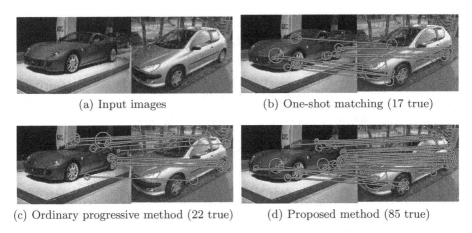

(a) Input images   (b) One-shot matching (17 true)

(c) Ordinary progressive method (22 true)   (d) Proposed method (85 true)

Fig. 1. An example of feature matching. The proposed method (d) enriches feature correspondence based on conventional graph matching and is superior to ordinary progressive method [10]. True matches are represented by cyan lines. (Best viewed in color) (Color figure online)

Recently, some progressive methods [10–12] have been proposed to deal with the above limitations of graph matching, attempting to increase the number of identified correct correspondences. These methods usually become sensitive to outliers and obtain low accuracy; Because they rely heavily on the initial matches which are pre-selected according to feature appearances.

In order to address the problems above, a progressive probabilistic model (PPM) is proposed in this paper, which combines an effective correspondence enrichment process with graph matching. The proposed method refines and enriches the candidate matches iteratively. In each iteration, the enrichment process propagates current graph matching results to their external matching space, then the dynamic update of the solution gradually promotes the graph matching objective. To satisfy the geometric consistency during the propagation, for each feature in reference domain, a corresponding location in target domain is optimized by imposing local invariance on affine transformations. Finally, a posterior probability is estimated as the matching confidence for each correspondence. In this way, the proposed approach abates the effects of outlier nodes or mismatches. PPM achieves robustness to appearance variation and outliers, and can be applied to various objects of the same class.

There are three main contributions in this paper: (1) A progressive model for graph matching is introduced to boost the number of true matches. (2) An effective probabilistic enrichment strategy is proposed to handle appearance variation and outliers. (3) The proposed model takes advantage of the cooperation between the refinement and enrichment, which outperforms the state-of-the-art progressive methods.

The rest of the paper is organized as follows. Section 2 describes PPM in detail, including graph matching and correspondence enrichment strategy. Section 3 presents the experiments on two datasets. Finally, main points of this paper are summarized in Sect. 4.

2   Progressive Matching Model

In this section, we introduce the progressive probabilistic model, consisting of two main steps: matches refinement by graph matching and probabilistic enrichment with local and geometric consistency constraints.

2.1   Graph Matching Formulation

Given two attributed graphs, $G^1 = (V^1, E^1, A^1)$ with n_1 nodes and $G^2 = (V^2, E^2, A^2)$ with n_2 nodes, where V, E and A denote a set of nodes, edges and attributes, respectively. The objective of graph matching is to find correct correspondences between the nodes of graphs G^1 and G^2. Generally, an assignment or permutation matrix $\mathbf{X} \in \{0, 1\}^{n_1 \times n_2}$ is used to represent the subset of $\chi \subset V^1 \times V^2$, where $\mathbf{X}_{ij} = 1$ if V_i^1 matches V_j^2 and 0 otherwise. In this paper, the column version of $\mathbf{X}$, defined as $\mathbf{x} \in \{0, 1\}^{n_1 n_2 \times 1}$, is used instead for real applications. In addition, relational similarity of the graph attributes are encoded in an affinity matrix $\mathbf{W}$, where the non-diagonal element $\mathbf{W}_{ij;ab}$ represents pairwise similarity between correspondences (V_i^1, V_j^2) and (V_a^1, V_b^2), and the diagonal element $\mathbf{W}_{ij;ij}$ represents unary similarity between nodes of the match (V_i^1, V_j^2).

Therefore, the graph matching problem can be formulated as an IQP problem with two-way constraints as follows:[1]

$$\mathbf{x}^* = \arg\min_{\mathbf{x}}(\mathbf{x}^T \mathbf{W} \mathbf{x}) \tag{1}$$

$$s.t. \begin{cases} \mathbf{x} \in \{0, 1\}^{n_1 n_2} \\ \sum_{i=1}^{n_1} \mathbf{x}_{ij} \leq 1, \sum_{i=1}^{n_2} \mathbf{x}_{ij} \leq 1. \end{cases}$$

Affinity matrix $\mathbf{W}$ is a key factor in determining the computational complexity of graph matching. The number of elements in $\mathbf{W}$ depends on the size of input graphs; Consequently, to input large graphs can cause high complexity and is impermissible. Moreover, another way to reduce complexity is to sparsify $\mathbf{W}$. In order to improve the efficiency of PPM, the approach used in [7] to sparsify $\mathbf{W}$ is adopted in this work.

Since the IQP problem is NP-hard, several efficient approximate algorithms for graph matching [4,6,7,9] have been proposed recently. In spite of the various strategies, any graph matching methods can be embedded into the proposed model and exert their capability as the refinement step.

2.2   Probabilistic Model for Correspondence Enrichment

Given a small number of candidate matches $M = \{m_1, m_2, \ldots, m_a, \ldots\}$ between two sets of nodes V^1 and V^2, our task in this phase is to find the other potential correspondences by exploiting M. In this probabilistic model, matching confidence for each correspondence between nodes V^1 and V^2 is represented as a posterior probability $p(v_i^1 \mapsto v_j^2 | M)$. The superscripts will be omitted in the equations below for brevity.

---

[1] Note that in this paper we use $\mathbf{x}_{ij}$ to denote $\mathbf{x}_{(i-1)n_2+j}$.

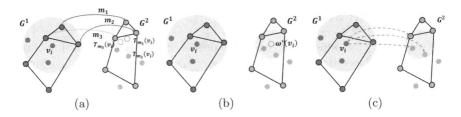

Fig. 2. Visualization of correspondence enrichment for $k_1 = 5$ and $k_2 = 3$. (a) For each feature v_i, potential correspondence locations $T_{m_a}(v_i)$ are estimated based on k_1 nearest neighbors(red area) correspondences m_a. (b) The geometric median of these potential locations $\omega^*(v_i)$ is optimized. (c) The posterior probabilities are evaluated for k_2 nearest neighbors(green area) of $\omega^*(v_i)$. (Best viewed in color) (Color figure online)

Since the feature appearances have been encoded in the affinity matrix of graph matching, the matching confidence is simply estimated based on local geometric consistency. In this way, PPM is insensitive to appearance variation among objects of the same class. The posterior probability of each correspondence is evaluated by geometric checking of neighboring candidates as illustrated in Fig. 2. For each feature in domain 1, a set of potential corresponding locations in domain 2 is obtained by imposing local invariance on affine transformations between neighboring candidates:

$$\Omega(v_i) = \{T_{m_a}(v_i)|m_a = (\hat{v}, m(\hat{v})), \hat{v} \in \text{kNN}(v_i, k_1)\}, \tag{2}$$

where $T_{m_a}(\cdot)$ is the affine homography transformation of the match m_a, and $m(\hat{v})$ is corresponding node of $\hat{v}$ determined by the best match. These projected locations $\Omega(v_i)$ are likely to be different and some of them may even suffer from outlier matches. To abate the effect of mismatches and obtain a more reliable location, geometric median [13] is used to regularize the potential locations:

$$\omega^*(v_i) = \arg\min_{\omega \in \mathbb{R}^2} \sum_{y \in \Omega(v_i)} ||\omega - y||_2, \tag{3}$$

which can be optimized by Weizfeid's algorithm [14]. Based on the corresponding location $\omega^*(v_i)$ optimized for each feature, the matching confidence is defined as:

$$p(v_i \mapsto v_j|M) = p(v_i \mapsto v_j|\omega^*(v_i)) \sum_{\substack{\hat{v} \in \text{kNN}(v_i, k_1) \\ m_a = (\hat{v}, m(\hat{v}))}} \text{mScore}(m_a), \tag{4}$$

which denotes that v_i is likely to match v_j if v_j is close to the optimized location $\omega^*(v_i)$ in the premise that v_i has many reliable neighboring matches, and mScore($\cdot$) is the confidence score of each correspondence derived from graph matching. The probabilistic of spatial support $p(v_i \mapsto v_j | \omega^*(v_i))$ is measured by the definition:

$$p(v_i \mapsto v_j | \omega^*(v_i)) = \begin{cases} 1 & \text{if } v_j = \text{NN}(\omega^*(v_i)) \\ \frac{exp(-d_{ij})}{S} & \text{if } v_j \in \text{ kNN}(\omega^*(v_i), k_2), \\ 0 & \text{otherwise} \end{cases} \qquad (5)$$

where NN($\cdot$) means the nearest neighbor node and d_{ij} is Euclidean distance between v_j and $\omega^*(v_i)$. S is a normalizing constant defined as

$$S = \sum_{v_j \in \text{kNN}(\omega^*(v_i), k_2)} exp(-d_{ij}). \qquad (6)$$

As a result, the probability score of false matches will be suppressed while more and more correct matches will be promoted in an iterative way.

The procedure of PPM is summarized in Algorithm 1.

Algorithm 1. Progressive probabilistic model (PPM)

Input: graphs G^R, G^T, # of candidates n_c
Output: matching results C
1 $(C, V^1, V^2) \leftarrow$ FindInitialMatch(V^R, V^T, n_c);
2 **while** *graph matching objective score S_{GM} increases* **do**
3     $(M, \text{mScore}, S_{GM}) \leftarrow$ GraphMatching(C);
4   Initialization: $p(v_i \mapsto v_j | M) \leftarrow 0 \ \forall v_i \in V^1, v_j \in V^2; \Omega \leftarrow \phi$;
5   **foreach** $v_i \in V^1$ **do**
6         $N_A \leftarrow \{v | v \in \text{kNN}(v_i, k_1), v \in M^1\}$;
7   **foreach** $\hat{v} \in N_A$ **do**
8             $m_a \leftarrow$ FindBestMatch($\hat{v}, M, \text{mScore}$);
9   compute $T_{m_a}(v_i)$;
10             $\Omega(v_i) \leftarrow [\Omega(v_i), T_{m_a}(v_i)]$;
11   **end**
12   compute $\omega^*(v_i)$ according to Eq.(4);
13         $N_B \leftarrow \{v | v \in \text{kNN}(\omega^*(v_i), k_2), v \in V^2\}$;
14   **foreach** $v_j \in N_B$ **do**
15             $p(v_i \mapsto v_j | M) \leftarrow p(l_i \mapsto l_j | \omega^*(v_i)) \sum \text{mScore}(m_a)$;
16   **end**
17   **end**
18     $C \leftarrow n_c$ best matches based on $p(v_i \mapsto v_j | M)$, which contains M;
19 **end**

---

3 Experiments and Evaluations

In this section, two sets of experiments are designed to evaluate PPM, showing the effect of the enrichment strategy and the robustness to outliers.

Table 1. Performance of PPM in accuracy (%) on cars and motorbikes dataset [6]

	Method	Cars	Motorbikes	avg
SM	w/o enrich.	52.14	74.28	63.21
	w/ enrich.	**79.81**	**78.47**	**79.14**
RRWM	w/o enrich.	57.08	59.71	58.40
	w/ enrich.	**80.06**	**85.30**	**82.68**
IPFP	w/o enrich.	50.05	64.41	57.23
	w/ enrich.	**81.08**	**74.07**	**77.58**
MPM	w/o enrich.	52.96	69.69	61.33
	w/ enrich.	**74.12**	**73.54**	**73.83**

Table 2. Performance of PPM in accuracy (%) on intra-class dataset [7]

Method	SM	RRWM	IPFP	MPM
w/o enrich.	78.77	82.58	80.64	79.87
w/ enrich.	**93.81**	**96.36**	**93.58**	**93.64**

The experiments are performed on two challenging datasets with deformation and intra-category variation: Cars and Motorbikes dataset [6] and Intra-class dataset [7]. Cars and Motorbikes dataset [6] contains two image pair sequences: 30 image pairs of cars and 20 motorbikes. Intra-class dataset [7] consists of 30 real image pairs of various objects collected from Caltech101 and MSRC datasets. These two datasets all provide some manually constructed correct correspondences for each image pair. More ground truth is derived by extending the original to its neighboring features in our experiments. Here, SIFT [15] is used to describe the features. Resultant performance is mainly presented with accuracy [4,5,7] defined as:

$$\text{accuracy} = \frac{\text{\# of true positive}}{\text{\# of ground truths}}, \qquad (7)$$

where true positive represent the correct correspondences identified by tested method.

The parameters of the step of enrichment are set as $k_1 = 20$ and $k_2 = 20$.

3.1 Effect of Enrichment

In fact, PPM can be considered as a combination of graph matching and correspondence enrichment. To evaluate the effect of the enrichment process, PPM is applied under two settings: with and without enrichment. Four conventional methods are employed as the graph matching module in the model: spectral matching (SM) [4], integer projected fixed-point matching (IPFP) [6], reweighted random-walk matching (RRWM) [7] and max-pooling graph matching (MPM)

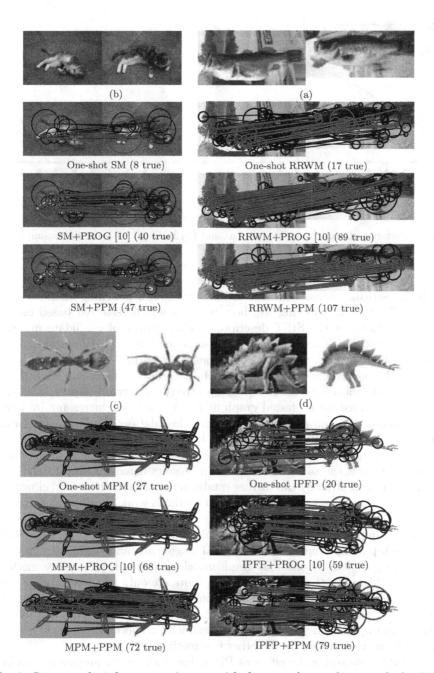

Fig. 3. Some results of our experiments with four graph matching methods. True matches are represented by red lines and false ones are shown by black lines. PPM performs better in the subtle parts of the objects like ears and rears and obtains more true matches. (Best viewed in color) (Color figure online)

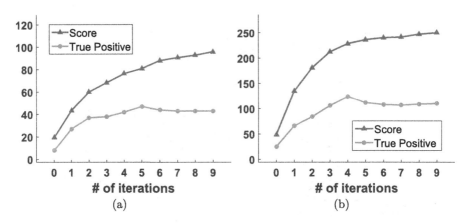

Fig. 4. Performance on the image pair (a) and (b) in Fig. 3 with the increasing iterations (Color figure online)

[9]. In the setting without enrichment, one-shot graph matching methods work individually. Before matching, N_c best matches will be selected based on the pairwise distance of the SIFT descriptors. The number of candidate matches selected by the enrichment process is the same as that of initial matches, fixed as $N_c = 1000$. The max number of iterations is set as 10.

Tables 1 and 2 show the performance in accuracy on Cars and Motorbikes dataset [6] and Intra-class dataset [7], respectively. Significant improvement can be clearly observed for all tested graph matching methods, increasing by 12%–24% for dataset [6] and about 14% for Intra-class dataset [7]. Reweighted random walk matching (RRWM) [7] performs best in PPM with enrichment among four graph matching methods, probably because the way to compute the affinity matrix here is the same as RRWM. The results demonstrate that the proposed enrichment strategy can further establish potential matches and eliminate the unreliable matches even when one-shot methods work poorly, which greatly boosts the true matches.

Some examples of the real image matching can be viewed in Figs. 1 and 3. In Fig. 3, the top images represent the original image pairs and the followings show the matching results by one-shot graph matching method, progressive method [10] and PPM, respectively. It's clear that there are only small-percentage true matches found by one-shot graph matching. After progressive matching [10], more correct matches are identified, which can be observed in the body of the dinosaur in Fig. 3. While, PPM performs well with respect to some subtle parts like the rears mismatched by the other two methods.

To further illustrate the effect of PPM, Fig. 4 shows its progressive performance with the increasing iterations on image pair (a) and (b) in Fig. 3. Objective scores of graph matching (red line) and the number of true matches (green line) both increase sharply in the early process and become almost convergent in the end. Both growth rates in scores reach about 500%. As for true matches, growth rates of image pair (a) and (b) are 588% and 629%, respectively. The performance growth of the other tested image pairs are similar.

All the results in this section demonstrate the significant improvement over one-shot graph matching and the effectiveness of our enrichment strategy.

3.2   Robustness to Outliers

During the progressive matching, outliers are first induced when selecting initial matches according to the feature appearances. In order to show the robustness to outliers, PPM is evaluated with varying number of initial matches N_c along with the state-of-the-art progressive graph matching method [10]. In this setting, N_c varies between 1000 and 3000 by increments of 500, while other parameters are fixed to the same as those of the first experiment. Since RRWM performs best in both [10] and PPM, it is used as the graph matching module in the comparison.

Figures 5 and 6 show the matching results on Cars and Motorbikes of dataset [6] and Intra-class dataset [7], respectively. PPM achieves high accuracy on both datasets of various objects with appearance variation. As shown in Fig. 5, on Cars and Motorbikes dataset [6], PPM (red line) shows remarkable tolerance to the number of initial matches, while method [10] (blue line) drops more sharply

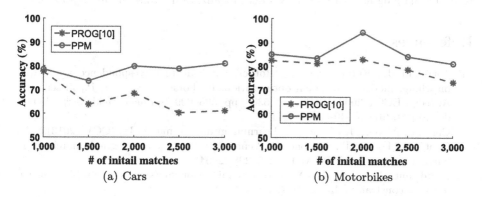

(a) Cars   (b) Motorbikes

Fig. 5. Performance on cars and motorbikes dataset [6] (Color figure online)

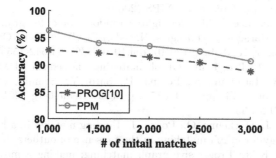

Fig. 6. Performance on intra-class dataset [7] (Color figure online)

with the increasing number. On Intra-class dataset [7], both methods descend with the varying number of initial matches, but PPM still achieves outstanding accuracy as shown in Fig. 6. The results demonstrate that the proposed PPM can handle a large number of outliers and some intra-class appearance variation in practical matching.

4   Conclusion and Future Work

In this paper, a progressive probabilistic graph matching model has been introduced to address the limitations of conventional graph matching method and improve the accuracy. The model combines graph matching with a probabilistic enrichment process to gradually boost the number of true candidates. In regard to the enrichment process, the geometric consistency with neighboring candidates, as well as the local regularization is taken into account. In turn, graph matching refines the candidates and eliminates some mismatches, leading to a more satisfactory performance. The experiment results demonstrate the effectiveness of the enrichment strategy. Moreover, PPM outperforms the state-of-the-art progressive method in the presence of appearance variation and outliers. In the future, the research will focus on applying the proposed model to other challenging tasks such as object localization and scene understanding.

References

1. Torresani, L., Kolmogorov, V., Rother, C.: Feature correspondence via graph matching: models and global optimization. In: Forsyth, D., Torr, P., Zisserman, A. (eds.) ECCV 2008. LNCS, vol. 5303, pp. 596–609. Springer, Heidelberg (2008). doi:10.1007/978-3-540-88688-4_44
2. Cho, M., Alahari, K., Ponce, J.: Learning graphs to match. In: ICCV (2013)
3. Conte, D., Foggia, P., Sansone, C., Vento, M.: Thirty years of graph matching in pattern recognition. IJPRAI **18**, 265–298 (2004)
4. Leordeanu, M., Hebert, M.: A spectral technique for correspondence problems using pairwise constraints. In: ICCV (2005)
5. Cour, T., Srinivasan, P., Shi, J.: Balanced graph matching. In: NIPS (2006)
6. Leordeanu, M., Herbert, M.: An integer projected fixed point method for graph matching and map inference. In: NIPS (2009)
7. Cho, M., Lee, J., Lee, K.M.: Reweighted random walks for graph matching. In: Daniilidis, K., Maragos, P., Paragios, N. (eds.) ECCV 2010. LNCS, vol. 6315, pp. 492–505. Springer, Heidelberg (2010). doi:10.1007/978-3-642-15555-0_36
8. Suh, Y., Cho, M., Lee, K.M.: Graph matching via sequential Monte Carlo. In: Fitzgibbon, A., Lazebnik, S., Perona, P., Sato, Y., Schmid, C. (eds.) ECCV 2012. LNCS, vol. 7574, pp. 624–637. Springer, Heidelberg (2012). doi:10.1007/978-3-642-33712-3_45
9. Cho, M., Sun, J., Duchenne, O., Ponce, J.: Finding matches in a haystack: a max-pooling strategy for graph matching in the presence of outliers. In: CVPR (2014)
10. Cho, M., Lee, K.M.: Progressive graph matching: making a move of graphs via probabilistic voting. In: CVPR (2012)

11. Wang, C., Wang, L., Liu, L.: Progressive mode-seeking on graphs for sparse feature matching. In: Fleet, D., Pajdla, T., Schiele, B., Tuytelaars, T. (eds.) ECCV 2014. LNCS, vol. 8690, pp. 788–802. Springer, Cham (2014). doi:10.1007/978-3-319-10605-2_51
12. Ham, B., Cho, M., Schmid, C., Ponce, J.: Proposal flow. In: CVPR (2016)
13. Lopuhaa, H.P., Rousseeuw, P.J.: Breakdown points of affine equivariant estimators of multivariate location and covariance matrices. Ann. Stat. 19(1), 229–248 (1991)
14. Chandrasekaran, R., Tamir, A.: Open questions concerning Weiszfeld algorithm for the Fermat-Weber location problem. Math. Program. 44, 293–295 (1989)
15. Lowe, D.G.: Object recognition from local scale-invariant features. In: ICCV (1999)

Automatic Detection of Utility Poles Using the Bag of Visual Words Method for Different Feature Extractors

Frank C. Cabello$^{(\boxtimes)}$, Yuzo Iano, Rangel Arthur, Abel Dueñas,
Julio León, and Diogo G. Caetano

Laboratory of Visual Communications,
School of Electrical and Computer Engineering,
University of Campinas, Av. Albert Einstein 400, Campinas 13083-852, Brazil
{fcc125,yuzo,aduenas,j.leon,diogo.gara}@decom.fee.unicamp.br,
rangel@ft.unicamp.br
http://lcv.fee.unicamp.br/

Abstract. One of the major problems in power distribution networks is abnormal heating associated with high resistance or excessive current flow, in which some of the affected components include three-phase transformers, switches, connectors, fuses, etc. Utility Pole detection aids in the classification of these affected components; thus, the importance of its study. In this work, we propose a method to detect the utility poles using a database of images obtained from Google Maps for the region of Campinas/SP. The Bag of Visual Words (BoVW) method was used to classify the two classes (those that are utility poles and those that are not utility poles), and know if the sub-image obtained belongs to a utility pole class.

Keywords: Hot-spots · Bag of Visual Words · Utility poles · Abnormal heating

1 Introduction

Image classification has been an active research topic in image processing. The research aims to classify images by extracting their features in order to detect, classify, or identify patterns. The recognition of utility poles will help the future detection of elements with overheating contained in the utility pole.

In this project, we have worked to develop an automatic detection method of utility poles in the city of Campinas, in the state of São Paulo - Brazil. Campinas is characterized as an urban city with buildings and single-story houses. However, while also located in a tropical region, Campinas is a city with many trees, which represent a problem when they are located near utility poles (utility pole detection is more difficult).

Thermal cameras are commonly used in order to know where the hot-spots are located. In this project, we use at least one thermal and one optical camera.

© Springer International Publishing AG 2017
M. Felsberg et al. (Eds.): CAIP 2017, Part II, LNCS 10425, pp. 116–126, 2017.
DOI: 10.1007/978-3-319-64698-5_11

It is necessary to calibrate the thermal and optical cameras in order to detect a hot region in the thermal camera, and to subsequently determine the exact location in the image from the optical camera. Once the location of the hot region is determined, the Region of Interest (ROI) is calculated. Finally, the ROI is classified to know whether it contains a utility pole or not.

This work begins at the detection of the region of interest, assuming a hotspot has been detected by the thermal camera and the system has correctly determined the ROI. In this region, the feature extraction and description are determined. Following the Bag of Visual Words (BoVW) and Space Pyramids models, the representation of each image is determined. This representation will be separated into two classes with the help of a Support Vector Machine (SVM) classifier. Therefore, each additional image will be classified using the pre-trained SVM.

The rest of this paper is structured as follows. In Sect. 2, the theory for feature extraction is introduced along with how to describe these features. Using those features, the BoVW method is explained in Sect. 3. Section 4 shows the Experimental results, comparisons, and discussion; and, finally, the conclusion is presented in Sect. 5. This work is part of the R&D project PD-0063-3014/2016 "ANEEL: DE3014 - Thermovision - Automatic hot spot detection system based on thermal imaging".

2   Feature Extraction

One of the objectives of this work is to classify an image by extracting and analyzing its features. For this purpose, it is necessary to detect the local features and then group them into a global representation within the image, and finally train a classifier in order to associate a label with an image. Thus, if a new image were to be classified, this classifier will know which label would be best to place on it.

The most useful information in the images is located around certain areas that normally correspond to the most important points and regions. In several object recognition applications, local processing around these most important points is sufficient as long as these points are stable and distinctive [1]. Local features are characterized by sudden changes of intensity in the region. These local features are classified as corners, borders, or blobs [2].

A feature extraction algorithm should, in a first step, extract, as effectively and robustly as possible, certain features that provide the maximum information possible. These features must meet the following conditions:

- The computational cost for feature extraction should not be excessive. The total extraction time must be the smallest possible.
- The location must be accurate, the error committed in the estimation of the characteristics should also be as small as possible.
- They must be robust and stable, they must remain along a sequence of similar images.

– They should contain as much information as possible of the scene, which means that the features must allow the extraction of information of a geometric type.

The most important features to extract are: corners, edges, and bobs; as the basic geometries in the feature extraction. Each feature stores the value representing a particular region around it, which is represented by a data vector. In the case of the SIFT algorithm [3], this vector has 128 real values (using $128 \times 32 = 4096$ bits of storage) and, in the case of the ORB algorithm [4], it uses 256 binary values (using 256 bits of storage).

The methods of feature detection and description most commonly used today in the literature are presented in the following subsection. For the visualization of the features, the image of the utility pole shown in Fig. 1a will be used.

2.1   Harris Corner Detector

The Harris corners detection algorithm is one of the simplest algorithms for extracting feature points in an image. The main idea is to locate points of interest where the neighborhood shows edges in more than one direction: these would be the corners of the image. An example of points of interest is shown in Fig. 1b.

(a)   (b)

Fig. 1. (a) One of the images of utility poles that will be used to obtain the local features of different algorithms that will be used in this work. (b) Feature points (red points in the image) extraction using the Harris algorithm. (Color figure online)

2.2   Features from Accelerated Segment Test (FAST)

The result of the detection of points of interest using the FAST [5] are represented as a circle in Fig. 2a. This algorithm only exists in the form of feature detection in a similar way to the Harris algorithm. The point of interest represents the information of the area contained in the green circle.

2.3 Scale-Invariant Feature Transform (SIFT)

The SIFT algorithm (developed by Lowe [6]) provides in one image a set of features that do not suffer from many of the complications experienced in other methods, such as scale or rotation changes. The SIFT algorithm transforms an image into a large collection of vectors of local features. Each of these characteristic vectors is invariant for any scale change or image rotation.

The SIFT algorithm has four steps: spatial-scaled extrema detection, location of feature points, assignment of orientation and generation of feature points descriptor [3]. The feature points and its representing region are shown in Fig. 2b.

2.4 Speeded up Robust Features (SURF)

The concept of SURF [7] is based on the fundamental principles of the SIFT algorithm. The main difference is a relatively efficient implementation that makes the SURF algorithm more suitable for near real-time applications and for processing a large amount of images. Figure 2c depicts the feature points and its representing region.

2.5 Binary Robust Invariant Scalable Keypoints (BRISK)

The BRISK algorithm was developed in [8] and stands out in the description of feature points. It motivated the ORB algorithm (created by Ethan Rublee [4]). The feature point and its representing region obtained by the BRISK algorithm are shown in Fig. 2d. The number of points and distribution throughout the image is similar to the SIFT algorithm.

2.6 Oriented-FAST and Rotated-BRIEF (ORB)

The ORB algorithm was developed as an alternative to the SIFT and SURF algorithms [4]. Additionally, it is a technique of detection and description of feature points that are invariant in scale and rotation, robust to noise and to affine transformations. The ORB algorithm is the combination of two known techniques, FAST (for feature extraction) and BRIEF (for feature description). The feature points and its representing region is shown are shown in Fig. 2e.

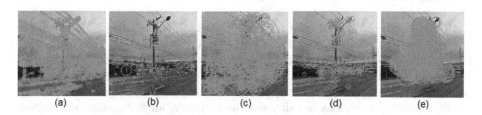

(a)   (b)   (c)   (d)   (e)

Fig. 2. Feature extraction methods. (a) FAST. (b) SIFT. (c) SURF. (d) BRISK. (e) ORB. (Color figure online)

3   Bag of Visual Words

The BoVW method [9] allows to represent an image as a single numerical vector, reducing the problems of noise sensitivity and variability, thus enabling the classifier training. Firstly, it is necessary to know what the most representative visual words could be that fit the concrete content of the images to be classified.

This means that is necessary to build a vocabulary of visual words. Therefore, the first step that we take is to select a set of images and extract the local features that we want to use for classification. Using the feature descriptor, each feature is a point in the coordinate space of the descriptor. This distribution of points (feature points) is divided into groups of similar features, applying some unsupervised learning algorithm. A visual word represents a group of similar features obtained with the previous algorithm. Once we have identified the visual words, it is possible to construct the histogram of visual words, as the final representation of the image.

It will match each local characteristic of the image to the most similar visual word and it will tell how often each word appears. Thus, each image that we want to classify will be represented by a vector of the same dimension. The BoVW method (Fig. 3 depicts the BoVW diagram) consists of 4 steps:

1. Extraction of local features of the set of training images using some feature extraction algorithm.
2. Construction of vocabulary of visual words using n random samples (for example, 30000 points of interest can be randomly chosen by the detector).
3. Features Quantization: it is necessary to obtain the description of the features obtained in step 1, with some algorithm of description, and to group all the similar features between them in order to find the visual words that are the representatives of each group of similar features (this grouping can be obtained using the unsupervised K-means classifier). Once the vocabulary is constructed, we can perform the representation and classification of images.
4. Constructing the histogram: it represents how often a visual word appears in the image. This representation is used to train the final classifier, and thus to label the new images.

4   Experiments and Results

Until this point, the spatial information of the image at the time of classification has not been taken into account. This is because the BoVW algorithm uses histograms of the characteristics of the entire image. In [10], spatial information was added, which consists of partitioning the image into a division of regions. In each region and at each level of the pyramid, the histogram of visual words that are found in these regions is calculated and, finally, all histograms are united where each histogram will have its own weight associated with its level in the space pyramid.

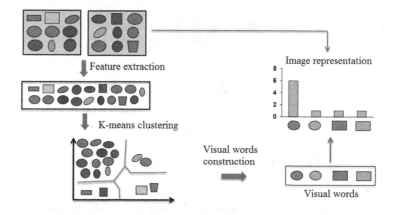

Fig. 3. Bag of Visual Words scheme.

The use of the Spatial Pyramids allows to find coincidences of characteristics at different levels of the pyramid. This method has a good performance with images containing an object in the center of the image as would be the case of the detection of utility poles. Four types of distributions of the Spatial Pyramid were designed; Fig. 4a represents the classical distribution of Spatial Pyramids. As utility poles generally have their main information in the upper part and in the middle of the image, the 3 remaining representations were drawn.

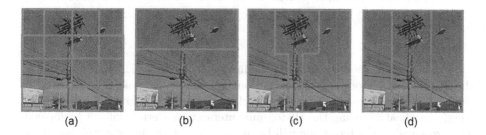

(a)   (b)   (c)   (d)

Fig. 4. Spatial Pyramid representations.

Among the 4 representations, the best results were obtained using the representation shown in Fig. 4d. A symmetric representation was chosen because as the detection of the hot-spots from the thermal image will generate a region of interest with the utility pole in the middle of the image. In the spatial distribution shown in Fig. 5, the dimensions of the representations are not known. Thus, in order to calculate them, it will be necessary to use a cross-validation technique. Likewise, the cross-validation technique will be applied to obtain the best weights of the internal regions.

Fig. 5. Distribution of the Spatial Pyramid.

Table 1 shows the tests that were performed for different values of a and b. These tests were obtained using 50000 samples with 256 centroids with the FAST–ORB combination. The best result is obtained using $a = 0.4$ e $b = 0.2$ (these fractions represent the relation with the width of the image) and the precision is 91.75%.

Table 1. Precision, sensitivity and specificity for some values for a and b

a	0.45	0.425	0.4	0.35	0.3	0.25	0.2
b	0.10	0.15	0.2	0.3	0.4	0.5	0.6
Precision (%)	90.66	91.66	91.75	90.916	88.75	86.25	85.5
Sensitivity (%)	87.078	89.724	92.95	87.47	81.4	75.839	74.72
Specificity (%)	97.239	97.69	98.622	97.448	97.26	96.349	95.68

This representation consists of 3 main regions (see Fig. 6): the whole image, the utility pole region, and the region outside the utility pole. The weights for the best results of precision for these regions were 0.4, 0.4, and 0.2 respectively. Here, classification using the Histogram Intersection Kernel will be performed. The cross-validation technique will be applied to obtain the best weights of the internal regions.

To obtain the results of the experiments, the remaining 120 utility poles and 120 non-utility poles images were used. The SVM [11] classifier is used solely to classify sets that are linearly separable and also completely separable without overlapping classes. In order for the SVM to be efficient in problems that are not linearly separable (in real-life problems, these cases usually happen) there are two alternatives. The first option is to allow certain overlapping between classes to relax the maximum margin condition. The second option, that is used in this work, is to transform the set of characteristics of all the classes to another space of characteristics where the classes are linearly separable.

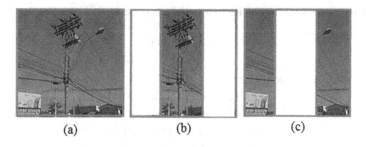

(a)   (b)   (c)

Fig. 6. Regions of the Spatial Pyramid, each region have its own weight.

The Kernel that we use for this transformation is the Histogram Intersection Kernel [12]. For the following tests, 50000 random samples with 256 centroids were chosen, and the cross-validation method was used to obtain the best parameters. Using the Histogram Intersection Kernel, without the use of Spatial Pyramids, the results of the precision of the classifier are shown in Table 2.

Table 2. Precision (%) using the Histogram Intersection Kernel without Spatial Pyramids.

		Description			
		SIFT	SURF	BRISK	ORB
Detection	Harris	91.33	86.16	90.58	89
	FAST	95.42	91.83	91.33	89.92
	SIFT	87.25	83.16	84.91	85.33
	SURF	91.75	91.91	84.25	78.75
	ORB	86.08	85.41	79.83	85.25
	BRISK	91.46	85.16	83.58	80.66
	Dense	84.58	83.08	88.91	84.41

All experiments were done using an Intel Core i7 4770k CPU (4GHz) with 8 GB DDR3 RAM at 1333 Mhz, and a Kingston v300 256GB SSD running 64-bit Microsoft Windows 8.1 with Visual Studio 2015 and OpenCV 3.1 library.

The average time of this Kernel was 5.25 s (to classify 240 images), and its average precision is 86.83%, being its maximum precision using the FAST–SIFT combination. Using the Histogram Intersection Kernel with spatial pyramids (the geometric distribution of the pyramid was shown in Fig. 6), the precision results are shown in Table 3 and the average processing time was 8 s.

The average precision using the Histogram Intersection Kernel with Spatial pyramids is 89.46%, the maximum precision is 96.16% and was obtained using the FAST–SIFT combination (see Table 3).

Table 3. Precision (%) using the Histogram Intersection Kernel with Spatial Pyramids.

		Description			
		SIFT	SURF	BRISK	ORB
Detection	Harris	94	88.33	92	90.08
	FAST	96.16	94.33	94.41	91.33
	SIFT	89.83	84.33	85.5	88.25
	SURF	93.66	93.83	87.41	81.83
	ORB	88.66	89.75	84.08	86.58
	BRISK	93.83	88	85	81
	Dense	93.25	89.25	93	87.25

Fig. 7. Images correctly classified as utility poles.

Table 4. Precision (%) of the classifiers using the combinations FAST–BRISK and FAST–SIFT.

Classification methods	FAST–BRISK	FAST–SIFT
Expectation maximization	51.66	60.42
Naive bayes	62.5	67.23
K – nearest neighbours	79.16	82.50
Random forest	79.58	83.72
Multi–layer perceptron	84.16	90.50
Decision tree	87.5	90.8
SVM Histogram Intersection Kernel without Spatial Pyramid	91.33	95.42
SVM Histogram Intersection Kernel with Spatial Pyramid	94.41	96.16
SVM Histogram Intersection Kernel with Spatial Pyramid and Weights	94.66	96.25

Since the two best combinations are FAST–SIFT and FAST–BRISK, they will be used in Table 4. Using the Histogram Intersection Kernel with Spatial Pyramids and with the weights obtained using Fig. 6, the precision of the classifier for FAST–BRISK is 94.66% and for FAST–SIFT, 96.25% (average processing time for both was around 23 s to classify 240 images). Figure 7 depicts some images that were correctly classified as poles; there are different shapes and sizes for the poles, and the classifiers studied in this work were able to perform their correct classification. The main classification results were gathered in Table 4, and the method using the Histogram Intersection Kernel and Spatial Pyramids(see Fig. 6) was the one that obtained the best results.

5 Conclusion

The best Spatial Pyramid division was shown in Fig. 6, this division given by the Pyramid helps to place a greater weight on the central part where the utility pole is generally located compared to the lateral regions. Using the Spatial Pyramids theory, was possible to increase the precision of the classifier, for FAST–SIFT combination, it increased from 95.42% to 96.25% and for FAST–BRISK combination increased from 91.33% to 94.66%. Although the classifier processing time increased from about 8 to about 23 s to classify 240 images.

Acknowledgments. The authors acknowledge the financial support from "Companhia Paulista de Força e Luz -"CPFL", "Companhia Piratininga de Força e Luz", "Rio Grande Energia S/A" and "Companhia Sul Paulista de Energia". Thanks also to the companies participating in this project: KascoSys P&D and RFerrarezi, both based in the city of Campinas–Brazil.

References

1. Suarez, O.D., del Carrobles, M.M.F., Enano, N.V., García, G.B., Gracia, I.S., Incertis, J.A.P., Tercero, J.S.: OpenCV Essentials. Packt Publishing Ltd., Birmingham (2014)
2. Kaspers, A.: Blob detection (2011)
3. Lowe, D.: Local feature view clustering for 3D object recognition. In: Proceedings of the 2001 IEEE Computer Society Conference on Computer Vision and Pattern Recognition, CVPR 2001, vol. 1. IEEE Computer Society, pp. I-682–I-688 (2001)
4. Rublee, E., Rabaud, V., et al.: ORB: an efficient alternative to SIFT or SURF. In: 2011 International Conference on Computer Vision, pp. 2564–2571. IEEE, November 2011
5. Rosten, E., Drummond, T.: Machine learning for high-speed corner detection. In: Leonardis, A., Bischof, H., Pinz, A. (eds.) ECCV 2006. LNCS, vol. 3951, pp. 430–443. Springer, Heidelberg (2006). doi:10.1007/11744023_34
6. Lowe, D.: Object recognition from local scale-invariant features. In: Proceedings of the Seventh IEEE International Conference on Computer Vision, vol. 2, pp. 1150–1157. IEEE (1999)
7. Bay, H., Ess, A.: Speeded-up robust features (SURF). Comput. Vis. Image Underst. J. **110**, 346–359 (2008). Elsevier

8. Leutenegger, S., Chli, M., Siegwart, R.Y.: BRISK: binary robust invariant scalable keypoints. In: Proceedings of the IEEE International Conference on Computer Vision, pp. 2548–2555 (2011)

9. Yang, J., Jiang, Y.-G.: Evaluating bag-of-visual-words representations in scene classification. In: Proceedings of the International Workshop on Workshop on Multimedia Information Retrieval, pp. 197–206 (2007)

10. Lazebnik, S., Schmid, C., Ponce, J.: Beyond bags of features: spatial pyramid matching for recognizing natural scene categories. In: Proceedings of the IEEE Computer Society Conference on Computer Vision and Pattern Recognition, vol. 2, pp. 2169–2178 (2006)

11. Boswell, D.: Introduction to support vector machines. http://dustwell.com/PastWork/IntroToSVM.pdf

12. Subhransu, M., Alexander, C., Jitendra, M.: Classification using intersection kernel support vector machines is efficient. In: IEEE Computer Vision and Pattern Recognition (2008)

Deep Objective Image Quality Assessment

Christopher Pramerdorfer[✉] and Martin Kampel

TU Wien, Vienna, Austria
{cpramer,kampel}@caa.tuwien.ac.at

Abstract. We present a generic blind image quality assessment method that is able to detect common operations that affect image quality as well as estimate parameters of these operations (e.g. JPEG compression quality). For this purpose, we propose a CNN architecture for multi-label classification and integrate patch predictions to obtain continuous parameter estimates. We train this architecture using softmax layers that support multi-label classification and simultaneous training on multiple datasets with heterogeneous labels. Experimental results show that the resulting multi-label CNNs perform similarly to multiple individually trained CNNs while being several times more efficient, and that common image operations and their parameters can be estimated with high accuracy. Furthermore, we demonstrate that the learned features are discriminative for subjective image quality assessment, achieving state-of-the-art results on the LIVE2 dataset via transfer learning. The proposed CNN architecture supports any multi-label classification problem.

Keywords: Convolutional neural networks · Image quality assessment · Multi-label classification · Deep learning

1 Introduction

The problem of assessing the quality of a given image is an active research field [8,10]. This work focuses on blind (no-reference) Image Quality Assessment (IQA), i.e. assessment without requiring "pristine" reference images. Existing methods in this field fall into two broad categories depending on whether they aim to predict quality scores as perceived by humans [1,15] or properties such as the amount of blur or compression artifacts [2,8]. Methods of the latter category have the advantage of being *objective* as they do not require ground-truth quality scores assigned by persons, which are *subjective* [10].

In this paper, we present a generic framework for blind objective IQA that is able to detect the presence of image *operations* that affect image quality as well as estimate the main *parameter* of every operation. For demonstration purposes, we consider three such operations that are popular in IQA [10]: JPEG and JP2K compression as well as image blur (and conversely sharpness). The corresponding parameters we aim to estimate are JPEG compression quality, JP2K compression ratio, and blur strength assuming a Gaussian kernel. However, our framework supports any number and combination of operations without modification; all adaptions are handled automatically via multi-task learning.

© Springer International Publishing AG 2017
M. Felsberg et al. (Eds.): CAIP 2017, Part II, LNCS 10425, pp. 127–138, 2017.
DOI: 10.1007/978-3-319-64698-5_12

We predict the parameter of a given operation by integrating patch classification results obtained using a Convolutional Neural Network (CNN). In order to be able to do so efficiently for multiple operations and to estimate which operations affect a given image, we propose a CNN architecture for multi-label classification. We train CNNs with this architecture using softmax layers that we adapt to support multi-label classification in a way that enables simultaneous training on multiple datasets with *heterogeneous* labels. This enables the networks to learn a shared set of features that are discriminative for detecting all individual operations and for estimating their parameters. The proposed architecture is flexible and can predict an arbitrary number of labels. It is applicable not only to IQA problems but to any multi-label classification problem.

We compare our multi-label CNNs to individually trained CNNs to demonstrate that the proposed architecture performs similarly while being several times more efficient. The results show that, by integrating patch predictions from these networks, we are able to reliably predict whether images are affected by the three operations considered, as well as estimate the JPEG and blur parameters at high accuracy. Furthermore, we connect our work to subjective IQA by showing that the learned features generalize to the LIVE2 dataset [10], achieving state-of-the-art results via transfer learning.

This paper is structured as follows. Section 2 summarizes related works in the fields of blind IQA and multi-label classification using CNNs. In Sect. 3, we present our method for regressing the parameter of a particular image operation and generalize this approach to multiple operations in Sect. 4, introducing the multi-label CNN architecture and generic IQA method. The experiments carried out to evaluate this architecture and method are discussed in Sect. 5.

2   Related Work

Blind IQA. Methods for subjective IQA require a dataset with image quality scores assigned by human labelers, such as LIVE2 [10]. Such methods usually extract handcrafted features, which are then mapped to quality scores using machine learning algorithms like neural networks [7,15]. Recent methods instead learn features from image data using CNNs in order to improve performance [1,6]. These methods differ mainly in terms of CNN architecture and integration of predicted patch quality scores. In both cases, the networks perform regression, whereas our CNNs perform multi-label classification.

Objective methods instead aim to detect image distortions that would cause human labelers to assign lower quality scores to images. This is done either directly by detecting specific image operations such as image blur [2], or indirectly by computing a general measure for image quality that is not directly related to human quality scores [8,14]. Methods of the latter type are generic and thus more flexible, while the former typically perform better on the specific operations they were designed for. Our proposed method is a compromise between both types. Similarly to [8,14], it is operation-agnostic (it does not include operation-specific algorithms) and supports any number and combination of operations. However, in contrast to these indirect methods, our method learns to discriminate between the individual operations and estimate their parameters. This information can be valuable in applications such as digital forensics.

Multi-label CNNs. Three loss functions for multi-label CNN classification are compared in [4], one of the first works on this topic. Two of these functions (pairwise and weighted approximate ranking) are ranking losses that are not applicable because no ordering exists between image operations. The third candidate generalizes the cross-entropy loss to support multiple labels by adapting the softmax layer. A limitation of this implementation with respect to our use case is that it does not discriminate between operations and parameters. [13] presents a similar approach but minimizes the ℓ_2 loss and requires pretraining on a single-label dataset. In contrast, our proposed softmax layers for multi-label classification distinguish between operations and their parameters (or generally between tasks and the classes of each task), and support end-to-end training on multiple datasets with heterogeneous labels. In a recent work [12], combine CNNs with LSTMs for multi-label classification, with the LSTM learning correlations between labels. However, in IQA there is little correlation because image operations often do not co-occur.

3   Operation Parameter Regression

We cast the problem of estimating the main parameter v of a *given* operation (e.g. knowing that an image is affected by JPEG compression artifacts) as a regression problem, $v \in \mathbb{R}$. This allows us to estimate both discrete and continuous parameters without modifications. To do so, we randomly sample M possibly overlapping patches $\{\mathcal{P}_1, \ldots, \mathcal{P}_M\}$ from every image. Instead of regressing $\hat{v}$ directly from these patches, we perform classification with C classes at the patch level, as classification problems are easier to optimize in a deep learning framework. The disadvantage is that the necessary discretization of the domain of v into C bins leads to a loss of precision. We mitigate this loss during patch-level fusion. This discretization is always possible in practice because v is always bounded. For instance, the JPEG quality must be between 1 and 100, and the amount of blurring typically encountered in practice is limited.

3.1   Deep Patch Classification

For patch classification we use a CNN that maps from single patches $\mathcal{P}$ to discrete probability distributions $\Pr(c|\mathcal{P})$. To obtain probabilistic outputs, which are required for patch-level fusion, we add a final softmax layer and minimize the cross-entropy loss during training.

Our CNN architecture is inspired by VGG-B [11] and is illustrated in Fig. 1. We chose this architecture because it is well-known and performs well in a variety of tasks [9,11]. All convolutions have a size of 3×3, a stride of 1, and apply zero-padding to preserve the spatial resolution of the input. All pooling layers perform max-pooling with size 2×2 and stride 2. All convolutional and the

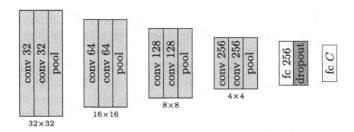

Fig. 1. Illustration of the CNN architecture used for patch classification. The label conv 32 denotes a convolutional layer with 32 feature maps. The dimensions given below each block are input dimensions assuming a patch size of 32×32 pixels.

first fully-connected (fc) layer use batch normalization [5] and ReLU activation functions. The final fc layer uses a softmax activation. The CNN has less than 1.5 million parameters at an input resolution of 32×32 pixels.

3.2   Patch-Level Fusion

Patch classification results in M estimated probability mass functions, which we integrate to obtain $\hat{v}$ as follows. Let $r(c) = \tilde{v}$ be a function that inverts the quantization (with a loss of precision). We compute

$$\hat{v} = \frac{1}{M} \sum_{m=1}^{M} \sum_{c=1}^{C} r(c) \, \Pr(c|\mathcal{P}_m). \tag{1}$$

As the result is a weighted average of $M \cdot C$ discrete values, $\hat{v}$ can take on arbitrary values within the original bounds of v. We later show empirically that this effectively mitigates quantization errors.

4   Deep Multi-label Classification for IQA

We now generalize the proposed patch classification method to support multiple image operations via multi-label classification, resulting in a generic framework for objective IQA. A simple approach to do so is to train one CNN per operation. However, this approach lacks efficiency in terms of computational and memory requirements; both increase linearly with the number of operations O, which can be a limiting factor on systems with limited resources.

To address this issue, we propose a *multi-label CNN* architecture with a shared feature extraction *frontend* and O classifier *backends*, one per operation. We present two types of softmax layers for training such networks, one that assumes that the operations applied to a given image are known at test time, and one that does not have this requirement. Both types support simultaneous training on multiple datasets with heterogeneous labels, enabling the CNN to optimize the joint performance on all datasets and operations.

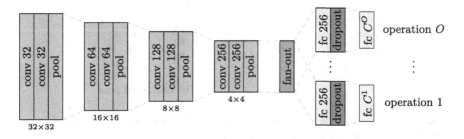

Fig. 2. Proposed architecture for multi-label classification.

4.1 Multi-label CNNs

Figure 2 illustrates the proposed multi-label CNN architecture. While the frontend is unchanged from the single-label version (Fig. 1), the backend starts with a *fan-out* layer that replicates the output of the final pooling layer O times during forward passes, and sums up the individual gradients during backward passes. Each copy is input to one of O separate backends. This architecture supports any frontend architecture and an arbitrary number of operations.

CNNs with this architecture have a shared frontend that extracts features that are discriminative for all operations and parameters, and O backends that perform classification on this basis. Compared to O individual CNNs, this leads to a significantly increased efficiency because the computationally demanding feature extraction is carried out only once. For instance, the multi-label CNN shown in Fig. 2 requires around 52 MFLOPs per forward pass at $O = 3$, whereas 3 CNNs as shown in Fig. 1 require around 154 MFLOPs.[1]

4.2 CNN Training

Our multi-label CNNs cannot be trained like single-label CNNs because all backends observe all samples and possibly heterogeneous ground-truth labels. To this end, we propose two layer types that support such data to replace the final softmax layers. Each layer contributes to a cross-entropy loss during training, which are summed to obtain the final scalar loss. Let $\mathcal{B} = \{(\mathbf{x}_b, t_b, w_b)\}_{b=1}^{B}$ be a minibatch of size B, with $\mathbf{x}_b$ being a feature vector computed by the final fc layer, t_b being the operation the corresponding sample belongs to, and w_b being the ground-truth parameter class of that operation.

Masking softmax layers. This layer type, when attached to backend o, discards all samples in $\mathcal{B}$ for which $t_b \neq o$, resulting in a subset $\mathcal{B}_o$, before proceeding like an ordinary softmax layer. Thus only samples of task o contribute to the loss of backend o during training. Due to the masking operation in the forward pass, the gradient matrix has $|\mathcal{B}_b| < B$ rows, preventing backpropagation. The layer thus keeps track which samples were discarded in the forward pass and assigns

---

[1] Assuming convolutions are implemented via matrix multiplication.

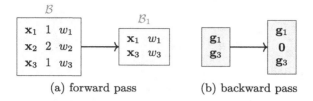

(a) forward pass   (b) backward pass

Fig. 3. Masking operations of the masking softmax layer assigned to backend 1. $\mathbf{g}_b$ is the gradient vector associated with sample $(\mathbf{x}_b, w_b)$.

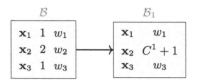

Fig. 4. Forward-pass of the remapping softmax layer assigned to backend 1.

a gradient of $\mathbf{0}$ to these samples in the backward pass. Figure 3 illustrates this operation. Due to the multiplicative effect of gradients, this causes samples for which $t_b \neq o$ to have no effect on the gradient updates in backend o.

This layer type is applicable if t_b is available at test time. If not, the layer can be replaced by an ordinary softmax layer after training. The resulting model is conceptually similar to O individually trained CNNs and shares one of their limitations: the individual backends were trained only on samples affected with a particular image operation, which has a negative impact on performance. We propose a second layer type that is more effective in this case.

Remapping softmax layers. This layer type, when attached to backend o, does not discard samples for which $t_b \neq o$ but instead remaps w_b to a new class $C^o + 1$ whose semantics is "sample not affected by this operation" (Fig. 4). During training, the CNN thus learns to detect whether a sample is affected by a given operation and, if so, to predict the correct parameter. In order to counteract the fact that the frequency v_0 of samples of class $C^o + 1$ is typically higher than those of any other class v_1 after remapping, samples contribute with factor $1/v_0$ respectively $1/v_1$ to the loss. The layer is otherwise identical to an ordinary softmax layer and is replaced by one after training.

5 Experiments

We test the proposed CNN architecture with both types of softmax layers; CNNs with masking layers are henceforth denoted as MML CNNs while those with remapping layers are referred to as RML CNNs.

5.1   Datasets

RAISE dataset. To our knowledge, there is no large public dataset available that includes images affected by JPEG and JP2K compression as well as blurring, all with a wide range of known parameter values. For this reason, we compile our own dataset using raw data from the RAISE dataset [3]. This dataset contains 8156 TIFF images that come directly from the camera and thus are unaffected by any image processing artifacts apart from demosaicing. We randomly select a subset of 1300 images and decimate every image to 20%, resulting in resolutions similar to other IQA datasets such as LIVE2. We randomly split this subset into a training set (1000 images), a validation set (100 images), and a test set (200 images). We refer to these images as being of the original class.

For every image, we then compute the following distorted versions: (i) JPEG compression at qualities $\{20, 35, 50, 65, 80, 95\}$ using `libjpeg`, (ii) JP2K compression at ratios $\{32, 24, 20, 16, 12, 8\}$ using `libopenjpeg`, and (iii) Gaussian blur with $\sigma \in \{0.5, 0.75, 1.0, 1.25, 1.5, 2.0\}$ using `OpenCV`. The parameter ranges encompass values that are typically encountered in practice. For instance, a JPEG compression quality of 20 leads to strong degradation whereas 95 produces images with no apparent compression artifacts (Fig. 5).

This results in 24700 images of 19 classes in total. We henceforth refer to this dataset as the RAISE dataset. Scripts for replicating the dataset are available at https://github.com/cpra/oiqa-cnns. The subset of original images and all JPEG images (7 classes) is called RAISE JPEG dataset. The RAISE JP2K and RAISE Blur datasets are defined analogously.

For patch-level training and validation, we randomly select 200 patch locations per original image, which we reuse for all corresponding distorted images. Validation and test patches are of size 32×32 whereas training patches are of size 40×40 to facilitate data augmentation during training.

LIVE2 dataset. We also test the generalizability of the MML CNN using the LIVE2 dataset [10]. LIVE2 is a popular dataset for evaluating subjective IQA methods and contains 29 reference images as well as distorted versions, including

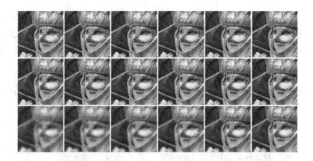

Fig. 5. Image operations and their effect. The top, middle, and bottom row correspond to JPEG compression, JP2K compression, and Gaussian blur, respectively

JPEG and JP2K compression, and blurring. We split the dataset into training, validation, and test sets at the image level. Following the practice in [1], we randomly select 17 reference images for training, 6 for validation, and 6 for testing. As this dataset is much smaller than RAISE, we randomly select 500 patch locations per reference image before proceeding as for the RAISE dataset.

5.2 Training

We train our multi-label CNN simultaneously for JPEG and JP2K compression, as well as blur strength classification, using the corresponding RAISE datasets. For data augmentation we randomly crop every training sample to 32×32 pixels and mirror it horizontally with a probability of 0.5.

For optimization we use stochastic gradient descent, an initial learning rate of 0.01, and Nesterov momentum. When training multi-label CNNs, we use a minibatch size of 32 per operation (96 in total). When training single-label CNNs for comparison, we use a minibatch size of 96. In any case, we train for a maximum of 200 epochs and halve the learning rate whenever the validation accuracy does not improve for 30 epochs. For regularization we use a weight decay of 0.0001 and dropout with a probability tuned using the validation set.

5.3 Evaluation Metrics

To assess the classification performance, we calculate confusion matrices and report the overall classification accuracy. The regression performance is measured using the Mean Absolute Error (MAE) and Mean Squared Error (MSE). For evaluating the subjective IQA performance, we additionally calculate the Pearson Correlation Coefficient (PCC) and Spearman's Rank Correlation Coefficient (SRCC) in order to facilitate comparison with existing work. These metrics measure the linear and rank correlation between true and predicted values, respectively, with the optimum being 1 in both cases.

5.4 Patch Classification Performance

We first compare multi-label CNNs to individually trained single-label CNNs using the RAISE JPEG, JP2K, and Blur datasets. Table 1 shows that the MML CNN performs similarly to individually trained CNNs. This confirms that the

Table 1. Average patch classification accuracies of different CNN architectures.

Dataset	Individual	MML	RML
RAISE JPEG	97.5%	96.4%	95.1%
RAISE JP2K	45.6%	45.2%	43.3%
RAISE Blur	90.4%	89.3%	89.1%

MML CNN is able to learn features that are discriminative for multiple image operations and, consequently, that it is a viable alternative to individual CNNs.

The RML CNN performs up to 2% worse because it must additionally predict whether a given operation is actually present. The relatively small performance decrease indicates that the network is able to do so reliably.

5.5   Image Classification Performance

We next study the image classification performance of MML and RML CNNs by averaging $\Pr(c|\mathcal{P})$ over all patches extracted from a given image.

MML CNN. Figure 6 shows that the MML CNN performs well on the JPEG quality and blur estimation tasks, as expected given the corresponding patch results. The network achieves an overall test accuracy of 100%, 56.0%, and 99.6% on the JPEG, JP2K, and Blur datasets, respectively. These scores are up to 10% higher than the corresponding patch scores (Table 1), highlighting the effectiveness of patch integration for compensating patch-level errors. While the network is able to determine whether images are affected by JP2K compression artifacts or not, it fails to predict reliable compression ratios. This is presumably because JP2K compression artifacts have no consistent block structure.

RML CNN. In order to assess the RML CNN performance, we test it on the full RAISE test set (i.e. every backend observes all JPEG, JP2K, and Blur test samples). Figure 7 summarizes the results. The network is able to determine the correct image operations for 3786 of 3800 test images, confirming that it performs well at this task. When considering the results for each operation, the overall test accuracies are 99.4%, 52.6%, and 99.3%. The differences compared to the MML CNN are in accordance with the patch classification results.

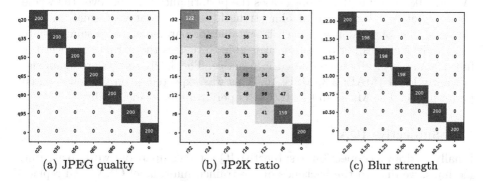

(a) JPEG quality   (b) JP2K ratio   (c) Blur strength

Fig. 6. MML CNN image classification performance on the individual RAISE testsets. The label o corresponds to the original class.

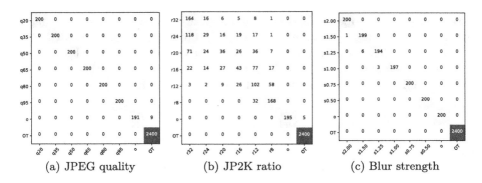

(a) JPEG quality   (b) JP2K ratio   (c) Blur strength

Fig. 7. RML CNN image classification performance on the full RAISE testset. The label OT corresponds to the "not affected by this operation" class $C^o + 1$.

5.6 Operation Parameter Regression Performance

We next evaluate the proposed IQA method in terms of parameter regression performance, using both MML and RML CNNs for patch classification. For this purpose, we perform a fine sampling of the individual parameter ranges (for instance, we consider JPEG compression ratios between 20 and 100 in steps of 1). For each parameter value, we randomly select 10 original images from the RAISE test set, apply the corresponding operation with the selected value, and extract 200 patches for classification. The resulting patch probabilities are then fused using the strategy explained in Sect. 3.2 to obtain quality scores $\hat{v}$. In case of the RML CNN, we consider only patches for which $\arg\max_c \Pr(c|\mathcal{P}) \neq C^o + 1$, i.e. those likely affected by the corresponding operation.

The results on the individual datasets are summarized in Fig. 8 and are consistent with the previous results; our IQA method performs well at estimating the JPEG compression quality (RML CNN MAE: 1.6, MSE: 3.9) and blur strength (MAE: 0.08, MSE: 0.01). In both cases the performance is stable over the whole domain, with the exception of blur strength estimation at very small σ. This confirms that our fusion approach is suitable for parameter regression; despite using only 7 classes per operation, it is able to regress the parameters over the full domains typically encountered in practice. The JP2K results are again significantly lower (MAE: 3.3, MSE: 16.4), as was expected given the lower patch performance. The MML and RML CNNs perform almost identically.

5.7 Subjective IQA

Finally, to draw a connection to subjective IQA and compare our work to existing methods, we discard the backends of the trained multi-label CNNs and replace them by a Multi-Label Perceptron (MLP) with a single hidden layer with 256 neurons. We then train this MLP to predict DMOS scores [10] using the LIVE2 dataset while fixing all frontend parameters, effectively performing DMOS score regression using multi-label CNN features at the patch level. These patch results are then averaged to obtain DMOS scores for images.

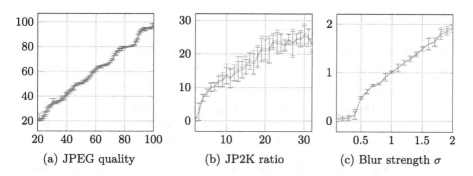

Fig. 8. Ground-truth (x-axis) and predicted (y-axis) parameter values using MML (blue) and RML CNNs (red). Error bars indicate the standard deviation of the 10 estimates per ground-truth value (Color figure online).

Table 2. Comparison of reported and our LCC and SRCC scores on the LIVE2 dataset.

Method	[7]	[14]	[6]	[15]	[1]	MML	RML
LCC	0.837	0.861	0.953	0.962	0.972	0.972	**0.975**
SRCC	0.827	0.886	0.956	0.964	0.960	0.966	**0.968**

As summarized in Table 2, the resulting subjective IQA method achieves state-of-the-art results in terms of both LCC and SRCC metrics. This shows that features that are discriminative for objective IQA are also discriminative for subjective IQA and indicates that pretraining on objective IQA datasets can be viable to overcome the comparatively small size of subjective IQA datasets. As the CNNs were trained on the RAISE dataset, the results confirm that the learned features generalize across datasets.

6   Conclusion

We have proposed a CNN architecture for multi-label classification and shown that corresponding multi-label CNNs perform similar to individually trained CNNs in IQA tasks. The proposed multi-label CNNs are several times more efficient due to their shared frontends. Additionally, the RML variant is able to reliably detect the operations by which a given image is affected. Furthermore, we have demonstrated that objective IQA parameters can be regressed by integrating patch classification results of these CNNs. The resulting IQA framework is generic and can be easily extended from the three operations considered.

While the proposed CNN architecture was developed for objective IQA, it is applicable to any multi-label classification problem. For the future, we plan to extend the architecture to support multiple operations per sample during training, to enable regression as well as classification, and to apply the architecture in other research fields such as face recognition.

References

1. Bosse, S., Maniry, D., Wiegand, T., Samek, W.: A deep neural network for image quality assessment. In: IEEE International Conference on Image Processing, pp. 3773–3777 (2016)
2. Chen, M.-J., Bovik, A.C.: No-reference image blur assessment using multiscale gradient. EURASIP J. Image Video Process. **2011**, 3 (2011)
3. Dang-Nguyen, D.-T., Pasquini, C., Conotter, V., Boato, G.: A raw images dataset for digital image forensics. In: ACM Multimedia Systems Conference, pp. 219–224 (2015)
4. Gong, Y., Jia, Y., Leung, T., Toshev, A., Ioffe, S.: Deep Convolutional Ranking for Multilabel Image Annotation. CoRR abs/1312.4894 (2013)
5. Ioffe, S., Szegedy, C.: Batch Normalization: Accelerating Deep Network Training by Reducing Internal Covariate Shift. CoRR abs/1502.03167 (2015)
6. Kang, L., Ye, P., Li, Y., Doermann, D.: Convolutional neural networks for no-reference image quality assessment. In: IEEE Conference on Computer Vision and Pattern Recognition, pp. 1733–1740 (2014)
7. Li, C., Bovik, A.C., Wu, X.: Blind image quality assessment using a general regression neural network. IEEE Trans. Neural Netw. **22**, 793–799 (2011)
8. Mittal, A., Soundararajan, R., Bovik, A.C.: Making a "Completely Blind" image quality analyzer. IEEE Signal Process. Lett. **20**, 209–212 (2013)
9. Parkhi, O.M., Vedaldi, A., Zisserman, A.: Deep face recognition. In: British Machine Vision Conference (2015)
10. Sheikh, H.R., Sabir, M.F., Bovik, A.C.: A statistical evaluation of recent full reference image quality assessment algorithms. IEEE Trans. Image Process. **15**, 3440–3451 (2006)
11. Simonyan, K., Zisserman, A.: Very Deep Convolutional Networks for Large-Scale Image Recognition. CoRR abs/1409.1556 (2014)
12. Wang, J., et al.: CNN-RNN: A unified framework for multi-label image classification. In: IEEE Conference on Computer Vision and Pattern Recognition, pp. 2285–2294 (2016)
13. Wei, Y., et al.: CNN: Single-label to Multi-label. CoRR abs/1406.5726 (2014)
14. Xue, W., Zhang, L., Mou, X.: Learning without human scores for blind image quality assessment. In: IEEE Conference on Computer Vision and Pattern Recognition, pp. 995–1002 (2013)
15. Zhang, P., Zhou, W., Wu, L., Li, H.: Semantic obviousness metric for image quality assessment. In: IEEE Conference on Computer Vision and Pattern Recognition, pp. 2394–2402 (2015)

Enhancing Textbook Study Experiences with Pictorial Bar-Codes and Augmented Reality

Huy Le$^{(\boxtimes)}$ and Minh Nguyen

Auckland University of Technology, 55 Wellesley Street East,
Auckland 1010, New Zealand
dsz8022@aut.ac.nz

Abstract. Augmented Reality (AR) could overlay computer-generated graphics onto the student's textbooks to make them more attractive, hence, motivate students to learn. However, most existing AR applications use either template (picture) markers or bar-code markers to conceal the information that it wants to display. The formal, being in a pictorial form, can be recognized easily but they are computationally expensive to generate and cannot be easily decoded. The latter displays only numeric data and are therefore cheap to produce and straightforward to decode. However, they look uninteresting and uninformative. In this paper, we present a way that combines the advantage of both the template and bar-code markers to be used in education, e.g. textbook's figures. Our method decorates on top of an original pictorial textbook's figure (e.g. historical photos, images, graphs, charts, maps, or drawings) additional regions, to form a single image stereogram that conceals a bar-code. This novel type of figure displays not only a realistic-looking picture but also contains encoded numeric information on students' textbooks. Students can turn the pages of the book, look at the figures, and understand them without any additional technology. However, if students observe the pages through a hand-held Augmented Reality devices, they see 3D virtual models appearing out of the pages. In this article, we also demonstrate that this pictorial bar-code is relatively robust under various conditions and scaling. Thus, it provides a promising AR approach to be used in school textbooks of all grades, to enhance study experiences.

Keywords: Augmented Reality · Computer vision · Image processing · Education · Textbook

1 Introduction and Motivation

1.1 Hard Copy Study Materials and AR Enhancement

We and our next generations are born and living in the digital world. Many schools, organizations and governments have been spending their time and money into new training technologies and methodologies to enhance today's student involved and engaged. There are many researchers, books and papers have been discussing the advantages and disadvantages of various learning methods

© Springer International Publishing AG 2017
M. Felsberg et al. (Eds.): CAIP 2017, Part II, LNCS 10425, pp. 139–150, 2017.
DOI: 10.1007/978-3-319-64698-5_13

such as e-learning, virtual learning, etc. However, there are still few key questions which have not been answered yet: Which training method is the most effectively? Which training method is easy for teaching and reaching by the new generations? Less than 20% of today's students prefer to study by reading a textbook, and more than 30% of those cannot live without Internet or cell phone for a day [12]. It shows that digital technologies are severely detriment to current students life and study effort. Students prefer to use digital learning tool such as the Internet, videos, virtual 3D as they are less boring and easier to understand while printed textbooks are not so attractive. On another hand, there are some would prefer to read printed materials as keeping their eyes away from eyestrain and other ocular symptoms [3]. For the solutions of those issues, blended learning (BL) has appeared. BL is a combination of traditional learning and e-learning which was defined in the research paper of Martin Oliver and Keith Trigwell in 2005 [16].

Augmented reality (AR) technology is acting as a revolutionary solution to make the idea of BL visible. AR is an ability to overlay 3D virtual objects into the real world in the real time [4]. Azuma in 1997 [1] also defined that AR is those technologies that combine real-world objects with virtual 3D objects and create interactivity between them (real and virtual) in the real time. In this case, the printed materials are real world environment while the rendered digital information are the virtual objects that are combined in the real time. In the recent decades, AR technologies have developed and reached a remarkably level in ICT industry. Those virtual objects represent certain information that users cannot detect directly through their senses. AR system was invented for the first time in the 1960s by Ivan Sutherland and his colleagues at Harvard University [22]. Since the information revolution started in the 1990s, AR has been exploited in several different areas of our life. With the advantage of new technologies, AR may just be the thing to change our daily lives. Probably, it is the next step in the way we collect, process and interacts with information [6]. The AR technology had many applications in different fields and kept expanding over time. The medical doctors can make use of AR for accurate positioning of the surgical site [9]. It also provides a better way for the patients to describe their symptoms. In military area, AR helps to identify any threats in the current location or receive commands from their commanders virtually in real time [15] as well as education [14] and tourism [8].

1.2   Current Technical Pros and Cons of Augmented Reality

Creating an effective AR experience requires to use different tools such as tracking, registration (for aligning the virtual objects with the real scene), and rendering (for displaying the virtual information). Those tools are easy to implement with the advantages of today's technologies. However, there are still few long term AR disadvantages which should be concerned such as system processing complexity and information orientation. The main principle behind AR technology is finding the target (could be a pictorial marker or a bar-code marker) and orientate the digital information on detected target.

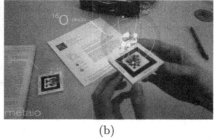

(a)   (b)

Fig. 1. (a) AR quiver application with its pictorial marker. (b) 2D bar codes are used to display Hydrogen and Oxygen elements in 3D.

Pictorial markers as shown in Fig. 1(a) are often used as they are convenient for detecting and displaying content. They also make more meaningful sense for the users, especially to younger users. However, image registration is required, and the recognition processing becomes unreliable sometimes due to the undesired similarity between images [24]. Therefore, this method is suitable for AR applications with smaller datasets such as Magic books [2] which is for children usage. This method will possibly create a processing complexity issue if it were applied for AR applications with large datasets such as AR Chemistry application as shown in Fig. 1(b). Chemistry education applications usually include a huge dataset which contains the information of over hundreds chemical elements and compounds. For those reasons, this application was designed to work with bar-code markers.

The bar-code markers are more popularly used within the robust and unambiguous applications [20]. These markers are normally designed with black and white data cells surrounding with a thick dark colored border. The data cells are encoded as a binary number which made them unique from each other. The decoded binary number on each marker will be compared with stored binary numbers in the dataset. The advantage of this method compared to pictorial markers is that the system processing complexity issue will be minimized. The other superior point is that besides encoding information, we can also make use of the binary bits for error detection and correction. Hamming codes [10] and Reed-Solomon [19] codes are two method which are often used for error detection and correction. Examples are ALVAR and ARTag [7]. However, these data cells are usually meaningless to the users if there was not another picture added alongside or other explanation information.

Both pictorial and bar-code makers have their advantages and disadvantages. The pictorial is giving better meaningful to the users in term of presentation. However, the system processing complexity and reliability is also concerned due to the comparing with huge stored pictorial database and undesired similarity between images. On the other hand, bar-code markers are containing unique hidden binary code which could solve the pictorial markers disadvantages.

Bar-code markers also provide a better way for error detection and correction. However, bar-code markers present non-useful information to users. Therefore, usability is not too easy.

2   Pictorial Marker with Hidden 3D Bar-Code (PMBC)

2.1   Overview

In this paper, we propose to use stereogram to conceal a multi-level bar-code optically. The end product is the proposed PMBC tag, its initial design is demonstrated in Fig. 2(a). The detail will be described further in this section; in short, our proposed tag presents some notable advantages:

- **Large range of data:** The multi-level bar-code can hold L^N different number, with L is the number of levels in each bar and N is number of bars.
- **Virtually Pictorial:** The image inside each PMBC tag is made from meaningful illustrations, rather than black bars, squares, or dots.
- **Flexibility of Pattern:** The decoded information is independent of image patterns, we can use a broad range of images to encode the same bar-code.

Each PMBC tag is a rectangle with a dimension $D = M \times N$ measured in pixels or millimeters; border thickness t is relatively small (we set $t = 10\%$ of D). The quadrilateral property of the rectangles can be used to detect their four straight lines and four corners. These are used for detecting the tag. The internal image is a stereogram (size $W \times H$) made of three regions. The central area is a fixed image (region A that fills up $\geq 50\%$ of the stereogram) and two repeated patterns on both sides of region A (region B and region C with $\leq 25\%$ of the stereogram each).

Hidden inside each stereogram is a bar-code with many horizontal bars with the same thickness. Each bar is coated with different grey levels between black and white. The grey level is used to represent different depth levels inside the stereogram. Figure 2(b) displays an example of 4-level binary bar-code with 10 horizontal bars. Each bar can hold four different levels: 0, 1, 2, 3; corresponding to black, dark grey, light grey, and white. Thus, this bar-code can store as many as $4^{10} = 1,048,576$ different numbers.

Figure 2(c) demonstrates basic steps of creating our proposed tag. As described, our PMBC Tag has a black border so that it is easily and reliably detectable under various circumstances. In theory, the internal stereogram of the PMBC tag can encode any 1D or 2D bar-code such as Code11, Code 32, Code 49, Code 93, Code 128, EAN-8, EAN-13, VSCode, Aztec Code, Data Matrix, Maxi Code, PDF417, and QR Code [13,18]. However, we opt to use a home-made Multi-level Bar-code for taking the advantages of being able to store depth at each bar.

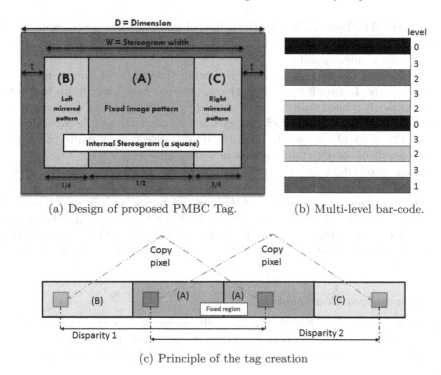

(a) Design of proposed PMBC Tag.   (b) Multi-level bar-code.

(c) Principle of the tag creation

Fig. 2. Design of PMBC tag that optically hides a multi-level bar-code.

2.2 Multi-level Bar-Code

This Multi-level Bar-code has a dimension of $W \times H$. It may have many vertical bars; each has thickness T, the data hold in each bar is one number, corresponding to depth levels (level 1, 2, 3, behind the surface). We represent it using grey levels (the darker, the lower depth, the brighter, the higher depth). Thus, black represents lowest level depth and white represents the highest level depth.

The number A presented by this bar-code is:

$$A = b_{N-1}L^{N-1} + b_{N-1}L^{N-2} + .. + b_1 L^1 + b_{k-1}L^0 \tag{1}$$

where b_i is depth level at i^{th} bar, L is number of depth levels at each bar, and N is the number of bars.

Controlled zone is a horizontal bar at the bottom of the bar-code, used to self-check the validity of the bar-code and it also specifies the bar-code version. The controlled bar has the same thickness as a vertical bar, and it is always having the lowest depth level (black color).

2.3 Bar-Code as a Disparity Map

A disparity map is needed to create stereogram; it holds disparity value of each pixel on the left stereo image compared to its corresponding pixel on the right stereo image [23]. Two pixels with the same color act as virtual conjugate pixel projections in 3D and relate to one pixel in the disparity map; these two pixels are the correspondent points.

A multi-level bar-code is capable of storing thousands of unique numbers using some bars. To turn it to a disparity map; the bar-code block needs to rotate 90° clockwise; it is then scaled, and converted to a corresponding disparity map D^* (a $H \times W$ matrix where each element is a number ranging between 0 and d_{MAX}). The map D^* is used to code to the distance in how many pixels are between a pair of corresponding points in a stereogram.

We can encode 0-level pixels in the bar-code block as a 0-level disparity point, and white colored pixels is coded as $4 \times N$-level disparity point ($N > 0$). A final disparity map can be presented as a grey-scaled image with intensity ranged between $[0, 3N]$. N represents the shift a point from its origin in stereogram image at each depth level. By this, if a bar is at a depth level 3, a pixel of the image will be copied to a location that is $3N$ pixels away to create the final stereogram.

2.4   Stereogram Construction

The internal image I_c of PMBC tag has a dimension of $W \times H$. First, we select a $W' \times H'$ illustration image I_i to be placed at the center of the stereogram. Let $s_H = H'/H$ is the ratio between the height of I_i and I_c; pixels of I_c is defined as:

$$I_c(x,y) = I_i(s_H \times x + \frac{(W - W')\%W'}{2}, s_H \times xy) \qquad (2)$$

Occasionally, image I_c may contains large blank monotone regions like the walls and tabletop, which could fail the stereo correlation. We opt to add artificial textures on I_c to eliminate all the blank regions if any. A texture image I_t is chosen, the new image I_s is made as:

$$I_s(x,y) = (1 - k) \times I_c(x,y) + k \times I_t(x,y) \qquad (3)$$

where k is the transmission coefficient ($0 < k < 1$).

The central 50% region of I_s is kept intact (region A - Fig. 2(a)) and will be used to repeat itself on the left and the right of the figure (regions B and C). In other words, pixels with identical/same colors to the left and right of a selected pattern are added with different horizontal shifts according to the disparity map D^*. Detail can be found in the pseudo-code below:

```
for h in range(0, H):
    for w in range(0, H/4):
        k = barcodeLevel[H/L]
        stereogram[h,w + k] = pattern[h,H/2+w]
        stereogram[h,3*H/4+w] = pattern[h,H/4+w+k]
```

The final image is then decorated with a thick black border to generate our proposed tag. As stated, the value stored is independent to look of our tags. Three examples of the PMBC tags are shown in Fig. 3. All of them hide the same Multi-level Bar-code storing number $29,984$.

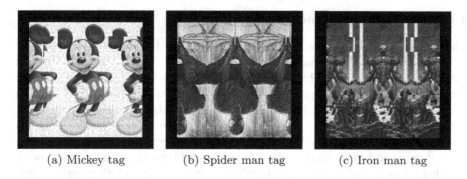

(a) Mickey tag   (b) Spider man tag   (c) Iron man tag

Fig. 3. Three examples of PMBC tags that conceal the same multi-level bar-code.

3 System Design and Implementation

We propose a new learning system which combines the traditional training method with the modern e-learning method. The system applies the advantages of AR technology to enhance perception and engagement of learners. However, we also provide students the interactivity with the normal pictorial markers to make sure that the users always receive the meaningful information. The system architecture is divided into five main modules as shown in Fig. 4. They are Data Acquisition, Preprocessing, Hidden Code Tracking, Information Rendering, and Client User Interface.

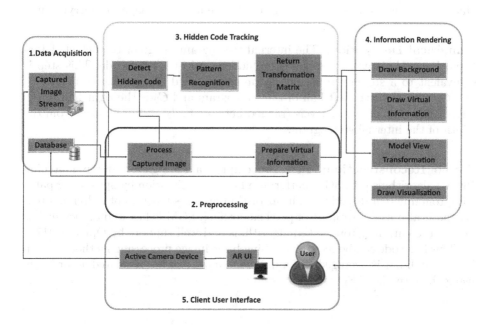

Fig. 4. Our system architecture

3.1 Data Acquisition and Storage

The system data acquisition is divided into two tires for internal database and input image stream data which captured by camera devices. The system internal database stores data of the hidden codes which are used for comparison with the decrypted code from the input marker. In this system, we use binary number (1 and 0) as stored comparable data to replace the original colored image data. Binary numbers are simple, and it has the limitation on how many items can be stored based on the number of available bits. They absorb fewer spaces in the database and also requires shorter processing time than the original data. Another advantage of using binary numbers is increasing processing performance as the manual image registration, and binary number indexing technique completely replaces the comparison. The similarity detecting failure is also minimized as each stored number is unique.

3.2 Pictorial Bar-Code Detection and Decryption

Detection: First, we need to find closed contours on the pictorial bar-code. Some basic image processing steps are required; they are outlined below:

- Convert the input image from RGB to greyscale
- Perform an adaptive binary thresholding method.
- Detect contours, if there are four vertices, it should be a quadrilateral.
- Apply Perspective Transform [17] to retrieve the internal image of the tag.

Once the border is detected, we can obtain the internal image for decryption.

Numerical Decryption: The internal stereogram image of the detected Pictorial Bar-code is used to rebuild the hidden multi-level bar-code. This step is equivalent to a stereo reconstruction process applied on two stereo images C_1 and C_2 with C_1 is the left half of the stereogram and C_2 is the right half of the stereogram. The disparity levels (ranged between 0, d_{MAX}) are known from the width of the internal stereogram.

Stereo Reconstruction: Here, we employ a semi-global stereo matching (SGM) algorithms [11]. SGM performs a fast approximation by optimizing pathwise from different directions. The algorithm matches blocks, not individual pixels and it also uses a support of pixel smoothing. SGM is fast and accurate; it is among the currently top-ranked algorithms and well support by OpenCV [17].

This bar-code can be recognized using basic image processing methods, depth level of each bar is averaged and a unique number can be calculated accordingly using the formula of Eq. 1.

3.3   Information Rendering

After pictorial bar-code and its hidden number are detected and determined. Virtual information is prepared according to the hidden code information. Then the camera position relative to marker should be calculated in order to blend the virtual information into the real world environment. The orientation of a virtual object on the top of fixed physical marker depends on the relationship between the camera coordinates and marker coordinates which shown in Fig. 5(a).

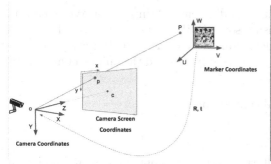

(a) Camera Pose Relationship.

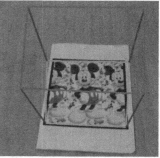

(b) Camera Pose 3D Sample.

Fig. 5. (a) Relationship between the marker coordinates and the camera coordinates. (b) Sample virtual 3D shape rendered on the top of the marker.

The relationship is also called pinhole camera model [21] or pose estimation in computer vision. There are three different coordinate systems presented in this situation. They are camera coordinates, camera screen coordinates and marker coordinates in the real world. A 3D object has only two kinds of motions on a static camera which is translation and rotation. Translation motion occurred when camera is moved from its current location (X, Y, Z) to a new location (X', Y', Z') in 3D space. It has three degrees of freedom and represented by vector t which can be calculated as in Eq. 4. Other motion is rotation which is absorbed when the camera is rotated about the X, Y and Z axes. Camera rotation motion is often represented by using Euler angles [5] (roll, pitch and yaw), a 3×3 matrix R or a direction of rotation and angle as shown in Eq. 5.

$$t = (X' - X, Y' - Y, Z' - Z) = \begin{bmatrix} t_x \\ t_y \\ t_z \end{bmatrix} \tag{4}$$

$$R = \begin{bmatrix} r_{00} & r_{01} & r_{02} \\ r_{10} & r_{11} & r_{12} \\ r_{20} & r_{21} & r_{22} \end{bmatrix} \tag{5}$$

To obtain the 3D pose estimation, we should have 2D coordinates of the object on the camera screen (x, y), 3D locations of the same 2D coordinates

(U, V, W) and intrinsic parameters of the camera or camera matrix. Let's assume that we want to present a 3D cube Fig. 5(b) and our camera is viewing from the top of the cube. The cube has eight vertices which are eight different 3D points. Therefore we should have four 2D point coordinates for the bottom of the cube and other four 2D point coordinates for the top. The same principle applies for 3D coordinates with an extra z value for each of those locations.

4   Experimental Results

Is our proposed PMBC tag sensitive to different lighting conditions, noises, and scaling? We test these criteria using some PMBC tags as shown in Fig. 6. The tag also optically hides the same unique number 29, 984. We carry out three experiments to observe the ability to correctly decoded the unique number 29984 after the following effects:

1. alternating the brightness and contrast of the tag.
2. scaling its original image.
3. adding noises and raindrops.

We use IrfanView tool[1] to alternate the sizes, apply effects, and add noises to the samples before checking the decoded results. Some IrfanView's output examples are shown in Fig. 6.

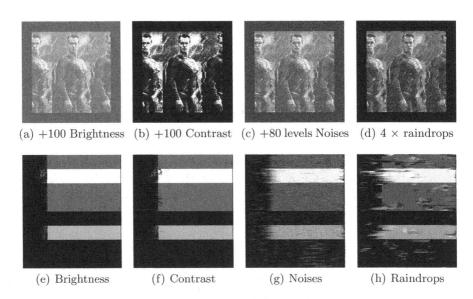

(a) +100 Brightness   (b) +100 Contrast   (c) +80 levels Noises   (d) 4 × raindrops

(e) Brightness   (f) Contrast   (g) Noises   (h) Raindrops

Fig. 6. Four examples of PMBC tags after various effects added by IrfanView (top row) and their corresponding decoded disparity maps (bottom row).

---

[1] http://www.irfanview.com/.

The above three effects have been tested; overall, our PMBC tag is found to be relatively robust under various changes in brightness, contrast, noises, and raindrop effects. The only failure is detected when image sample is smaller than 150×150 pixels. This could be improved by employing other image processing methods such as histogram equalization, to identify different levels of depths more reliably.

5   Conclusion

We propose a new learning system which combines the traditional training method with the modern e-learning method. The system employs our proposed Pictorial Marker with Hidden 3D Bar-code (PMBC) and current AR technology to enhance perception and engagement of learners PMBC is an AR tag that is capable of hiding multi-level optically (3D) bar-codes. It is a single image stereogram concealing a depth-coded bar-code while maintaining a complete color picture of what it is presenting. Most today's similar applications are using either template (picture) markers or bar-code markers. Template markers are pictorial, but they are computationally expensive and uncertainly identified by computers. On the other hand, bar-code markers contain decodable numeric data, but they look uninteresting and uninformative. Our PMBC tag is not only presenting a realistic-looking image but also encoding a broad range of numeric data. Moreover, PMBC tag is relatively robust under various conditions; thus, it could be a promising approach to be used in this new learning system future and future AR applications.

References

1. Azuma, R.T.: A survey of augmented reality. Presence Teleoperators Virtual Environ. **6**(4), 355–385 (1997)
2. Billinghurst, M., Kato, H., Poupyrev, I.: The magicbook: a transitional AR interface. Comput. Graphics **25**(5), 745–753 (2001)
3. Blehm, C., Vishnu, S., Khattak, A., Mitra, S., Yee, R.W.: Computer vision syndrome: a review. Surv. Ophthalmol. **50**(3), 253–262 (2005)
4. Carmigniani, J., Furht, B.: Augmented reality: an overview. In: Furht, B. (ed.) Handbook of Augmented Reality, pp. 3–46. Springer, New York (2011). doi:10. 1007/978-1-4614-0064-6_1
5. Diebel, J.: Representing attitude: euler angles, unit quaternions, and rotation vectors. Matrix **58**(15–16), 1–35 (2006)
6. Feiner, S.K.: Augmented reality: a new way of seeing. Sci. Am. **286**(4), 48–55 (2002)
7. Fiala, M.: ARTag, a fiducial marker system using digital techniques. In: IEEE Computer Society Conference on Computer Vision and Pattern Recognition, CVPR 2005, vol. 2, pp. 590–596. IEEE (2005)
8. Fritz, F., Susperregui, A., Linaza, M.T.: Enhancing cultural tourism experiences with augmented reality technologies. In: Proceedings of the 6th International Symposium on Virtual Reality, Archaeology and Cultural Heritage (VAST) (2005)

9. Fuchs, H., Livingston, M.A., Raskar, R., Colucci, D., Keller, K., State, A., Crawford, J.R., Rademacher, P., Drake, S.H., Meyer, A.A.: Augmented reality visualization for laparoscopic surgery. In: Wells, W.M., Colchester, A., Delp, S. (eds.) MICCAI 1998. LNCS, vol. 1496, pp. 934–943. Springer, Heidelberg (1998). doi:10.1007/BFb0056282

10. Hamming, R.W.: Error detecting and error correcting codes. Bell Labs Tech. J. **29**(2), 147–160 (1950)

11. Hirschmüller, H.: Stereo processing by semiglobal matching and mutual information. IEEE Trans. Pattern Anal. Mach. Intell. **30**(2), 328–341 (2008)

12. Ivanova, A., Ivanova, G.: Net-generation learning style: a challenge for higher education. In: Proceedings of the International Conference on Computer Systems and Technologies and Workshop for PhD Students in Computing, p. 72. ACM (2009)

13. Kato, H., Tan, K.T.: Pervasive 2d barcodes for camera phone applications. IEEE Pervasive Comput. **6**(4), 76–85 (2007)

14. Lee, K.: Augmented reality in education and training. TechTrends **56**(2), 13–21 (2012)

15. Livingston, M.A., Rosenblum, L.J., Julier, S.J., Brown, D., Baillot, Y., Swan, I., Gabbard, J.L., Hix, D., et al.: An augmented reality system for military operations in urban terrain. Technical report, DTIC Document (2002)

16. Oliver, M., Trigwell, K.: Can blended learning be redeemed? E-learn. Digit. Media **2**(1), 17–26 (2005)

17. Bradski, G., Kaebler, A.: Learning OpenCV: Computer Vision with the OpenCV Library. O'Reilly, California (2008)

18. Palmer, R.C., Eng, P.: The Bar Code Book: A Comprehensive Guide to Reading, Printing, Specifying. Evaluating and Using Bar Code and Other Machine-readable Symbols. Trafford Publishing (2007)

19. Reed, I.S., Solomon, G.: Polynomial codes over certain finite fields. J. Soc. Ind. Appl. Math. **8**(2), 300–304 (1960)

20. Siltanen, S.: Theory and applications of marker-based augmented reality (2012). http://www.vtt.fi/inf/pdf/science/2012/S3.pdf

21. Sturm, P.: Pinhole camera model. In: Ikeuchi, K. (ed.) Computer Vision, pp. 610–613. Springer, New York (2014). doi:10.1007/978-0-387-31439-6_472

22. Sutherland, I.E.: A head-mounted three dimensional display. In: Proceedings of the Fall Joint Computer Conference, Part I, 9–11 December 1968, pp. 757–764. ACM (1968)

23. Szeliski, R.: Computer Vision: Algorithms and Applications. Springer Science & Business Media, Heidelberg (2010)

24. Tikanmäki, A., Röning, J.: Markers-toward general purpose information representation. In: IROS 2011 Workshop: Knowledge Representation for Autonomous Robots (2011)

Stacked Progressive Auto-Encoders for Clothing-Invariant Gait Recognition

TzeWei Yeoh[✉], Hernán E. Aguirre, and Kiyoshi Tanaka

Faculty of Engineering, Shinshu University,
4-17-1 Wakasato, Nagano 380-8553, Japan
yeohtzewei@gmail.com,
{ahernan,ktanaka}@shinshu-u.ac.jp

Abstract. Gait recognition has been considered as an unique and useful biometric for person identification at distance. However, variations in covariate factors such as view angles, clothing, and carrying condition can alter an individual's gait pattern. These variations make the task of gait analysis much more complicated. Recognizing different subjects under clothing variations remains one of the most challenging tasks in gait recognition. In this paper, we propose a Stacked Progressive Auto-encoders (SPAE) model for clothing-invariant gait recognition. A key contribution of this work is to directly learn clothing-invariant gait features for gait recognition in a progressive way by stacked multi-layer auto-encoders. In each progressive auto-encoder, our SPAE is designed to transform the Gait Energy Images (GEI) with complicated clothing types to ones of normal clothing, while keeping the GEI with normal clothing type unchanged. As a result, it gradually reduces the effect of appearance changes due to variations of clothes. The proposed method is evaluated on the challenging clothing-invariant gait recognition OU-ISIR Treadmill dataset B. The experimental results demonstrate that the proposed method can achieve a far better performance compared to existing works.

Keywords: Gait recognition · Gait Energy Image (GEI) · Clothing-invariant · Stacked progressive auto-encoders (SPAE)

1 Introduction

There is a growing demand for robust and reliable human identification systems in various significant applications, such as video surveillance and access control [6]. A wide variety of human identification based biometric have been proposed by using physiological or behavioral traits of a person such as face, iris, fingerprint and gait [10]. In recent trends, gait recognition has attracted increasing attention as a novel biometric due to its capability to identify a person at a far distance by the way they walk [15]. However, it is still a far way to go towards automated visual surveillance for identity recognition using gait on those uncontrolled scenarios due to the effect caused by various covariate factors

© Springer International Publishing AG 2017
M. Felsberg et al. (Eds.): CAIP 2017, Part II, LNCS 10425, pp. 151–161, 2017.
DOI: 10.1007/978-3-319-64698-5_14

such as viewing angle, clothing condition, carrying condition, walking surface, elapsed time and walking speed [4].

Many promising works of human identification based on gait approaches have been introduced (for a recent review see [10,11]). These works demonstrate the feasibility of using gait signature for human identification at a distance. However, gait recognition is still a complicated task because of the variations in view angles, clothing, carring condition and other covariate factors. These factors can degrade the overall performance of gait recognition. Amongst all covariate factors, clothing type has been demonstrated to be the most challenging one [5]. It will drastically alter the individuals appearance with the variation of different clothing types, such as baggy pants, skirt, down jacket, and coats. The task of gait analysis becomes much more challenging [13]. In recent conventional methods, such as that in [5], proposed a technique by introducing a Random Subspace Method (RSM) framework for clothing-invariant gait recognition by setting up multiple inductive biases in a random manner. Islam et al. [7] conducted a study on human gait by dividing it into small window chunks and developed a Random Window Subspace Method (RWSM) for clothing invariant gait recognition. More recently, in [16], they developed a CNN-based method that can automatically learn to extract the most discriminative changes of gait features from a low-level input data (i.e. GEI) for clothing-invariant gait recognition. Although most of the conventional approaches have shown promising results, the overall performance of these approaches often got degraded greatly by the challenging clothing types.

In gait recognition, when the combinations between the clothing types is simple, it is easier to obtain an individual's gait patterns. However, if the type of clothing differs between the gallery and the probe, parts of the body seen in the silhouettes are likely to change and the ability to identify subjects decreases with respect to these body parts [6]. In this paper, we proposed to extract clothing-invariant gait feature by learning the complex non-linear transform from the most challenging combinations of clothing types to normal condition. Specifically, our proposed method is inspired by the one [9] where a stacked progressive auto-encoders network is proposed to deal with face recognition across different poses. In this work, we propose a solution based on the extension of this principle to deal with the effect of challenging combinations of different clothing covariates. We summarize the contributions of this study as follows.

- We present a model Stacked Progressive Auto-Encoders for clothing-invariant gait recognition named as SPAEGait. It's designed to transform the input gait images from challenging combinations of clothing types to a normal one. The method tries to leverage gradually the information about the different clothing combinations in order to achieve clothing-invariant identification.
- We evaluate the performance of the proposed method on the challenging clothing-invariant of the OU-ISIR Treadmill-B dataset, achieving improved performance compared to other conventional approaches.

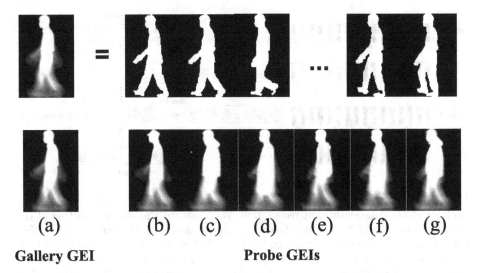

(a)   **(b)**   **(c)**   **(d)**   **(e)**   **(f)**   **(g)**

Gallery GEI   **Probe GEIs**

Fig. 1. Samples of GEIs of a similar person are computed by averaging the silhouettes over one gait cycle taken from OU-ISIR Treadmill B dataset [6]. The leftmost GEI image (a) is the gallery GEI in normal clothing condition, while the rightmost GEI image (b)–(g) are probe GEIs with different clothing combinations.

The rest of this paper is organized as follows. Section 2 presents the stacked progressive auto-encoders network for clothing-invariant gait recognition. Experimental results and discussion are given in Sect. 3, and conclusion and future works of this paper are highlighted in Sect. 4.

2 Proposed Approach

In this section, we propose a solution for a clothing-invariant feature learning based gait recognition framework, which is especially effective for dealing with clothing variations. The proposed framework for gait recognition is represented in Fig. 2. We will describe our proposed framework in the following sections.

2.1 Input Data (GEI)

In this work, we employ a spatio-temporal gait representation called Gait Energy Image (GEI) [12] as the input raw data of our method. Example GEIs belonging to one subject are shown in Fig. 1. GEI is constructed by averaging the silhouette in one complete gait cycle. Given a size-normalized and horizontal aligned human walking binary silhouette sequenece $B(x, y, t)$, the grey-level GEI $G(x, y)$ is defined as follows

$$G(x, y) = \frac{1}{N} \sum_{t=1}^{N} B(x, y, t), \tag{1}$$

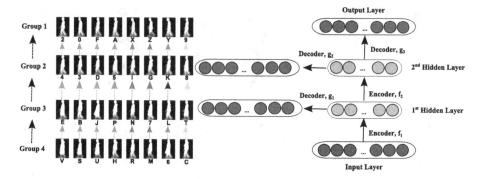

Fig. 2. The schema of the proposed Stacked Progressive Auto-Encoders (SPAE) network for clothing-invariant gait recognition. We illustrate a model architecture of the stacked network with two hidden layers, which can deal with the variations in different combination of clothing types ranging from challenging ones to normal conditions. In training stage of our SPAE, each progressive auto-encoder aims at mapping the GEI images at challenging combinations of clothing types to normal condition, while keeping the GEI images with normal combination of clothing type unchanged. In the testing stage, given a GEI image, it is pass into the SPAE network, and the outputs of the topmost hidden layers with easy combination of simple clothing types are used as the clothing-invariant features for gait recognition.

where N is the number of frames in complete cycles of the sequence, x and y are values in the 2D image coordinate.

2.2   Auto-Encoder (AE)

In recent years, auto-encoder [1,3] played an important role in unsupervised learning and in deep architectures for transfer learning and other tasks. In general, a shallow auto-encoder neural network usually contains two parts: encoder and decoder. In addition, it is usually made up of a input layer, a hidden layer and followed by an output layer [14]. The encoder, denoted as $f(.)$ can transform the input data into a new representation in the hidden layer. It usually comprises of a linear transform and a nonlinear transformation as follows:

$$z = f(x) = s(Wx + b), \tag{2}$$

where W is the linear transform, b is the basis and $s(.)$ is the nonlinear transform, which is also called a element-wise "activation function", such as sigmoid function or tanh function:

$$s(x) = \frac{1}{1 + e^{-x}} \tag{3}$$

or

$$s(x) = \frac{e^x - e^{-x}}{e^x + e^{-x}}. \tag{4}$$

The decoder, denoted as g, tries to transform the hidden representation z back to input data x, i.e.,

$$x' = g(z) = s(W'z + b'),$$ (5)

where $g(.)$ denotes the decoder, w' and b' denote the linear transformation and basis in decoder and x' is the output data. In general, we usually employ least square error as the cost function to optimize the parameter of w, b, W' and b'.

$$\left[\hat{W}, \hat{b}, \hat{W}', \hat{b}'\right] = \arg\min_{W,b,W',b'} \sum_{i=1}^{N} \left\|x_i - x_i'\right\|^2$$ (6)

$$= \arg\min_{W,b,W',b'} \sum_{i=1}^{N} \|x_i - g(f(x))\|^2,$$ (7)

where x_i denotes the ith training sample of N samples and x_i' means the corresponding output of x_i. In our experiments, we train the autoencoder with Stochastic Gradient Descent (SGD). Our implementation for this network is based on the Caffe framework [8].

2.3 SPAEGait for Clothing Variations

In this paper, we propose a method by stacking multiple progressive auto-encoders together to deal with the effect caused by clothing variations. During the model training, the output is synthesized in a progressive way. For gait recognition, a combination of simple clothing types contains more dynamic information about one person's gait pattern because the appearance changes are very minimal. Hence, we try to transform all the gait energy images to normal conditions. The first layer of auto-encoders is employed to handle the most challenging combination of clothing types. Then, after some layer of auto-encoders, all GEI images would gradually narrow down the clothing variation to normal condition as shown in Fig. 2, improving the accuracy for gait recognition.

Here, we assumed that there are $2 \times L + 1$ clothing variations in the dataset. The combination of clothing types are $\{0,2,3,4,\ldots,X,Y,Z\}$. In the first layer, the auto-encoder would transform the GEI images from challenging combinations of clothing types in Group 4 to Group 3. In other words, this progressive AE narrows down the clothing variations. Meanwhile it keeps the GEI images with normal condition unchanged. Similarly, the auto-encoder in the second layer is designed transform the GEI image which is slightly complicated in Group 3 to Group 2. Finally, the last progressive AE layer would transform all the GEI images to normal conditions but maintain the simple clothing types unchanged.

We train each progressive AE layer separately and the output of a hidden layer is the input of the next successive layer. The whole network is fine tuned by optimizing all layers together after training all the auto-encoders, as bellow

$$\left[\hat{W}_k|_{k=1}^{L}, \hat{b}_k|_{k=1}^{L}, \hat{W}_L', \hat{b}_L'\right] = \arg\min_{W_k|_{k=1}^{L}, b_k|_{k=1}^{L}, W_L', b_L'} \sum_{i=1}^{N} \|x_i' - g_L(f_L(f_{L-1}(\ldots(f_1(x_i)))))\|^2,$$ (8)

where k is the kth layer in all L layers.

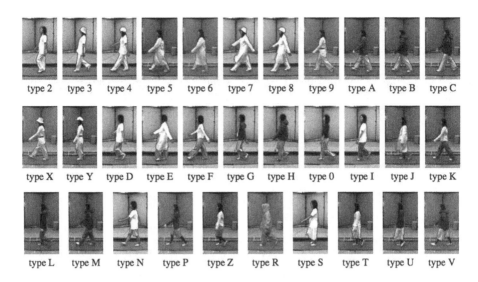

type 2   type 3   type 4   type 5   type 6   type 7   type 8   type 9   type A   type B   type C

type X   type Y   type D   type E   type F   type G   type H   type 0   type I   type J   type K

type L   type M   type N   type P   type Z   type R   type S   type T   type U   type V

Fig. 3. A sample of images for each clothing variations taken from the OU-ISIR-B dataset [6].

2.4   Clothing-Invariant Feature Extraction and Recognition

As the GEIs with clothing variations are transformed to normal ones, the output representation of the topmost layer f_L should be almost reduced down to normal condition. However, the representation emdedded in the lower hidden layers do not contain clothing-invariant features, but only contain very small clothing variations. Therefore, we accumulate the output representation of multiple hidden layers as the clothing-invariant features as follows:

$$F = [f_{L-i}, f_{L-i+1}, \ldots, f_L], \tag{9}$$

where $0 \leq i \leq L - 1$. The resulting features, F learnt from SPAEGait model may not be class discriminative. We then employ Principal Component Analysis (PCA) for dimensionality reduction and the nearest neighbor for classification.

3   Experiments and Results

In this section, we firstly describe the experimental settings for the evaluations including the datasets and implementation details; then investigate our proposed SPAEGait model on the effect of variation of different combination of clothing types; finally, compare it with the existing approaches on the OU-ISIR Treadmill dataset B [6].

3.1   Dataset Description

The OU-ISIR Treadmill dataset B [6] is one of the largest clothing variations gait database. It consists of 68 subjects in total with 15 to 32 different clothing

Table 1. Different combinations of clothing variation in the OU-ISIR-B dataset [6].

Type	S_1	S_2	S_3	Type	S_1	S_2	Type	S_1	S_2
3	RP	HS	Ht	0	CP	CW	F	CP	FS
4	RP	HS	Cs	2	RP	HS	G	CP	Pk
6	RP	LC	Mf	5	RP	LC	H	CP	DJ
7	RP	LC	Ht	9	RP	FS	I	BP	HS
8	RP	LC	Cs	A	RP	Pk	J	BP	LC
C	RP	DJ	Mf	B	RP	DJ	K	BP	FS
X	RP	FS	Ht	D	CP	HS	L	BP	Pk
Y	RP	FS	Cs	E	CP	LC	M	BP	DJ
N	SP	HS	-	P	SP	Pk	R	RC	-
S	Sk	HS	-	T	Sk	FS	U	Sk	Pk
V	Sk	DJ	-	Z	SP	FS	-	-	-

combinations as listed in Table 1, while the list of clothes available in the dataset is shown in Table 2. The original dataset is divided into three subsets, i.e., training set, gallery set and probe set. During the training, same as in conventional approaches, we used a subset containing 446 sequences of 20 subjects with all possible clothing types for training purpose. For testing, we used a gallery set consisting gait sequences of the remaining 48 subjects in standard clothing combination. The probe set consists of 856 gait sequences for these 48 subjects with all types of different clothing combinations excluding the standard one in the gallery set. Figure 3 shows the sample images of different combinations of clothing variation in the dataset.

Table 2. List of variations in clothing types used in the dataset (Abbreviation: Clothes type ID).

RP: Regular pants	HS: Half shirt	CW: Casual wear
BP: Baggy pants	FS: Full shirt	RC: Rain coat
SP: Short pants	LC: Long coat	Ht: Hat
Sk: Skirt	Pk: Parker	Cs: Casquette cap
CP: Casual pants	DJ: Down jacket	Mf: Muffler

3.2   Implementation Details

In the experiments, the structure of our proposed autoencoder with 2 fully-connected hidden layers as shown in Fig. 2 are used. During the training phase, we used the Caffe software [8], a very popular open source deep learning framework to train the model. Particularly, each progressive auto-encoder needs to be trained separately. The weights are initialized to random values using Gaussian distribution with zero mean and standard deviation of 0.01 for all trainable layers. All the bias terms are initialized with the constant zero. We set our base learning rate to 0.1. We reduced the learning rate for all layers by a factor of 10 every 1000 iterations prior to termination. We trained the model for 10,000 iterations and the activation function is sigmoid for all experiments. After this, the two hidden layers have been trained then combined in a stacked way, and fine-tuned as a whole in the model. In the fine-tuning, the base learning rate was set to 0.01 and the maximum number of iterations was 5000. One important parameter to the proposed model is the numbers of hidden layer neurons. Following the work in [2], our model is evaluated with different numbers of neurons from 500 to 3000. From the results (tables not reported), we noticed that when the number is 2000 the model achieves the best recognition rate. So, we set this as the number of neurons for each hidden layer in the model. Besides the feature from the topmost layer, we can also select features from lower layers as the invariant gait feature for gait recognition. From the results (tables not reported), we observed that the feature consisting of the multiple hidden layers did not help to improve recognition rates. However, in our case, the features from the topmost hidden layer achieved the best performance with a significant improvement. The final step of the invariant gait feature extraction was to reduce the dimension using PCA, same as in [9]. The feature dimension was reduced from 2000 to 100. The value 100 was chosen according the experiments. The features obtained from the topmost encoding layer are then compressed using PCA and used as input to a nearest-neighbour classifier (NN).

3.3   On the Effect of Clothing Variations

We conducted all our experiments on the OU-ISIR Treadmill B dataset to examine the performance of our proposed approach with the effect of different clothing types. The performance for the proposed method can be observed in Fig. 4. From the experiment results, we have the following observations:

In experiment 1 (Exp. 1), consisting of walking sequences that only involve normal clothing combinations (e.g., Y:regular pants+full shirt, 2:regular pants+half shirt, etc.), we can see that the average CCR is above 98.5%. It is also clear from Fig. 4 that the gait pattern is not affected by an appearance change caused by easy combinations of clothing types. Apart from the basic clothing combinations in Experiment 1, we can observe that in Experiment 2 (Exp. 2), the clothing combinations consists of walking sequences that involve some headwear covariates (e.g., 3:regular pants+half shirt+hat, 4:regular pants+half shirt+casquette cap, etc.). In this case, we are still able to achieve relatively high performance. The average CCR is above 94.2%. The headwear covariates does not affect the performance of gait recognition significantly.

In Experiment 3 (Exp. 3), we analyze the scenario of clothing combination type B and E when a complicated clothing type (e.g., down jacket or long coat) is combined with a basic clothing type (e.g., casual or regular pants), the recognition task becomes much more complicated. However, based on results obtained by the proposed method, the performance still yields competitive results, with

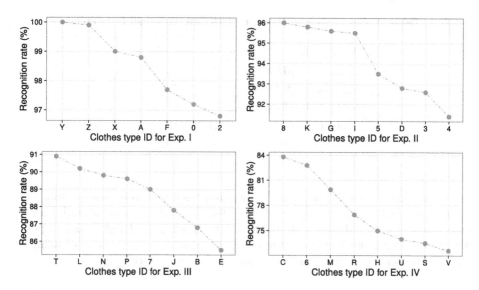

Fig. 4. Performance evaluation on the OU-ISIR Treadmill B dataset under the effect of different clothing combinations.

an average recognition rate above than 85.5%. The combinations between the challenging clothing types shown in Experiment 4 (Exp. 4) lead to a decrease in gait recognition accuracy performance. We can observe the average CCR is reduced further to 77.3%. For instance, given a query gait pattern under the effect of difficult combinations of clothing types S, the recognition rate decreased to around 73.5%. Similarly, for clothing type V, the recognition rate decreased to around 72.6%. Notice that for the same task a human would also have difficulties as clothes of this type occlude many discriminative features, nevertheless our proposed method is still able to recognize in most cases the individual's gait.

In general, these experiments revealed that the proposed approach is able to provide high classification accuracy even when the combination of clothing types get more complicated. The performance of our method in identifying subjects based on gait under the effect of clothing variations shows that the proposed stacked progressive AE is able to learn clothing-invariant features by converting challenging combinations clothing types to basic ones progressively which, in turn, means the clothing variations are narrowed down eventually.

3.4   Comparison with Conventional Approaches

The experimental results have been compared with the ones obtained by other approaches available in the literature. Table 3 summarizes the comparison of results with respect to other conventional approaches on the OU-ISIR Treadmill B dataset. It can been seen that our method displays good results, especially for the challenging clothing types. This can be explained by the fact that our method gradually narrows down the clothing variations which, in turn, improves

Table 3. Comparison of results on OU-ISIR Treadmill B dataset.

Algorithms	CCR(%)
Guan et al. (RSM) [5]	80.44
Islam et al. (RWSM) [7]	78.54
Yeoh et al. (CNN+SVM) [16]	91.38
Yeoh et al. (CNN+Softmax) [16]	87.80
Proposed method (SPAEGait)	94.80

the recognition performance in the challenging clothing combination conditions when these features are considerably affected. In addition, the overall performance outperforms all the state-of-the-art methods considered in our experiments [5,7,16]. This further demonstrates the specific effectiveness of using a stacked auto-encoder on gait recognition with clothing variations.

4   Conclusions and Future Works

In this paper, we have presented a stacked progresssive auto-encoders model to extract clothing-invariant gait feature for gait recognition under various clothing covariates. The proposed framework can tranform gradually GEI images from a challenging combination of clothing types to the normal ones by multiple shallow progressive auto-encoders. These features extracted from few topmost layers of stacked AE network only contain very small clothing variations, that are further integrated with PCA for clothing-invariant gait recognition. The experiments were performed on a well-known and the largest clothing variation gait dataset, namely the OU-ISIR Treadmill B dataset. Experimental results show that our model can effectively improve recognition rate by reducing the clothing variations especially when there is a large clothing variation and achieves a far better performance over existing approaches.

As a future work, we intend to extend our approach to cope with more challenging covariate variations. Moreover, it will be interesting to see how our approach can be modified to accomodate the case of gait identification under the effect of various covariate factors. Besides, using another large gait database can also be investigated to evaluate the effectiveness of our method.

Acknowledgment. The authors would like to express our sincere thanks to Institute of Scientific and Industrial Research, Osaka University for providing access to the OU-ISIR Gait Treadmill-B dataset for the use in this work.

References

1. Baldi, P.: Autoencoders, unsupervised learning, and deep architectures. In: ICML Unsupervised and Transfer Learning, vol. 27(37–50), p. 1 (2012)
2. Bengio, Y.: Practical recommendations for gradient-based training of deep architectures. In: Montavon, G., Orr, G.B., Müller, K.-R. (eds.) Neural Networks: Tricks of the Trade. LNCS, vol. 7700, pp. 437–478. Springer, Heidelberg (2012). doi:10.1007/978-3-642-35289-8_26
3. Bengio, Y., et al.: Learning deep architectures for ai. Foundations and trends® . Mach. Learn. **2**(1), 1–127 (2009)
4. Bouchrika, I., Carter, J.N., Nixon, M.S.: Towards automated visual surveillance using gait for identity recognition and tracking across multiple non-intersecting cameras. Multimedia Tools Appl. **75**(2), 1201–1221 (2016)
5. Guan, Y., Li, C.T., Hu, Y.: Robust clothing-invariant gait recognition. In: 2012 Eighth International Conference on Intelligent Information Hiding and Multimedia Signal Processing (IIH-MSP), pp. 321–324. IEEE (2012)
6. Hossain, M.A., Makihara, Y., Wang, J., Yagi, Y.: Clothing-invariant gait identification using part-based clothing categorization and adaptive weight control. Pattern Recogn. **43**(6), 2281–2291 (2010)
7. Islam, M.S., Islam, M.R., Akter, M.S., Hossain, M., Molla, M.: Window based clothing invariant gait recognition. In: 2013 International Conference on Advances in Electrical Engineering (ICAEE), pp. 411–414. IEEE (2013)
8. Jia, Y., Shelhamer, E., Donahue, J., Karayev, S., Long, J., Girshick, R., Guadarrama, S., Darrell, T.: Caffe: convolutional architecture for fast feature embedding. In: Proceedings of the 22nd ACM International Conference on Multimedia, pp. 675–678. ACM (2014)
9. Kan, M., Shan, S., Chang, H., Chen, X.: Stacked progressive auto-encoders (spae) for face recognition across poses. In: Proceedings of the IEEE Conference on Computer Vision and Pattern Recognition, pp. 1883–1890 (2014)
10. Lee, T.K., Belkhatir, M., Sanei, S.: A comprehensive review of past and present vision-based techniques for gait recognition. Multimed. Tools Appl. **72**(3), 2833–2869 (2014)
11. Makihara, Y., Matovski, D.S., Nixon, M.S., Carter, J.N., Yagi, Y.: Gait recognition: Databases, representations, and applications. Wiley Encyclopedia of Electrical and Electronics Engineering (2015)
12. Man, J., Bhanu, B.: Individual recognition using gait energy image. IEEE Trans. Pattern Anal. Mach. Intell. **28**(2), 316–322 (2006)
13. Nandy, A., Chakraborty, R., Chakraborty, P.: Cloth invariant gait recognition using pooled segmented statistical features. Neurocomputing **191**, 117–140 (2016)
14. Vincent, P., Larochelle, H., Lajoie, I., Bengio, Y., Manzagol, P.A.: Stacked denoising autoencoders: Learning useful representations in a deep network with a local denoising criterion. J. Mach. Learn. Res. **11**, 3371–3408 (2010)
15. Yeoh, T.W., Zapotecas-Martínez, S., Akimoto, Y., Aguirre, H.E., Tanaka, K.: Feature selection in gait classification using geometric PSO assisted by SVM. In: Azzopardi, G., Petkov, N. (eds.) CAIP 2015. LNCS, vol. 9257, pp. 566–578. Springer, Cham (2015). doi:10.1007/978-3-319-23117-4_49
16. Yeoh, T., Aguirre, H.E., Tanaka, K.: Clothing-invariant gait recognition using convolutional neural network. In: 2016 International Symposium on Intelligent Signal Processing and Communication Systems (ISPACS), pp. 1–5. IEEE (2016)

Biometrics

An Improved Scheme of Local Directional Pattern for Texture Analysis with an Application to Facial Expressions

Abuobayda M. Shabat and Jules-Raymond Tapamo[(✉)]

School of Engineering, University of KwaZulu-Natal, Howard College Campus,
Durban 4041, South Africa
abshabat@gmail.com, tapamoj@ukzn.ac.za

Abstract. In this paper, several extensions and modifications of Local Directional Pattern (LDP) are proposed with an objective to increase its robustness and discriminative power. Typically, Local Directional pattern generates a code based on the edge response value for the eight directions around a particular pixel. This method ignores the center value which can include important information. LDP uses absolute value and ignores sign of the response which carries information about image gradient and may contain more discriminative information. The sign of the original value carries information about the different trends (positive or negative) of the gradient and may contain some more data. Centered Local Directional Pattern (CLDP), Signed Local Directional Pattern (SLDP) and Centered-SLDP (CSLDP) are proposed in different conditions. Experimental results on 20 texture types using 5 different classifiers in different conditions shows that CLDP in both upper and lower traversal and CSLDP substantially outperforms the formal LDP. All the proposed methods were applied to facial expression emotion application. Experimental results show that SLDP and CLDP outperform original LDP in facial expression analysis.

Keywords: Texture features · Local Directional Pattern · Centered Local Directional Pattern · Signed Local Directional Pattern · Centered-SLDP · Classification · Facial expression

1 Introduction

In computer vision, texture is a very significant aspect. It presents information about the spatial properties like color or pixel intensity which can be extracted from an image. It describes the relationship between the value of the gray value in a particular pixel and its neighbors. Several textural features methods have been proposed, including Gray Level co-occurrence Matrix (GLCM) [1], Grey-Level Run Length Matrix (GLRLM) [2], Gabor Filter [3], Local Binary pattern [4] and Local Directional Pattern [5] and many others. Local Binary Pattern (LBP) was an eye catching for its simplicity and excellent accuracy in extracting

© Springer International Publishing AG 2017
M. Felsberg et al. (Eds.): CAIP 2017, Part II, LNCS 10425, pp. 165–177, 2017.
DOI: 10.1007/978-3-319-64698-5_15

data from images. It has been used in various applications, such as face and hand recognition [6]. The LBP operator generates binary digits, from the binary derivation that describes the neighboring pixels, which is utilized as an integral measure for regional image contrast. It takes the center value as a threshold for the regional 3×3 neighboring pixels, hence generating one binary digit if the neighbor pixel is larger or equal to the threshold, otherwise it generates zero binary digit. Inspired by the success of LBP, many scholars have proposed several adjustments of LBP [7–10]. However, LBP is sensitive to illumination changes and noise.

A stabler feature method, based on the computation of the directional information, Local Directional Pattern was proposed by Jabid et al. [5]. Unlike LBP which is based on computing the pixel intensity, LDP computes the directional information around the pixel using a gradient operator. It has been used in various applications; signature verification [11], face recognition [12], and facial expression recognition [5]. Despite the great achievement of LDP in pattern recognition and computer vision, its fundamental working mechanism still needs more investigation. Zhong [13] proposed an Enhanced Local Directional Pattern (ELDP) by taking the two most prominent directional edge response value. ELDP code is then generated by converting the two values into an octal digit. The result established the robustness of ELDP against non-illumination changes. LDP codes generation is based on the edge response values in the eight directions around the central pixel, but it doesn't take into account the center pixel value. Due to the fact that the center pixel is a very significant factor, ignoring it may lead to a critical lost of information. Centered Local Directional Pattern (CLDP) is proposed with an aim to include the center pixel value based on its relation with the neighboring pixels. Another issue with the classical LDP is that it is encoded using the absolute value, however, the sign of the original value indicates a trend (positive or negative) which may hold more information and it is applied in the proposed Signed Local Directional Pattern (SLDP) method.

The remainder of this study is organised as follows. In section two, Local Directional Pattern, CLDP, SLDP and CSLDP, for texture analysis are presented. Section three, is devoted to the evaluation of discrimination performances of feature methods presented. Section four presents conclusion and recommendations of the study.

2   Local Features for Texture Analysis

In this section, the original LDP is presented together with the proposed methods, CLDP, SLDP and CSLDP.

2.1   Local Directional Pattern

LBP is considered unstable because it extensively depends on the neighboring pixels intensity which makes it vulnerable and sensitive to random noise. To overcome these problems, Jabid et al. [5] proposed the Local Directional Pattern

(LDP). Unlike LBP which considers the intensities of the neighboring pixels, LDP considers edge response values in eight different directions. To calculate the eight directional edge response values of a particular pixel, Kirsch masks, shown in Fig. 1, are used.

$$
\begin{bmatrix} -3 & -3 & 5 \\ -3 & 0 & 5 \\ -3 & -3 & 5 \end{bmatrix}
\begin{bmatrix} -3 & -3 & 5 \\ -3 & 0 & 5 \\ -3 & -3 & 5 \end{bmatrix}
\begin{bmatrix} -3 & 5 & 5 \\ -3 & 0 & 5 \\ -3 & -3 & -3 \end{bmatrix}
\begin{bmatrix} 5 & 5 & 5 \\ -3 & 0 & -3 \\ -3 & -3 & -3 \end{bmatrix}
$$

M_0(East) M_1(North East) M_2(North) M_3(North West)

$$
\begin{bmatrix} 5 & 5 & -3 \\ 5 & 0 & -3 \\ -3 & -3 & -3 \end{bmatrix}
\begin{bmatrix} 5 & -3 & -3 \\ 5 & 0 & -3 \\ 5 & -3 & -3 \end{bmatrix}
\begin{bmatrix} -3 & -3 & -3 \\ 5 & 0 & -3 \\ 5 & 5 & -3 \end{bmatrix}
\begin{bmatrix} -3 & -3 & -3 \\ -3 & 0 & 5 \\ -3 & 5 & 5 \end{bmatrix}
$$

M_4 (West) M_5(South West) M_6(South) M_6(South Est)

Fig. 1. Kirsch masks

Given a pixel (x, y) of an image, **I**. For each direction i and using the corresponding mask M_i, the ith directional response $m_i(x, y)$ can be computed as

$$
m_i(x, y) = \sum_{k=-1}^{1} \sum_{l=-1}^{1} M_i(k + 1, l + 1) \times I(x + k, y + l) \tag{1}
$$

A vector $(m_0, \dots, m_7)$ is derived from each of the eight directions, where for each pixel (x, y) m_i represents $m_i(x, y)$. The k most significant responses are selected from the directional response vector. Thus, placing the corresponding positions to 1 bit code, leaving the remaining $(8 - k)$ to 0 bit code. The LDP code, $LDP_{x,y}$, of the pixel (x, y) base on the directional response $(m_0, \dots, m_7)$, is derived using Eq. 2.

$$
LDP_{x,y}(m_0, m_1, ..., m_7) = \sum_{i=0}^{7} s(m_i - m_k) \times 2^i \tag{2}
$$

where m_k is the kth most significant response and $s(x)$ is defined as

$$
s(x) = \begin{cases} 1 & x \geq 0 \\ 0 & otherwise \end{cases} \tag{3}
$$

From the LDP transformed image, an histogram, H, is extracted. H is defined as

$$
H_i = \sum_{x=0}^{M-1} \sum_{y=0}^{N-1} p(LDP_{x,y}, C_i) \tag{4}
$$

where C_i is the ith LDP pattern value, $i = 1, ..., ^8C_3$ and p is given as

$$
p(x, a) = \begin{cases} 1 \text{ if } x = 0 \\ 0 \text{ otherwise} \end{cases}
$$

2.2   Centered Local Directional Pattern

The original LDP codes are generated based on the value of the edge response in eight directions around pixel, but this method ignores the center pixel value. However, the center pixel may contain more information. In LDP, central pixels are not included; as a result only $^8C_3 = 56$ patterns can be generated. If the central picxel is considered, $^9C_4 = 126$ patterns will be generated. Consequently, CLDP enables the extraction of more information, that will potentially lead to better characterization of visual artefacts. The CLDP feature method is calculated in three steps;

1. Calculate of the eight directional responses. This remains the same as in the first and the second steps in LDP calculation (see Eq. 2).
2. Compute the average m_{AVG} of the 8 neighbouring pixels (see Eq. 5) as a threshold, Generate 1 binary code if the center pixel is greater or equal to the threshold, otherwise generates 0 binary code (see Eq. 6).

$$m_{AVG} = \frac{1}{8} \sum_{i=0}^{7} m_i \tag{5}$$

$$CLDP_{threshold} = s(m_c - m_{AVG}) \tag{6}$$

where m_c is the center pixel value and $s(x)$ is defined in Eq. 3.
3. Using the response values computed in the first step, and the threshold obtained in the previous step, the CLDP can be calculated as

$$CLDP = \sum_{i=0}^{7} s(m_i - m_k) \times 2^i + CLDP_{threshold} \tag{7}$$

There are two ways to generate the binary code in CLDP, depending on the direction:

– CLDP-UP: in this option, the computation of the binary code begins from the center pixels and walks up (anticlockwise) through all the neighbouring pixels, in the following order $(m_c, m_0, ..., m_7)$ as shown in Fig. 2. The $CLDP_{UP}$ can be calculated as

$$CLDP_{UP} = \sum_{i=0}^{7} s(m_i - m_k) \times 2^i + s(m_c - m_{AVG})$$

– $CLDP_{Down}$: in this other option, the computation of the binary code begins from the central pixels and walk down (clockwise) through all the neighbouring pixels, in the following order $(m_c, m_0, m_7, \ldots, m_1)$ as shown in Fig. 2. $CLDP_{UP}$, as represented, starts from the center pixel m_0 and walks up through all neighboring pixels; then $CLDP_{Down}$ starts from the center pixel m_0 and walks down through all neighboring pixels. $CLDP_{Down}$ can be calculated as

$$CLDP_{Down} = \sum_{i=0}^{7} s(m_{7-i} - m_k) \times 2^i + s(m_c - m_k) \times 2 + s(m_c - m_{AVG})$$

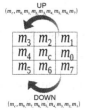

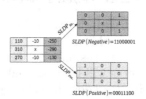

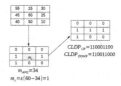

Fig. 2. Representation of $CLDP_{UP}$ and $CLDP_{Down}$

Fig. 3. A computing example for Signed Local Directional Pattern

Fig. 4. A computing example for Centered Local Directional Pattern

Figure 4 shows an example of computing the Centered Local Directional Pattern. To construct the CLDP descriptor, the calculation of the CLDP code for each pixel (x, y) is necessary. The histogram H obtained from the transformed Image can be defined as

$$H_i = \sum_{x=0}^{M-1} \sum_{y=0}^{N-1} p(CLDP_{(x,y)}, C_i) \tag{8}$$

2.3 Signed Local Directional Pattern

LDP and SLDP are for most of the steps similar; the only difference thing is the way the most prominent edges are chosen. On the LDP, the selection of the most prominent edges is based on the absolute values, while for the SLDP the signs (positive or negative) of edges are considered. Typically, LDP is encoded using the absolute value, however the sign of the original value indicates the trends (positive or negative) of the gradient and may hold more data. SLDP features are based on eight bit binary codes assigned to each pixel of an input image. In case of the positive trends, three of the most prominent edges are chosen as calculated in Eq. 9.

$$SLDP(Pos)_{x,y} = \sum_{i=0}^{7} s(m_i - m_k) \times 2^i \tag{9}$$

On the other hand, in the negative trends are calculated as:

$$SLDP(Neg)_{x,y} = \sum_{i=0}^{7} s((-m_i) - m_k) \times 2^i \tag{10}$$

where s is defined a in Eq. 3. Figure 3 shows an example of computing SLDP. Histogram H for both directions are obtained from the transformation has 56 bins and can be defined as

$$H_i(P/N) = \sum_{x=0}^{M-1} \sum_{y=0}^{N-1} p(SLDP_{(x,y)}(P/N), C_i) \tag{11}$$

2.4   Centered Signed Local Directional Pattern

The SLDP codes are generated based on two different directions (positive or negative) of the gradients. However, this method ignores the center pixel value. In the original SLDP, $^8C_3 = 56$ patterns only are generated, but if the center pixel is included $^9C_4 = 126$ patterns will be generated, which will include more information. Center-SLDP (CSLDP), adds the center pixel for both directions (positive or negative). Their computation is done in two steps (Fig. 5):

1. Compute SLDP as in the Subsect. 2.3 above.
2. CSLDP coding of the directional response is generated based on the first step and the calculation of the center pixel obtained in Subsect. 2.2. The CSLDP can be calculated as

$$CSLDP = SLDP(Pos/Neg) + CLDP_{threshold} \qquad (12)$$

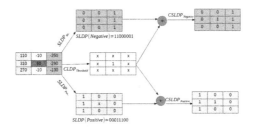

Fig. 5. A computing example for Centered Signed Local Directional Pattern

3   Experimental Analysis

The main steps used for texture classification are:

1. *Feature operator selection*: This step consists of picking a features method (LDP, CLDP and SLDP) and determining the descriptions for each pixel, creating a suitable scale for the textual description of the image.
2. *Local feature extraction*: The resulting description of the image is formed through a concatenation of sub-regional histogram of local pattern.
3. *Classification*: Match the unknown image description (test set) with all known images (training set) using different classifiers in different conditions.

Python-Fortran, OpenCV and scikit-learn frameworks [15] are used for experiments.

3.1 Classification Evaluations

The effectiveness of these methods were evaluated using different classification measures, learning curve, accuracy, precision, recall, F-score and Cohen's kappa. The learning curve is a very useful algorithm that evaluates the sanity of an algorithm. It plots the relation between the training set size and the performance. In a basic manner it shows the starting point where the classifier begins to learn. Accuracy is the number of samples classified correctly, for example if the classifier accuracy is 50% it means that the classifier manages to classify correctly 50% of the dataset

$$Accuracy = \# \text{ of samples correctly classified}/ \# \text{ of samples} \qquad (13)$$

Precision is the ratio of positive predictions to all positive classes value predicted.

$$precision = \text{True positive}/(\text{True positive} + \text{False positive}) \qquad (14)$$

Sensitivity is the ratio of positive predictions to all positive classes in test data.

$$recall = \text{True positive}/(\text{True positive} + \text{False negative}) \qquad (15)$$

F-score conveys the balance between the precision and the recall.

$$F - score = 2 * ((precision * recall)/(precision + recall)) \qquad (16)$$

Cohen's kappa is a very good measure that can handle both multi-class and imbalanced class problem very well. It calculates the agreement between categorical data. If the value is less than or equal 0, it indicates that the classifier is useless. Table 1 shows the interpretation of Kappa value.

Table 1. Interpretation of Kappa(k)

Strength of argument	Value of κ
Poor agreement	$\kappa \leq 0.20$
Fair agreement	$0.20 < \kappa \leq 0.40$
Moderate agreement	$0.40 < \kappa \leq 0.60$
Good agreement	$0.60 < \kappa \leq 0.80$
Very good agreement	$0.80 < \kappa \leq 1.00$

Each classifier is trained using different parameters as shown in Table 2.

3.2 Textural Application

In this section, all the presented method are evaluated for texture analysis using Kylberg dataset.

Table 2. Classifiers parameters

Classifiers	Parameters
SVM	Polynomial linear kernel, configuration parameter c = 0.025
k-NN	k = 5
DT	Entropy, the minimum number of split is 10
RF	The number of the trees is 10, the maximum depth of the tree is 5
Adboost	The maximum number of estimator is 50, learning rate = 1
Gaussian NB	Autoselected
Perceptron	The number of passes over the training data = 100, constant eta = 0.1

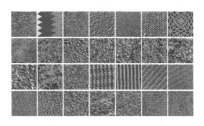

Fig. 6. The sample images of texture from Kylberg

Kylberg Datasets. Kylberg dataset [14] consists of 28 categories, each categories has 160 images. All selected images have the size of 576×576. In Fig. 6 a sample of each category is shown.

Experimental Results and Discussion for Kylberg Dataset. In this paragraph, we evaluate the power of the proposed descriptor for texture analysis using 5 different classifiers (K-neareast neighbor algorithm(KNN), Support Vector Machine (SVM), Perceptron, Naive-Bayes(NB), Decision Tree(DT)), under two different conditions:

1. Classification without preprocessing: In this option, raw feature vectors generated are fed into classifiers.
2. Classification with preprocessing:
   - Standardization (Z-score normalization): Re-scale the features so that they'll have the attributes of a standard normal distribution with the mean (μ) equals to 0 and the standard deviation (σ) equals to 1.
   - Min-Max scaling: The data is normalized to a specified range - usually 0 to 1. A Min-Max scaling is typically performed using Eq. 17.

$$X_{norm} = \frac{X - X_{min}}{X_{max} - X_{min}} \tag{17}$$

The Kylberg dataset is split into two: 80% of the dataset as a training set and 20% as test set, and 10 cross-validation is used. The performance of LDP, CLDP (up or down), SLDP (Positive or negative) and CSLDP(positive or negative) will

Table 3. $F - score$ of LDP, CLDP, SLDP and CSLDP using six classifiers

Classifiers	LDP	$CLDP_{UP}$	$CLDP_{Down})$	SLDP(Pos.)	SLDP(Neg.)	CSLDP(Pos.)	CSLDP(Neg.)
k-NN	0.928	0.953	0.954	0.911	0.897	0.954	0.957
SVM	0.992	0.999	0.999	0.988	0.978	0.988	0.985
DT	0.897	0.924	0.916	0.85	0.848	0.916	0.87
RF	0.688	0.807	0.752	0.745	0.716	0.741	0.769
NB	0.867	0.91	0.911	0.868	0.869	0.929	0.929
Perceptron	0.756	0.888	0.891	0.559	0.568	0.83	0.808

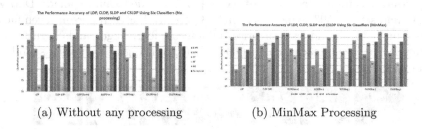

(a) Without any processing   (b) MinMax Processing

Fig. 7. The classification performance of LDP, CLDP, SLDP and CSLDP using six classifiers

be compared using 6 different classifiers (K-neareast neighbor algorithm(KNN), Support Vector Machine (SVM), Perceptron, Naive-Bayes (NB), Decision Tree (DT)) under different conditions. Regarding the length of the proposed descriptor, the basic LDP and SLDP has 56 bins and both CLDP and CSLDP have 126 bins. F-score and the performance accuracy are used to evaluate the effectiveness of our proposed methods. We begin by describing the performance of the proposed method without any preprocessing for the feature vectors, Fig. 10a shows the performance accuracy of LDP, SLDP, CLDP and CSLDP in six different classifiers. The results establish that the addition of the center pixel to the LDP in CLDP has a substantial impact on the performance, where there is an increase in the performance ranging from 1% to 9% according to each classifiers. It was noted that the trend (up or down) to calculate the edge responses vector in CLDP has no effect; as both have proximity the same accuracy. The result also shows that the sign of the gradient (positive or negative) in SLDP has improved the performance compared to LDP which use only the absolute value. For example, the accuracy for SLDP is equal to 95% and 100% compare to 93% and 99% in LDP using NB and SVM, respectively. For CSLDP, the two properties were merged, adding the center pixel value and the sign of the gradient. The $F - score$ in Table 3 shows that the value for CSLDP is always greater than LDP except for DT classifier. It is clear that the best accuracies are achieved for the Kylberg texture dataset was 100% for both CLDP and SLDP using SVM classifier. Table 3 shows that the best performance is distributed among CLDP and CSLDP for all the classifier, however, LDP did not provide the best performance in any of the classifiers.

When features extracted are preprocessed using standardization or MinMax before the classification, it can be noticed that there is an improvement in performance by 1% to 7% in all the classifiers except SVM which decreases by 1% (see Figs. 10a and 7b). For example, the best performance of the perceptron classifier was 92% but when we process the data it improved the performance to 99%.

3.3   Facial Expression Application

Extended Cohn-Kada Dataset (CK+). Extended Cohn-Kanade Dataset (CK+) dataset has 593 sequences from 123 persons. For each person seven facial expression neutral, sadness, surprise, happiness, fear, anger and disgust were captured. The size of each image is 640 × 490 pixel. Figure 8 shows a sample of each expression.

Fig. 8. A sample of face expression images from Cohn-Kanade dataset

Experimental Results and Discussion for CK+. Usually in facial applications the data size is fixed, so we calculated the learning curve to determine how much training dataset is sufficient to teach the classifiers. Note that in each of the two methods LDP and CLDP, the SVM began to be learned when the sample size was between 0 and 500 with an accuracy around 85% as shown in Figs. 9a and b. On the other hand, we also noticed that SVM learned faster on SLDP with an accuracy around 90% utilizing the same sample size. These results demonstrate the strength and effectiveness of the methods despite the small sample size. It is found that the learning curve starts to flatten when the sample size is around 1500 with an accuracy of $\simeq$ 98% with all the features methods which indicate that the classifiers are gaining less knowledge.

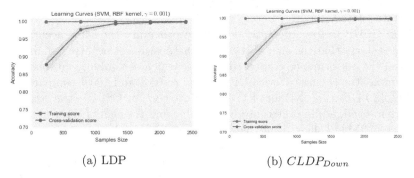

(a) LDP   (b) $CLDP_{Down}$

Fig. 9. Learning curve of LDP, $CLDP_{Down}$ using SVM

In Table 4 the average kappa scores when applying each classifier to 40% of dataset as a training dataset and reminder as a test dataset are shown. It was

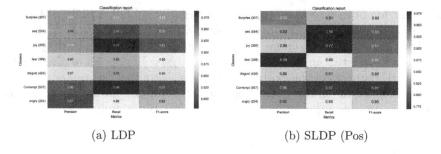

(a) LDP   (b) SLDP (Pos)

Fig. 10. Classification report of LDP and SLDP (Pos) using SVM

Table 4. The average kappa scores of LDP, CLDP and SLDP

Features	DT	SVM	NB	k-NN	Perceptron
LDP	0.64	0.98	0.61	0.82	0.98
CLDP (UP)	0.62	0.99	0.58	0.84	0.99
CLDP (Down)	0.62	0.99	0.62	0.83	0.99
SLDP (Pos)	0.75	0.99	0.58	0.83	0.99
SLDP (Neg)	0.73	0.99	0.61	0.83	0.99

noticed that LDP performance was not the best in any of the classifiers. On the contrary, for instance, its performance with (DT) was 0.64 compare to the best performance 0.75 using CLDP. Which establishes the strength and efficiency of the presented methods. The best classifier performance was for both SVM and perceptron with accuracy ranging from 0.98 to 0.99 in all feature methods. This is why we chose SVM to plot a learning curve and the classification report as it will be seen later. For DT classifier there was a 10% improvement in the SLDP method compared to LDP, LDP accuracy was 0.64 compared to SLDP at 0.75. The weak performance of all classifiers was when NB was used to judge the methods.

In Fig. 10 a classification report is plotted for each facial expression emotions class, where the x-axis shows the performance for precision, recall and F-score and y-axis shows the facial expression classes and their images number. Once again, our features methods achieved the best score on almost every facial expression class. The F-score of our SLDP outperforms the original LDP in every facial expression class.

4   Conclusion

Several extensions and modifications of Local Directional Pattern (LDP) have been proposed, CLDP, which takes into account the center value, SLDP, which considers the two different trends of the gradient and CSLDP, which includes the center for both directions (positive or negative). LDP, CLDP, SLDP and CSLDP were tested using Kylberg dataset of 3200 images using six different classifiers in different conditions. The performance of the proposed operators was also investigated on facial expression analysis. Results show that CLDP and CSLDP outperform the existing LDP. This shows that the center pixel and the directions of the gradient are very important in the extraction of textural features.

References

1. Eleyan, A., Demirel, H.: Co-occurrence matrix and its statistical features as a new approach for face recognition. Turkish J. Electr. Eng. Comput. Sci. **19**(1), 97–107 (2011)
2. Galloway, M.M.: Texture analysis using gray level run lengths. Comput. Graph. Image Process. **4**(2), 172–179 (1975)
3. Bovik, A.C., Gopal, N., Emmoth, T., Restrepo, A.: Localized measurement of emergent image frequencies by Gabor wavelets. IEEE Trans. Inf. Theory **38**(2), 691–712 (1992)
4. Ojala, T., Pietikinen, M., Harwood, D.: A comparative study of texture measures with classification based on featured distributions. Pattern Recogn. **29**(1), 51–59 (1996)
5. Jabid, T., Kabir, M.H., Chae, O.: Local directional pattern (LDP) for face recognition. In: Proceedings of the IEEE International Conference on Consumer Electronics (ICCE), pp. 329–330, January 2010
6. Wang, X., Gong, H., Zhang, H., Li, B., Zhuang, Z.: Palmprint identification using boosting local binary pattern. In: Proceedings of 18th International Conference on Pattern Recognition (ICPR 2006), vol. 3, pp. 503–506, August 2006
7. Ahonen, T., Hadid, A., Pietikainen, M.: Face description with local binary patterns: application to face recognition. IEEE Trans. Pattern Anal. Mach. Intell. **28**(12), 2037–2041 (2006)
8. Zhang, G., Huang, X., Li, S.Z., Wang, Y., Wu, X.: Boosting local binary pattern (LBP)-based face recognition. In: Li, S.Z., Lai, J., Tan, T., Feng, G., Wang, Y. (eds.) Advances in Biometric Person Authentication. LNCS, vol. 3338, pp. 179–186. Springer, Berlin Heidelberg (2004). doi:10.1007/978-3-540-30548-4_21
9. Ojala, T., Pietikainen, M., Maenpaa, T.: Multiresolution gray-scale and rotation invariant texture classification with local binary patterns. IEEE Trans. Pattern Anal. Mach. Intell. **24**(7), 971–987 (2002)
10. Pietikinen, M., Hadid, A., Zhao, G., Ahonen, T.: Local binary patterns for still images. Computer Vision Using Local Binary Patterns. Computational Imaging and Vision, pp. 13–47. Springer, London (2011)
11. Jabid, T., Kabir, M.H., Chae, O.: Robust facial expression recognition based on local directional pattern. ETRI J. **32**(5), 784–794 (2010)
12. Kabir, M.H., Jabid, T., Chae, O.: A local directional pattern variance (LDPv) based face descriptor for human facial expression recognition. In: Seventh IEEE International Conference Proceedings of Advanced Video and Signal Based Surveillance (AVSS), pp. 526–532, August 2010
13. Zhong, F., Zhang, J.: Face recognition with enhanced local directional patterns. Neurocomputing **119**, 375–384 (2013)

14. Kylberg, G.: Kylberg Texture Dataset v. 1.0. Centre for Image Analysis, Swedish University of Agricultural Sciences and Uppsala University (2011)
15. Pedregosa, F., Varoquaux, G., Gramfort, A., Michel, V., Thirion, B., Grisel, O., Blondel, M., Prettenhofer, P., Weiss, R., Dubourg, V., Vanderplas, J.: Machine learning in Python. J. Mach. Learn. Res. **12**, 2825–2830 (2011)

A Missing Singular Point Resistant Fingerprint Classification Technique, Based on Directional Patterns

Kribashnee Dorasamy[1,2]([✉]), Leandra Webb-Ray[1],
and Jules-Raymond Tapamo[2]

[1] CSIR, Modelling and Digital Science, P.O. Box 395, Pretoria 0001, South Africa
KDorasamy@csir.co.za, leeleewebb@gmail.com
[2] School of Engineering, UKZN, King George V Avenue, Durban 4041, South Africa
tapamoj@ukzn.ac.za

Abstract. Biometric fingerprint scanners that are integrated into numerous electronic devices, are compact. Commonly, individuals place their fingers on these compact scanners incorrectly causing loss of Singular Points (SPs). This has a severe impact on Exclusive Fingerprint Classification due to small inter-class variability amongst fingerprint classes. Directional Patterns (DPs) have recently shown potential in classifying fingerprints with missing SPs. However, the recent technique is designed to classify frequently occurring cases of missing SPs. In this paper the rules for complex cases where most of the key information has not been captured and tends to be extremely difficult to classify, are proposed to develop a complete classification algorithm using DPs. The proposed algorithm is tested on the FVC 2002 DB1 and 2004 DB1 and achieves an overall accuracy of 92.48%.

1 Introduction

Fingerprint biometrics which can be used to automatically determine or verify the identities of people, has played a key role in government and commercial organisations. Biometric systems used by government organisation in particular, capture and process millions of fingerprint data. Owing to the large-sized databases (DBs), searching the entire DB to find a fingerprint match is computationally expensive. Therefore, the input fingerprint is often compared to other fingerprints in the DB of the same class pattern inorder to reduce the number of comparisons. Exclusive fingerprint classification is the technique used to identify the input fingerprints' class pattern which can either be a Right Loop (RL), Left Loop (LL), Tented Arch (TA), Plain Arch (PA) or a Whorl (W). Visually these classes can be identified by the number and position of the Singular Points (SPs) (i.e. loops and deltas) and the structural layout of the ridges around these SPs. RLs, LLs, and TAs that are fully captured, have one loop and one delta. Complete W fingerprints have two loops and two deltas, whereas complete PAs have no SPs. Figure 1 shows the complete fingerprint classes, as well as fingerprint

© Springer International Publishing AG 2017
M. Felsberg et al. (Eds.): CAIP 2017, Part II, LNCS 10425, pp. 178–189, 2017.
DOI: 10.1007/978-3-319-64698-5_16

classes where certain SPs have not been captured as a result of the incorrect placement of the finger on the scanner. Challenges in exclusive fingerprint classification arises as a result of small inter-class variability between classes that have the same number of loops and deltas, especially when the fingerprint class is not completely captured with missing SPs as seen in Fig. 1(i) to (k). As a result they appear so similar that even a slight rotation can make it difficult to even visually determine the correct class. When the fingerprint is incorrectly classified, there is a higher risk of a false fingerprint match. Most existing fingerprint classification techniques have not developed a complete algorithm that includes all rule-sets which classifies fingerprints with missing SPs [1,2]. Many techniques generally address frequently occurring cases where fingerprints are not fully captured [2–5], like fingerprints with a single loop [2,5] or Ws with only two loops or two deltas [2–4]. Complex cases, such as cases where fingerprints with single delta or differentiating a Partial Fingerprint (PF) where all SPs are not captured from a PA are extremely difficult to classify since most of the key information about the class is lost. PFs can be RLs, LLs, TAs or even Ws that have been cut off before the class pattern resulting in having no SPs. Since both PAs and PFs have no SPs, most classification techniques commonly misclassify PFs as PAs or remove them from the testing DBs. Recent local orientation classification methods [5,6] do attempt to address complex cases. However, the rule-set that classifies PAs from PFs is only dependent on the size of the fingerprint which makes the rule is inefficient, as individuals have various finger sizes. In addition, the single delta rule-set presented by Guo *et al.* [6] was dependent on the fingerprints being upright, therefore the rule did not preserve rotation. Furthermore, it is much harder to distinguish between classes using local orientation

Fig. 1. Various fingerprint classes containing different number of SPs based on the way the finger was placed on the scanner (a) complete W with two loops and two deltas, (b) W with two loops, (c) W with one loop and one delta, (d) RL with one loop and one delta, (e) LL with one loop and one delta, (f) TA with one loop and one delta, (g) RL with a single delta, (h) LL with a single delta, (i) RL with a single loop, (j) LL with a single loop, (k) TA with a single loop, (l) PA with no SPs, (m) PF of PA with no SPs, (n) PF of LL with no SPs, and (o) PF of LL with no SPs

fields, where most of the key information is not captured, especially when there is noise. Even slight amounts of noise changes the orientation fields.

Grouping the orientation fields to form a DP can be advantageous when there is minimal information about a class. In our earlier works [7] we show that unique patterns can be consistently formed for complex cases, namely fingerprints with single deltas, Ws with single loops and deltas, and fingerprints with no SPs. However, the work only evaluates the DPs under rotation to establish which alignment technique proposes a unique DP pattern for each class. Furthermore, it only presents the manual classification to determine whether the method of alignment produces consistent patterns. The manual classification results can not be used to benchmark against automated systems and therefore the true level of accuracy of the algorithm was not established. In addition, the automated classification rule-sets for complex cases using DPs have not been presented nor have there been a complete automated DP classification technique that includes rules for both the complex cases and frequently occurring cases covered in our earlier works [8].

Therefore, this paper proposes the classification rules for fingerprints with single deltas, fingerprints with no SPs and Ws with a single loop and delta. In addition, the paper will benchmark the complete automated classification that includes the proposed rules and the rules of frequently occurring fingerprint cases covered in [8]. Section 2 introduces the set-up used for the fingerprint classification scheme, the details of the DP formation and SP detection. The implementation of the proposed classification rules is covered in Sect. 3. Section 4 discusses the experimental results.

2   Classification Set-Up

The complete classification method is presented in a decision tree in Fig. 2. In the pre-processing step the background of the fingerprint image is segmented using the work of Wang *et al.* [9]. The output image should contain only the ridges and valleys of the fingerprint. Orientation fields of the segmented fingerprint can be estimated using the method proposed by Hong *et al.* [10]. To generate the DP, these orientation fields are grouped into n regions shown in Table 1. Each group is refereed to a specific region number ($region_{num}$) with a specific range value ($range_i$) and interval ($\triangle\phi$).

Isolated regions that are not connected to SPs are considered as visible noise and eliminated by assigning its region number to the surrounding region. On the DPs, the SPs are located by detecting point where all 3 regions intersect. For each pixel in the entire matrix, the neighbourhood (ND) of 24 pixels will be used to search for a possible intersection and the type of SP (i.e., loops and deltas). When the regions at the intersecting point move from 1 to n in an anti-clockwise direction, it is a loop as shown in Fig. 3. When the regions at the intersecting point move from 1 to n in a clockwise direction, it is a delta. Each classification rule is evoked based on the number of loops and deltas found on the fingerprints' directional pattern.

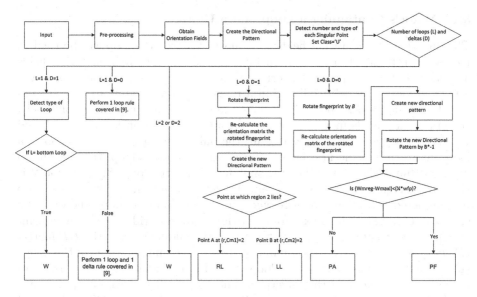

Fig. 2. Brief overview of the complete classification process of the proposed algorithm

Table 1. Calculated values that produce a 3-region DP

n	$\triangle\phi$	i	$range_i$	$region_{num}$
3	60°	1	0°:60°	1
		2	61°:90°	2
		3	91°:180°	3

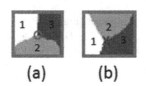

(a)   (b)

Fig. 3. Two types of SPs, namely (a) loop and (b) delta [8]

3   Implementation of Classification Rules

The classification rules presented in this section is the extension of our previous works [7,8]. The fingerprint is first rotated a specific way to form unique DPs for each class based on the number of loops and deltas. Thereafter, the rules are evoked. The class output is initialised to an unclassifiable class U.

3.1   Whorls Containing a Loop and a Delta

The direction of orientation fields as it moves towards the loop can be used to indicate a top loop from a bottom loop. Based on how the orientation fields are calculated in this paper, the orientation moves downwards as it approaches a bottom loop. The fields fall within a range of approximately 61° to 90° and therefore region 2 lies above the detected loops' coordinate points. Conversely, the orientation fields moves upwards as it approaches a top loop. RL, LL and TA have a top loop. However, Ws are the only class that can have a bottom loop due to their structural layout. To identify a bottom loop, the rule shown in Algorithm 1 checks whether region 2 lies above the loop. Since region 2 is the largest region on a W and covers the entire fingerprint width, it can also be considered as a reliable feature when detecting a bottom loop. The variables shown in Fig. 4(e) used to execute algorithm 1 are the width of the fingerprint (w_{fp}), the width of region 2 (w_{r2}), the row of the midpoint (r_m), the DP matrix ($Region$) which contains region numbers 1 to n, the minimum row of the image (y_{min}), and the coordinates of the loop (r, c). The midpoint referred to here, lies between the top edge of the fingerprint and loop L. If the class output is still U, the rule for a single loop and delta RL, LL and TA is evoked. The details of this rule is covered in [8].

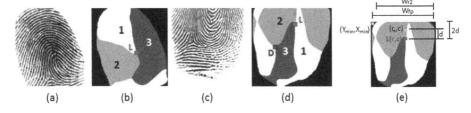

Fig. 4. Fingerprint with a single loop and single delta of (a) a complete RL fingerprint with a single bottom loop (L) and its' (b) DP, (c) a W with a bottom loop and its (d) DP, and (e) symbols illustrated on a W DP that will be used for this rule-set

---

Algorithm 1. Classify a W with a bottom loop and a delta

Input: $w_{fp}, w_{r2}, r_m, Region, y_{min}, r, c$
Output: $class$
1: $r_m = (r + y_{min})/2$
2: **if** $(Region(c, r_m) == 2)\&(3/4 * w_{fp} < w_{r2})$ **then**
3:     $class = 'W'$
4: **end if**

3.2   Right Loop and Left Loop with a Single Delta

When a fingerprint with a single delta is detected, it is rotated so that the
orientation fields below the delta are horizontal, as seen in Fig. 5(b).

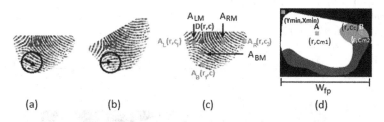

(a)   (b)   (c)   (d)

Fig. 5. Fingerprint with single delta D, (a) where the orientation fields below D flow
approximately $35°$, (b) where the orientation fields below D flow horizontal, (c) containing the symbols used to perform the classification rule

The average orientation fields around points A_{Lm}, A_{Rm} and A_{Bm} with coordinates (r, c_{1mid}), (r, c_{2mid}) and (r_{1mid}, c), respectively, are used to compute the
angle (ρ_R and ρ_L) in which the fingerprint is rotated by. The orientation fields
considered are around the midpoints between the delta D at (r, c), and the side
and bottom limits of the fingerprint (i.e., A_L, A_R and A_B), as shown in Fig. 5(c).
The orientation fields are obtained from matrix O.

$$\rho_R = \frac{\sum_{i=c_{2mid}}^{i=c_{2mid}+10} \lceil O(r_{1mid}, i) \rceil}{10} * (180/\pi) \tag{1}$$

$$\rho_L = \frac{\sum_{i=c_{1mid}}^{i=c_{1mid}+10} \lceil O(r_{1mid}, i) \rceil}{10} * (180/\pi) \tag{2}$$

The orientation matrix O is re-calculated from the rotated fingerprint image
and used to create the new DP. This approach is independent from the amount
of ridges captured. When the centroid of region 2 lies on the right of the fingerprint, it is a LL. Conversely, if the centroid of region 2 lies on the left of
the fingerprint, it is a RL. This procedure is described in Algorithm 2, using
symbols w_{fp}, c_c, c_{m1}, c_{m2}, matrix $Region$ which represents the DP, x_{min} and r
as depicted in Fig. 5(d). The initial step is to locate the centroid (r, c_c) of the
region 2. Points A and B are the midpoints taken from column c_c to the x_{min}
and from column c_c to $x_{min} + w_{fp}$, respectively. The limits are found using the
entire fingerprint width w_{fp} and fingerprint minimum column value x_{min}.

3.3   PA and PF with No Loops and Deltas

When no SPs are detected on the DP, it is either a PA or PF of RLs, LLs,
TAs, Ws or even a PA. It was established that regions in a PF DP have a
narrower width near the highest curvature point than PA DP, since each region

---

Algorithm 2. Classify a DP with a single delta

Input: $w_{fp}, c_c, c_{m1}, c_{m2}, Region, x_{min}, r$
Output: $class$
1: $c_{m1} = (c_c/2 + x_{min})$
2: $c_{m2} = (w_{fp} - c_c)/2 + (x_{min})$
3: **if** $(Region(c_{m1}, r) == 2)$ **then**
4:      $class='RL'$
5: **else if** $(Region(c_{m2}, r) == 2)$ **then**
6:      $class='LL'$
7: **end if**

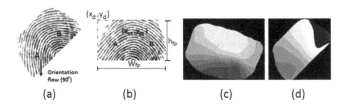

(a)   (b)   (c)   (d)

Fig. 6. Illustrations of a (a) rotated PF where the orientation flow near point A is 90°, (b) points of interest on a PF fingerprint, (c) 10-region DP of a rotated PA, (d) 10-region DP of a rotated PF

is approaching the point which would have been the loop. This attribute can be used to differentiate these classes. To ensure the innermost region of the converging point is made more pronounced, the fingerprint is rotated so that it is in a diagonal position, as shown in Fig. 6. The initial step to achieve this is to locate the centroid at column x_{ct} and row y_{ct}, as shown in Fig. 6(b). The height h_{fp} and width w_{fp} of the fingerprint are used to obtain the orientation fields around the regions of interest, A and B. Point A is located at $(y_2, x_{ct} + x)$ and B lies at position $(y_2, x_{ct} - x)$. Equations 3 and 4 are formulae used to compute the coordinates of A and B:

$$x = w_{fp}/4 \tag{3}$$

$$y_2 = y_{ct} + h_{fp}/4 \tag{4}$$

The average orientation field is taken around the reference points, A and B. Given the following equations:-

$$\sigma_R = \frac{\sum_{i=y_2}^{i=y_2+10} \lceil O(i, x_{ct} - x) \rceil}{10} * (180/\pi) \tag{5}$$

$$\sigma_L = \frac{\sum_{i=y_2}^{i=y_2+10} \lceil O(i, x_{ct} + x) \rceil}{10} * (180/\pi) \tag{6}$$

The smaller value between σ_R and σ_L is selected to compute the angle α by which the fingerprint is rotated, so that it points downwards. Once the fingerprint is rotated the orientation matrix is re-calculated to form the new DP where the

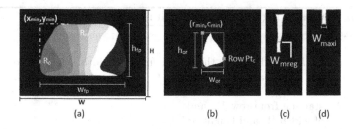

Fig. 7. Illustration of (a) a *PF DP* containing the symbols used for the algorithm, (b) the widths used to classify a *PA* and a *PF* where, (c) denotes the smallest width w_{mreg} in region R_m, and (d) denotes the maximum width w_{maxi} in region R_{m2} (cropped R_m)

orientation fields are partitioned into 10 regions. The interval $\triangle\phi$ between each $range_i$ is 18° (i.e., 180°/10°), so that the converging point is pronounced with the least amount of visible noise.

Region Rm is the region that has the smallest width, which normally appears at the point where all regions converge. The rule compares the width of region Rm at the converging point and its width at the edge of the fingerprint to determine whether the class is a PA or a PF class. The parameters used for this rule are the minimum row y_{min} and minimum column x_{min}, height h_{fp}, leftmost region R_o and width w_{fp} of the fingerprint, as shown in Fig. 7. H and W are the height and width of the entire image, which includes the fingerprint and the black background. The furthest region on the left of the DP is referred to as R_o and is used as a guideline to achieve a rough estimate of the row of the possible point of convergence. The leftmost region R_o is depicted in Fig. 7(b), having a minimum row r_{min}, minimum column c_{min}, height h_{or}, and width w_{or}. It appears that region R_o has a longer width at a specific row Pt_c, since all regions are advancing towards the point of convergence. This row is also the row at which the maximum convergence occurs. Hence it can be used to detect the region Rm, by searching for the region with the smallest width w_{mreg} on row Pt_c. The pixel value representing region R_m is initialised to 255 and the other pixel values are set to 0. Region R_m is cropped from row P_{tc} to the edge of the fingerprint. The cropped image is stored in matrix R_{m2}. The maximum width of R_{m2} is denoted by w_{maxi}. Figure 7(c) and (d) depicts the widths w_{mreg} and w_{maxi}, that is of regions R_m and R_{m2}, respectively. The width w_{maxi} is compared to width w_{mreg} for classification, as presented in Algorithm 3.

4   Experimental Results

To illustrate the efficiency of the proposed algorithm shown in Fig. 2, it was tested on the Fingerprint Verification Competition (FVC) 2002 $DB1$ and 2004 $DB2$, since it contains a variety of fingerprint classes with different number loops and deltas. The ground truths were manually created based on the definitions in [11]. There will be seven classes, namely: RL, LL, TA, PA, PF, W and Unclassifiable classes U. A small percentage of 0.03% of the entire set of images, was removed

---

Algorithm 3. Classifying a fingerprint with no SPs

Input: P_{tc}, w_{fp}
Output: $class$
 1: Find the region R_m by detecting the region smallest width w_{mreg} on row P_{tc}
 2: Set pixels in R_m to 255 and other pixels to 0.
 3: $height_C = H\text{-}r_c$
 4: // Crop R_m starting from row P_{tc} and
 5: // column 0 of width W and height $height_C$
 6: R_{m2} = Crop R_m
 7: Determine the maximum width w_{maxi} of R_{m2}
 8: **if** $(w_{mreg} - w_{maxi}) < 1/4 * w_{fp}$ **then**
 9:     $class=$'PF'
10: **else**
11:     $class=$'PA'
12: **end if**

Table 2. Confusion matrix of the exclusive classification using the proposed algorithm on 871 images from 880 images from the FVC 2002 $DB1$

Actual	Assigned						
	W	RL	LL	TA	PA	PF	U
W	**185**	7	2	0	0	0	0
RL	7	**310**	4	2	0	0	2
LL	10	4	**288**	0	0	0	2
TA	1	10	2	**6**	0	0	0
PA	0	1	1	0	**8**	6	0
PF	0	0	0	0	0	**9**	0
U	0	0	1	0	0	1	**1**
Accuracy	**93%**						

from the testing DB, since the error was not related to classification rule-sets and was a result of the intersecting regions not forming the DPs.

Tables 2 and 3 depict the confusion matrix of the classification results. The proposed method was tested on a total of 871 and 845 flat fingerprint images from the FVC 2002 $DB1$ and FVC 2004 $DB1$ that achieved an accuracy of 93% and 91.95%, respectively. Even when the testing was conducted on all images in the DBs, the algorithm still achieved high accuracies of 91.93% and 87.5% on the FVC 2002 $DB1$ and FVC 2004 $DB1$. The results depict that using DPs and the remaining SPs can allow successful exclusive classification of fingerprints with missing SPs. Globally representing the fingerprint class aids in classifying fingerprints with minimal key class information, especially when differentiating difficult cases like PA and PF. In addition, the algorithm can even differentiate between complete PAs and partial PAs shown in Fig. 1. It was also found that

Table 3. Confusion matrix of the exclusive classification using the proposed algorithm on 845 images from the 880 images from the *FVC* 2004 *DB*1

Actual	Assigned						
	W	*RL*	*LL*	*TA*	*PA*	*PF*	*U*
W	**183**	18	9	0	0	0	0
RL	3	**246**	4	0	0	0	0
LL	13	2	**283**	0	0	0	0
TA	1	7	2	**0**	1	0	0
PA	0	0	1	0	**52**	0	0
PF	0	1	0	0	6	**11**	0
U	0	0	0	0	0	2	**2**
Accuracy	**91.95%**						

Table 4. Comparison of accuracy results against literature

Author	*DB*	Average accuracy	No. of classes
Karu and Jain [1]	*FVC* 2002 *DB*1, 2004 *DB*1	50.53% (867/1716)	5
Msiza *et al.* [2]	*FVC* 2002 *DB*1, 2004 *DB*1	68.88% (1182/1716)	5
Webb and Mathekga [5]	*FVC* 2002 *DB*1, 2004 *DB*1	85.83% (1473/1716)	7
Dorasamy *et al.* [8]	*FVC* 2002 *DB*1, 2004 *DB*1	90.67% (1556/1716)	7
Jung and Lee [13]	*FVC* 2002 *DB*1, 2004 *DB*1	80.1% (650/812)	4
Guo *et al.* [6]	FVC 2002 DB1, 2004 DB1 & DB2	92.7% (7345/7920)	4
Proposed method	**FVC 2002 DB1, 2004 DB1**	**92.48% (1587/1716)**	7

errors mostly occurred for *W*s with single loops and single deltas which was misclassified has an *RL* or an *LL*, or vice versa. This is caused by over smoothing of the orientation fields in noisy images, when there is no information above the loop and the image is rotated enough to appear as an *LL* or an *RL*.

The average exclusive classification accuracy of 92.48% was benchmarked against previous algorithms that are capable of classifying cases of missing *SP*s using local orientation fields (but not *DP*s) and the remaining *SP*s. To compare the accuracy level of the proposed technique directly, methods by Karu and Jain, Webb and Mathekga *et al.*, Msiza *et al.* and Dorasamy *et al.* [1,5,8,12] have been re-implemented and tested on the same set of fingerprints. The proposed method was also benchmarked against the work by Guo *et al.* [6] and work by Jung and Lee [13]. Table 4 shows that the proposed algorithm outperforms several methods presented in literature. It is also fairly close to the accuracy achieved by Guo *et al.* [6] which classified 4 classes, with a difference of 0.22%. However, classification for a smaller number of classes automatically increases the accuracy since fingerprints have less probability of falling into another class. The benefit of the proposed technique is that it achieves a good accuracy level while classifying 7 different classes. The more classes implies that there will be less fingerprints to compare against when conducting fingerprint matching.

5   Conclusion

Exclusive classification DPs rule-sets that can classify more challenging cases of fingerprints with minimal information about a class were presented within this paper. It is an extension of our previous works [8] that are carried to accomplish a complete DP algorithm that is resistant to fingerprints with missing SPs. The complete automated classification algorithm was tested on the FVC 2002 DB1 and FVC 2004 DB1 which produced an average accuracy of 92.48%. The DPs provided a global representations of the classes with missing SPs which makes it easier to distinguish. The method was even able to classify partial PAs (i.e. PFs) from complete PAs. Fingerprint matching also experiences challenges with PF fingerprints and therefore the exclusive classification rule-set for PFs can be extended further and applied in the matching process, as a future work.

References

1. Karu, K., Jain, A.K.: Fingerprint classification. Pattern Recogn. **29**, 389–404 (1996)
2. Msiza, I.S., Leke-Betechuoh, B., Nelwamondo, F.V., Msimang, N.: A fingerprint pattern classification approach based on the coordinate geometry of singularities. In: Proceedings of the 2009 IEEE International Conference on Systems, Man, and Cybernetics, San Antonio, TX, USA, pp. 510–517. IEEE Computer Society (2009)
3. Hong, L., Jain, A.K.: Classification of fingerprint images. In: Proceedings of the Scandinavian Conference on Image Analysis, East Lansing, pp. 665–672 (1999)
4. Zhang, Q., Huang, K., Yan, H.: Fingerprint classification based on extraction and analysis of singularities and pseudoridges. Pattern Recogn. **37**, 2233–2243 (2004)
5. Webb, L., Mathekga, M.: Towards a complete rule-based classification approach for flat fingerprints. In: 2014 Second International Symposium on Computer Networks, South Africa, Pretoria, pp. 549–555. IEEE (2014)
6. Guo, J., Liu, Y., Chang, J., Lee, J.: Fingerprint classification based on decision tree from singular points and orientation field. Expert Syst. Appl. **41**, 752–764 (2014)
7. Dorasamy, K., Webb-Ray, L., Tapamo, J.-R.: Evaluating the change of directional patterns for fingerprints with missing singular points under rotation. In: Bebis, G., et al. (eds.) ISVC 2016. LNCS, vol. 10073, pp. 555–565. Springer, Cham (2016). doi:10.1007/978-3-319-50832-0_54
8. Dorasamy, K., Webb, L., Tapamo, J., Khanyile, N.: Fingerprint classification using a simplified rule-set based on directional patterns and singularity features. In: 8th IAPR International Conference on Biometrics, Phuket, Thailand, vol. 2, pp. 400–407. IEEE (2015)
9. Wang, L., Bhattacharjee, N., Gupta, G., Srinivasen, B.: Adaptive approach to fingerprint image enhancement. In: Proceedings of 8th International Conference on Advances in Mobile Computing and Multimedia, pp. 42–49. ACM (2010)
10. Hong, L., Wan, Y., Jain, A.: Fingerprint image enhancement: algorithm and performance evaluation. IEEE Trans. Pattern Anal. Mach. Intell. **20**, 777–789 (1998)

11. Maltoni, D., Maio, D., Jain, A.K., Prabhakar, S.: Handbook of Fingerprint Recognition, 2nd edn. Springer, London (2009). doi:10.1007/978-1-84882-254-2
12. Msiza, I.S., Mistry, J., Leke-Betechuoh, B., Nelwamondo, F.V., Marwala, T.: On the introduction of secondary fingerprint classification. In: Yang, J., Nanni, L. (eds.) Sate Art Biometrics, pp. 104–120. InTech, China (2009)
13. Jung, H., Lee, J.: Fingerprint classification using the stochastic approach of ridge direction information. In: International Conference on Fuzzy System, pp. 169–174. IEEE (2009)

Indexing of Single and Multi-instance Iris Data Based on LSH-Forest and Rotation Invariant Representation

Naser Damer[1(✉)], Philipp Terhörst[1,2], Andreas Braun[1], and Arjan Kuijper[1,3]

[1] Fraunhofer Institute for Computer Graphics Research (IGD), Darmstadt, Germany
{naser.damer,philipp.terhoerst,andreas.braun,
arjan.kuijper}@igd.fraunhofer.de
[2] Physics Department, Technische Universität Darmstadt, Darmstadt, Germany
[3] Mathematical and Applied Visual Computing,
Technische Universität Darmstadt, Darmstadt, Germany

Abstract. Indexing of iris data is required to facilitate fast search in large-scale biometric systems. Previous works addressing this issue were challenged by the tradeoffs between accuracy, computational efficacy, storage costs, and maintainability. This work presents an iris indexing approach based on rotation invariant iris representation and LSH-Forest to produce an accurate and easily maintainable indexing structure. The complexity of insertion or deletion in the proposed method is limited to the same logarithmic complexity of a query and the required storage grows linearly with the database size. The proposed approach was extended into a multi-instance iris indexing scheme resulting in a clear performance improvement. Single iris indexing scored a hit rate of 99.7% at a 0.1% penetration rate while multi-instance indexing scored a 99.98% hit rate at the same penetration rate. The evaluation of the proposed approach was conducted on a large database of 50k references and 50k probes of the left and the right irises. The advantage of the proposed solution was put into prospective by comparing the achieved performance to the reported results in previous works.

1 Introduction

Large-scale biometric systems are spreading to facilitate security and service goals worldwide. An example of such a system is the e-Aadhaar project of the Unique IDentification Authority of India [10] with a goal of enrolling 1.2 billion citizens. Building such a system requires trillions of biometric comparisons within duplicate enrollment checks to insure single enrollment per person. Duplicate checks and identification queries in such a system are challenged by the limitation of computational efficiency as they require exhaustive comparisons.

This work was supported by the German Federal Ministry of Education and Research (BMBF) as well as by the Hessen State Ministry for Higher Education, Research and the Arts (HMWK) within CRISP.

M. Felsberg et al. (Eds.): CAIP 2017, Part II, LNCS 10425, pp. 190–201, 2017.
DOI: 10.1007/978-3-319-64698-5_17

Iris texture is rich in information like spots, rifts, colors, filaments, minutia, and other details. It has about 10^{72} possible pattern [19] that makes it very unique and generally results in one of the smallest false-matching rates of all biometric traits [1]. These properties make the iris an outstanding candidate for large-scale-biometric systems.

Indexing techniques aim at reducing the number of comparisons (candidate identities) required by an identification system with a large number of biometric references. The fuzziness of biometric data makes the indexing task quite challenging.

The majority of the previously proposed solutions suffer from the accuracy-efficiency tradeoff, which leads to low performance when achieving fast searches [17,18]. The data storage is another challenge as some indexing structures grow very large with an expanding database [22]. The maintenance of such an indexing structure, insertions and deletions, have to be performed with minimum costs, which is not the case in some of previously proposed solutions [20].

In this work, we propose an iris indexing approach based on the rotation invariant iris representation (RIR) [8]. This iris representation is fed into a Locality-Sensitive Hashing Forest (LSH-Forest) [2], which provides accurate projection into low dimensional space with minimum parametrization requirement. Moreover, we present a multi-instance iris indexing technique that illustrates the advantages of using multiple biometric sources.

Our proposed approach achieved high accuracy at very low search space (e.g. 99.8% hit rate at 0.2% penetration rate for single iris) compared to the reported performances in related works. The complexity of query was kept at $\mathcal{O}(l \log_2(N))$, where N is the database size and l is the number of trees in the LSH-Forest. Moreover, the required storage grows linearly with the database size N and the complexity of insertion or deletion is limited to the same complexity of a query.

In Sect. 2, a detailed look into related works is presented. Section 3 discusses the proposed solution including its theoretical background. The experiment setup and the achieved results are detailed in Sects. 4 and 5. Finally a conclusion is drawn in Sect. 6.

2   Related Work

Different features extraction approaches were proposed for biometric iris representation. However, some of the most accurate and widely-used approaches, such as the Daugman iris codes [9] and the ordinal measures (OM) [24], suffer from rotational-inconsistency inherited from the sensitiveness to eye tilt. This has limited the possibilities of developing accurate and fast indexing structures for iris databases. Recently, a number of rotation-invariant feature transformations were proposed with an aim to enable iris indexing [8,21].

An indexing structure is a data structure that is used to quickly locate where an index value occurs. To perform fast identification, only the neighborhood of the query index has to be searched to reduce the search space drastically.

Unfortunately, biometric data has no natural order by which one can sort it and thus indexing biometric data is a challenging task.

Driven by the demand for large-scale biometric systems, different approaches were proposed to reduce the response time for iris identification. Daugman et al. [14] proposed a fast search algorithm based on Beacon Guided Search on iris codes using the multiple colliding segment principle, which results in low query times but has the need of a complex memory management. Mehrotra et al. [17] propose an indexing algorithm that divides the iris image into subbands, then create a histogram of transform coefficients for each subband. A key is created based on these histograms, then organized into a search tree achieving a hit rate 98.5% at a penetration rate of 41%. Mukherjee and Ross [18] proposed two indexing techniques for both iris codes and iris textures, which achieve hit rate of 84% at 30% of the search space. One of the most accurate tree based approaches was presented by Jayaraman et al. [16]. By using principal component analysis in combination with B+ trees, a hit rate of 93.2% was achieved along with a penetration rate of 66.3%.

Gadde et al. proposed a technique based on Burrows Wheeler transformation reaching a good hit rate of 99.8% while reducing the search space to 12.3% [11] on an evaluation database containing only 249 subjects. In an indexing method proposed by Rathgeb and Uhl [22], the search space could be reduced to only 3% while reaching a hit rate around 90%. This is achieved by generating 4-bit biometric keys from the iris image to use it as a start position in a Karnaugh map, however, these results go along with a high storage cost. More recently, Rathgeb et al. proposed an iris indexing approach based on Bloom filters achieving a hit rate of 93.5% at 6.2% penetration rate [20]. However, this approach uses all samples at every tree level and it requires a full tree replacement for any deletion operation in the database, with a complexity of $\mathcal{O}(N \log(N))$.

It can be noticed that accurate indexing structures suffer from computational and structural challenges. Therefore, This work focus both on the accuracy and computational efficiency of such a structure.

Fewer works addressed the promising aspect of multi-biometric indexing. Gyaourova and Ross proposed solutions for multi-biometric indexing of face and fingerprint biometrics [12,13]. They created index codes based on the comparison scores of an input with a small constant set of references. Three approaches were evaluated, namely the union and intersection of both candidate lists, as well as the concatenation of both index codes. The union of the candidate lists performed the best with a hit rate of around 99, 5% at a penetration rate of 5%, compare to around 92.5% and 90% hit rate for face and fingerprint respectively at the same penetration rate. However, this was achieved with the large index code of 256 dimensions.

3   Methodology

In a typical content-based image retrieval process, a submitted image query is analyzed for discriminant features related to color, shape, or texture. These features are compared to images in a reference database resulting in a set of similarity measures. Based on these measures, images with a certain level of similarity to the query image are retrieved.

Similarly, in biometric systems, identification describes the problem of finding the corresponding identity for a given query q in a database. To tackle this problem an $m-$nearest neighbor search (also called similarity search) is performed. Here, m points are returned for a query q, these points have the closest distance to q according to a given distance function D. Unfortunately, finding the exact solution to an m-nearest neighbor problem is not a trivial one [15], especially in higher dimensions where commonly used grid techniques fail when facing the curse of dimensionality. However, good solutions can be found by relaxing the problem. The $m-$approximate nearest neighbor problem describes the task of finding m neighbors for a fixed value $\epsilon > 0$, such that the distance between them and q is at most $(1 + \epsilon)R$, where R is the true nearest neighbor distance.

For small datasets, it is common to compare each object to the query q in order to find the most similar ones. However, this becomes infeasible for bigger datasets due to linear query cost. The solution introduced here is based on indexing. The idea is to develop an index for each entry in the database. Then, for a given query q, the index will be computed and instead of searching the whole database, only the neighborhood of the index will be searched. In short, the given query will select only a small subset of candidates, which will be compared against the query [2]. In general, a good indexing structure should meet high standards for accuracy, query and maintenance efficiency, data independence, and minimum storage cost.

In the following, our proposed approach will be presented by initially discussing the Locality-Sensitive Hashing and its derivative coupled with a rotation invariant iris representation. These algorithms will be deployed to build the highly accurate and efficient indexing approach proposed in this work.

3.1   Locality-Sensitive Hashing

The main goal of Locality-Sensitive Hashing (LSH) is to perform a projection into a lower dimensional space where if similar high-dimensional vectors are projected, their projections to the lower dimensional subspace are similar too [23]. Hence, objects are hashed to different buckets such that similar objects have a higher probability to get hashed to the same bucket than dissimilar objects. Hashing to the same value and thus to the same bucket is called *collision*.

In order to create an LSH index, the family of LSH-functions $\mathcal{H}$ is used to construct a set of hash tables which are used collectively to refer to such an index.

1. In the LSH approach k functions $h_1, h_2, \ldots, h_k$ are chosen randomly and with replacement from $\mathcal{H}$. Then every point $p \in S$ in the database is placed in the bucket with the label $g(p) = (h_1(p), h_2(p), \ldots, h_k(p))$.
2. Construct l separate hash tables with the hash functions $g_1, g_2, \ldots, g_l$ by independently performing step 1 l times.

The first step ensures that far away points are less probable to collide, however, it has similar effect on nearby points. Therefore, the second step insures that the nearby points restores a high collision probability.

LSH offers a great solution for indexing problems and it was previously used for indexing of biometric fingerprint data [3]. However, it suffers of some theoretical and practical problems:

– Due to the dependence of the parameters k and l on the size of the database N and the distance to the true nearest neighbor R, the index may need to be reconstructed with different parameters whenever N changes by a sufficient large amount or when the data characteristic changes.
– Expensive storage and preprocessing cost are given if an adequate ϵ-nearest neighbor search is required for all queries.

Practically, the first problem can be solved by setting l and k to constant values. This works well if l is large enough (e.g. $l = 10$) so that the performance does not depend crucially on l. By contrast, k is critically dependent on data characteristics and strongly affected by N. As a result k needs to be tuned carefully and re-tuned periodically.

To tackle the second problem, R can be chosen such that most of the query distances to the true nearest neighbors are captured, which avoids large number of indices, save storage and preprocessing costs. However, if that is done and since R can critically affect the index performance, the overall performance for every query will decrease [2].

3.2   LSH-Forest

So far the LSH index suffers of the need of tuning, the continuous retuning and the lack of a quality guarantee for all queries. This can be solved by some adoptions leading to LSH-Forest [2].

While in the LSH index each point p is placed in a bucket with label $g(p) = (h_1(p), \ldots, h_k(p))$, now labels can have a variable length. The only sufficient condition is that each label must be distinct and therefore long enough. Thereby $g(p, x) = (h_1(p), \ldots, h_x(p))$ indicates to the label of the point p with length x. To avoid very long labels, which can happen by hashing very similar points, a maximum label length k_m can be introduced.

Data Structure. Instead of building hash tables as in LSH, logical prefix trees are constructed in LSH-Forest. Such a tree is called an LSH-Tree and is constructed on the set of all labels, where each leaf corresponds to a point. The whole indexing structure consists of l LSH-Trees and is called an LSH-Forest, which is constructed with independently drawn random sequences of hash functions from $\mathcal{H}$.

Query. Given an LSH-Forest consisting of l LSH-Trees created of a dataset and an $m-$NN query q. Then the query can be answered by traversing the LSH-Trees in two phases:

- *Top-Down Phase:* Descend each LSH-Tree T_i until the leaf with the highest prefix match is found. Then maintain $x := \max_i^l\{x_i\}$ as the maximum level of leaf nodes across all l trees.
- *Bottom-Up Phase:* Collect M points from the LSH-Forest by moving up from level x towards the root synchronously across all LSH-Trees. Here, M must satisfy $M \leq cl$, where c is a small constant, to ensure that there are at least m distinct points to return since M includes duplicates.

After these two phases, the M points are ranked in order of decreasing approximate similarity and the top m distinct points are returned. The approximate similarity is based on the distance between the hash labels.

Inserts. The insertions of a point p are done independently on each LSH-Tree. For this, the top-down search is performed until a leaf node p' is reached. Afterward, the labels of p and p' are extended with additional digits until they become distinct and unique leaf nodes.

Deletes. For a deletion, the top-down search is performed to reach the node, which then can be removed from the tree. Afterward, the labels of different nodes in the tree are contracted with a bottom-up traversal.

In conclusion, the LSH-Forest offers a great indexing framework. Using regular Prefix B-Trees requires only $\mathcal{O}(l \log_B(N))$ disk accesses for a query and can be parallelized easily by separating the LSH-Trees or by using range partitioning. Due to its self-tuning nature, a fluent management of the database without the need of continuous retuning can be guaranteed. Furthermore, this approach offers a probabilistic quality guarantee for all queries at the same time. It is fast, since it has a speed-up factor of around $\frac{N}{M}$ compared to exhaustive search, and every tree can be stored with linear storage costs.

3.3 Rotation Invariant Representation

To build an accurate biometric indexing structure, indices have to be extracted from discriminant and compact representations of the biometric characteristics. Iris biometrics inherently suffers from rotation inconsistency, which effect indexing efforts. Based on this, our indexing scheme will be based on the recently proposed rotation invariant, accurate, and compact RIR transformation [8]. This transformed representation will be based on OM binary features extracted from iris images [24].

The RIR transformation consists of a linear combination of two basic transformations $\hat{u}(v)$ and $\hat{v}(v)$. Given a binary iris code $v \in \{0,1\}^n$ of length n, the $\hat{u}_k$ basic transformation describes how many pairs of 0's have a distance of k within v, while $\hat{v}_k$ specifies the same for pairs of 1's. Now, the RIR transformation is defined component-wise as

$$\text{RIR}_k(v) = \hat{u}_k(v) + \hat{v}_k(v) = \sum_{i<j} \delta(d(i,j) - k)\, \delta_{v_i, v_j} \tag{1}$$

where δ is the Kronecker delta and the distance $d(i, j)$ between the ith and j^{th} location of a vector $\mathbf{v} \in \{0, 1\}^n$ is defined as

$$d(i, j) = \min \{|i - j|, n - |i - j|\}. \tag{2}$$

As a result, the kth component of RIR(v) is the number of same labeled pairs with a distance of k within v. This leads to a rotation invariant representation of the iris that enables our proposed indexing structure.

3.4   Iris Indexing

The proposed solution is based on the OM iris codes [24] and its transformation into the RIR domain [8] discussed earlier. This iris representation was deployed as an input to the LSH-Forest to create trees that contain the input labels and allow performing fast search for these labels.

Building an LSH-Forest based on RIR iris codes is performed by multiple insertions. Before each insertion, a top-down search is done to find a leaf node (with the same label as the input). This leaf node becomes an inner node for two new leaf nodes, one is the old leaf and the second is the input. Here, the labels are extended with additional digits until they become distinct. This insertion is done to create l LSH-Trees with a maximum label length of 32, thus a dynamic index length (maximum 32). The hash functions used to create the labels are generated by using the random projection principle on cosine similarity measure.

To retrieve a candidate list from an RIR based query, a top-down search is performed in each tree. From the lowest level of the trees, candidates are collected by synchronously moving upward in the trees reaching the required penetration rate (bottom-up phase). Moving upwards, duplicate candidates are neglected.

This structure allows for a search with the complexity $\mathcal{O}(l \log_2(N))$. This is highly scalable especially when compared to previous works, e.g. Gyaourova and Ross [12,13] achieved a complexity of $\mathcal{O}(N)$.

From maintenance point of view, inserting or removing a sample into the indexing structure have the same complexity as candidate retrieval. This enables the practical use of this approach in large-scale and constantly changing databases. Moreover, the proposed approach requires linear storage costs in relation to the number of entries in the database. This is achieved by storing chains of internal nodes (of degree 2) as a single node, which limits the storage requirements for each tree to $\mathcal{O}(N)$.

To extend the proposed approach to multi-instance iris indexing (left and right iris), a simple concatenation on the feature level is deployed. This is done by concatenating the RIR codes of both irises and building LSH-Forest based on the concatenated codes. This will largely improve the indexing accuracy as will be seen later in Sect. 5.

4   Experimental Setup

The ISYN1 iris synthetic images database [4–6,24,25] was used in this work to develop and evaluate the proposed solution. This database is generated by CASIA [7] using its synthetic generator software [25]. In this work, 50,000 reference and 50,000 probe iris images for each of the left and right irises were used.

The ordinal iris codes (OM) were extracted from each image in the database. These codes are transformed into a compact and rotation invariant space using the RIR transformation [8] then each element in the resulting vectors is normalized using z-score normalization.

For single iris indexing, LSH-Forest was used, as described earlier, to build an indexing structure based on the normalized RIR codes. The number of tree used in the LSH-Forest $l = 10$ and a maximum 32 labels (index dimensionality). The hash functions to create the labels are generated by using the random projection principle on cosine similarity measure.

The multi-instance iris indexing used the concatenated and normalized RIR codes of the left and right irises to build the indexing structure using LSH-Forest. As for single iris indexing, 10 LSH-trees are used and the maximum label length is 32. For further comparison, the multi-biometric indexing approach presented by Gyaourova and Ross [12,13] was implemented and evaluated based on the RIR iris codes. The implementation used an index code dimensionality of 32 for direct comparison and the index code candidate list union, as the best performing multi-biometric approach presented in [12,13].

The achieved performance is presented as a relation between the penetration rate and hit rate. Here, the hit rate is the portion of the searches where the correct identity is found within the considered percentage of the references out of the complete database (penetration rate). Small penetration rates results in smaller number of in-depth comparisons required to make the final identification decision. Beside the tradeoff between the penetration and hit rates, the computational time required for a query is presented. The achieved query time is discussed to put the efficiency of the proposed approach in prospective as this implementation does not utilize the high possibility of computational parallelization and is measured on a desktop PC running on an Intel®Core™i5-4590 3.30 GHz CPU.

5   Results

The achieved performance of the proposed single iris indexing approach is demonstrated in Fig. 1. To view the achieved accuracy in broader prospective, performances reported in previous related works are pointed out in the plot. Figure 1 also shows more details of the hit rate scored at very low penetration rates, with 99.85% hit rate at 0.4% penetration rate.

To give a prospective on the query speed of the proposed approach, Fig. 3 presents the average query time with relation to different sizes of the retrieved

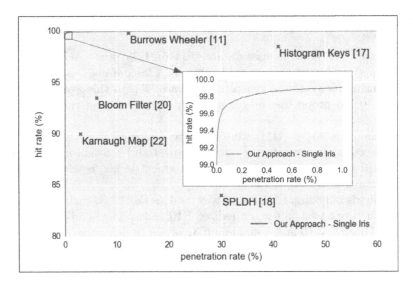

Fig. 1. Single iris indexing performance of the proposed approach in comparison to reported performances in the related work.

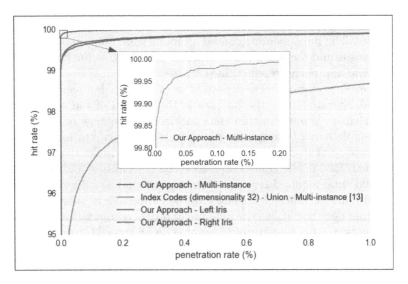

Fig. 2. Multi-instance iris indexing performance of the proposed approach in comparison to single iris indexing and multi-instance approach based on the iris codes candidate lists union [12,13].

candidate list and the reference database. As mentioned earlier, this query time is achieved on a desktop PC without any computational parallelization, which if implemented, would greatly decrease the query time. This plot indicates that a

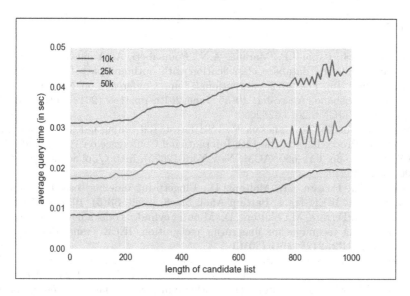

Fig. 3. Query time needed to determine a number of candidate matches in a database of 10, 25, and 50 thousand references.

query in a 50k users database would require slightly more than 0.03 s to retrieve 200 candidates (0.4% penetration rate) with a hit rate around 99.85%.

Figure 2 shows the indexing performance of the proposed multi-instance solution in comparison to single iris indexing. A comparison is also made to the multi-instance index codes candidate lists union method [12,13] applied on the same RIR features and with the same index dimensionality. The benefit of including both irises in to perform the fast search is clear by increasing the hit rate from 99.8% for the single iris to 99.99% for the multi-instance solution at 0.2% penetration rate. The proposed approach also outperforms the union of index codes indexing method. It should be mentioned that the query time for the multi-instance database does not differ significantly from the single iris one as it uses the same indexing structure.

6 Conclusion

Driven by the need to replace the impractical exhaustive search in large databases, biometric data indexing presents a chance to limit this search to a small subset of the database. This work presented an iris indexing approach based on rotation invariant iris representation and LSH-Forest. The proposed approach achieved high accuracy while maintaining logarithmic query complexity along with simple maintainability and limited storage requirements. The presented indexing structure was extended to a multi-index one, with a clear positive effect on the accuracy. An evaluation was performed on a large iris image database and the results were put into prospective by comparing to performances reported in the state-of-the-art.

References

1. Afonso, L.C.S., Papa, J.P., Marana, A.N., Poursaberi, A., Yanushkevich, S.N.: A fast large scale iris database classification with optimum-path forest technique: a case study. In: The 2012 International Joint Conference on Neural Networks (IJCNN), Brisbane, Australia, 10–15 June 2012, pp. 1–5 (2012). http://dx.doi.org/10.1109/IJCNN.2012.6252660

2. Bawa, M., Condie, T., Ganesan, P.: LSH forest: self-tuning indexes for similarity search. In: Proceedings of the 14th International Conference on World Wide Web, WWW 2005, pp. 651–660. ACM, New York (2005). http://doi.acm.org/10.1145/1060745.1060840

3. Cappelli, R., Ferrara, M., Maltoni, D.: Fingerprint indexing based on minutia cylinder-code. IEEE Trans. Pattern Anal. Mach. Intell. **33**(5), 1051–1057 (2011)

4. Cappelli, R., Ferrara, M., Maltoni, D.: Minutia cylinder-code: a new representation and matching technique for fingerprint recognition. IEEE Trans. Pattern Anal. Mach. Intell. **32**, 2128–2141 (2010)

5. Cappelli, R., Ferrara, M., Maltoni, D., Tistarelli, M.: MCC: a baseline algorithm for fingerprint verification in FVC-ongoing. In: ICARCV, pp. 19–23. IEEE (2010)

6. Cappelli, R., Maio, D., Maltoni, D.: Synthetic fingerprint-database generation. In: Proceedings of the 16th International Conference on Pattern Recognition (ICPR 2002) Volume 3, ICPR 2002, vol. 3, p. 30744. IEEE Computer Society, Washington, DC (2002). http://dl.acm.org/citation.cfm?id=839291.842894

7. Chinese Academy of Sciences, Institute of Automation (2013). http://english.ia.cas.cn/

8. Damer, N., Terhörst, P., Braun, A., Kuijper, A.: Efficient, accurate, and rotation-invariant iris code. IEEE Signal Process. Lett. **24**(8), 1233–1237 (2017). http://ieeexplore.ieee.org/document/7956157/

9. Daugman, J.: How iris recognition works. IEEE Trans. Circuits Syst. Video Technol. **14**(1), 21–30 (2004)

10. e-Aadhaar - Unique Identification Authority of India (2015). https://eaadhaar.uidai.gov.in/

11. Gadde, R.B., Adjeroh, D., Ross, A.: Indexing iris images using the burrows-wheeler transform. In: 2010 IEEE International Workshop on Information Forensics and Security, pp. 1–6, December 2010

12. Gyaourova, A., Ross, A.: A coding scheme for indexing multimodal biometric databases. In: 2009 IEEE Computer Society Conference on Computer Vision and Pattern Recognition Workshops, pp. 93–98, June 2009

13. Gyaourova, A., Ross, A.: Index codes for multibiometric pattern retrieval. IEEE Trans. Inf. Forensics Secur. **7**(2), 518–529 (2012)

14. Hao, F., Daugman, J., Zielinski, P.: A fast search algorithm for a large fuzzy database. IEEE Trans. Inf. Forensics Secur. **3**(2), 203–212 (2008). http://dx.doi.org/10.1109/TIFS.2008.920726

15. Indyk, P., Motwani, R.: Approximate nearest neighbors: towards removing the curse of dimensionality. In: Proceedings of the Thirtieth Annual ACM Symposium on Theory of Computing, STOC 1998, pp. 604–613. ACM, New York (1998). http://doi.acm.org/10.1145/276698.276876

16. Jayaraman, U., Prakash, S., Devdatt, B., Gupta, P.: An indexing technique for biometric database. In: 2008 International Conference on Wavelet Analysis and Pattern Recognition, vol. 2, pp. 758–763, August 2008

17. Mehrotra, H., Srinivas, B.G., Majhi, B., Gupta, P.: Indexing iris biometric database using energy histogram of DCT subbands. In: Ranka, S., et al. (eds.) IC3 2009. CCIS, vol. 40, pp. 194–204. Springer, Heidelberg (2009). doi:10.1007/978-3-642-03547-0_19

18. Mukherjee, R., Ross, A.: Indexing iris images. In: 19th International Conference on Pattern Recognition (ICPR 2008), 8–11 December 2008, Tampa, Florida, USA, pp. 1–4. IEEE Computer Society (2008). http://dx.doi.org/10.1109/ICPR.2008.4761880

19. Proenca, H., Alexandre, L.A.: Toward noncooperative iris recognition: a classification approach using multiple signatures. IEEE Trans. Pattern Anal. Mach. Intell. 29(4), 607–612 (2007). http://dx.doi.org/10.1109/TPAMI.2007.1016

20. Rathgeb, C., Breitinger, F., Baier, H., Busch, C.: Towards bloom filter-based indexing of iris biometric data. In: 2015 International Conference on Biometrics (ICB), pp. 422–429, May 2015

21. Rathgeb, C., Breitinger, F., Busch, C., Baier, H.: On application of bloom filters to iris biometrics. IET Biom. 3(4), 207–218 (2014). http://dx.doi.org/10.1049/iet-bmt.2013.0049

22. Rathgeb, C., Uhl, A.: Iris-biometric hash generation for biometric database indexing. In: 2010 20th International Conference on Pattern Recognition (ICPR), pp. 2848–2851, August 2010

23. Slaney, M., Casey, M.: Locality-sensitive hashing for finding nearest neighbors. IEEE Sig. Process. Mag. 25(2), 128–131 (2008). http://dx.doi.org/10.1109/msp.2007.914237

24. Sun, Z., Tan, T.: Ordinal measures for iris recognition. IEEE Trans. Pattern Anal. Mach. Intell. 31(12), 2211–2226 (2009). http://dx.doi.org/10.1109/TPAMI.2008.240

25. Wei, Z., Tan, T., Sun, Z.: Synthesis of large realistic iris databases using patch-based sampling. In: 19th International Conference on Pattern Recognition, ICPR 2008, pp. 1–4, December 2008

Machine Learning

Clustering-Based, Fully Automated Mixed-Bag Jigsaw Puzzle Solving

Zayd Hammoudeh$^{(\boxtimes)}$ and Chris Pollett

Department of Computer Science, San José State University, San José, CA, USA
{zayd.hammoudeh,chris.pollett}@sjsu.edu

Abstract. The jig swap puzzle is a variant of the traditional jigsaw puzzle, wherein all pieces are equal-sized squares that must be placed adjacent to one another to reconstruct an original, unknown image. This paper proposes an agglomerative hierarchical clustering-based solver that can simultaneously reconstruct multiple, mixed jig swap puzzles. Our solver requires no additional information beyond an unordered input bag of puzzle pieces, and it significantly outperforms the current state of the art in terms of both the reconstructed output quality as well the number of input puzzles it supports. In addition, we define the first quality metrics specifically tailored for multi-puzzle solvers, the Enhanced Direct Accuracy Score (EDAS), the Shiftable Enhanced Direct Accuracy Score (SEDAS), and the Enhanced Neighbor Accuracy Score (ENAS).

1 Introduction

The first jigsaw puzzle was introduced over 250 years ago. Despite being considered a hobby for children, puzzle solving is strongly NP-complete when inter-piece compatibility is an unreliable metric for determining adjacency [1]. Jigsaw puzzle techniques have been applied to a variety of disciplines including: archaeological artifact reconstruction [8], deleted file analysis [5], image editing [3], shredded document reconstruction [15], and DNA fragment reassembly [9].

Most recent automated puzzle solving research has focused on the jig swap puzzle, which is similar to a traditional jigsaw puzzle except that all pieces are equal-sized squares. This makes them significantly more challenging to solve since piece shape cannot be used. In addition, the original "ground-truth" solution image is generally unknown by the solver.

The jig swap puzzle problem is subclassified into three different categories based on the level of difficulty [4]. The simplest variety is the *Type 1* puzzle, which fixes piece orientation by disallowing their rotation. While the puzzle's image contents are unknown, the overall dimensions are known as well as potentially the correct location of one or more pieces. In contrast, *Type 2* jig swap puzzles allow piece rotation, which for puzzles of n pieces increases the number of possible solutions by a factor of 4^n; the dimensions for this type of puzzle may be unknown. *Mixed-bag* puzzles contain pieces from multiple input images as shown in Fig. 1. Puzzle piece orientation may be provided, but image dimensions are unknown and may vary. Most current mixed-bag solving algorithms require the specification of the number of ground-truth inputs.

© Springer International Publishing AG 2017
M. Felsberg et al. (Eds.): CAIP 2017, Part II, LNCS 10425, pp. 205–217, 2017.
DOI: 10.1007/978-3-319-64698-5_18

Mixed 6,255 Piece Input

540 Pieces, SEDAS=1   805 Pieces, SEDAS=1

805 Pieces,   805 Pieces,   3,300 Pieces, SEDAS=0.998
SEDAS=0.990   SEDAS=0.990

Fig. 1. Fully-automated mixed-bag puzzle solving: Our solver generated these results without any external information, including the number of input puzzles. The average, weighted EDAS and ENAS scores were 0.997 and 0.993 respectively.

In 2011, Pomeranz *et al.* developed a greedy, Type 1 jig swap puzzle solver that has been foundational for much of the subsequent research. They introduced the concept of *best buddies*, which are two puzzle piece sides (e.g., left, right, top, bottom) that are mutually more similar to each other than they are to any other piece's side. Pomeranz *et al.* also defined multiple test datasets, some of which are used in this paper.

Paikin and Tal [11] advanced the current state of the art in 2015 with their greedy solver that supports both missing pieces and mixed-bag puzzles. Their approach has two primary limitations. First, the solver must be provided the number of ground-truth inputs. In addition, seed piece selection is based on very localized information (i.e., only 13 pieces), which often results in poor runtime decisions such as multiple puzzles spawning from the same ground-truth input. These suboptimal selections can catastrophically degrade solution quality.

This paper's primary contribution is a novel, clustering-based, mixed-bag puzzle solver that significantly outperforms the state of the art both in terms of solution quality and the number of supportable puzzles. Unlike previous work, our approach requires no externally supplied, "oracle" information including the number of ground-truth inputs.

In addition, previously proposed, single-puzzle-solver performance metrics [2] are unusable for mixed-bag puzzles since they do not account for the presence of pieces from different images in a single output nor for the dispersion of one

---

Algorithm 1. The Mixed-Bag Solver

1: **function** MIXEDBAGSOLVER(*pieces*)
2:     *segments* ← SEGMENTATION(*pieces*)
3:     *overlap_matrix* ← STITCH(*segments, pieces*)
4:     *clusters* ← CLUSTER(*segments, overlap_matrix*)
5:     *seeds* ← SELECTSEEDS(*clusters*)
6:     *solved_puzzles* ← FINALASSEMBLY(*seeds, pieces*)
7:     **return** *solved_puzzles*

---

input's pieces across multiple outputs. As such, we introduce the first quality metrics for mixed-bag puzzles. We also enhance an existing metric to correct for the potential to be misleadingly punitive when puzzle dimensions are unknown.

2   Overview of the Mixed-Bag Solver

Humans commonly solve jigsaw puzzles by correctly assembling subregions and then iteratively merge those smaller reconstructions to form larger ones. This strategy forms the basis of our *Mixed-Bag Solver* shown in Algorithm 1. Its only input is the combined bag of pieces. The number of puzzles, their dimensions, and piece orientation are all unknown.

The first Mixed-Bag Solver stage identifies disjoint sets of pieces (i.e., segments) where there is strong confidence of correct placement. Next, the solver quantifies inter-segment relationships via the stitching process; agglomerative hierarchical clustering uses these quantified similarity scores to group related segments. Each resulting segment cluster represents what the solver identified as a single ground-truth input. A seed piece is selected from each cluster for use in the final assembly stage, which generates the reconstructed puzzle output(s).

Although not shown in Algorithm 1, the Mixed-Bag Solver requires a placer, which organizes (i.e., places) the individual pieces. Our architecture is independent of the specific placer used, granting it significant flexibility. For all experiments in this paper, we used the placer algorithm proposed by Paikin and Tal [11] as it is the current state of the art and due to its multiple puzzle support.

3   Segmentation

Segmentation provides basic structure to the unordered bag of pieces by partitioning it into disjoint, ordered sets, known as *segments*, which are partial puzzle assemblies where there is a high degree of confidence of correct piece placement.

Segmentation is performed across one or more rounds. Initially, pieces have no segment assignment. In each round, all unassigned pieces are assembled together as though they belong to the same input image as shown in Fig. 2; this eliminates the need to make any assumptions regarding the number of input puzzles. Once the pieces have been placed, the single, reconstructed puzzle is segmented as

(a) Ground-Truth   (b) Reconstruction as a
       Images   Single Puzzle   (c) Segmented Output

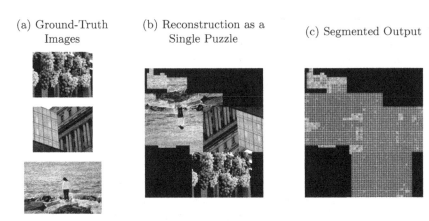

Fig. 2. Segmentation example: Three ground-truth inputs of two different sizes are shown in (a). All pieces are placed in the single, reconstructed output puzzle in (b). Segmented output in (c) is shown with any contiguous group of matching colored pieces belonging to the same segment. Stitching pieces are denoted with a white "+" mark.

described in Algorithm 2, which is partially based on the approach originally proposed by Pomeranz *et al.* in [13].

Segments in the single, reconstructed output are found iteratively, with all pieces eventually assigned to a single segment. Each segment's growth starts by adding one *seed* piece from the *unassigned* pool to an empty queue. Pieces are popped from the queue and added to the current, expanding segment. If the popped piece's neighbor in the reconstructed output is both in *unassigned* and also its best buddy, then that neighbor is added to the queue. A segment's growth terminates once no pieces remain in the queue to be popped.

As mentioned previously, two puzzle pieces, p_i and p_j, are *best buddies* on their respective sides, s_x and s_y, if they mutually more similar to each other than they are to a side, s_z, of any other piece, p_k. Given a metric, C, that quantifies inter-piece similarity, we define the best buddy relationship as:

$$\forall p_k \neq p_j \forall s_z, C(p_i, s_x, p_j, s_y) > C(p_i, s_x, p_k, s_z)$$
$$\text{and} \tag{1}$$
$$\forall p_k \neq p_i \forall s_z, C(p_j, s_y, p_i, s_x) > C(p_j, s_y, p_k, s_z).$$

This approach differs slightly from that of [11,13,14] by limiting best buddies to between exclusively two piece sides. This change is required because images with very low variation (e.g., those generated by a computer) often have large numbers of "best buddy cliques" that significantly degrade segmentation performance.

Correctly assembled regions from multiple ground-truth inputs commonly merge into a single segment via very tenuous linking. Our segmentation algorithm trims each segment by removing all articulation points, which is any piece whose removal increases the number of connected segment components. Also removed are any pieces disconnected from the segment's seed after articulation

---

Algorithm 2. Pseudocode for segmenting the single, reconstructed puzzle

---

```
 1: function SEGMENT(puzzle)
 2:     puzzle_segments ← {}
 3:     unassigned ← {all pieces in puzzle}
 4:     while |unassigned| > 0 do
 5:         segment ← new empty segment
 6:         seed ← next piece in unassigned
 7:         queue ← [seed]
 8:         while |queue| > 0 do
 9:             piece ← next piece in queue
10:             add piece to segment
11:             for each neighbor in NEIGHBORS(puzzle, piece) ∩ unassigned do
12:                 if ISBESTBUDDY(neighbor, piece) then
13:                     add neighbor to queue
14:                     remove neighbor from unassigned
15:         remove segment articulation pieces
16:         remove segment pieces disconnected from seed
17:         add removed pieces back to unassigned
18:         add segment to puzzle_segments
19:     return puzzle_segments
```

---

point deletion. All pieces no longer part of the segment are returned to the unassigned pool. Once this is completed, the segment is in its final form.

At the end of a segmentation round, only segments meeting a set of criteria are saved. First, all segments must exceed a minimum size. In our experiments, a minimum segment size of seven resulted in the best solution quality. If the largest segment exceeds this minimum size, it is automatically saved. Any other segment is saved if its size exceeds both the minimum and some fraction, α (where $0 < \alpha \leq 1$), of the largest segment. We found that $\alpha = 0.5$ provided appropriate balance between finding the largest possible segments and reducing segmentation's execution time.

The only change in subsequent segmentation rounds is the exclusion of all pieces already assigned to a saved segment. Segmentation terminates once either all pieces are assigned to a saved segment or when no segment in a given round exceeds the minimum savable size.

4   Identifying Related Segments

Traditional image stitching involves combining multiple overlapping photographs to form a single panoramic or higher resolution image. The Mixed-Bag Solver's *Stitching* stage uses a similar technique to identify segments that originate from the same ground-truth input.

Ground-Truth	Segment Images	Segment Grid Partitioning with Stitching Pieces	Mini-Assembly

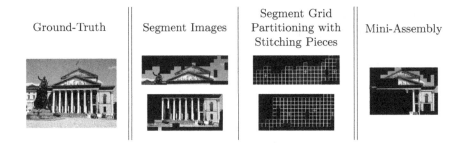

Fig. 3. An input image split into two disjoint segments that are sub-partitioned into a grid of (colored) cells. Stitching pieces are denoted with a white "+" mark. The mini-assembly, which uses a stitching piece from the upper segment, is composed of pieces from both segments (e.g., the building's roof and columns).

4.1 Stitching

Segmentation commonly partitions a single image into multiple disjoint segments. If a pair of such segments are adjacent in an original input, it is expected that they would eventually overlap if allowed to expand. A larger intersection between these two expanded segments (i.e., puzzle piece sets) indicates a stronger relationship. In contrast, if a ground-truth image consists of only a single, saved segment, then that segment generally resists growth. Since inter-segment spatial relationships, if any, are unknown by the solver, segment growth must be allowed, but never forced, to proceed in all directions.

Rather than attempt to grow a segment in its complete form, the Mixed-Bag Solver tests for localized expansion through the use of *grid cells*, which are non-overlapping subregions of a segment. These grid cells are defined by placing a bounding rectangle around the entire segment. Then starting from the upper left corner, this rectangle is partitioned into a grid of a target width (e.g., the equivalent of 10 puzzle pieces wide as used in this paper). This process is shown in Fig. 3 where an image split into two segments, both of which are further subdivided into three grid cells. If a segment's dimensions are not evenly divisible by the target width, then any grid cells along the segment's bottom and rightmost boundaries will be narrower than this ideal target.

Intuitively, it is obvious that expansion can only occur along a segment's edges. This is done by focusing on those grid cells that contain at least one piece next to an *open location*, which is any puzzle slot not occupied by a member of the segment including both the segment's external perimeter and any internal voids. For each such grid cell, localized expansion is done via a *mini-assembly* (MA). Unlike traditional placement, the MA places only a fixed number of pieces (e.g., 100 for all experiments in this paper). This placement size partially dictates the solver's inter-segment relationship sensitivity.

The MA's placement seed is referred to as a *stitching piece* and must be a member of the candidate grid cell. The selection of an appropriate stitching piece is critical; for example, if a piece too close to a boundary is selected, erroneous coupling with unrelated segments may occur. As such, the algorithm finds the set of pieces, if any, within the candidate grid cell whose distance to the nearest open location equals a predefined target (we used a distance of 3 for our experiments). If no pieces satisfy that distance criteria, the target value is decremented until at least one satisfying piece is identified. Then from this pool of possible stitching pieces, the one closest to the grid cell's center is used for stitching.

By selecting the stitching piece closest to the grid cell's center, the solver is able to enforce an approximate maximum inter-stitching piece spacing. This ensures that stitching pieces are not too far apart, which would hinder the detection of subtle inter-segment relationships. It also prevents multiple near-identical mini-assemblies, caused by stitching pieces being too close together, that contribute little added value.

4.2   Quantifying Inter-Segment Relationships

A mini-assembly is performed for each stitching piece, ζ_x, in segment, Φ_i, where $\zeta_x \in \Phi_i$. If the mini-assembly output, MA_{ζ_x}, is composed of pieces from multiple segments, there is a significantly increased likelihood that those segments come from the same ground-truth input.

Equation (2) defines the overlap score between a segment, Φ_i, and any other segment, Φ_j. The intersection between mini-assembly output, MA_{ζ_x}, and segment Φ_j is normalized with respect to the size of both, since the smaller of the two dictates the maximum possible overlap. Also, this score must use the maximum intersection across all of the segment's mini-assemblies as two segments may only be adjacent along a small portion of their boundaries.

$$Overlap_{\Phi_i,\Phi_j} = \max_{\zeta_x \in \Phi_i} \frac{|MA_{\zeta_x} \cap \Phi_j|}{\min(|MA_{\zeta_x}|, |\Phi_j|)} \tag{2}$$

Each segment generally has different mini-assembly outputs, meaning the overlap scores for each permutation of segment pairs is usually asymmetric. All overlap scores are combined into the m by m, square *Segment Overlap Matrix*, whose order, m, is the total number of saved segments.

5   Segment Clustering and Final Assembly

After stitching, the solver performs agglomerative hierarchical clustering of the saved segments to determine the number of ground-truth inputs. This necessitates that the overlap matrix be triangularized into the *Cluster Similarity Matrix*. Each element, $\omega_{i,j}$, in this new matrix represents the similarity (bounded between 0 and 1 inclusive) of segments, Φ_i and Φ_j; it is calculated via:

$$\omega_{i,j} = \frac{Overlap_{\Phi_i,\Phi_j} + Overlap_{\Phi_j,\Phi_i}}{2}. \tag{3}$$

In each clustering round, the two most similar segment clusters, Σ_x and Σ_y, are merged if their similarity exceeds a specified threshold. Based on a dozen random samplings of between two to five images from the dataset in [12], we observed a minimum similarity of 0.1 provided the best clustering accuracy.

Inter-cluster similarity, $r_{x \cup y, z}$, with respect to any other remaining segment cluster, Σ_z, is updated according to the single-linkage paradigm as shown in Eq. (4), wherein the similarity between any pair of clusters equals the similarity of their two most similar members. Solely the maximum similarities are considered as two clusters may only be adjacent along two of their member segments. The number of segment clusters remaining at the end of hierarchical clustering is the Mixed-Bag Solver's estimate of the ground-truth input count.

$$r_{x \cup y, z} = \max_{\Phi_i \in (\Sigma_x \cup \Sigma_y)} \left(\max_{\Phi_j \in \Sigma_z} \omega_{i,j} \right). \qquad (4)$$

Some modern jigsaw puzzle placers including [11,13,14] use a kernel-growing technique. If the placer used by the Mixed-Bag Solver requires this additional step, we select a single seed from each segment cluster. This approach leads to better seed selection since most other placers make their seed decisions either randomly or greedily at runtime. Once this is completed, final piece placement begins simultaneously across all puzzle seeds. The resulting fully reconstructed puzzles, with all pieces placed, are the Mixed-Bag Solver's final outputs.

6   Quality Metrics for Mixed-Bag Puzzles

The direct and neighbor accuracy metrics for quantifying the quality of single puzzle reconstructions were defined in [2] and used by [4,11,13,14]. However, both measures are unusable for mixed-bag puzzles since neither account for two complications unique to this problem, specifically that pieces from multiple ground-truth inputs may be placed in the same generated output and that pieces from a single input image can be spread across different outputs [7].

6.1   Enhanced and Shiftable Direct Accuracy

Puzzle solving involves generating a set of output puzzles, S, from a set of inputs, P. Each $P_i \in P$ is composed of n_i pieces. $c_{i,j}$ is the number of pieces in the same location in both P_i and output, $S_j \in S$. In contrast, $m_{i,j}$ is the total number of pieces from P_i in S_j, making $0 \leq c_{i,j} \leq m_{i,j} \leq n_i$.

Standard direct accuracy (where $|P| = |S| = 1$ and $n_1 = m_{1,1}$) is the fraction of pieces that are correctly placed in the reconstructed output. It is defined as:

$$DA = \frac{c_{1,1}}{n_1}. \qquad (5)$$

A solved image is *perfectly reconstructed* if the location of all pieces exactly match the original image (i.e., $DA = 1$) [4].

Our *Enhanced Direct Accuracy Score* (EDAS) in Eq. (6) addresses standard direct accuracy's deficiencies for mixed-bag puzzles in three primary ways. First, since pieces from P_i may be in multiple reconstructed puzzles, EDAS uses the maximum score across all of S so as to focus on the best overall reconstruction of P_i. Second, dividing by n_i marks as incorrect any piece from P_i that is not in S_j. Lastly, the summation of all $m_{k,j}$ penalizes for the placement of any pieces from inputs other than P_i.

$$EDAS_{P_i} = \max_{S_j \in S} \frac{c_{i,j}}{n_i + \sum_{k \neq i}(m_{k,j})} \tag{6}$$

Both standard and enhanced direct accuracy can be misleadingly punitive for shifts in the output, in particular when the solved puzzle's boundaries are not fixed/known. As noted in [7], even a single misplaced piece can cause these metrics to drop to zero. Direct accuracy more meaningfully quantifies output quality if the comparison reference location, l, is allowed to shift within a fixed set of possible locations, L_j, in S_j. As such, our *Shiftable Enhanced Direct Accuracy Score* (SEDAS) in Eq. (7) updates the term $c_{i,j}$ to $c_{i,j,l}$ to denote the use of this variable reference when determining the correctly placed piece count.

$$SEDAS_{P_i} = \max_{S_j \in S} \left(\max_{l \in L_j} \frac{c_{i,j,l}}{n_i + \sum_{k \neq i}(m_{k,j})} \right) \tag{7}$$

For this paper, L_j was the set of all puzzle locations within the radius defined by the Manhattan distance between the upper left corner of S_j and the nearest puzzle piece, inclusive. An alternative approach is for L_j to be the set of all locations in S_j, but that can be computationally prohibitive for large puzzles.

6.2 Enhanced Neighbor Accuracy

Standard neighbor accuracy (where $|P| = |S| = 1$) is the fraction of puzzle piece sides with the same neighbors in both the input and output puzzles. If $a_{i,j}$ is the number of puzzle piece sides with matching neighbors in both P_i and S_j, then for square pieces, this single-puzzle metric is formally defined as:

$$NA = \frac{a_{1,1}}{4n_1}. \tag{8}$$

Similar to the reasons described for EDAS, our *Enhanced Neighbor Accuracy Score* (ENAS), which is defined as:

$$ENAS_{P_i} = \max_{S_j \in S} \frac{a_{i,j}}{4(n_i + \sum_{k \neq i}(m_{k,j}))} \tag{9}$$

addresses standard neighbor accuracy's limitations for mixed-bag puzzles. Neighbor accuracy is immune to shifts [2] making a shiftable version of it unnecessary.

Table 1. Number of solver experiments for each puzzle input count

#Puzzles	2	3	4	5
#Iterations	55	25	8	5

7 Experimental Results

Our experiments followed the standard puzzle parameters established by previous work including [2,4,11,13,14]. All of the square puzzle pieces were 28 pixels wide. We also used the three, 20 image datasets of sizes 432, 540, and 805 pieces from [2,10,12]. Only the more challenging Type 2 mixed-bag puzzles were investigated, meaning piece rotation and puzzle(s) dimensions were unknown.

The current state of the art, Paikin and Tal's algorithm, was used as the comparative performance baseline. In each test, two to five images were randomly selected, without replacement, from the 805 piece dataset [12] and input into the two solvers. Table 1 shows the number of tests performed for each input count.

7.1 Determining the Number of Input Puzzles

Most previous solvers including [2,11,13,14] either assumed or were provided the number of input images. In contrast, the Mixed-Bag Solver determines this information via hierarchical clustering.

Clustering a Single Input Image: The solver's accuracy determining the number of inputs when passed only a single image represents its overall performance ceiling. For the 432 [2], 540 [10], and 805 piece [12] datasets, the solver's accuracy determining that the pieces came from a single puzzle was 100%, 80%, and 85% respectively. While there was a degradation in performance for larger puzzles, it was not significant. In all cases where an error was made, the solver reported that there were two input images.

Input puzzle count errors are more likely for images with large areas of little variation (e.g., a clear sky, smooth water, etc.). These incorrectly classified images have on average lower numbers of best buddies (by 8% and 12% for the 540 and 805 piece datasets respectively), which adversely affected segmentation.

Clustering Multiple Input Images: Figure 4 shows the Mixed-Bag Solver's performance identifying the number of input puzzles when randomly selecting, without replacement, multiple images from the 805 piece dataset. The number of input images was correctly determined in 65% of tests. Likewise, the solver overestimated the number of inputs by more than one in less than 8% tests, with a maximum overestimation of three. Across all experiments, it never underestimated the input puzzle count. This indicates the solver can over-reject cluster mergers due to clusters being too isolated to merge with others.

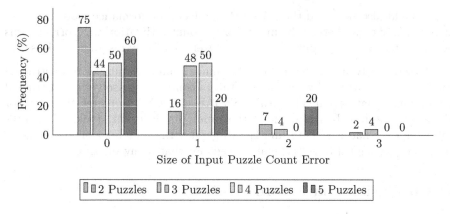

Fig. 4. Multiple input puzzle clustering Accuracy: A correct estimation of the input puzzle count is an error of "0." An overestimation of a single puzzle is an error of "1."

7.2 Comparison of Solver Output Quality for Multiple Input Images

Table 2 contains the comparative results when both solvers were supplied multiple input images. The values for each of the three metrics, namely SEDAS, ENAS, and percentage of puzzles reassembled perfectly, are averaged. The Mixed-Bag Solver (MBS) results are subdivided between when the number of input puzzles was correctly determined (denoted with a "†") versus all combined results ("‡"); the former value represents the performance ceiling had our solver been provided the input puzzle count like Paikin and Tal's algorithm.

Despite receiving less information, the quality of our results exceeded that of Paikin and Tal by between 2.5 to 8 times for SEDAS and up to four times ENAS. The Mixed-Bag Solver was also substantially more likely to perfectly reconstruct the images. Furthermore, unlike Paikin and Tal, our algorithm showed no significant performance degradation as the number of input puzzles increased.

Table 2. Solver performance comparison for multiple input puzzles. Results with "†" indicate the Mixed-Bag Solver (MBS) correctly estimated the input puzzle count while "‡" values include all MBS results.

Puzzle count	Average SEDAS			Average ENAS			Perfect reconstruction		
	MBS†	MBS‡	Paikin	MBS†	MBS‡	Paikin	MBS†	MBS‡	Paikin
2	0.850	0.757	0.321	0.933	0.874	0.462	29.3%	23.6%	5.5%
3	0.953	0.800	0.203	0.955	0.869	0.364	18.5%	18.8%	1.4%
4	0.881	0.778	0.109	0.920	0.862	0.260	25.0%	15.6%	0%
5	0.793	0.828	0.099	0.868	0.877	0.204	20.0%	24%	0%

It should also be noted the Mixed-Bag Solver's performance scores are similar irrespective of whether the input puzzle count estimation was correct. This indicates that any extra puzzles generated were relatively insignificant in size.

Ten Puzzle Solving: The previous maximum number of puzzles reconstructed simultaneously was five by Paikin and Tal. In contrast, our solver reconstructed the 10 image dataset in [6], with a SEDAS and ENAS greater than 0.9 for all images. Despite being provided the input puzzle count, Paikin and Tal's algorithm only had a SEDAS and ENAS greater than 0.9 for a single image as their solver struggled to select quality seeds for that many puzzles.

8   Conclusion and Future Work

We presented an algorithm for simultaneous reassembly of multiple jig swap puzzles without prior knowledge. Despite the current state of the art requiring specification of the input puzzle count, our approach still outperforms it in terms of both the output quality and the supportable number of input puzzles.

Potential improvements to our solver remain that merit further investigation. First, rather than performing segmentation through placement, it may be faster and yield better, larger segments if the entire set of puzzle pieces were treated as nodes in an undirected graph with edges being the best buddy relationships. This would enable segment identification through the use of well-studied graph partition techniques. In addition, our approach requires that stitching pieces be members of a saved segment. Superior results may be achieved if pieces not assigned to a segment are also used, as they may help bridge inter-segment gaps.

References

1. Altman, T.: Solving the jigsaw puzzle problem in linear time. Appl. Artif. Intell. **3**(4), 453–462 (1990)
2. Cho, T.S., Avidan, S., Freeman, W.T.: A probabilistic image jigsaw puzzle solver. In: CVPR, pp. 183–190 (2010)
3. Cho, T.S., Butman, M., Avidan, S., Freeman, W.T.: The patch transform and its applications to image editing. In: CVPR, pp. 1489–1501 (2008)
4. Gallagher, A.C.: Jigsaw puzzles with pieces of unknown orientation. In: CVPR, pp. 382–389 (2012)
5. Garfinkel, S.L.: Digital forensics research: The next 10 years. Digit. Invest. **7**, S64–S73 (2010)
6. Hammoudeh, Z.S.: Ten puzzle dataset. http://www.cs.sjsu.edu/faculty/pollett/masters/Semesters/Spring16/zayd/?10_puzzles.html
7. Hammoudeh, Z.S.: A Fully Automated Solver for Multiple Square Jigsaw Puzzles Using Hierarchical Clustering. Master's thesis, San José State University (2016)
8. Koller, D., Levoy, M.: Computer-aided reconstruction and new matches in the forma urbis romae. Bullettino Della Commissione Archeologica Comunale di Roma **2**, 103–125 (2006)
9. Marande, W., Burger, G.: Mitochondrial DNA as a genomic jigsaw puzzle. Science **318**(5849), 415 (2007)

10. Olmos, A., Kingdom, F.A.A.: McGill calibrated colour image database. http://tabby.vision.mcgill.ca/

11. Paikin, G., Tal, A.: Solving multiple square jigsaw puzzles with missing pieces. In: CVPR, pp. 4832–4839 (2015)

12. Pomeranz, D., Shemesh, M., Ben-Shahar, O.: Computational jig-saw puzzle solving. https://www.cs.bgu.ac.il/~icvl/icvl_projects/automatic-jigsaw-puzzle-solving/

13. Pomeranz, D., Shemesh, M., Ben-Shahar, O.: A fully automated greedy square jigsaw puzzle solver. In: CVPR, pp. 9–16 (2011)

14. Sholomon, D., David, O., Netanyahu, N.S.: A genetic algorithm-based solver for very large jigsaw puzzles. In: CVPR, pp. 1767–1774 (2013)

15. Zhu, L., Zhou, Z., Hu, D.: Globally consistent reconstruction of ripped-up documents. Trans. Pattern Anal. Mach. Intell. **30**, 1–13 (2008)

A Violence Detection Approach Based on Spatio-temporal Hypergraph Transition

Jingjia Huang, Ge Li[✉], Nannan Li, Ronggang Wang, and Wenmin Wang

School of Electronic and Computer Engineering, Shenzhen Graduate School,
Peking University, Lishui Road 2199, Nanshan District, Shenzhen 518055, China
jjhuang@pku.edu.cn, {geli,wangwm}@ece.pku.edu.cn,
{linn,rgwang}@pkusz.edu.cn

Abstract. In the field of activity recognition, violence detection is one of the most challenging tasks due to the variety of action patterns and the lack of training data. In the last decade, the performance is getting improved by applying local spatio-temporal features. However, geometric relationships and transition processes of these features have not been fully utilized. In this paper, we propose a novel framework based on spatio-temporal hypergraph transition. First, we utilize hypergraphs to represent the geometric relationships among spatia-temporal features in a single frame. Then, we apply a new descriptor called Histogram of Velocity Change (HVC), which characterizes motion changing intensity, to model hypergraph transitions among consecutive frames. Finally, we adopt Hidden Markov Models (HMMs) with the hypergraphs and the descriptors to detect and localize violence in video frames. Experiment results on BEHAVE dataset and UT-Interaction dataset show that the proposed framework outperforms the existing methods.

Keywords: Violence detection · Action recognition · Hypergraph · Spatio-temporal feature · HMM

1 Introduction

Violence detection is one of the most challenging works in video processing. Considerable efforts have been done for its essential applications in video surveillance and smart camera systems. Since violence is a kind of aggressive interaction among multiple people with an unstable movement patterns, the major challenge for violence detection is to distinguish violent events from other normal human group activities. Therefore, in literature, violence detection has been considered to be a problem of abnormal detection [10,11,21] or group activity recognition [12,13].

In order to automatically characterize violence in videos, some impressive works have been done. Most of these methods are based on spatio-temporal features [10,11,14,15,21] of human objects. Spatio-temporal features are always extracted from the areas called spatial temporal interest points (STIP), proposed by Laptev and Lindeberg [16]. Spatio-temporal features are able to efficiently

© Springer International Publishing AG 2017
M. Felsberg et al. (Eds.): CAIP 2017, Part II, LNCS 10425, pp. 218–229, 2017.
DOI: 10.1007/978-3-319-64698-5_19

retain the information of motion variation in spatio-temporal domain. However, few researches concern a full expression of the relationships among the detected features. More specifically, there is a lack of expression on geometric relationships as well as the transition of these features. Geometric relationship is the representation of a posture in a single frame. Transition description is the representation of the posture variation process in consecutive frames. We then propose a framework based on spatio-temporal hypergraph transition to improve the performance of violence detection. On one hand, we propose an effective hypergraph model which can represent the concrete geometric relationships of the spatio-temporal features. On the other hand, we propose a descriptor called Histogram of Velocity Changing (HVC) to express the transition process among consecutive hypergraphs. HVC can efficiently distinguish normal actions such as walking, meeting and hugging, which have a relatively stable as well as gentle movement patterns, from violence activities such as fighting which change drastically. The major contributions of our work are listed in the following:

(1) Introduce an innovative hypergraph to model the intra-frame relationships of local spatio-temporal features.
(2) Propose a novel descriptor called HVC to model the transition of spatio-temporal hypergraphs in consecutive frames for violence detection.
(3) Adopt a new spatio-temporal hypergraph transition based framework, which concatenates the hypergraph model and HVC to a Hypergraph-Transition Chain (H-T Chain) based on a redundant trajectory extraction algorithm.

Experiment results on UT-Interaction dataset and BEHAVE dataset demonstrate that the performance of our method outperforms the existing methods.

2   Related Work

Action recognition has been extensively studied in the last decade [1,4–6]. However, violence detection is still a challenge problem due to the variety of patterns and the lack of training data. According to different application environments, various approaches are developed to solve the problem.

On the purpose of violence detection in movies, most of the works in the literature use audio features as an additional resource to represent the video elements [17]. [18] is one of the first proposals to characterize and index violent scenes in general TV drama and movies, in which characteristic sounds of violent events were used.

However, in surveillance systems, which is a more general application, people encounter the problems of no audio signal, low video resolution and relatively small scale of objects in a vast majority of cases. Therefore, many researchers have made efforts to figure out more suitable approaches to handle the problems. Hassner et al. propose a low-level descriptor for violence detection in crowded scenes called Violence Flow based on the magnitude changes of optical flow over successive frames [19]. Based on socio-psychological studies, Helbing et al. in [20] originally introduce a Social Force model to investigate the pedestrian movement dynamics. In the model, they treat the moving particles as individuals. How the individuals react to energy potentials caused by others and static obstacles through a repulsive force, is considered as an important clue for crowds activity recognition. Inspired by this work, Mehran *et al.* obtain Force Flow from Social Force Model and apply it to violence detection [11]. Based on STIP [16] trajectory, Cui *et al.* evaluate the anomaly in crowd with an interaction energy potential function [10]. Similarly to Cui *et al.*'s work, the approach proposed in [21] consider trajectories too. Hossein *et al.* develop a spatio-temporal descriptor called Histograms of Oriented Tracklets, which simultaneously captures the magnitude and orientation information of a set of STIP trajectories passing through a spatio-temporal volume.

Ryoo and Aggarwal [12] have demonstrated that the similarity between two videos can be well reflected by the structural similarity between sets of features extracted from them. On the other hand, graph is an effective tool for modeling complex structured data [15]. Consequently, some researchers adopt graphs to represent the structure of features in their works [14,15,22–24]. Brende and Sinisa propose a volumetric-based approach to learn spatiotemporal graphs of activities from videos [22]. In [14], Wu *et al.* construct two directed and attributed graphs based on intra-frame relationships and inter-frame relationships of the local features. Similarly, Aoun *et al.* construct Spatial Frame Graphs and Temporal Video Graphs for action recognition in [15]. Yi and Lin [23] construct undirected spatio-temporal graphs for each detected instance and attribute the spatio-temporal interactive relationship to the edges. However, all of them [14,15,23] are second-order matching methods which are limited to the affinity matrix embedding pairwise relationships between feature points with unary information, where the pairwise relationships are rotation-invariant but not scale-invariant as well as affine-invariant. Comparing to them, hypergraph based methods are more expressive with better integration of geometric information [25]. Therefore, Ta *et al.* propose a model for recognizing and localizing human activities through hypergraph matching in [24]. In this paper, we also apply hypergraphs in our work. However, Ta *et al.'s* work is aiming at the recognition of individual activities while ours is able to analysis interactions among multiple people. Furthermore, we not only apply the posture information represented by hypergraphs but also the transition process between consecutive hypergraphs. Hypergraphs and corresponding HVCs are arranged in a H-T Chain to simulate the motion sequences.

3   Methodology

We propose a new framework to model group interactions for violence detection based on spatio-temporal hypergraph transition. Figure 1 is an illustration to the framework. We first extract STIPs and track them. Meanwhile, we use a foreground segmentation algorithm to detect the moving objects and assign STIPs to respective segmentations. Then, we construct hypergraphs for the segmentations and extract the trajectories of them. Transition descriptors are calculated

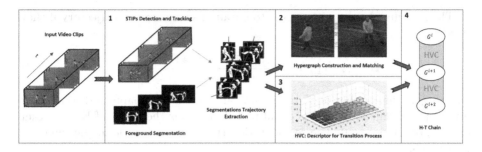

Fig. 1. A review for our framework. **Step 1:** pre-processing and trajectory extraction. **Step 2:** hypergraph model construction. **Step 3:** HVC descriptor construction. **Step 4:** H-T Chain construction based on the trajectory extracted in step 1. HMM is used to deal with the chains.

between consecutive hypergraphs in the trajectory. Finally, for each trajectory, a H-T Chain is constructed and Hidden Markov Model (HMM) is used to detect violent events.

3.1   Pre-processing: Coarse Trajectory Extraction

During violence detection, interactions of a group of people instead of the individuals are the major concern, such as a group of people in fighting. Thus, a coarse object detection and tracking pre-processing step are employed in our frame work firstly. This will provide two benefit: (1) It helps to maintain the geometric information of the group. (2) It helps to analysis the motion transition in the following steps. As a result, we develop a simple but robust object trajectory extraction algorithm based on STIPs tracking in the framework.

As described in [16], Space-Time Interest Point (STIP) is an extension of the Harris corner detection operator to spacetime which is able to efficiently retain the information of motion variation in spatio-temporal domain. We use Laptev's release of STIP code for detection task and KLT tracker to track interest points. Then, we use Gaussian Mixture Model to get a set of foreground segmentations in tth frame:

$$S_t = \{s_{t,i}\}(i = 1...n) \tag{1}$$

where n is the number of segmentations in tth frame and the $s_{t,i}$ is

$$s_{t,i} = \{p_{t,j}\}(j = 1...m) \tag{2}$$

where $p_{t,j}$ is the jth STIPs in the segmentation and m is the total number of STIPs in it.

Comparing to the methods based on overlapping spatial sectors or sliding windows which break the original spatial relationships of STIPs [12,19,21], our method clusters the STIPs belonging to the same segmentation as one analysis unit to benefit the further hypergraph construction.

Then we match s_{t,i_1} with s_{t+1,i_2} to obtain a segmentation trajectory if they satisfy:

$$\psi(s_{t,i_1}, s_{t+1,i_2}) = \frac{\sum_{j_1,j_2} \phi(p_{t,j_1}, p_{t+1,j_2})}{m_1} \geq thd \qquad (3)$$

where m_1 is the number of STIPs in s_{t,i_1} and thd is a probability threshold for objects matching. $\phi(p_{j_1}, p_{j_2})$ is equal to 1 if p_{t+1,j_2} is the result of KLT tracker for p_{t,j_1}, otherwise to 0. The segmentation trajectory is for the further analysis in Sect. 3.3.

3.2   Constructing and Matching Hypergraph Model

A hypergraph is a generalization of a graph in which an edge can join any number of vertices. Hypergraph is initially introduced to computer vision area for its ability of encoding higher order geometric invariants such as scale and affine invariants. In this paper, we use hypergraph to represent the posture information for segmentations by modeling the intra-frame relationships of STIPs. We define a hypergraph for a segmentation s_i as:

$$G^i = (V^i, E^i, F^i) \qquad (4)$$

where V^i is a set of vertices $i.e$ STIPs belonging to G^i. E^i is a set of hyperedges correspond to a 3-tuple of vertices. F^i is a set of feature vectors correspond to each vertices.

We consider two segmentations s_1 and s_2, and assume that N_1 and N_2 are the number of STIPs tracked in s_1 and s_2, respectively. Then we defined the hypergraph model for them as G^1 and G^2. A matching between G^1 and G^2 is equivalent to looking for an $N_1 \times N_2$ assignment matrix X such that $X_{i,j}$ is equal to 1 when p_i in G^1 matched to p_j in G^2, and to 0 otherwise. We follow the constrain in [26] that one node in G^1 can be matched to exactly one node in G^2 but no constrain to the nodes in G^2. Thus, we can get a set of assignment matrices:

$$A = \{X \in \{0,1\}^{N_1 \times N_2}, \Sigma_i X_{i,j} = 1\} \qquad (5)$$

The measurement of similarity between two graphs can be formulated as the maximization of the following score on A:

$$score(A) = \sum_{i,i',j,j',k,k'} H_{i,i',j,j',k,k'} X_{i,i'} X_{j,j'} X_{k,k'} \qquad (6)$$

where $H_{i,i',j,j',k,k'}$ is a similarity measure for the sets of features between hyperedges $E = \{i, j, k\}$ and $E' = \{i', j', k'\}$. The higher the value, the greater the similarity is. We utilize Duchenne $et\ al.$'s tensor-based power-iteration algorithm [26] for the optimization problem.

In order to adapt the hypergraph model to our work, we construct the similarity measurement $H_{i,i',j,j',k,k'}$ as follows. First, to measure the geometric similarity of two hypergraphs, we use the properties of the triangle formed by three

STIPs. We denote a vector of the sine of the 3-tuple computed from their spatial coordinates as a_{ijk}:

$$a_{ijk} = [sin(\overrightarrow{ij}, \overrightarrow{ik}), \; sin(\overrightarrow{ji}, \overrightarrow{jk})] \qquad (7)$$

Then, averaged optical flows, extracted from 3D patches around the corresponding STIPs, are used as motion descriptors which are denoted as F earlier in this section. As a result, we get the feature vector for a hyperedge E as:

$$f_{E_{ijk}} = [a_{ijk}, \; F_i, \; F_j, \; F_k] \qquad (8)$$

and calculate the similarity score with a Gaussian kernel:

$$H_{i,i',j,j',k,k'} = exp\{-\frac{\|f_{E_{ijk}} - f_{E_{i'j'k'}}\|_{l_2}}{\sigma}\} \qquad (9)$$

where σ is the parameter of the kernel which governs the intra class variations of the features.

For the further analysis in the following section, we construct a codebook for the hypergraphs based on the proposed hypergraph similarity measurement. Spectral cluster is used for the unsupervised clustering.

3.3 Transition Description

If we consider the intra-frame relationships of local spatio-temporal features modeled with hypergraph as the postures in a motion sequence, then the transition of hypergraphs refer to the changing process from one posture to the next one. We produce Histogram of HVC for the transition description. A formation process is shown in Fig. 2. In Sect. 3.1 we extract trajectories of segmentations in which multiple STIP trajectories existed. The HVC computation for the transition from G^k to G^{k+1} starts with the portions of STIP trajectories passing through these two graphs. We denote this portion of the trajectory as $tr = \{p_1, ...p_i, ...p_n\}$, where i indicates the ith point in the STIP trajectory and

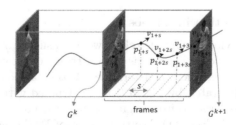

Fig. 2. Formation process for HVC. HVC for the transition from G^k to G^{k+1} is extracted from the portion of STIPs trajectory (red curve) passing through them. ν_i is the optical flow on the STIP p_i and s is the step length for velocity sampling. (Color figure online)

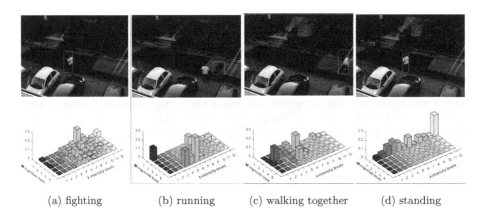

(a) fighting   (b) running   (c) walking together   (d) standing

Fig. 3. HVC examples for fighting, running, walking together and standing instances from BEHAVE dataset.

n is the number of frames between G^k and G^{k+1}. Then, we define the intensity of velocity changing from p_i to p_{i+s} as:

$$I_{i,i+s} = \frac{\|\nu_i - \nu_{i+s}\|_{l_2}}{\|\nu_i\|_{l_2}} \tag{10}$$

where ν_i is the optical flow of p_i and s is the step length of sampling. Considering to get a more smooth intensity changing process, the sampling range can be expanded to G^{k-1} and G^{k+2} dependently. Besides, we compute the average velocity magnitudes of the sub-trajectory $\overrightarrow{p_i p_{i+s}}$ as:

$$M_{i,i+s} = \frac{\sum_{t=i}^{i+s} \|\nu_t\|_{l_2}}{s+1} \tag{11}$$

Finally, the intensities and magnitudes of all the sub-trajectories involving in the trajectories passing through G^k and G^{k+1} are quantized in I intensities and M magnitudes bins, respectively. Each of the sub-trajectories gives a contribution to the HVC histogram. Some HVC examples for different activities from BEHAVE dataset are shown in Fig. 3.

3.4   Violence Detection in Videos

In order to locate the violence in video clips, we use a sliding observation window to traverse the whole video. For segmentation trajectories within the window, a HMM is used as a classifier to determine whether a violence exist or not. A HMM sequence illustrated by Fig. 4(2) is constructed with hypergraphs and corresponding HVCs extracted from segmentation trajectories within the observation window shown in Fig. 4(1). We train a set of HMM models for different events including one for violence and the others for normal events. Given a behavior, we calculate probabilities of the HMM models, and we declare the behavior as a violence if the violence HMM model has the highest probability among all the HMMs.

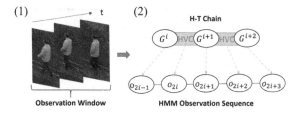

Fig. 4. (1) a motion sequence in an observation window. In (2), the top layer of graph is the H-T Chain for the motion sequence where G^i is the ith hypergraph in it. The second layer in (2) is the observation sequence for HMMs where o_{2i} is the $2i$th observation in it. The arrows, which connect two layers, indicate the correspondence between elements in the H-T Chain and observations for HMMs.

4 Experiments and Analysis

4.1 The Behave Dataset

Dataset. BEHAVE is a video dataset for multi-person behaviour analysis. There are a set of complex group activities annotated in the video, including meeting, splitting up, loitering, walking together, escaping, fighting and other interaction behaviours.

Parameters. We fixed the $thd = 0.2$ for trajectory extraction in Eq. 3. For the hypergraphs, spectral cluster is used and a codebook with 25 vocabularies is constructed. For the HVC, we fixed $M = 5$ and $I = 12$. Meanwhile, a codebook with 15 vocabularies is trained for HVC. To construct a H-T Chain for a motion sequence, we set the size for observation window as 80 frames while the step length is 20 frames. For the portion of segmentation trajectory within the window, the interval size between two hypergraphs is set to 3 frames and the sampling range for the HVC is 9 frames with a step length of 3 frames.

Evaluation. The detection task of BEHAVE dataset is a violence/non-violence detection. The comparison methods include the optical flow based method, social force model [11] and interaction energy potential [10]. Following settings in [10], we used half of normal and abnormal videos for training and the rest for testing. According to the past practice, the results are evaluated by the means of ROC as shown in Fig. 5.

Analysis. According to the ROC curves in Fig. 5, our method outperforms the Interaction Energy Potentials [10], Social Force Model [11] and Optical Flow, which demonstrates that our framework is competitive with the existing methods on BEHAVE dataset.

4.2 The UT-Interaction Dataset

Dataset. The UT-Interaction dataset includes two sets. UTI[#1] was captured on a parking lot with mostly static background, while UTI[#2] was captured on a

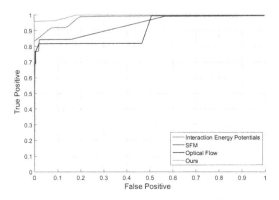

Fig. 5. ROC curves on BEHAVE dataset. Our method (green line) is compared with Interaction Energy Potentials [10], Social Force Model [11] and Optical Flow. (Color figure online)

lawn with slight background dynamics and camera jitters. Both the sets contain 60 videos of 6 categories of human interactions: push, kick, punch, shake hands, point and hug(10 videos for each category).

Parameters. We keep the same settings for UTI$^{\#1}$ and UTI$^{\#2}$. We fixed the $thd = 0.3$ for trajectory extraction in Eq. 3. Since the events to recognize in UT dataset have a relatively short duration, we set the size of observation window as the length of each clips. For the other parameters, We follow the same setting as that in the experiment for BEHAVE.

Evaluation. We evaluate the performance of our framework via the leave-one-out cross validation strategy proposed in [12], *i.e*, for each category, 9 sequences of segmented videos (54 training videos in total) are used for training and the remaining one video (6 testing videos in total) is used for testing. The confusion matrixes of our recognition results on UTI$^{\#1}$ and UTI$^{\#2}$ are shown in Fig. 6(a) and (b). Besides, since our method address on the problem of violence detection, we also perform the violence/non-violence recognition experiments on this dataset. First, we divide the action categories into two groups, where push, kick and punch are labeled as violent actions; while shake hands, point and hug are labeled as non-violent actions. Such classification is shown with different colors (gray for violent group and blue for non-violent group) in Fig. 6. Then, for each group, we choose the maximum log probability of the three actions' HMMs as the score for the entire group. For instance, if a clip to recognize has the log probabilities of HMMs for kick, push and punch as -0.43, -1.30, and -2.71, the score of the violent group is selected as -0.43 (the maximum one). Finally, we classify the actions as violence or non-violence via the normalized group scores. Table 1 shows our experiment results comparing to the approaches of Ryoo and Aggarwal [12], [7], Ryoo [9], Ke [3], Yu and Yun [2] and Xu *et al.* [8]. In the table, Accuracy$^{\#1}$-α and Accuracy$^{\#2}$-α mean the recognition accuracy for the six

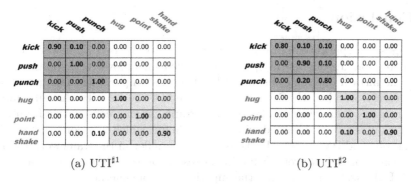

(a) UTI[#1]   (b) UTI[#2]

Fig. 6. Confusion matrix of our method on UT-Interaction dataset (Color figure online)

Table 1. Experiments results on UT-Interaction dataset. '-' means no corresponding experiment results.

Methods	Accuracy[#1]-α	Accuracy[#1]-β	Accuracy[#2]-α	Accuracy[#2]-β
Ryoo and Aggarwal [12]	70.80%	-	-	-
MSSC [7]	83.33%	-	81.67%	-
Ryoo [9]	85.00%	-	70.00%	-
Ke [3]	93.33%	-	-	-
Yu and Yun [2]	93.33%	96.67%	**91.67%**	95.00%
Xu *et al.* [8]	**96.67%**	96.67%	90.00%	-
Ours	**96.67%**	**98.33%**	90.00%	**100.00%**

categories of actions in UTI[#1] and UTI[#2]; Accuracy[#1]-β and Accuracy[#2]-β are the violence/non-violence recognition accuracy in the two sets.

Analysis. On the dataset, [12] is the benchmark for action recognition task. As it illustrated in column 1 and 3 of Table 1, for action recognition task, our method achieves the same performance as state-of-the-art methods [8] on UTI[#1] and has a competitive performance on UTI[#2]; while for the violence/non-violence recognition task, which are presented in column 2 and 4, we outperform state-of-the-art methods on both sets. We notice that though [2] has a better performance than ours on UTI[#2], we get a higher accuracy for violence recognition. It is because that [2] is confused by hug and punch that belong to non-violence and violence group respectively. Thanks to HVC, which is sensitive to the motion changing intensity, our method is able to distinguish the gentle movement from vigorous movement. Therefore, less confusion exists between non-violent and violent actions in our methods. The results demonstrate that our method achieves superior performance over these comparison methods.

5   Conclusion

We proposed a spatio-temporal hypergraph transition based framework for violence detection. In our method, We introduced an innovative hypergraphs to model the intra-frame relationships of local spatio-temporal features. Besides, a novel descriptor called HVC is proposed to model the velocity changing intensity during the transition process of hypergraphs among consecutive frames. A motion sequence is represented with a H-T Chain, which was constructed with hypergraphs and HVCs. HMM was used as a classifier in the framework to indicate violent/nonviolent event. Experiment results on UT-Interaction dataset and BEHAVE dataset demonstrated the superiority of our method.

Acknowledgments. This work was supported by National Science Foundation of China (No. U1611461), National Natural Science Foundation of China (61602014), Shenzhen Peacock Plan (20130408-183003656), and Science and Technology Planning Project of Guangdong Province, China (No. 2014B090910001).

References

1. Wang, L., Xiong, Y., Wang, Z., Qiao, Y., Lin, D., Tang, X., Gool, L.: Temporal segment networks: towards good practices for deep action recognition. In: Leibe, B., Matas, J., Sebe, N., Welling, M. (eds.) ECCV 2016. LNCS, vol. 9912, pp. 20–36. Springer, Cham (2016). doi:10.1007/978-3-319-46484-8_2
2. Kong, Y., Yun, F.: Close human interaction recognition using patch-aware models. IEEE Trans. Image Process. Publ. IEEE Signal Process. Soc. **25**, 167–178 (2015)
3. Ke, O., Bennamoun, M., An, S., Boussaid, F., Sohel, F.: Human interaction prediction using deep temporal features. In: 2016 European Conference on Computer Vision (2016)
4. Du, T., Bourdev, L., Fergus, R., Torresani, L., Paluri, M.: Learning spatiotemporal features with 3D convolutional networks. In: IEEE International Conference on Computer Vision, pp. 4489–4497 (2015)
5. Wang, L., Qiao, Y., Tang, X.: Action recognition with trajectory-pooled deep-convolutional descriptors. In: 2015 IEEE Conference on Computer Vision and Pattern Recognition, pp. 4305–4314 (2015)
6. Zhang, B., Wang, L., Wang, Z., Qiao, Y., Wang, H.: Real-time action recognition with enhanced motion vector CNNs. In: 2016 IEEE Conference on Computer Vision and Pattern Recognition (2016)
7. Lan, T., Chen, T.-C., Savarese, S.: A hierarchical representation for future action prediction. In: Fleet, D., Pajdla, T., Schiele, B., Tuytelaars, T. (eds.) ECCV 2014. LNCS, vol. 8691, pp. 689–704. Springer, Cham (2014). doi:10.1007/978-3-319-10578-9_45
8. Xu, Z., Qing, L., Miao, J.: Activity auto-completion: predicting human activities from partial videos. In: International Conference on Computer Vision, pp. 3191–3199 (2015)
9. Ryoo, M.S.: Human activity prediction: early recognition of ongoing activities from streaming videos. In: 2011 International Conference on Computer Vision, pp. 1036–1043, November 2011

10. Cui, X., Liu, Q., Gao, M., Metaxas, D.N.: Abnormal detection using interaction energy potentials. In: The IEEE Conference on Computer Vision and Pattern Recognition, CVPR 2011, 20–25 June 2011, pp. 3161–3167. Colorado Springs Co, USA, June 2011

11. Mehran, R., Oyama, A., Shah, M.: Abnormal crowd behavior detection using social force model. In: IEEE Conference on Computer Vision and Pattern Recognition, pp. 935–942 (2009)

12. Ryoo, M.S., Aggarwal, J.K.: Spatio-temporal relationship match: video structure comparison for recognition of complex human activities. In: IEEE International Conference on Computer Vision, pp. 1593–1600 (2009)

13. Blunsden, S.J., Fisher, R.B.: The BEHAVE video dataset: ground truthed video for multi-person. Ann. BMVA **4**, 1–11 (2009)

14. Wu, B., Yuan, C., Hu, W.: Human action recognition based on context-dependent graph kernels. In: IEEE Conference on Computer Vision and Pattern Recognition, pp. 2609–2616 (2014)

15. Ben Aoun, N., Mejdoub, M., Ben Amar, C.: Graph-based approach for human action recognition using spatio-temporal features. J. Vis. Commun. Image Represent. **25**(2), 329–338 (2014)

16. Laptev, I., Lindeberg, T.: On space-time interest points. Int. J. Comput. Vision **64**, 107–123 (2005)

17. De Souza, F.D.M., Chavez, G.C., Do Valle, E.A., De A. Araujo, A.: Violence detection in video using spatio-temporal features. In: 2012 Proceedings of the 25th SIBGRAPI Conference on Graphics, Patterns and Images, pp. 224–230 (2010)

18. Nam, J.H., Alghoniemy, M., Tewfik, A.H.: Audio-visual content-based violent scene characterization. In: Proceedings of the International Conference on Image Processing, ICIP 1998, pp. 353–357 (1998)

19. Hassner, T., Itcher, Y., Kliper-Gross, O.: Violent flows: real-time detection of violent crowd behavior. In: Computer Vision and Pattern Recognition Workshops, pp. 1–6 (2012)

20. Helbing, D., Molnár, P.: Social force model for pedestrian dynamics. Phys. Rev. E Stat. Phys. Plasmas Fluids Relat. Interdiscip. Top. **51**(5), 4282–4286 (1995)

21. Mousavi, H., Galoogahi, H.K., Perina, A., Murino, V.: Detecting abnormal behavioral patterns in crowd scenarios. In: Esposito, A., Jain, L.C. (eds.) Toward Robotic Socially Believable Behaving Systems - Volume II. ISRL, vol. 106, pp. 185–205. Springer, Cham (2016). doi:10.1007/978-3-319-31053-4_11

22. Brendel, W., Todorovic, S.: Learning spatiotemporal graphs of human activities. In: IEEE International Conference on Computer Vision, pp. 778–785 (2011)

23. Yi, Y., Lin, M.: Human action recognition with graph-based multiple-instance learning. Pattern Recogn. **53**(C), 148–162 (2016)

24. Ta, A.P., Wolf, C., Lavou, G., Baskurt, A.: Recognizing and localizing individual activities through graph matching. In: Seventh IEEE International Conference on Advanced Video and Signal Based Surveillance, pp. 196–203 (2010)

25. Park, S., Park, S., Hebert, M.: Fast and scalable approximate spectral matching for higher order graph matching. IEEE Trans. Pattern Anal. Mach. Intell. **36**(3), 479–492 (2014)

26. Duchenne, O., Bach, F., In So, K., Ponce, J.: A tensor-based algorithm for high-order graph matching. IEEE Trans. Pattern Anal. Mach. Intell. **33**(12), 2383–95 (2011)

Blur Parameter Identification Through Optimum-Path Forest

Rafael G. Pires[1], Silas E.N. Fernandes[1], and João Paulo Papa[2(✉)]

[1] Department of Computing, Federal University of São Carlos (UFSCar),
Rodovia Washington Luís, Km 235 - SP 310, São Carlos, SP 13565-905, Brazil
{rafael.pires,silas.fernandes}@dc.ufscar.br
[2] Department of Computing, São Paulo State University (Unesp),
Av. Eng. Luiz Edmundo Carrijo Coube, 14-01, Bauru, SP 17033-360, Brazil
papa@fc.unesp.br

Abstract. Image acquisition processes usually add some level of noise and degradation, thus causing common problems in image restoration. The restoration process depends on the knowledge about the degradation parameters, which is critical for the image deblurring step. In order to deal with this issue, several approaches have been used in the literature, as well as techniques based on machine learning. In this paper, we presented an approach to identify blur parameters in images using the Optimum-Path Forest (OPF) classifier. Experiments demonstrated the efficiency and effectiveness of OPF when compared against some state-of-the-art pattern recognition techniques for blur parameter identification purpose, such as Support Vector Machines, Bayesian classifier and the k-nearest neighbors.

Keywords: Image restoration · Machine learning · Optimum-Path Forest

1 Introduction

During the image acquisition process, some level of noise is usually added to the real data mainly due to physical limitations of the acquisition sensor, and also regarding imprecisions during the data transmission and manipulation. Therefore, the resultant image needs to be processed in order to attenuate its noise without loosing details present at high frequencies regions, being the field of image processing that addresses such issue called "image restoration".

However, one of the main problems in image restoration is to restore the image details smoothed by the blurring process (the image can get blurred due to the sensor's movement, lens defocusing and physical limitations during the image acquisition process), which is modeled by the point spread function (PSF), but with the compromise of keeping the noise at acceptable levels. Such concern has oriented the development of iterated image restoration techniques, in which the amount of image restoration can be controlled among the iterations until some convergence criterion is met [1].

© Springer International Publishing AG 2017
M. Felsberg et al. (Eds.): CAIP 2017, Part II, LNCS 10425, pp. 230–240, 2017.
DOI: 10.1007/978-3-319-64698-5_20

A classical problem in image restoration is to obtain a suitable estimation of the blur parameters encoded by the PSF. In this context, one can face two distinct problems: *blind deconvolution* and *image deconvolution*. While the former deals with the problem of smoothing the noise, but without any kind of prior knowledge about the blur model, the latter approach makes use of the knowledge concerning the blurring process. Roughly speaking, both approaches belong to the well-known *image denoising* research area [2]. Usually, three blur models are often addressed in image restoration, since most part of the problems are related to them: defocus, Gaussian and motion blur.

In the last years, one can find a number of image restoration and deblurring techniques such as inverse and Wiener filter regularization, projection-based [3–5] and *Maximum a Posteriori* probability techniques [6]. Despite of machine learning being a well consolidated research field dating back to the 60's, only in the last years it has been employed to address the problem of image restoration. Zhou et al. [7], Paik e Katsaggelos [8] and Sun and Yu [9] are among the first ones to propose image restoration by means of neural networks. Later on, the number of machine learning-driven works related to image restoration has increased considerably [10–12]. Deep learning techniques, for instance, have been considered a page-turner due to the outstanding results in a number of computer vision-related problems, such as face and object recognition, just to name a few. Some works have addressed the problem of image restoration using such approaches, such as Tang et al. [13], which proposed the Robust Boltzmann Machine (RoBM), that usually allows Boltzmann Machines to be more robust to image corruptions. The model is trained in an unsupervised fashion with unlabeled noisy data, and it can learn the spatial structure of the occluders. Recently, Zhang et al. [14] employed the well-known Support Vector Machines (SVMs) and neural networks for image denoising, and Dalong et al. [15] modeled the problem of blind deconvolution by means of Support Vector Regression.

In short, the main idea is to model the problem of blur parameter identification as a pattern recognition task, in which phantom images are blurred with different parameters to design a labeled training set, being the main task to identify the blur parameters in a set of unseen images. Basically, one can perform two distinct approaches concerning image deconvolution by means of machine learning techniques: (i) given a specific blur model, to identify its parameters (*blur parameter identification*), or (ii) given a blurred image, to identify its blur formulation among some models learned from a set of training images (*blur identification*). This paper focuses on the first approach.

Some years ago, Papa et al. [16–19] proposed the Optimum-Path Forest (OPF) classifier, which is a graph-based technique that models the problem of pattern recognition as a graph partition task, being the dataset samples encoded as graph nodes, and connected to each other by means of an adjacency relation. Roughly speaking, OPF rules a competition process among some key samples (*prototypes*) in order to partition the graph into optimum-path trees rooted at each prototype node. The competition process concerns offering to the non-prototype samples optimum-path costs (a sort or *reward*), being a sample associated to the same label of its conqueror (the competition process starts at the prototype nodes, which can not be conquered by any node). The OPF has demonstrated promising results in a number of applications, being usually similar to SVMs, but faster for training, since it is parameterless and does not assume any kind of feature space separability.

However, to the best of our knowledge, OPF has never been applied to the context of blur parameter identification so far. Therefore, the main contribution of this paper is to evaluate OPF effectiveness for blur parameter identification in natural images against some well-known supervised machine learning techniques, such as SVMs and the k-nearest neighbors (k-NN) classifier. The remainder of this paper is organized as follows. Section 2 presents a brief theoretical background about OPF-based classification, and Sect. 3 states the methodology employed in this work. Sections 4 and 5 discuss the experimental results and conclusions, respectively.

2   Optimum-Path Forest

The OPF framework is a recent highlight to the development of pattern recognition techniques based on graph partitions. The nodes are the data samples, which are represented by their corresponding feature vectors, and are connected according to some predefined adjacency relation. Given some key nodes (prototypes), they will compete among themselves aiming at conquering the remaining nodes. Thus, the algorithm outputs an optimum path forest, which is a collection of optimum-path trees (OPTs) rooted at each prototype. This work employs the OPF classifier proposed by Papa et al. [17,18], which is explained in more details as follows.

Let $\mathcal{D} = \mathcal{D}_1 \cup \mathcal{D}_2$ be a labeled dataset, such that $\mathcal{D}_1$ and $\mathcal{D}_2$ stands for the training and test sets, respectively. Let $\mathcal{S} \subset \mathcal{D}_1$ be a set of prototypes of all classes (i.e., key samples that best represent the classes). Let $(\mathcal{D}_1, A)$ be a complete graph whose nodes are the samples in $\mathcal{D}_1$, and any pair of samples defines an arc in $A = \mathcal{D}_1 \times \mathcal{D}_1$ Additionally, let π_s be a path in $(\mathcal{D}_1, A)$ with terminus at sample $s \in \mathcal{D}_1$.

The OPF algorithm proposed by Papa et al. [17,18] employs the path-cost function $f_{\max}$ due to its theoretical properties for estimating prototypes (Sect. 2.1 gives further details about this procedure):

$$f_{\max}(\langle s \rangle) = \begin{cases} 0 & \text{if } s \in \mathcal{S} \\ +\infty & \text{otherwise,} \end{cases}$$
$$f_{\max}(\pi_s \cdot \langle s, t \rangle) = \max\{f_{\max}(\pi_s), d(s, t)\}, \tag{1}$$

where $d(s, t)$ stands for a distance between nodes s and t, such that $s, t \in \mathcal{D}_1$. Therefore, $f_{\max}(\pi_s)$ computes the maximum distance between adjacent samples in π_s, when π_s is not a trivial path. In short, the OPF algorithm tries to minimize $f_{\max}(\pi_t), \forall t \in \mathcal{D}_1$.

2.1   Training Phase

We say that S^* is an optimum set of prototypes when minimizes the classification errors for every $s \in \mathcal{D}_1$. We have that S^* can be found by exploiting the theoretical relation between the minimum-spanning tree and the optimum-path tree for $f_{\max}$ [20]. The training essentially consists of finding S^* and an OPF classifier rooted at S^*. By computing a minimum spanning tree (MST) in the complete graph $(\mathcal{D}_1, A)$, one obtain a connected acyclic graph whose nodes are all samples of $\mathcal{D}_1$ and the arcs are undirected and weighted by the distances d between adjacent samples. In the MST, every pair of samples is connected by a single path, which is optimum according to $f_{\max}$. Hence, the minimum-spanning tree contains one optimum-path tree for any selected root node.

The optimum prototypes are the closest elements of the MST with different labels in $\mathcal{D}_1$ (i.e., elements that fall in the frontier of the classes). By removing the arcs between different classes, their adjacent samples become prototypes in S^*, and can define an optimum-path forest with minimum classification errors in $\mathcal{D}_1$.

Classification Phase. For any sample $t \in \mathcal{D}_2$, we consider all arcs connecting t with samples $s \in \mathcal{D}_1$, as though t were part of the training graph. Considering all possible paths from S^* to t, we find the optimum path $P^*(t)$ from S^* and label t with the class $\lambda(R(t))$ of its most strongly connected prototype $R(t) \in S^*$. This path can be identified incrementally, by evaluating the optimum cost $C(t)$ as follows:

$$C(t) = \min_{\forall s \in \mathcal{D}_1} \{\max\{C(s), d(s, t)\}\}. \tag{2}$$

Let the node $s^* \in \mathcal{D}_1$ be the one that satisfies Eq. 2 (i.e., the predecessor $P(t)$ in the optimum path $P^*(t)$). Given that $L(s^*) = \lambda(R(t))$, the classification simply assigns $L(s^*)$ as the class of t. An error occurs when $L(s^*) \neq \lambda(t)$.

Another interesting point to be considered concerns with the relation between OPF and the nearest neighbor classifier (NN). Although OPF uses the distance between samples to compose the cost to be offered to them, the path-cost function encodes the power of connectivity of the samples that fall in the same path, being much more powerful than the sole distance. Therefore, this means OPF is not a distance-based classifier. Additionally, Papa et al. [17] showed that OPF is quite different than NN, being those techniques exactly the same only when all training samples become prototypes. The complexity for the OPF training step is given by $\theta(|\mathcal{D}_1|^2)$, while the complexity for classification is given by $O(p|\mathcal{D}_2|)$, in which $p \in O(|\mathcal{D}_1|)$, i.e., $0 < p \leq |\mathcal{D}_1|$.

3   Methodology

In this section, we present the methodology employed to evaluate the robustness of OPF classifier for blur parameter identification.

3.1   Blur Identification as a Pattern Recognition Task

According to Dash et al. [21], if one considers the variance over blurred image patches according to different severities of blur, we shall notice a variability of such variance values. This means the greater the blur severity, the more homogeneous the image becomes (smaller variance). However, the variance value itself is not sufficient to guarantee a good discriminative power among different blur severity models. Similarly to the work of Dash et al. [21], we used the variance of blurred image patches as the criterion to select the ones that will to compose the final dataset. Therefore, each sample (blurred patch) whose variance is greater than a threshold T will be represented by the brightness of its 8-neighborhood system (3×3 patches obtained without overlapping), as depicted in Fig. 1.

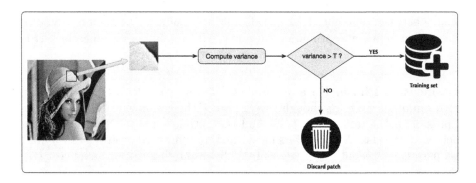

Fig. 1. Methodology used to design the dataset samples.

3.2   Experimental Setup

In order to fulfil the main goal of this work, we employed two distinct blur models, i.e., motion and Gaussian blur. While the former has the *motion length* (L) as the main parameter (we fixed the direction of the motion blur to 45°), the latter model has the *variance* (σ) of the Gaussian distribution as the sole parameter. Figures 2a and 3a display the original images used in this work.

The Lena image is motion-blurred (MB) with four different blur lengths $S \in \{1, 15, 30, 45\}$ (Figs. 2b–e), being the $14,447$ samples taken by considering a patch of size 3×3 in which the variance is larger than $T = 0.04$. The Cameraman image is degraded with Gaussian blur (GB) and $\sigma \in \{1, 4, 7, 10\}$ (Figs. 3b–e), being the $13,209$ samples from the blurred image taken by considering the same patch size and variance value of Lena image[1]. Therefore, the feature vector for each sample (patch) is represented by the brightness of the pixels contained on it.

In regard to the pattern recognition techniques, we compared OPF against SVM with Radial Basis Function, a Bayesian classifier (BAYES), and k-NN for

---

[1] All these ranges for both L and σ were empirically chosen.

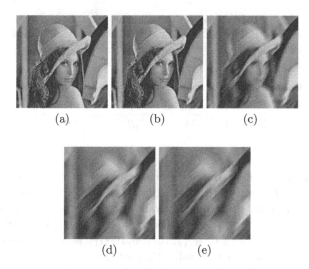

Fig. 2. (a) Original Lena image, and (b) blurred images with motion blur length = 1, (c) blur length = 15, (d) blur length = 30, (e) and blur length = 45.

Fig. 3. (a) Original Cameraman image, and (b) blurred images with Gaussian blur $\sigma = 1$, (c) $\sigma = 4$, (d) $\sigma = 7$, and (e) $\sigma = 10$.

blur parameter identification. The experiments were conducted over four classification datasets, whose main characteristics are presented in Table 1. Notice each class stands for a different blur parameter, and the features are obtained over a 3×3 path[2].

---

[2] The experiments were conducted on a computer with a Pentium Intel Core $i5^{®}$ 650 3.2 Ghz processor, 4 GB of memory RAM and Linux Ubuntu Desktop LTS 12.04 as the operational system.

Table 1. Description of the datasets.

Dataset	# samples	# features	# classes
Cameraman (MB)	13, 191	9	4
Cameraman (GB)	13, 209	9	4
Lena (MB)	14, 447	9	4
Lena (GB)	14, 403	9	4

For each dataset, we conducted a cross-validation procedure with 20 runnings, being each of them partitioned as follows: 40% of the samples were used to compose the training set, being the validation and testing sets ranged from 10%–50%, 20%–40%, ..., 40%–20%. Both k-NN ($k \in \{3, 5, 7, 9\}$) and SVM ($\gamma \in \{0.001, 0.01, 0.1, 1\}$ and $C \in \{1, 10, 100, 1000\}$) were optimized through a cross-validation procedure over a validation set using a grid-search. As BAYES and OPF are parameterless, they were trained using both the training and validating sets (i.e., with the merged sets). These percentages have been empirically chosen, being more intuitive to provide a larger validation set to fine-tune SVM and k-NN parameters.

In addition, we conducted an experiment where Cameraman image was used to train the techniques, for further classification of the Lena image, as well as the opposite situation (we denote such experiments as "cross-training"). The mean computational load was also considered, being the k-NN and SVM execution times computed together with their fine-tuning parameter step.

4   Experiments

In this section, we present the experimental results concerning the task of blur parameter identification. Table 2 presents the mean accuracy rates over the test set considering all classification techniques, being such results the average over all possible configurations of validating and test sets. These results were evaluated through the Wilcoxon signed-rank test with significance of 0.05 [22].

With respect to the Table 2, we can draw three important conclusions: (i) OPF classifier has been much more accurate than all compared techniques, (ii) clearly, the cross-training experiment seemed to add an extra complexity when learning the blur parameter models, and (iii) OPF appeared to be less sensitive to the cross-training procedure when compared to the remaining techniques.

Table 3 shows the mean computational load of all compared techniques with respect to the training step. The values in bold stand for the faster techniques concerning the Wilcoxon signed-rank test. For k-NN and SVM, the training time includes the training phase + learning step to fine-tune parameters. Clearly, OPF has been much faster than SVM technique, since their bottleneck concerns the fine-tuning step. However, the BAYES approach has been considered the fastest one in all of datasets concerning the training phase.

Table 2. Mean accuracy results: the bolded values stand for the most accurate techniques according Wilcoxon test. The recognition rates were computed according to Papa et al. [17], which consider unbalanced datasets.

Dataset	BAYES	k-NN	OPF	SVM
Cameraman (MB)	29.72 ± 0.56	42.06 ± 0.93	$\mathbf{62.37 \pm 0.64}$	49.88 ± 1.17
Cameraman (GB)	28.16 ± 0.62	40.32 ± 0.81	$\mathbf{61.02 \pm 0.61}$	43.29 ± 1.31
Lena (MB)	29.56 ± 0.50	41.02 ± 0.79	$\mathbf{61.94 \pm 0.58}$	51.43 ± 1.78
Lena (GB)	30.10 ± 0.50	44.67 ± 0.67	$\mathbf{63.24 \pm 0.53}$	50.53 ± 1.35
Train Cameraman-Test Lena (MB)	30.82 ± 0.00	31.95 ± 0.40	$\mathbf{54.27 \pm 0.07}$	41.37 ± 0.44
Train Cameraman-Test Lena (GB)	28.85 ± 0.00	29.45 ± 0.35	$\mathbf{53.47 \pm 0.12}$	35.87 ± 0.79
Train Lena-Test Cam (MB)	29.61 ± 0.00	34.68 ± 0.46	$\mathbf{57.31 \pm 0.07}$	43.50 ± 0.86
Train Lena-Test Cam (GB)	28.28 ± 0.00	29.35 ± 0.26	$\mathbf{53.28 \pm 0.15}$	33.38 ± 0.87

Table 3. Mean computational load (in seconds) with respect to the training time.

Dataset	BAYES	k-NN	OPF	SVM
Cameraman (MB)	$\mathbf{0.004 \pm 0.001}$	0.422 ± 0.228	3.623 ± 1.225	390.85 ± 285.35
Cameraman (GB)	$\mathbf{0.004 \pm 0.001}$	0.283 ± 0.102	3.507 ± 1.187	77.352 ± 52.496
Lena (MB)	$\mathbf{0.005 \pm 0.001}$	0.429 ± 0.217	4.293 ± 1.461	427.77 ± 326.43
Lena (GB)	$\mathbf{0.005 \pm 0.001}$	0.298 ± 0.104	4.362 ± 1.477	99.799 ± 70.646
Train Cameraman-Test Lena (MB)	$\mathbf{0.006 \pm 0.001}$	0.391 ± 0.112	8.283 ± 0.059	142.31 ± 42.982
Train Cameraman-Test Lena (GB)	$\mathbf{0.006 \pm 0.000}$	0.340 ± 0.072	7.986 ± 0.083	136.39 ± 59.032
Train Lena-Test Cam (MB)	$\mathbf{0.007 \pm 0.001}$	0.416 ± 0.148	9.903 ± 0.067	226.96 ± 67.303
Train Lena-Test Cam (GB)	$\mathbf{0.006 \pm 0.001}$	0.393 ± 0.113	10.00 ± 0.079	171.40 ± 90.530

In regard to the testing phase, Table 4 presents the mean computational load in seconds with respect to the testing phase. As one can observe, the BAYES approach has been considered the fastest one concerning all datasets for the testing phase. In addition, the non-parametric Friedman test was performed to rank the algorithms for each dataset separately. In case of Friedman test provides meaningful results to reject the null-hypothesis (h_0: all techniques are equivalent), we can perform a post-hoc test further. For this purpose, we conducted the Nemenyi test, proposed by Nemenyi [23] and described by Demšar [24], which allows us to verify whether there is a critical difference (CD) among techniques or not. The results of the Nemenyi test can be represented in a simple diagram, in which the average ranks of the methods are plotted on the horizontal axis, where the lower average rank is, the better the technique is. Moreover, the groups with no significant difference are then connected with a horizontal line.

Figure 4 depicts the statistical analysis considering the accuracy results for all classification techniques. As one can observe, the OPF approach can be considered the most accurate one. Such point reflects the OPF technique achieved the best accuracy rates in all datasets. Figures 5a and b depict the statistical analysis considering the training (training+validating) and testing time with the Nemenyi test, respectively. On average, a comparison between OPF and SVM showed the

Table 4. Mean computational load (in seconds) with respect to the testing time.

Dataset	BAYES	k-NN	OPF	SVM
Cameraman (MB)	**0.004** ± 0.001	0.062 ± 0.020	1.756 ± 0.293	0.715 ± 0.240
Cameraman (GB)	**0.004** ± 0.001	0.042 ± 0.014	1.579 ± 0.277	0.818 ± 0.280
Lena (MB)	**0.005** ± 0.002	0.067 ± 0.023	2.265 ± 0.374	0.880 ± 0.321
Lena (GB)	**0.004** ± 0.001	0.053 ± 0.016	2.149 ± 0.357	0.917 ± 0.330
Train Cameraman-Test Lena (MB)	**0.012** ± 0.001	0.236 ± 0.021	10.08 ± 0.035	3.522 ± 0.583
Train Cameraman-Test Lena (GB)	**0.012** ± 0.000	0.162 ± 0.023	10.27 ± 0.047	4.003 ± 0.706
Train Lena-Test Cam (MB)	**0.011** ± 0.000	0.225 ± 0.028	9.795 ± 0.031	3.510 ± 0.595
Train Lena-Test Cam (GB)	**0.011** ± 0.000	0.161 ± 0.022	9.136 ± 0.083	3.803 ± 0.749

OPF has been about 32.19 times faster than SVM in training+validating phase, since OPF has no parameters to be fine-tuned. Even considering the SVM without grid-search, the OPF are still faster (around 5.29 times). In regard to the testing phase, the OPF stands for the slowest technique, being about 2.58 times slower than SVM.

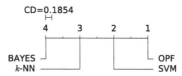

Fig. 4. Comparison of all techniques with the Nemenyi test regarding the accuracy results. Groups of techniques that are not significantly different (at p = 0.05) are connected.

(a)   (b)

Fig. 5. Nemenyi statistical test regarding the computational load for: (a) training (training+validating) and (b) testing phases. Groups that are not significantly different (at $p = 0.05$) are connected to each other.

5 Conclusions

In this paper, we introduced the OPF classifier in the context of blur parameter identification. We used two different blurring models and a cross-training approach to assess the robustness of supervised pattern recognition techniques to identify the blur parameters. The well-known Lena and Cameraman images were degraded with different blur severities in order to compose a training set with different classes, being each dataset sample represented by the brightness of the pixels that fall in 3×3 patches. The experiments showed OPF is much more accurate for blur parameter identification than all compared techniques, as well as it has a very suitable computational load.

Acknowledgment. The authors are grateful to FAPESP grants #2014/16250-9 and #2014/12236-1, CNPq grant #306166/2014-3, as well as CAPES.

References

1. Mammone, R.J.: Computational Methods of Signal Recovery and Recognition. Wiley, New York (1992)
2. Shacham, O., Haik, O., Yitzhaky, Y.: Blind restoration of atmospherically degraded images by automatic best step-edge detection. Pattern Recogn. Lett. **28**(15), 2094–2103 (2007)
3. Papa, J.P., Fonseca, L.M.G., de Carvalho, L.A.S.: Projections onto convex sets through particle swarm optimization and its application for remote sensing image restoration. Pattern Recogn. Lett. **31**(13), 1876–1886 (2010)
4. Pires, R.G., Pereira, D.R., Pereira, L.A.M., Mansano, A.F., Papa, J.P.: Projections onto convex sets parameter estimation through harmony search and its application for image restoration. Nat. Comput. **15**(3), 493–502 (2016)
5. Papa, J.P., Mascarenhas, N.D.A., Fonseca, L.M.G., Bensebaa, K.: Convex restriction sets for cbers-2 satellite image restoration. Int. J. Remote Sens. **29**(2), 443–458 (2008)
6. Katsaggelos, A.K.: Digital Image Restoration. Springer-Verlag New York, Inc., Secaucus (1991)
7. Zhou, Y.-T., Chellappa, R., Vaid, A., Jenkins, B.K.: Image restoration using a neural network. IEEE Trans. Acoust. Speech Sig. Process. **36**(7), 1141–1151 (1988)
8. Paik, J.K., Katsaggelos, A.K.: Image restoration using a modified hopfield network. IEEE Trans. Image Process. **1**, 49–63 (1992)
9. Sun, Y., Yu, S.Y.: A modified hopfield neural network used in bilevel image restoration and reconstruction. In: International Symposium on Information Theory Application, vol. 3, pp. 1412–1414 (1992)
10. Qiao, J., Liu, J.: A SVM-based blur identification algorithm for image restoration and resolution enhancement. In: Gabrys, B., Howlett, R.J., Jain, L.C. (eds.) KES 2006. LNCS, vol. 4252, pp. 28–35. Springer, Heidelberg (2006). doi:10.1007/11893004_4
11. Li, D.: Support vector regression based image denoising. Image. Vis. Comput. **27**(6), 623–627 (2009)
12. Xia, Y., Sun, C., Zheng, W.X.: Discrete-time neural network for fast solving large linear l1 estimation problems and its application to image restoration. IEEE Trans. Neural Netw. Learn. Syst. **23**(5), 812–820 (2012)

13. Tang, Y., Salakhutdinov, R., Hinton, G.: Robust boltzmann machines for recognition and denoising. In: IEEE Conference on Computer Vision and Pattern Recognition, 2012, Providence, Rhode Island, USA (2012)
14. Zhang, G.-D., Yang, X.-H., Xu, H., Lu, D.-Q., Liu, Y.-X.: Image denoising based on support vector machine. In: 2012 Spring Congress on Sarvajanik College of Engineering & Technology, pp. 1–4. IEEE (2012)
15. Li, D., Mersereau, R.M., Simske, S.: Blind image deconvolution through support vector regression. IEEE Trans. Neural Netw. **18**(3), 931–935 (2007)
16. Papa, J.P., Falcão, A.X.: A new variant of the optimum-path forest classifier. Int. Symp. Vis. Comput. **47**(1), 935–944 (2008)
17. Papa, J.P., Falcão, A.X., Suzuki, C.T.N.: Supervised pattern classification based on optimum-path forest. Int. J. Imaging Syst. Technol. **19**(2), 120–131 (2009)
18. Papa, J.P., Falcão, A.X., Albuquerque, V.H.C., Tavares, J.M.R.S.: Efficient supervised optimum-path forest classification for large datasets. Pattern Recogn. **45**(1), 512–520 (2012)
19. Papa, J.P., Fernandes, S.E.N., Falcão, A.X.: Optimum-path forest based on k-connectivity: theory and applications. Pattern Recogn. Lett. **87**, 117–126 (2017). Advances in Graph-based Pattern Recognition
20. Allène, C., Audibert, J.-Y., Couprie, M., Keriven, R.: Some links between extremum spanning forests, watersheds and min-cuts. Image Vis. Comput. **28**(10), 1460–1471 (2010)
21. Dash, R., Sa, P.K., Majhi, B.: Blur parameter identification using support vector machine. In: Proceedings of the International Conference on Advances in Computer Science, pp. 89–92 (2011)
22. Wilcoxon, F.: Individual comparisons by ranking methods. Biom. Bull. **1**(6), 80–83 (1945)
23. Nemenyi, P.: Distribution-free Multiple Comparisons. Princeton University (1963)
24. Demšar, J.: Statistical comparisons of classifiers over multiple data sets. J. Mach. Learn. Res. **7**, 1–30 (2006)

Real-Time Human Pose Estimation via Cascaded Neural Networks Embedded with Multi-task Learning

Satoshi Tanabe$^{(\boxtimes)}$, Ryosuke Yamanaka, Mitsuru Tomono, Makiko Ito, and Teruo Ishihara

FUJITSU LABORATORIES LTD., Kawasaki, Japan
tanabe.s@jp.fujitsu.com

Abstract. Deep convolutional neural networks (DCNNs) have recently been applied to Human pose estimation (HPE). However, most conventional methods have involved multiple models, and these models have been independently designed and optimized, which has led to sub-optimal performance. In addition, these methods based on multiple DCNNs have been computationally expensive and unsuitable for real-time applications. This paper proposes a novel end-to-end framework implemented with cascaded neural networks. Our proposed framework includes three tasks: (1) detecting regions which include parts of the human body, (2) predicting the coordinates of human body joints in the regions, and (3) finding optimum points as coordinates of human body joints. These three tasks are jointly optimized. Our experimental results demonstrated that our framework improved the accuracy and the running time was 2.57 times faster than conventional methods.

Keywords: Human pose estimation · Neural networks · Multi-task learning · End-to-end learning

1 Introduction

Human pose estimation (HPE) from images is a challenging task in computer vision. It predicts the coordinates of human body joints in images. It has many applications, such as those in gesture recognition, clothing parsing, and human tracking. This task is still challenging due to camera viewpoints, complicated backgrounds, occlusion, and running time. Image recognition has recently been improved with deep convolutional neural networks (DCNNs) [1]. Krizhevsky et al. [1] achieved the best recognition rate and attracted a great deal of attention. State-of-the-art performance of HPE has also been achieved with DCNNs [2–10]. However, because the computational cost of DCNNs is very high, the number of calculations should be reduced as much as possible. Furthermore, in order to achieve state-of-the-art performance, end-to-end learned models should be used.

This paper proposes a novel end-to-end framework for HPE implemented with cascaded neural networks. Figure 1 overviews the architecture of our framework, which includes three tasks: (1) detecting region proposals [14] which

© Springer International Publishing AG 2017
M. Felsberg et al. (Eds.): CAIP 2017, Part II, LNCS 10425, pp. 241–252, 2017.
DOI: 10.1007/978-3-319-64698-5_21

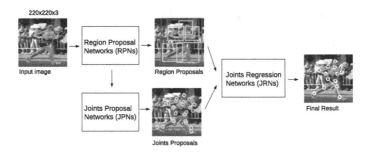

Fig. 1. Overview of the proposed framework

include parts of the human body via region proposal networks (RPNs), (2) predicting the coordinates of human body joints in region proposals via joints proposal networks (JPNs), and (3) finding optimum points as the coordinates of human body joints via joints regression networks (JRNs). These three tasks are jointly optimized. We demonstrated the efficiency of our framework on the Leeds sports pose (LSP) dataset [11]. Our experiments revealed that our framework improved accuracy and reduced the running time compared to conventional methods. The remainder of the paper discusses related works in Sect. 2, and then introduces our framework in Sect. 3. The experimental results are presented in Sect. 4. Section 5 concludes the paper.

2   Related Work

A number of different approaches using DCNNs have been proposed for HPE. DeepPose [2] proposes a cascade of DCNNs-based pose predictors. Such a cascade allows for increased precision of joint localization, which achieves very high levels of accuracy. However, this model includes multiple DCNNs that are computationally expensive, and each pose predictor is independently designed and optimized. Chen et al. [10] use DCNNs to learn conditional probability for the presence of parts. The conditional probability is also called a heat map. The human pose is predicted using graphical models with prior knowledge such as geometric relationships among body parts. However, the DCNNs and the graphical models are independently optimized. Yang et al. [12] propose a model, which combines the DCNNs for generating a heat map with the graphical models, and these models are jointly optimized. This approach also achieves high levels of accuracy. However, generating a heat map requires the use of many DCNNs, which leads to large computational costs. Wang et al. [13] propose a model which handles two tasks: (1) it generates a heat map from depth images via a fully convolutional network (FCN) [15] and (2) it seeks an optimal configuration of body parts via an inference built-in MatchNet [16]. However, MatchNet imposes large computational costs due to the use of chains in multiple convolutional layers [17].

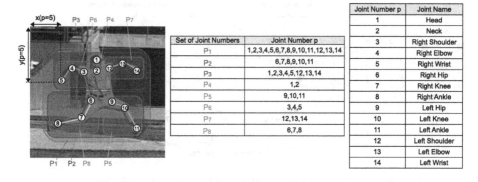

Fig. 2. Representing a human body as a graph

3   Our Framework

This section presents our framework, which consists of three stages. The first stage is region proposal networks (RPNs), the second stage is joints proposal networks (JPNs), and the third stage is joints regression networks (JRNs). These are described in Subsects. 3.1, 3.2, and 3.3, respectively. In Subsect. 3.4, the multi-task learning procedure of our model is described.

3.1   Region Proposal Networks (RPNs)

Our model predicts region proposals $\mathbf{R}$ via RPNs in the first stage. Region proposals $\mathbf{R}$ denote a vector that consists of bounding boxes, which include multiple parts of the human body. Region proposals $\mathbf{R}$ are obtained as follows:

$$\mathbf{R}(\mathbf{I}) = (\mathbf{B}_1, \, \mathbf{B}_2, \, \ldots, \, \mathbf{B}_K) \tag{1}$$

$$\mathbf{B}_k = (b_1^k, \, b_2^k, \, b_3^k, \, b_4^k) = \left(\min_{p \in P_k} x(p), \, \min_{p \in P_k} y(p), \, \max_{p \in P_k} x(p), \, \max_{p \in P_k} y(p)\right), \tag{2}$$

where $\mathbf{I}$ denotes an input image, p denotes a joint number, and P_k denotes a set of joint numbers. Here, $1 \leq k \leq K$, K denotes the number of bounding boxes which is set to eight, and $x(p)$ and $y(p)$ denote the coordinates of human body joints with joint number p in the input image. Figure 2 outlines the relationship between a joint number p and P_k. Figure 3 shows an example of the architecture for RPNs, where feature map 1, 2, and 3 are used in joints proposal networks (JPNs) as input.

We adopted two architectures which have been widely used for image classification. The first was VGG-16 [18], and the second was GoogLeNet [23], because these architectures have provided outstanding results for image classification. For example, Faster-RCNN [14] predicts region proposals via VGG-16 [18] for object detection. The performance impact of each RPNs architecture is described in Sect. 4.

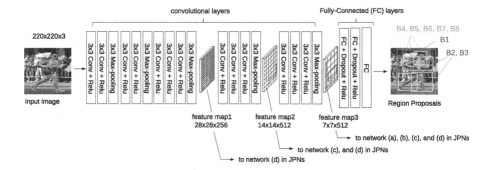

Fig. 3. The architecture of RPNs by using VGG-16 [18]

3.2   Joints Proposal Networks (JPNs)

Our model takes feature maps and region proposals $\mathbf{R}$ as input in the second stage, and it predicts joints proposals $\mathbf{J}$ via JRNs, where joints proposals $\mathbf{J}$ are defined as the coordinates of human body joints in region proposals $\mathbf{R}$. Joints proposals $\mathbf{J}$ are obtained as follows:

$$\mathbf{J}(\mathbf{I}) = (\mathbf{J}_0, \mathbf{J}_1, \mathbf{J}_2 \ldots, \mathbf{J}_K) \tag{3}$$

$$\mathbf{J}_0 = (x(1), \ y(1), \ \ldots, \ x(L), \ y(L)) \tag{4}$$

$$\mathbf{J}_k = (j_1^k(p_1), \ j_2^k(p_1), \ldots, \ j_1^k(p_{s(k)}), \ j_2^k(p_{s(k)})) \tag{5}$$

$$j_1^k(p) = (x(p) - b_1^k)/(b_3^k - b_1^k) \tag{6}$$

$$j_2^k(p) = (y(p) - b_2^k)/(b_4^k - b_2^k), \tag{7}$$

where $1 \leq k \leq K$, $p_1, \ldots, p_{s(k)} \in P_k$, b_1^k, b_2^k, b_3^k, and b_4^k denote vertices of bounding box $\mathbf{B}_k$ defined in Eq. (2), and L denotes the number of human body parts, which is set to 14 as described in Fig. 2.

Figure 4 shows the architectures for JPNs, which consists of four types of networks. Network (a) takes a feature map from the middle layer in RPNs as input, and predicts $\mathbf{J}_0$. Networks (b), (c), and (d) take feature maps from multiple middle layers in RPNs and region proposals $\mathbf{R}$ as input, and predict $\mathbf{J}_1$, $\mathbf{J}_2$, $\ldots$, and $\mathbf{J}_K$.

The purpose of the region-of-interest (RoI) pooling layers [14] is to extract the area indicated by region proposals $\mathbf{R}$ from feature maps and to produce a fixed-length feature vector because full-connected (FC) layers [1] require it as input. DeepPose [2] extracts the area from an input image. However, such an approach requires many calculations using convolutional layers for the feature extraction, which leads to large computational costs. In our model, the calculation is performed only once because the area is extracted from feature maps. This approach is computationally efficient and suitable for real-time applications. Moreover, our model takes feature maps from multiple middle layers as input to increase the resolution of feature maps. For example, feature maps with

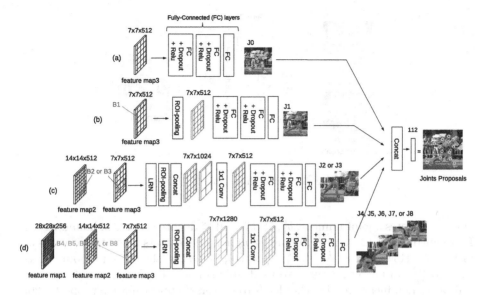

Fig. 4. The architecture of JPNs.

high resolution are required to calculate $\mathbf{J}_4$, $\mathbf{J}_5$, $\mathbf{J}_6$, $\mathbf{J}_7$, and $\mathbf{J}_8$, because $\mathbf{B}_4$, $\mathbf{B}_5$, $\mathbf{B}_6$, $\mathbf{B}_7$, and $\mathbf{B}_8$ have a small area (see Fig. 3). On the other hand, feature maps with high resolution are not required for calculating $\mathbf{J}_1$ because $\mathbf{B}_1$ has a large area. Our model changes the number of feature maps depending on the size of the bounding box. As shown in Fig. 4, Network (d) takes feature map 3, 2, and 1 as input. Network (c) takes feature map 3 and 2 as input. Network (b) and (a) only take feature map 3 as input.

The purpose of the 1×1 convolutional layers is to reduce the channel dimensions of feature maps. They enable the training time to shorten by reducing the dimensions. The Local Response Normalization (LRN) layers are used to align the amplitude of feature maps in each convolutional layer.

3.3   Joints Regression Networks (JRNs)

Our model takes region proposals $\mathbf{R}$ and joints proposals $\mathbf{J}$ as input in the third stage, and it predicts the coordinates of human body joints via JRNs. Figure 5 shows the architectures for JRNs, whose purpose is finding the optimum points as human body joints.

These layers can be replaced with a linear function under ideal conditions, in the case that region proposals $\mathbf{R}$ and joints proposals $\mathbf{J}$ are correct values. The coordinates of human body joints are obtained as follows:

$$x(p) = (b_3^k - b_1^k)\, j_1^k(p) + b_1^k \tag{8}$$

$$y(p) = (b_4^k - b_2^k)\, j_2^k(p) + b_2^k, \tag{9}$$

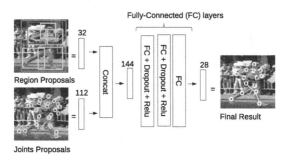

Fig. 5. The architecture of JRNs

where p denotes a joint number, b_1^k, b_2^k, b_3^k, and b_4^k denote vertices of bounding box $\mathbf{B}_k$ defined in Eq. (2), and $j_1^k(p)$ and $j_2^k(p)$ denote elements of joints proposals $\mathbf{J}_k$ in Eq. (5). However, as b_1^k, b_2^k, b_3^k, b_4^k, $j_1^k(p)$, and $j_2^k(p)$ fluctuate randomly, Eqs. (8) and (9) do not work well. We used fully-connected (FC) layers [1] because they are a nonlinear function that leads to universal approximation property [20].

3.4 Multi-task Learning

We define the loss function of the entire network as:

$$l(\mathbf{w}_1, \mathbf{w}_2, \mathbf{w}_3) = l_1(\mathbf{w}_1) + l_2(\mathbf{w}_1, \mathbf{w}_2) + l_3(\mathbf{w}_1, \mathbf{w}_2, \mathbf{w}_3), \qquad (10)$$

where $\mathbf{w}_1$, $\mathbf{w}_2$, and $\mathbf{w}_3$ denote weight parameters in RPNs, JPNs, and JRNs. $l_1(\mathbf{w}_1)$, $l_2(\mathbf{w}_1, \mathbf{w}_2)$, and $l_3(\mathbf{w}_1, \mathbf{w}_2, \mathbf{w}_3)$ correspond to the mean-squared-error (MSE) [21] for RPNs, JPNs, and JRNs.

The loss function, $l(\mathbf{w}_1, \mathbf{w}_2, \mathbf{w}_3)$, is minimized with respect to $\mathbf{w}_1$, $\mathbf{w}_2$, and $\mathbf{w}_3$. We employ an Adaptive Moment Estimation (Adam) [27] to optimize $\mathbf{w}_1$, $\mathbf{w}_2$, and $\mathbf{w}_3$. The entire algorithm for multi-task learning is summarized in Algorithm 1. First, $\mathbf{w}_1$, $\mathbf{w}_2$, and $\mathbf{w}_3$ are initialized randomly in step 1. Then, RPNs, JPNs and JRNs are independently optimized in step 2, 3, and 4. Finally, the entire network is trained in step 5. Step 2, 3, and 4 are pre-training techniques [22] to shorten training time in step 5. The end-to-end learning is performed in step 5. The details about values of the several parameters are described in Sect. 4.

4 Experiments

4.1 Experimental Settings

Datasets. We evaluated the proposed methods on well-known public pose estimation benchmarks: Leeds sports poses (LSP) dataset [11] and Leeds sports pose extended training (LSPET) dataset [12]. The LSP dataset consists of 1,000 training and 1,000 testing images, and the LSPET dataset consists of 10,000 training images. However, a lot of data are required for training DCNNs. We

---

Algorithm 1. Optimize $\mathbf{w}_1$, $\mathbf{w}_2$, and $\mathbf{w}_3$

Input: Training Samples $(\mathbf{I}_1, \mathbf{R}(\mathbf{I}_1), \mathbf{J}(\mathbf{I}_1)) \dots (\mathbf{I}_N, \mathbf{R}(\mathbf{I}_N), \mathbf{J}(\mathbf{I}_N))$
Output: $\mathbf{w}_1$, $\mathbf{w}_2$ and $\mathbf{w}_3$
1: Initialize $\mathbf{w}_1$, $\mathbf{w}_2$ and $\mathbf{w}_3$ randomly.
2: $\mathbf{w}_1 \leftarrow \arg \min_{\mathbf{w}_1} l_1(\mathbf{w}_1)$
3: $\mathbf{w}_2 \leftarrow \arg \min_{\mathbf{w}_2} l_2(\mathbf{w}_1, \mathbf{w}_2)$
4: $\mathbf{w}_3 \leftarrow \arg \min_{\mathbf{w}_3} l_3(\mathbf{w}_1, \mathbf{w}_2, \mathbf{w}_3)$
5: $\mathbf{w}_1, \mathbf{w}_2, \mathbf{w}_3 \leftarrow \arg \min_{\mathbf{w}_1, \mathbf{w}_2, \mathbf{w}_3} l_1(\mathbf{w}_1) + l_2(\mathbf{w}_1, \mathbf{w}_2) + l_3(\mathbf{w}_1, \mathbf{w}_2, \mathbf{w}_3)$

---

also used our 3D-CAD models to increase the amount of data. The 3D-CAD models consist of 11,000 training images that are automatically generated by using open source 3D-CAD tools [28,29], and the human motions are created by using the motion capture database [30]. We combined these datasets with the LSP and the LSPET datasets. As a result, the combination contained 22,000 training images. Peng et al. [19] augment the training images with synthetic images generated from 3D-CAD models for image classification. We used this approach.

We augmented the training images to reduce overfitting by horizontally mirroring the images, rotating them through 360 degrees for every 9 degrees, cropping them randomly, and injecting white noise into them. The final amount of training samples was 5,000,000.

Metrics. We used widely accepted evaluation metrics called the percent of detected joints (PDJ) [12], which calculates the detection rate of human body joints, where a joint is considered as being detected if the distance between the predicted joint and the correct joint is less than a fraction of the torso diameter. The torso diameter is defined as the distance between the left shoulder and the right hip. We also computed an Area-Under-the-Curve (AUC) to compare our work with other approaches (see Figs. 6, 7, and 8).

Person-Centric/Observer-Centric. In person-centric [24] annotations, right/ left body parts are marked according to the viewpoint of the person in an image. For example, the right wrist of a person facing the camera is left in the image. However, if the person faces away from the camera, it is right in the image. On the other hand, in observer-centric [24] annotations, right/left body parts are marked regardless of the viewpoint. Person-centric annotations are more difficult than observer-centric annotations because it is necessary to recognize the viewpoint. The information of the viewpoint is important for action recognition. Therefore, we used person-centric annotations.

DCNN Architectures. We investigated two DCNN architectures in RPNs. The first was VGG-16 [18] that consists of 16 convolutional layers and FC layers.

Table 1. Running time on an Intel Xeon CPU at 3.50 GHz and a NVIDIA Tesla K40 GPU. Note that the Heat Map [12] only indicates the running time for generating a heat map, and it does not include the processing time of other tasks.

Method	Running time [s]
DeepPose [2]	0.18
Heat map [12]	0.50
Our work (VGG-16 [18])	**0.094**
Our work (GoogLeNet [23])	**0.070**

The second was GoogLeNet [23] that consists of three convolutional layers, nine inception layers, and FC layers.

Implementation Details. All of our experiments were carried out on an Intel Xeon CPU at 3.50 GHz and a NVIDIA Tesla K40 GPU. Our model was implemented on the Chainer library [31]. In order to optimize our model, we used pre-training models in Model Zoo [32] for fine-tuning [22]. The learning rate and the batch size were set to 0.0001 and 24, respectively, for training RPNs, JPNs and JRNs. On the other hand, the learning rate and the batch size were set to 0.00001 and 20 for the end-to-end learning. The total training time was about 2 weeks.

4.2   Experimental Results

Table 1 lists the running time results. Our model was 2.57 times faster than DeepPose [2]. As described in Sect. 3.2, the conventional methods use a lot of DCNNs that are computationally expensive. However, our model does not use them, so the running time of our model was fast.

Figure 6 shows the PDJ results on the LSP dataset. We used person-centric annotations for fair comparison with related work [2,25]. Our model achieved the best performance compared with the conventional methods. Our results were particularly better in the low precision domain. The AUC of our model was 7.34% – 29.66% higher than that of DeepPose [2].

We analyzed how different RPNs architectures affected performance. Figure 7 shows the PDJ results with different RPNs architectures. Our best performance was achieved by using the architecture of VGG-16 [18] in RPNs. The AUC of VGG-16 [18] was 8.65% – 19.62% higher than that of GoogLeNet [23].

Figure 8 compares JPNs with JRNs. The PDJ of JRNs was higher than that of JPNs, especially for ankle. The AUC was improved from 0.66% to 12.56%. Here, we can observe that JRNs have an effect on improving accuracy. Figure 9 shows some pose estimation results.

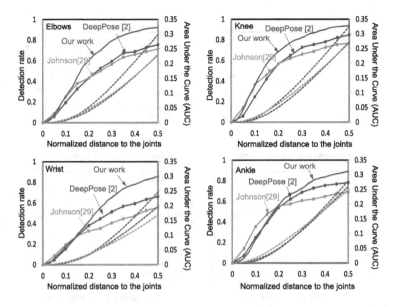

Fig. 6. PDJ comparison of our work and other approaches on the LSP dataset. The solid lines and the dashed lines represent PDJ and AUC on the LSP dataset, respectively. All results are from author's papers, and these are the person-centric results. The architecture of RPNs was VGG-16 [18]. Note that we adopted only person-centric results as related work. For example, Chen and Yuille [10] and Yang et al. [12] used observer-centric annotations, therefore these were excluded from comparisons.

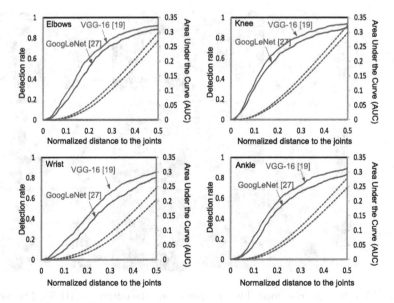

Fig. 7. Influences of different RPNs architectures are plotted. The solid lines and the dashed lines represent PDJ and AUC on the LSP dataset, respectively.

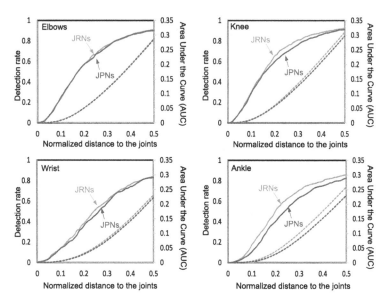

Fig. 8. Figures compare JPNs with JRNs. Results for JPNs were calculated by using output of network (a) in Fig. 4. The architecture of RPNs was VGG-16 [18]. The solid lines and the dashed lines represent PDJ and AUC on the LSP dataset, respectively.

Fig. 9. HPE results we obtained. The first row shows the outputs of RPNs. The second row shows the outputs of JPNs. The third row shows the outputs of JRNs.

5 Conclusion and Future Work

We proposed a novel end-to-end framework for HPE implemented with cascaded neural networks. We demonstrated the efficiency of our framework on the LSP dataset [11]. As a result, our model achieved accuracy that was higher than that of conventional models, and the running time was 2.57 times faster than conventional methods.

As a future work, we plan to evaluate our method on other datasets, for example, Frames Labeled In Cinema dataset (FLIC) [33], Kinect2 Human Gesture Dataset (K2HGD) [13], and MPII Human Pose Dataset [34]. Furthermore, we apply other methods of speeding up HPE to our model, such as binarized weights [26] or low rank approximation [17].

Acknowledgments. The authors would like to thank Professor Hironobu Fujiyoshi at Chubu University for his forthright comments and valuable suggestions.

References

1. Krizhevsky, A., Sutskever, I., Hinton, G.E.: ImageNet classification with deep convolutional neural networks. In: NIPS (2012)
2. Toshev, A., Szegedy, C.: DeepPose: human pose estimation via deep neural networks. In: CVPR (2014)
3. Tompson, J., Jain, A., LeCun, Y., Bregler, C.: Joint training of a convolutional network and a graphical model for human pose estimation. In: NIPS (2014)
4. Tompson, J., Goroshin, R., Jain, A., LeCun, Y., Bregler, C.: Efficient object localization using convolutional networks. In: CVPR (2015)
5. Jain, A., Tompson, J., LeCun, Y., Bregler, C.: MoDeep: a deep learning framework using motion features for human pose estimation. In: Cremers, D., Reid, I., Saito, H., Yang, M.-H. (eds.) ACCV 2014. LNCS, vol. 9004, pp. 302–315. Springer, Cham (2015). doi:10.1007/978-3-319-16808-1_21
6. Jain, A., Tompson, J., Andriluka, M., Taylor, G.W., Bregler, C.: Learning human pose estimation features with convolutional networks. In: ICLR (2014)
7. Fan, X., Zheng, K., Lin, Y., Wang, S.: Combining local appearance and holistic view: dual-source deep neural networks for human pose estimation. In: CVPR (2015)
8. Chu, X., Ouyang, W., Li, H., Wang, X.: Structured feature learning for pose estimation. In: CVPR (2016)
9. Chen, X., Yuille, A.: Parsing occluded people by flexible compositions. In: CVPR (2015)
10. Chen, X., Yuille, A.L.: Articulated pose estimation by a graphical model with image dependent pairwise relations. In: NIPS (2014)
11. Johnson, S., Everingham, M.: Clustered pose and nonlinear appearance models for human pose estimation. In: BMVC (2010)
12. Yang, W., et al.: End-to-end learning of deformable mixture of parts and deep convolutional neural networks for human pose estimation. In: CVPR (2016)
13. Wang, K., et al.: Human pose estimation from depth images via inference embedded multi-task learning. In: Multimedia Conference (2016)

14. Ren, S., et al.: Faster R-CNN: towards real-time object detection with region proposal networks. In: NIPS (2016)
15. Long, J., et al.: Fully convolutional networks for semantic segmentation. In: CVPR (2015)
16. Han, X., et al.: MatchNet: unifying feature and metric learning for patch-based matching. In: CVPR (2015)
17. Jaderberg, M., et al.: Speeding up convolutional neural networks with low rank expansions. BMVA Press (2014)
18. Simonyan, K., et al.: Very deep convolutional networks for large-scale image recognition. In: ICLR (2015)
19. Peng, X., et al.: Learning deep object detectors from 3D models. In: CVPR (2015)
20. Sonoda, S., et al.: Neural network with unbounded activation functions is universal approximator. In: Applied and Computational Harmonic Analysis (2015)
21. Bishop, C.M.: Pattern Recognition and Machine Learning. Information Science and Statistics, 2nd edn. Springer, New York (2006)
22. Reyes, A., Caicedo, J., Camargo, J.: Fine-tuning Deep Convolutional Networks for Plant Recognition. In: Proceedings of CEUR Workshop, CLEF (Working Notes), vol. 1391 (2015). CEUR-WS.org
23. Szegedy, C., et al.: Going deeper with convolutions. In: CVPR (2014)
24. Eichner, M., Ferrari, V.: Appearance sharing for collective human pose estimation. In: Lee, K.M., Matsushita, Y., Rehg, J.M., Hu, Z. (eds.) ACCV 2012. LNCS, vol. 7724, pp. 138–151. Springer, Heidelberg (2013). doi:10.1007/978-3-642-37331-2_11
25. Johnson, S., et al.: Learning effective human pose estimation from inaccurate annotation. In: CVPR (2011)
26. Courbariaux, M., et al.: BinaryConnect: training deep neural networks with binary weights during propagations. In: NIPS (2015)
27. Kingma, D.P., et al.: Adam: a method for stochastic optimization. In: ICLR (2015)
28. Website of MakeHuman. http://www.makehuman.org/. Accessed 18 Apr 2017
29. Website of Blender. https://www.blender.org. Accessed 18 Apr 2017
30. Website of motion capture database. http://mocap.cs.cmu.edu/. Accessed 18 Apr 2017
31. Website of Chainer. http://chainer.org/. Accessed 18 Apr 2017
32. Website of Model Zoo. https://github.com/BVLC/caffe/wiki/Model-Zoo. Accessed 18 Apr 2017
33. Benjamin, S., et al.: MODEC: multimodal decomposable models for human pose estimation. In: CVPR (2013)
34. Andriluka, M., et al.: 2D human pose estimation: new benchmark and state of the art analysis. In: CVPR (2014)

Image Restoration I

Multi-view Separation of Background and Reflection by Coupled Low-Rank Decomposition

Jian Lai$^{(\boxtimes)}$, Wee Kheng Leow$^{(\boxtimes)}$, Terence Sim$^{(\boxtimes)}$, and Guodong Li$^{(\boxtimes)}$

Department of Computer Science,
National University of Singapore, Singapore, Singapore
{laij,leowwk,tsim}@comp.nus.edu.sg, lguodong@u.nus.edu

Abstract. Images captured by a camera through glass often have reflection superimposed on the transmitted background. Among existing methods for reflection separation, multi-view methods are the most convenient to apply because they require the user to just take multiple images of a scene at varying viewing angles. Some of these methods are restricted to the simple case where the background scene and reflection scene are planar. The methods that handle non-planar scenes employ image feature flow to capture correspondence for image alignment, but they can overfit resulting in degraded performance. This paper proposes a multiple-view method for separating background and reflection based on robust principal component analysis. It models the background and reflection as rank-1 matrices, which are decomposed according to different transformations for aligning the background and reflection images. It can handle non-planar scenes and global reflection. Comprehensive test results show that our method is more accurate and robust than recent related methods.

Keywords: Reflection removal · Non-planar scenes · Robust PCA

1 Introduction

When an image is captured by a camera through a piece of glass, reflection is often superimposed on the transmitted background scene. Reflection is annoying because it corrupts the image content, and it is difficult to remove them manually. Therefore, separation of background and reflection from the superimposed image has attracted research interests over the years. Existing methods for reflection separation can be broadly grouped into three categories: single-image, single-view multiple-image, and multi-view multiple-image. Single-image methods [12,13] in general require user inputs or training images, which are inconvenient to apply. Single-view methods [1,6,10,21,22] take multiple images of a scene with a fixed camera under varying imaging conditions such as polarization, flash/no-flash, and focusing, and use the resulting differences between background and reflection to separate them. These methods require additional

© Springer International Publishing AG 2017
M. Felsberg et al. (Eds.): CAIP 2017, Part II, LNCS 10425, pp. 255–268, 2017.
DOI: 10.1007/978-3-319-64698-5_22

accessories, professional setting skill, and fixed camera position, and so are not suitable for images taken by non-professional users.

Multi-view methods take multiple images of a scene at varying viewing angles. Thus, the images are not aligned. The reflection of objects as seen by a camera, which we call *virtual scene*, is located on the same side of the reflecting glass as the background scene. When the background scene and the virtual scene are not coplanar, the background image and reflection image can be separated by multi-view methods. When the objects in the background are on a plane, and similarly for the reflection, the background and reflection scenes are planar and can be aligned by single homography. Several methods [7, 8, 23] assume this simplifying condition. Some recent methods [14, 24, 25] try to solve the more complex case of non-planar scenes. They employ nonlinear flows of image features to capture correspondence for image alignment, but they can overfit resulting in degraded performance [16].

This paper proposes a multiple-view method for separating reflection and background by coupled low-rank decomposition (CLORD). It is based on the method of robust principal component analysis (robust PCA) [4], which has been successfully applied to various computer vision problems [11, 18–20]. It models the background and reflection as rank-1 matrices, which are decomposed according to different transformations for aligning the background and reflection images. It can handle non-planar scenes and global reflection. Comprehensive test results show that CLORD is more accurate and robust than recent related methods.

2   Coupled Low-Rank Decomposition

2.1   Overview

Given a set of superimposed images of stationary objects, which are captured over a small range of viewing angles, CLORD is proposed to recover the background and foreground simultaneously. CLORD is inspired by but differs from that of Guo et al. [8]. In comparison, Guo et al. [8], Szeliski et al. [23], and Gai et al. [7] assume planar scenes, and Guo et al. [8] also assume sparse reflection. So, their methods solve a simpler version of the reflection separation problem. Like CLORD, Li and Brown [14], Xue et al. [24] and Yang et al. [25] also handle non-planar scenes and global reflection. However, CLORD is based on robust PCA [4], which has been successfully applied to various challenging computer vision tasks, whereas [14, 24, 25] apply iteratively reweighted least squares.

2.2   Problem Formulation

For a set of superimposed images $\mathbf{f}_i'$ with m pixels, $i = 1, \ldots, n$, captured at different viewing angles, each image $\mathbf{f}_i'$ is a linear combination of the transmitted background $\mathbf{b}_i'$ and the reflection $\mathbf{r}_i'$, i.e., $\mathbf{f}_i' = \mathbf{b}_i' + \mathbf{r}_i'$. By arranging each image as a column in an $m \times n$ matrix, the above relationship can be written as $\mathbf{F}' = \mathbf{B}' + \mathbf{R}'$, where $\mathbf{F}'$, $\mathbf{B}'$, and $\mathbf{R}'$ denote the matrices of the unaligned superimposed images, transmitted background, and reflection, respectively. The corresponding matrices $\mathbf{B}$ and $\mathbf{R}$ of aligned background and reflection are low-rank, specifically, rank-1. They are related to the unaligned matrices by transformation functions T and T':

$$\mathbf{B} = T_b(\mathbf{B}'), \quad \mathbf{R} = T_r(\mathbf{R}'), \quad \mathbf{B}' = T_b'(\mathbf{B}), \quad \mathbf{R}' = T_r'(\mathbf{R}).$$

So, $\mathbf{F}' = T_b'(\mathbf{B}) + T_r'(\mathbf{R})$. The transformations T and T' differ for different $\mathbf{b}_i'$ and $\mathbf{r}_i'$. They are nonlinear in general, and T' is a good approximation of T^{-1}.

Distinctive features such as edges correspond to large gradients in the images. Image gradients are computed by convolving gradient filter kernel $\mathbf{g}$ with the background and the reflection giving $\mathbf{g} * \mathbf{b}_i$ and $\mathbf{g} * \mathbf{r}_i$. Since the gradient filter is linear, these convolutions can be written in matrix form as $\mathbf{DB}$ and $\mathbf{DR}$, where $\mathbf{D}$ is gradient operator.

To enforce mutual exclusion of edges, two coupled weight matrices $\mathbf{W}_b$ and $\mathbf{W}_r$ are applied to the gradients giving $\mathbf{W}_b \circ \mathbf{DB}$ and $\mathbf{W}_r \circ \mathbf{DR}$, where $\circ$ denotes element-wise multiplication. If an edge belongs to the background, its corresponding entry in $\mathbf{W}_b$ is small whereas its entry in $\mathbf{W}_r$ is large. The converse is true if edge belongs to reflection.

Now, the reflection separation problem can be formulated as follows:

$$\min_{\mathbf{B},\mathbf{R},\mathbf{E}',\mathbf{N}',T_b',T_r'} \|\mathbf{B}\|_* + \|\mathbf{R}\|_* + \lambda_1\|\mathbf{E}'\|_1 + \frac{\lambda_2}{2}\|\mathbf{N}'\|_F^2 + $$
$$\lambda_3\|\mathbf{W}_b \circ \mathbf{DB}\|_1 + \lambda_3\|\mathbf{W}_r \circ \mathbf{DR}\|_1 \tag{1}$$
$$\text{subject to } \mathbf{F}' = T_b'(\mathbf{B}) + T_r'(\mathbf{R}) + \mathbf{E}' + \mathbf{N}'.$$

The first two terms are the nuclear norms of background and reflection, which are small for low-rank matrices. The third term uses l_1-norm to model sparse noise $\mathbf{E}'$ with possibly large amplitudes whereas the fourth term uses Frobenius norm to model non-sparse, small-amplitude noise $\mathbf{N}'$. The last two terms model mutually exclusive sparse gradients. This formulation handles the aligned background $\mathbf{B}$ and aligned reflection $\mathbf{R}$ in a uniform manner. This uniformity results in a simpler objective function compared to that of Guo et al. [8], which models low-rank background with sparse reflection instead. Moreover, they model mutual exclusion of features with $\|\mathbf{DB} \circ \mathbf{DR}\|_1$. The minimization of $\|\mathbf{DB} \circ \mathbf{DR}\|_1$ can be achieved by small $\mathbf{DB}$ and small $\mathbf{DR}$, which is ambiguous where the edge belongs. In comparison, CLORD is more specific and less prone to such ambiguity (Sect. 2.5).

In general, the background scene and the virtual scene are not coplanar. So, they have different transformations T_b and T_r, which permit $\mathbf{B}$ and $\mathbf{R}$ to be separated. In the degenerate case that the background and virtual scenes are coplanar, separation of background and reflection is impossible regardless of the method used.

To solve Problem 1, an alternating optimization algorithm is applied to minimize various parts of the objective function alternatively. The details of our algorithm CLORD is derived in a similar manner as [8,9,11,15,19,20].

2.3 Optimization of Background

First, let us consider the optimization of $\mathbf{B}$, $\mathbf{E}'$, and $\mathbf{N}'$ while keeping $\mathbf{R}$, T_b', and T_r' fixed. As the superimposed images are captured at a small range of varying viewing angles, the structures of the images are essentially the same. So, minimizing $\mathbf{E}'$ and $\mathbf{N}'$ is equivalent to minimizing $\mathbf{E} = T_b(\mathbf{E}')$ and $\mathbf{N} = T_b(\mathbf{N}')$. Denoting $\mathbf{F} = T_b(\mathbf{F}') - T_b(T_r'(\mathbf{R}))$ and applying T_b to both sides of the constraint equation of Problem 1 yields equivalent constraint $\mathbf{F} = \mathbf{B} + \mathbf{E} + \mathbf{N}$. With $\mathbf{R}$, T_b', and T_r' fixed, Problem 1 reduces to

$$\min_{\mathbf{B},\mathbf{E},\mathbf{N}} \|\mathbf{B}\|_* + \lambda_1\|\mathbf{E}\|_1 + \frac{\lambda_2}{2}\|\mathbf{N}\|_F^2 + \lambda_3\|\mathbf{W}_b \circ \mathbf{DB}\|_1 \tag{2}$$
$$\text{subject to } \mathbf{F} = \mathbf{B} + \mathbf{E} + \mathbf{N}.$$

To match Problem 2 to robust PCA's formulation, we introduce two auxiliary variables $\mathbf{A}$ and $\mathbf{C}$ such that $\mathbf{B} = \mathbf{A}$ and $\mathbf{DB} = \mathbf{C}$. Then, Problem 2 becomes

$$\min_{\mathbf{B},\mathbf{E},\mathbf{N},\mathbf{A},\mathbf{C}} \|\mathbf{A}\|_* + \lambda_1\|\mathbf{E}\|_1 + \frac{\lambda_2}{2}\|\mathbf{N}\|_F^2 + \lambda_3\|\mathbf{W}_b \circ \mathbf{C}\|_1 \tag{3}$$
$$\text{subject to } \mathbf{F} = \mathbf{B} + \mathbf{E} + \mathbf{N}, \ \mathbf{B} = \mathbf{A}, \ \mathbf{DB} = \mathbf{C}.$$

Now, Problem 3 matches robust PCA, which can be solved using augmented Lagrange multiplier (ALM) method [15]. With ALM, Problem 3 is reformulated as

$$\min_{\mathbf{B},\mathbf{E},\mathbf{N},\mathbf{A},\mathbf{C}} \|\mathbf{A}\|_* + \lambda_1\|\mathbf{E}\|_1 + \frac{\lambda_2}{2}\|\mathbf{N}\|_F^2 + \lambda_3\|\mathbf{W}_b \circ \mathbf{C}\|_1 +$$
$$\frac{\mu}{2}\|\mathbf{F} - \mathbf{B} - \mathbf{E} - \mathbf{N}\|_F^2 + \frac{\mu}{2}\|\mathbf{B} - \mathbf{A}\|_F^2 + \tag{4}$$
$$\frac{\mu}{2}\|\mathbf{DB} - \mathbf{C}\|_F^2 + \langle \mathbf{Y}_1, \mathbf{F} - \mathbf{B} - \mathbf{E} - \mathbf{N} \rangle +$$
$$\langle \mathbf{Y}_2, \mathbf{B} - \mathbf{A} \rangle + \langle \mathbf{Y}_3, \mathbf{DB} - \mathbf{C} \rangle,$$

where $\mathbf{Y}_1, \mathbf{Y}_2, \mathbf{Y}_3$ are the Lagrange multipliers, μ is a penalty parameter, and $\langle \cdot, \cdot \rangle$ denote the sum of product of corresponding matrix elements.

An important operator used in various solutions of RPCA is the *soft thresholding* operator [4,15], which is applied to each matrix element individually:

$$S_\varepsilon(x) = \begin{cases} x - \varepsilon, & \text{if } x > \varepsilon, \\ x + \varepsilon, & \text{if } x < -\varepsilon, \\ 0, & \text{otherwise.} \end{cases} \tag{5}$$

With this operator, [3] show that, for matrix $\mathbf{M}$ with SVD $\mathbf{USV}^\top$, the minimizations of nuclear norm and l_1-norm are given by

$$\mathbf{U}\,S_\varepsilon(\mathbf{S})\mathbf{V}^\top = \arg\min_{\mathbf{X}} \varepsilon\|\mathbf{X}\|_* + \frac{1}{2}\|\mathbf{M} - \mathbf{X}\|_F^2, \tag{6}$$

$$S_\varepsilon(\mathbf{M}) = \arg\min_{\mathbf{X}} \varepsilon\|\mathbf{X}\|_1 + \frac{1}{2}\|\mathbf{M} - \mathbf{X}\|_F^2. \tag{7}$$

Applying Eqs. 6 and 7 to the optimization of $\mathbf{A}$, $\mathbf{E}$, and $\mathbf{C}$ in Problem 4 give

$$\mathbf{A} = \mathbf{U} S_{1/\mu}(\mathbf{S}) \mathbf{V}^\top, \tag{8}$$

$$\mathbf{E} = S_{\lambda_1/\mu} \left(\mathbf{F} - \mathbf{B} - \mathbf{N} + \mathbf{Y}_1/\mu \right), \tag{9}$$

$$\mathbf{C} = S_{\lambda_3 \mathbf{W}_b/\mu} \left(\mathbf{D}\mathbf{B} + \mathbf{Y}_3/\mu \right), \tag{10}$$

where $\mathbf{U}\mathbf{S}\mathbf{V}^\top$ is the SVD of $\mathbf{B} + \mathbf{Y}_2/\mu$. In our problem, the ranks of the static background and reflection are known to be 1. Therefore, Eq. 8 is equivalent to

$$\mathbf{A} = \mathbf{U} S_1 \mathbf{V}^\top, \tag{11}$$

where S_1 is a diagonal matrix that contains only the first singular value of $\mathbf{S}$.

The optimization of $\mathbf{N}$, with other variables fixed, involves only Frobenius norm, and can be derived directly as

$$\mathbf{N} = \frac{\mu}{\lambda_2 + \mu} \left(\mathbf{F} - \mathbf{B} - \mathbf{E} + \frac{1}{\mu} \mathbf{Y}_1 \right). \tag{12}$$

Now, with $\mathbf{A}$, $\mathbf{E}$, $\mathbf{C}$, and $\mathbf{N}$ fixed, Problem 4 reduces to

$$\min_{\mathbf{B}} \frac{\mu}{2} \| \mathbf{B} - (\mathbf{F} - \mathbf{E} - \mathbf{N} + \frac{1}{\mu} \mathbf{Y}_1) \|_F^2 +$$
$$\frac{\mu}{2} \| \mathbf{B} - (\mathbf{A} - \frac{1}{\mu} \mathbf{Y}_2) \|_F^2 + \frac{\mu}{2} \| \mathbf{D}\mathbf{B} - (\mathbf{C} - \frac{1}{\mu} \mathbf{Y}_3) \|_F^2,$$

which can be rearranged as

$$\min_{\mathbf{B}} \frac{\mu}{2} \left(\mathbf{B}^\top (\mathbf{D}^\top \mathbf{D} + 2\mathbf{I}) \mathbf{B} - \mathbf{T}^\top \mathbf{B} \right), \tag{13}$$

where

$$\mathbf{T} = \mathbf{F} - \mathbf{E} - \mathbf{N} + \frac{1}{\mu} \mathbf{Y}_1 + \mathbf{A} - \frac{1}{\mu} \mathbf{Y}_2 + \mathbf{D}^\top \left(\mathbf{C} - \frac{1}{\mu} \mathbf{Y}_3 \right).$$

Problem 13 is a quadratic optimization problem whose solution is given by [5]

$$(\mathbf{D}^\top \mathbf{D} + 2\mathbf{I}) \mathbf{B} = \mathbf{T}. \tag{14}$$

Equation 14 can be efficiently solved using fast Fourier transform (FFT) [5] as follows: Let $\mathbf{b}_i$ and $\mathbf{t}_i$ denote the 2D forms of column i of $\mathbf{B}$ and $\mathbf{T}$ respectively. Then,

$$\mathbf{b}_i = \mathcal{F}^{-1} \left(\frac{\mathcal{F}(\mathbf{t}_i)}{\overline{\mathcal{F}(\mathbf{D})} \circ \mathcal{F}(\mathbf{D}) + 2} \right), \tag{15}$$

where $\mathcal{F}$, $\mathcal{F}^{-1}$, and $\overline{\mathcal{F}}$ are the 2D FFT, 2D inverse FFT and the complex conjugate of 2D FFT, respectively, and $\mathbf{2}$ is a matrix whose elements are all 2. The division in Eq. 15 is performed element-wise.

Finally, the Lagrange multipliers and parameter μ are updated according to ALM method [15] as follows:

$$
\begin{aligned}
\mathbf{Y}_1 &= \mathbf{Y}_1 + \mu(\mathbf{F} - \mathbf{B} - \mathbf{E} - \mathbf{N}), \\
\mathbf{Y}_2 &= \mathbf{Y}_2 + \mu(\mathbf{B} - \mathbf{A}), \\
\mathbf{Y}_3 &= \mathbf{Y}_3 + \mu(\mathbf{DB} - \mathbf{C}), \\
\mu &= \rho\mu, \text{ for } \rho > 1.
\end{aligned}
\tag{16}
$$

The variables discussed in this section are updated iteratively according to ALM until convergence (Algorithm 2).

2.4 Updating of Reflection Transformations

In our alternating optimization scheme, updating of transformations T_r and T_r' of reflection is performed before updating of the aligned reflection $\mathbf{R}$. After the aligned background $\mathbf{B}$ is obtained (Sect. 2.3), the unaligned reflection is computed as $\mathbf{R}' = \mathbf{F}' - T_b'(\mathbf{B})$. Among the reflection images $\mathbf{r}_i'$ in $\mathbf{R}'$, the one that corresponds to the neutral viewing angle, denoted as $\mathbf{r}_0'$, is selected as reference. All other reflection images $\mathbf{r}_i'$ are aligned to $\mathbf{r}_0'$. That is, the transformation of $\mathbf{r}_0'$ is identity function, giving $\mathbf{r}_0' = \mathbf{r}_0$.

The transformation between $\mathbf{r}_i'$ and $\mathbf{r}_0' = \mathbf{r}_0$ is computed as follows. First, matching pairs of feature points are extracted from $\mathbf{r}_i'$ and $\mathbf{r}_0$ based on scale-invariant feature transform (SIFT) [17] and speeded up robust features (SURF) [2]. The matching pairs may contain undesirable features that belong to the background instead of the reflection. Consider a matching pair (p_i, p_0). If their positions $T_b(p_i)$ and $T_b(p_0)$ after aligning to the background $\mathbf{B}$ are close to each other, then they belong to the background instead of the reflection. So, they are removed from matching pairs for computing T_r and T_r'.

As the reflection scene is non-planar, a single homography may not be sufficient to model the transformation. So, RANSAC is applied to find the largest set of consistent matching pairs that fits a homography. Then, the consistent set is removed and RANSAC is applied to find the next largest set of consistent matching pairs. This procedure is repeated until there are too few remaining matching pairs. If the number of consistent set is one, then a pair of homographies can be computed from the consistent set as T_r and T_r'. Otherwise, thin-plate spline is used to derive T_r and T_r' from the union of the consistent sets.

This procedure is repeated for each $\mathbf{r}_i'$. The transformations for different $\mathbf{r}_i'$ are different. Initially, the transformations computed are between $\mathbf{r}_i'$ and the reference $\mathbf{r}_0$. As the algorithm iterates, the rank-1 matrix $\mathbf{R}$ converges such that $\mathbf{r}_i = \mathbf{r}_0$ for all i. Then, desired transformations between $\mathbf{r}_i'$ and $\mathbf{r}_i$ are obtained.

2.5 Updating of Background Weights

There are three kinds of edges in the images: (1) edges that belong to the background, (2) edges that belong to the reflection, and (3) phantom edges that result from separating non-edge pixels into background and reflection. To discriminate between these edges, three gradient fields are computed from the aligned images, namely $\mathbf{G}_f = DT_b(\mathbf{F}')$, $\mathbf{G}_r = DT_b(T_r'(\mathbf{R}))$, and $\mathbf{G}_b = \mathbf{G}_f - \mathbf{G}_r$. Note that $T_b(\mathbf{F}')$ and $T_b(T_r'(\mathbf{R}))$ make up $\mathbf{F}$ that is used for optimization of background (Sect. 2.3). Salient edges are extracted by thresholding the gradient fields into corresponding binary matrices $\mathbf{H}_f$, $\mathbf{H}_b$, and $\mathbf{H}_r$ at threshold τ. Then, element $[j, k]$ of background weight $\mathbf{W}_b$, denoted as $\mathbf{W}_b[j, k]$, is updated as follows:

$$\mathbf{W}_b[j, k] = \begin{cases} 0, \text{ if majority of } \mathbf{H}_f[j, l], \text{ for } l = 1, \ldots, n, \text{ satisfies} \\ \quad \mathbf{H}_f[j, l] = 1 \text{ and } |\mathbf{G}_b[j, l]| > \alpha|\mathbf{G}_r[j, l]|, \\ 5, \text{ if } \mathbf{H}_f[j, k] = 1 \text{ and } |\mathbf{G}_r[j, k]| > \alpha|\mathbf{G}_b[j, k]|, \\ 5, \text{ if } \mathbf{H}_f[j, k] = 0 \text{ and } (|\mathbf{G}_b[j, k]| > \tau \text{ or } |\mathbf{G}_r[j, k]| > \tau), \\ 1, \text{ otherwise.} \end{cases} \quad (17)$$

The first three conditions correspond, respectively, to background edges, reflection edges, and phantom edges. In other words, background edges are associated with zero weight whereas non-background edges are associated with large weights. The first condition fills the whole row j of $\mathbf{W}_b$ with 0 if the majority of the elements in row j of $\mathbf{H}_f$ satisfies the condition. This condition overrides the 2nd and 3rd conditions. This method facilitates the convergence of $\mathbf{B}$ into a rank-1 matrix. The last condition provides default weight of 1 for the other pixels. In the current implementation, $\tau = 0.05$ and $\alpha = 1.2$.

2.6 Initialization

The initialization of our algorithm CLORD is similar to the min-max alternation method of [23] for reflection separation. Among the superimposed images $\mathbf{f}_i'$ in $\mathbf{F}'$, the one at the neutral viewing angle, denoted as $\mathbf{f}_0'$, is selected as the reference. Since the image intensity is dominated by background intensity, we can apply the transformation update algorithm described in Sect. 2.4 on $\mathbf{f}_i'$ to estimate the transformations T_b and T_b' of the background. Next, an initial estimate of the background image $\mathbf{b}$ is obtained by finding the row-minima of $T_b(\mathbf{F}')$, which is inserted into each column of $\mathbf{B}$. Next, $\mathbf{R}'$ is computed as $\mathbf{F}' - T_b'(\mathbf{B})$. Then, the transformation update algorithm is applied on $\mathbf{r}_i'$ to estimate the transformations T_r and T_r' of the reflection. Finally, an initial estimate of the reflection image $\mathbf{r}$ is obtained by finding the row-maxima of $T_r(\mathbf{R}')$, which is inserted into each column of $\mathbf{R}$.

2.7 Summary

Our CLORD method handles background $\mathbf{B}$ and reflection $\mathbf{R}$ in a uniform manner. So, the optimization of $\mathbf{R}$, updating of background transformations T_b and T_b', and updating of reflection weights $\mathbf{W}_r$ are achieved by applying the same algorithms described in the preceding sections, with $\mathbf{B}$ and $\mathbf{R}$ swapped. The complete algorithm is summarized in Algorithm 1, and the optimization of rank-1 matrix $\mathbf{B}$ (as well as $\mathbf{R}$, with $\mathbf{B}$ and $\mathbf{R}$ swapped) is summarized in Algorithm 2. In the current implementation, λ_1, λ_2, λ_3 are set to 50λ, 2000λ, and 10λ respectively, where $\lambda = 1/\sqrt{\max(m, n)}$ [4].

Algorithm 1. Coupled Low-Rank Decomposition (CLORD)

Input: $\mathbf{F}'$
1 Initialize $\mathbf{B}, \mathbf{R}, T_b, T_b', T_r, T_r'$ (Sect. 2.6).
2 **repeat**
3   Update the weight matrix $\mathbf{W}_b$ (Sect. 2.5).
4   Optimize background $\mathbf{B}$ (Algorithm 2, Sect. 2.3).
5   Estimate the transformations of reflection T_r and T_r' (Sect. 2.4).
6   Update the weight matrix $\mathbf{W}_r$ (Sect. 2.5).
7   Optimize reflection $\mathbf{R}$ (Algorithm 2, Sect. 2.3).
8   Estimate the transformations of background T_b and T_b' (Sect. 2.4).
9 **until** *convergence*;
   Output: B, R.

Algorithm 2. Updating of Rank-1 Matrix

Input: $\mathbf{F}$
1 Initialize $\mathbf{Y}_1, \mathbf{Y}_2, \mathbf{Y}_3, \mu > 0, \rho > 1$.
2 Initialize $\mathbf{B}, \mathbf{A}, \mathbf{E}, \mathbf{C}, \mathbf{N}$ to 0.
3 **repeat**
4      $\mathbf{U}, \mathbf{S}, \mathbf{V} = \text{SVD}(\mathbf{B} + \mathbf{Y}_2/\mu)$.
5      $\mathbf{A} = \mathbf{U}\mathbf{S}_1\mathbf{V}^\top$.
6      $\mathbf{E} = S_{\lambda_1/\mu}(\mathbf{F} - \mathbf{B} - \mathbf{N} + \mathbf{Y}_1/\mu)$.
7      $\mathbf{C} = S_{\lambda_3 \mathbf{W}_b/\mu}(\mathbf{DB} + \mathbf{Y}_3/\mu)$.
8      $\mathbf{N} = \frac{\mu}{\lambda_2+\mu}(\mathbf{F} - \mathbf{B} - \mathbf{E} + \mathbf{Y}_1/\mu)$.
9   Update $\mathbf{B}$ according to Eq. 15.
10      $\mathbf{Y}_1 = \mathbf{Y}_1 + \mu(\mathbf{F} - \mathbf{B} - \mathbf{E} - \mathbf{N})$.
11      $\mathbf{Y}_2 = \mathbf{Y}_2 + \mu(\mathbf{B} - \mathbf{A})$.
12      $\mathbf{Y}_3 = \mathbf{Y}_3 + \mu(\mathbf{DB} - \mathbf{C})$.
13      $\mu = \rho\mu$.
14 **until** *convergence*;
   Output: B

---

3 Experiments and Discussions

This section compares the performance of CLORD and those of recent existing methods denoted as LI [14], GUO [8], and XUE [24]. As the program for XUE is not publicly available, we compare it only with the results published in [24] (Sects. 3.1 and 3.2). To handle color images, the methods are applied to each of the R, G and B channels and their outputs are combined into the final results. Test programs were implemented in Matlab, and ran on a PC with Intel Core 3.5 GHZ CPU and 32 GB RAM.

3.1 Planar Background and Reflection Scenes

This test evaluates the methods' base-case performance on images of planar background and reflection scenes. This scenario is valid when all objects are far

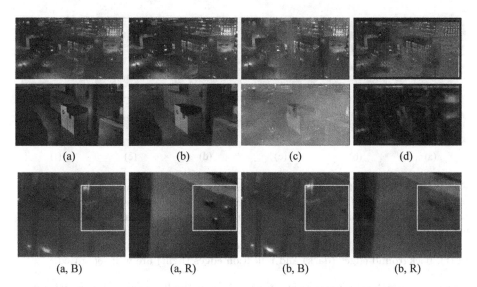

| (a) | (b) | (c) | (d) |
| (a, B) | (a, R) | (b, B) | (b, R) |

Fig. 1. Test results of night scene images with planar background and reflection. (Row 1) Recovered background, (row 2) separated reflection, (row 3) zoom-in views of bottom-right corner of the images. White boxes show that XUE's background has remnants of reflection. (a) CLORD, (b) XUE, (c) LI, (d) GUO. (B) Background, (R) reflection.

away from the camera. 5 images of a night scene captured through a window, and 5 images of indoor scene with glass reflection were used for the test. These images were obtained from [24] and ground truth was not available.

For the night scene, Fig. 1 shows that the results of CLORD and XUE are visually accurate whereas GUO and LI cannot separate the background and the reflection well. This could be because GUO and LI explicitly optimize the background but not the reflection, which implies that better reflection recovery can improve background recovery. There is a subtle difference between the results of CLORD and XUE (Fig. 1, row 3). The background recovered by XUE has some remnants of reflection. On the other hand, the background recovered by CLORD does not have reflection, and its recovered reflection is much more complete than that of XUE. This difference could be due to XUE's non-locally linear flow filed.

The indoor results of CLORD and XUE are shown in Fig. 2(b) and (c). Consistent with the previous test, XUE's background has more remnants of reflection compared to that of CLORD. Figure 2(d) and (e) show that input images transformed by the computed T_b align very well to the reference background. Figure 2(f) visualizes the weight W_b of the reference image in Fig. 2(a). It shows that the edge of background and reflection objects are well separated. In this test, CLORD takes 8 min to process 5 images of size 1152×648 whereas XUE takes about 20 min.

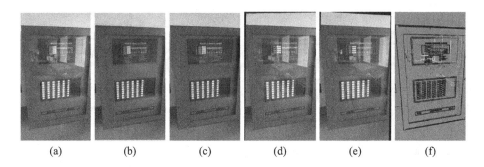

(a)   (b)   (c)   (d)   (e)   (f)

Fig. 2. Test results of indoor scene images with planar background and reflection. (a) Reference image, (b) CLORD, (c) XUE, (d, e) input images transformed by the computed T_b align very well to the reference background in (a), (f) W_b associated with (a). As shown in (b) and (c), XUE's background has more remnants of relection compared to that of CLORD, their recovered reflections are similar and are omitted. In (f), background edges (dark) and reflection edges (light) are well separated.

Table 1. Normalized cross correlation of various methods tested on images with non-planar background and reflection scenes.

Method	CLORD	XUE	LI	GUO
Background	**0.94**	0.90	0.79	0.77
Reflection	**0.83**	0.75	0.61	0.69

3.2   Non-planar Background and Reflection Scenes

This test evaluates the methods' performance on images of non-planar background and reflection scenes. This scenario is valid when the objects are located at different depths near the camera. 5 images from [24] captured at different viewing angles were used, and ground truths of the background and reflection were available. The degree of match between the recovered images and the ground truths were measured as normalized cross correlation (NCC).

Figure 3 shows that the results of CLORD and XUE are visually accurate. On the other hand, GUO and LI cannot separate the background and the reflection well because they do not explicitly optimize the reflection. In addition, GUO is not designed to handle non-planar scenes. In comparison, CLORD's recovered background is more accurate, and XUE's background has some distortions around the head of the toy frog. Table 1 confirms that CLORD is more accurate than the other methods for non-planar scenes.

3.3   Scene Distance

This experiment investigates the effect of the distances of background objects and reflection objects from the reflecting glass on the methods' performance. A background object was placed behind a piece of glass, and a reflection object

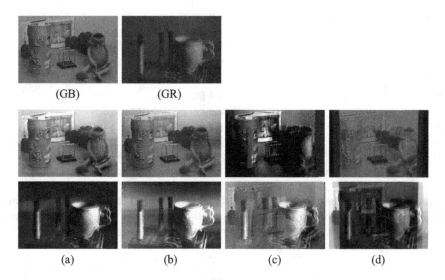

Fig. 3. Test results on images of non-planar background and reflection scenes. (GB) Ground truth background, (GR) ground truth reflection. (Row 2) Recovered background, (row 3) separated reflection. (a) CLORD, (b) XUE, (c) LI, (d) GUO.

was placed in front. The ratio d_r of the distances of the reflection object and the background object from the glass was varied from 4/8 to 8/8 = 1 in steps of 1/8. For each distance ratio d_r, ground truths and 5 superimposed images were captured at different viewing angles. The methods were tested, and NCC between ground truths and recovered images were measured.

Figure 4 shows that CLORD is more accurate than LI and GUO. CLORD's background NCC exceeds 0.9 for $d_r \leq 6/8$. On the other hand, LI's background NCC is less than 0.9, and GUO's background NCC hovers around 0.6. The accuracies of CLORD and LI decrease gradually with increasing d_r as expected because the background scene and virtual scene become more coplanar. The accuracy of GUO remains low for all d_r. The methods' reflection NCC shows a similar trend as their background NCC.

3.4   Degree of Non-planarity

This experiment investigates the effect of the degree of non-planarity on the methods' performance. Two background objects were placed behind a piece of glass, and a reflection object was placed in front. The ratio of the distance of the reflection object and the nearer background object from the glass was set at 1/2. The further background object was placed behind the nearer object, and its distance ratio d_b from the nearer object varied from 0 to 4/8 in steps of 1/8. For each distance ratio d_b, ground truths and 5 superimposed images were captured at different viewing angles. The methods were tested, and NCC between ground truths and recovered images were measured.

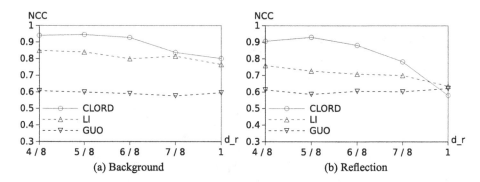

Fig. 4. Normalized cross correlation (NCC) of various methods at varying distance ratio d_r of reflection with respect to background. Smaller d_r means greater depth disparity between background and reflection objects.

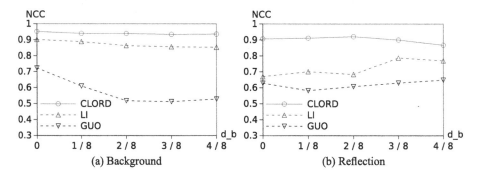

Fig. 5. Normalized cross correlation (NCC) of various methods at varying distance ratio d_b of the further background object from the nearer background object. Larger d_b means greater non-planarity.

Figure 5 shows that CLORD's background NCC exceeds 0.9 for all d_b, whereas LI's background NCC is less than 0.9. GUO's background NCC is slightly above 0.7 at $d_b = 0$ when the background scene is planar, and it decreases with increasing d_b, which makes the background scene more and more non-planar. CLORD's reflection NCC is around 0.9 for all d_b. LI's reflection NCC is slightly below 0.7 at $d_b = 0$, and it increases with increasing d, suggesting that LI can identify reflection more accurately for non-planar scenes. GUO's reflection NCC is slightly above 0.6. The test results show that CLORD is more accurate than LI and GUO for the non-planar scenes.

4   Conclusion

This paper presented a multiple-view method for separating reflection and background in unaligned images by coupled low-rank decomposition (CLORD). It is based on the method of robust PCA. It models the background and reflection as rank-1 matrices in a uniform manner, which are decomposed according to different transformations that align the background images and the reflection images separately. Comprehensive test results show that CLORD is more accurate and robust than recent related methods, especially for images with non-planar scenes and global reflection.

References

1. Agrawal, A., Raskar, R., Nayar, R.K., Li, Y.Z.: Removing photography artifacts using gradient projection and flash-exposure sampling. ACM Trans. Graph. **24**(3), 828–835 (2005)
2. Bay, H., Ess, A., Tuytelaars, T., Gool, L.V.: Speeded-up robust features (SURF). Comput. Vis. Image Underst. **110**(3), 346–359 (2008)
3. Cai, J.F., Candès, E.J., Shen, Z.: A singular value thresholding algorithm for matrix completion. SIAM J. Optim. **20**(4), 1956–1982 (2010)
4. Candès, E.J., Li, X., Ma, Y., Wright, J.: Robust principal component analysis? J. ACM **58**(3), 1–37 (2011)
5. Chaudhuri, S., Velmurugan, R., Rameshan, R., Deconvolution, B.I.: Methods and Convergence. Springer, New York (2014)
6. Farid, H., Adelson, E.H.: Separating reflections from images by use of independent component analysis. J. Optical Soc. Am. A **16**(9), 2136–2145 (1999)
7. Gai, K., Shi, Z.W., Zhang, C.S.: Blindly separating mixtures of multiple layers with spatial shifts. In: Proceedings of CVPR (2008)
8. Guo, X., Cao, X., Ma, Y.: Robust separation of reflection from multiple images. In: Proceedings of CVPR (2014)
9. Jiang, X., Lai, J.: Sparse and dense hybrid representation via dictionary decomposition for face recognition. IEEE Trans. PAMI **37**(5), 1067–1079 (2015)
10. Kong, N., Tai, Y., Shin, J.S.: A physically-based approach to reflection separation: From physical modeling to constrained optimization. IEEE Trans. PAMI **36**(2), 209–221 (2014)
11. Leow, W.K., Cheng, Y., Zhang, L., Sim, T., Foo, L.: Background recovery by fixed-rank robust principal component analysis. In: Wilson, R., Hancock, E., Bors, A., Smith, W. (eds.) CAIP 2013. LNCS, vol. 8047, pp. 54–61. Springer, Heidelberg (2013). doi:10.1007/978-3-642-40261-6_6
12. Levin, A., Weiss, Y.: User assisted separation of reflections from a single image using a sparsity prior. IEEE Trans. PAMI **29**(9), 1647–1654 (2007)
13. Levin, A., Zomet, A., Weiss, Y.: Separating reflections from a single image using local features. In: Proceedings of CVPR (2004)
14. Li, Y., Brown, M.S.: Exploiting reflection change for automatic reflection removal. In: Proceedings of ICCV (2013)
15. Lin, Z., Chen, M., Wu, L., Ma, Y.: The augmented Lagrange multiplier method for exact recovery of corrupted low-rank matrices. Technical report, UIUC (2009)
16. Z. Lou and T. Gevers. Image alignment by piecewise planar region matching. IEEE Trans. Multimedia, 16(7), 2014
17. Lowe, D.G.: Distinctive image features from scale-invariant keypoints. Int. J. Comput. Vis. **60**(2), 91–110 (2004)
18. Oh, T.H., Lee, J.Y., Tai, Y.W., Kweon, I.S.: Robust high dynamic range imaging by rank minimization. IEEE Trans. PAMI **37**(6), 1219–1232 (2015)

19. Oh, T.H., Tai, Y.W., Bazin, J.C., Kim, H., Kweon, I.S.: Partial sum minimization of singular values in robust pca: Algorithm and applications. IEEE Trans. PAMI **38**(4), 744–758 (2016)

20. Peng, Y., Ganesh, A., Wright, J., Xu, W., Ma, Y.: Rasl: robust alignment by sparse and low-rank decomposition for linearly correlated images. IEEE Trans. PAMI **34**(11), 2233–2246 (2012)

21. Schechner, Y.Y., Kiryati, N., Basri, R.: Separation of transparent layers using focus. Int. J. Comput. Vis. **39**(1), 25–39 (2000)

22. Schechner, Y.Y., Shamir, J., Kiryati, N.: Polarization-based decorrelation of transparent layers: the inclination angle of an invisible surface. In: Proceedings of ICCV (1999)

23. Szeliski, R., Avidan, S., Anandan, P.: Layer extraction from multiple images containing reflections and transparency. In: Proceedings of CVPR (2000)

24. Xue, T.F., Rubinstein, M., Liu, C., Freeman, W.T.: A computational approach for obstruction-free photography. ACM Trans. Graph. **34**(4), 1–11 (2015)

25. Yang, J.L., Li, H.D., Dai, Y.C., Tan, R.T.: Robust optical flow estimation of double-layer images under transparency or reflection. In: Proceedings of CVPR (2016)

Fast and Easy Blind Deblurring Using an Inverse Filter and PROBE

Naftali Zon, Rana Hanocka, and Nahum Kiryati[(✉)]

School of Electrical Engineering, Tel Aviv University, Tel Aviv, Israel
nk@eng.tau.ac.il

Abstract. PROBE (**P**rogressive **R**emoval **o**f **B**lur R**e**sidual) is a recursive framework for blind deblurring. PROBE is neither a functional minimization approach, nor an open-loop sequential method where blur kernel estimation is followed by non-blind deblurring. PROBE is a feedback scheme, deriving its unique strength from the closed-loop architecture. Thus, with the rudimentary modified inverse filter at its core, PROBE's performance meets or exceeds the state of the art, both visually and quantitatively. Remarkably, PROBE lends itself to analysis that reveals its convergence properties.

1 Introduction

Non-blind deblurring, *i.e.* the recovery of a latent image u given a blurred image g and the blur kernel h, can be formulated as a functional minimization problem

$$\hat{u} = \min_{u} \frac{\gamma}{2}||h * u - g||_2^2 + Q(u). \tag{1}$$

The data term $||h * u - g||_2^2$ reflects Gaussian noise, and the regularization term $Q(u)$ represents *a-priori* information about image structure. This formulation is equivalent to a maximum a-posteriori (MAP) statistical estimation problem.

Mainstream *blind* deblurring schemes fall into two major categories: *sequential* approaches first estimate the blur kernel then employ non-blind deblurring; *parallel* approaches simultaneously estimate the latent image and the blur kernel. Sequential approaches which utilize functional minimization for the non-blind deblurring phase, do so as in Eq. (1). Following the same energy minimization paradigm, a parallel approach should supposedly minimize a joint functional

$$(\hat{u}, \hat{h})_{\text{MAP}} = \min_{u,h} \frac{\gamma}{2}||h * u - g||_2^2 + Q(u) + R(h), \tag{2}$$

where the blur regularizer $R(h)$ represents *a-priori* information about the blur point spread function (PSF). This is equivalent to maximizing the posterior

$$(\hat{u}, \hat{h})_{\text{MAP}} = \max_{u,h} p(u, h|g) = \max_{u,h} p(g|u, h)p(h)p(u), \tag{3}$$

N. Zon, R. Hanocka—equal contributors.

© Springer International Publishing AG 2017
M. Felsberg et al. (Eds.): CAIP 2017, Part II, LNCS 10425, pp. 269–281, 2017.
DOI: 10.1007/978-3-319-64698-5_23

and is referred to as (parallel) $\mathrm{MAP}_{u,h}$. The variational problem (2) is solved via alternate minimization (AM).

Classic parallel true $\mathrm{MAP}_{u,h}$ approaches [2,18] are elegant, but known [1, 4,10,14] to fail in practice. Levin *et al.* [10] demonstrated that simultaneous estimation of u and h using parallel $\mathrm{MAP}_{u,h}$ with common image priors $Q(u)$, actually favors the trivial solution $\hat{u} = g$ and $\hat{h} = \delta$. This means that the observed blur in g is assumed to be due to blur in the latent image u, rather than due to the blur process h. Such assumption obviously results in a blurred solution $\hat{u}$. Therefore, the true $\mathrm{MAP}_{u,h}$ minimizers $(\hat{u}, \hat{h})_{\mathrm{MAP}}$ are *not* the desired solution.

Works which claim to follow the true $\mathrm{MAP}_{u,h}$ paradigm, bypass the (undesired) true minimizers using ad-hoc steps to minimize a non-equivalent functional, referred to as parallel ad-hoc $\mathrm{MAP}_{u,h}$. They usually increase the likelihood weight (γ in Eq. (2)) during minimization and apply the blur kernel constraints via sequential-projection (see [14]). Perrone and Favaro [14] showed that in practice [2] used ad-hoc steps, which do not minimize the true $\mathrm{MAP}_{u,h}$. Levin *et al.* [10] demonstrated the same for [15]. These ad-hoc steps avoid the true $\mathrm{MAP}_{u,h}$ minimizers [10,14], and often lead to the desired solution [10].

While parallel ad-hoc $\mathrm{MAP}_{u,h}$ is a viable alternative to true $\mathrm{MAP}_{u,h}$, modern works have had better success utilizing the sequential approach [6,7,9]. Levin *et al.* [10] showed that the sequential approach is more stable and a lower dimensional problem. We refer to the sequential class of algorithms which use the MAP paradigm for kernel estimation as ad-hoc $\mathrm{MAP}_{u_c,h}$. It usually applies ad-hoc steps as in parallel ad-hoc $\mathrm{MAP}_{u,h}$, but additionally uses unnatural, exceedingly sparse priors [3,7,17] to obtain a *cartoon image* u_c. The cartoon image u_c is then discarded, and a separate non-blind deblurring process is invoked.

The cartoon image emphasizes salient edges and suppresses weak details in flat regions [16]. The cartoon image is useful for accurate kernel estimation, conceivably since it directs the kernel estimate to stronger step-edges, diverting away from the weak edges possibly related to noise. Thus, the cartoon image functions as a regularizer. Additionally, Levin *et al.* [10] showed that increasing the likelihood weight results in a cartoon-like image, with a higher prior energy $Q(u) = \|\nabla u\|_p^p$, and actually increases the true $\mathrm{MAP}_{u,h}$ energy.

From another perspective, ad-hoc $\mathrm{MAP}_{u_c,h}$ exploits the ill-posedness of the problem by focusing on an equivalent but easier problem: the outcome of blurring a cartoon image u_c is similar to blurring the true latent image u. Formally, $g_c \approx g$ where $g_c = u_c * h$ and $g = u * h$. This has been observed in [6] for discrete images. The cartooning effect can simply be achieved via shock filtering [13]. This can be used to steer $\mathrm{MAP}_{u_c,h}$ towards the desired minimum.

The empirical failure of the parallel $\mathrm{MAP}_{u,h}$ approach, and its explanation by Levin *et al.* [10], have driven the blind deblurring field towards employing the sequential deblurring framework, where blur kernel estimation is followed by non-blind deblurring. However, the sequential framework is limited, since any inaccuracy in either PSF estimation or non-blind deblurring leads to irreparable deblurring errors. Thus, successful application of the sequential framework requires accurate, sophisticated, complex and computationally expensive PSF estimation and non-blind deblurring algorithms. The proposed PROBE framework is a superior alternative to both the parallel $\text{MAP}_{u,h}$ and sequential approaches.

2   PROBE Framework

Recently, Hanocka and Kiryati [5] presented a successful blind deblurring algorithm, based on sophisticated and computationally expensive algorithmic components. For blur-kernel estimation they extracted an ad-hoc $\text{MAP}_{u_c,h}$ algorithm from the implementation of Kotera et al. [7]. For non-blind deblurring they adopted the non-blind version of Bar et al. [1], based on the Γ-convergence approximation of Mumford-Shah regularization. We claim that the successful result of [5] is primarily due to its special architecture, rather than due to the sophisticated algorithmic components used. We specify this architecture and call it PROBE, shorthand for **P**rogressive **R**emoval **o**f **B**lur R**e**sidual. We show that [5] is a needlessly expensive version of PROBE. We replace the non-blind deblurring module by a trivial modified inverse filter, show that it yields results as good as those of [5] or better, and are similar or superior to the state of the art. Furthermore, we show that by simplifying [5], PROBE not only yields faster processing and better experimental results, but unlike previous works lends itself to analysis. Specifically, we obtain analytic results characterizing the convergence of PROBE. Thus, the PROBE architecture makes the blind deblurring scheme better than its non-blind deblurring component.

PROBE is illustrated in Fig. 1(left). Initially, the switch is at the lower position, and the system coincides with the familiar sequential deblurring scheme, consisting of PSF estimation followed by non-blind deblurring. After the first iteration, the switch is thrown to the upper position. Then, the imperfect outcome of the current iteration is fed back as input to the next iteration, i.e., $g^l \leftarrow \hat{u}^{l-1}$ where superscripts denote iteration numbers. The PSF estimator identifies the *residual blur* remaining in the current input image g^l. The non-blind deblurring module removes some of the residual blur, leading to an even better deblurring result $\hat{u}^l$. Recursion continues until a stopping criterion is met. The feedback is the key to PROBE's superior performance: unlike an open-loop system, feedback can potentially correct errors due to imperfect system components.

PROBE is fundamentally different than the parallel $\text{MAP}_{u,h}$ approach, illustrated in Fig. 1(right). In PROBE, the input-image for the non-blind deblurring module varies between iterations. In contrast, in parallel $\text{MAP}_{u,h}$ the input to the deblurring module is fixed throughout the iterative process. Furthermore, in PROBE the outcome of the PSF estimation module should be the *residual* blur, gradually coming close to an impulse function. In parallel $\text{MAP}_{u,h}$, the output of the blur estimation module should approach the original blur kernel. Thus, ideally, $\lim_{l\to\infty} \hat{h}^l_{\text{PROBE}} = \delta$ while $\lim_{l\to\infty} \hat{h}^l_{\text{MAP}} = h$.

The ability of PROBE's feedback scheme to compensate for component inaccuracies is a key observation. We replace the intricate non-blind deblurring algorithm of [1] that is used in [5] by the simplest, fastest and crudest practical alternative: the modified inverse filter, given by the transfer function

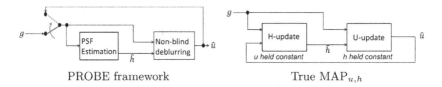

PROBE framework   True MAP$_{u,h}$

Fig. 1. Left: PROBE framework adds a feedback loop to the familiar sequential approach to progressively correct for errors. Right: parallel MAP$_{u,h}$ approach which jointly minimizes a unified energy functional.

$$H_{\text{RI}} = \frac{H^*}{|H|^2 + C}, \tag{4}$$

where H is the assumed (current, residual) blur. H_{RI} is a smooth approximation of the pseudo-inverse filter and is also a crude approximation of the Wiener filter with unknown noise statistics, approximated by the constant C. Before demonstrating PROBE's superior experimental performance, we mathematically analyze its convergence.

3   PROBE Convergence

Using the modified inverse filter simplifies PROBE substantially. However, effective PSF estimation algorithms are intricate. To facilitate convergence analysis, we model PROBE's blur estimation module by an *oracle*, providing the exact blur kernel that relates the input image g^l to the latent image u.

PROBE recursively estimates the residual blur kernel and employs non-blind deblurring to reduce the residual blur. The input for each PROBE iteration l is a different blurred image g^l given by

$$g^l = h_o^l * u + n^l, \tag{5}$$

where the blur kernel h_o^l provided by the oracle, relates the current image g^l to the latent image u. The blur kernel is forwarded to the modified inverse filter for the non-blind deblurring phase. In the current iteration, the transfer function of the modified inverse filter becomes

$$H_{\text{RI}}^l = \frac{(H_o^l)^*}{|H_o^l|^2 + C}, \tag{6}$$

where C is a regularization parameter and H_o^l is the Fourier transform of h_o^l.

Equations (5) and (6) are the basis for convergence analysis. In the frequency domain, the input G^{l+1} for the coming iteration $l+1$ is a filtered version of G^l:

$$G^{l+1} = H_{\text{RI}}^l G^l = H_o^l H_{\text{RI}}^l U + N^l H_{\text{RI}}^l = \frac{|H_o^l|^2}{|H_o^l|^2 + C} U + N^{l+1} \tag{7}$$

where N^l and U are the Fourier transforms of n^l and u respectively. The blur kernel provided by the oracle in iteration $l + 1$ is therefore simply related to the blur kernel the oracle had given in the previous iteration l by

$$H_o^{l+1} = \frac{|H_o^l|^2}{|H_o^l|^2 + C}. \tag{8}$$

PROBE convergence is indicated by convergence of the blur kernel provided by the oracle, *i.e.*, when $H_o^{l+1} = H_o^l$. Equation (8) implies that H_o^l is strictly positive for any $l > 0$. The final filter H_o^∞ can obtained by solving Eq. (8) with $H_o^{l+1} = H_o^l = H_o^\infty$. The resulting equation has three solutions,

$$H_o^\infty = 0 \quad or \quad H_o^\infty \approx C \quad or \quad H_o^\infty \approx 1 - C$$

Note that the large and small roots of $\frac{1 \pm \sqrt{1-4C}}{2}$ are approximated by $1 - C$ and C, respectively. Dynamically, for $C > \frac{1}{4}$ the filter converges to zero, while for $C < \frac{1}{4}$ it can be shown that:

$$\begin{cases} H_o^{l+1} < H_o^l & H_o^0 \in (0, C) \\ H_o^{l+1} > H_o^l & H_o^0 \in (C, 1 - C) \\ H_o^{l+1} < H_o^l & H_o^0 \in (1 - C, +\infty) \end{cases} \tag{9}$$

The final PROBE blur kernel H_o^∞ at each spatial frequency converges to one of the two stable solutions 0 and $1 - C$. As seen in Eq. (9), the third solution C is unstable. In the case of defocus blur, where the original PSF H_o^0 that blurred the latent image can be assumed to be a monotonically decreasing low-pass filter, the final output H_o^∞ of the oracle is an ideal low-pass filter assuming values of either 0 or $1 - C$. For small values of C, H_o^∞ approaches 1 at a wide range of spatial frequencies, and in the limit $\lim_{C \to 0} H_o^\infty = 1$ and $\lim_{C \to 0} h_o^\infty = \delta$. This is the ideal outcome, since the oracle indicates that all the blur has been removed. It is not surprising, since for $C \approx 0$, the modified inverse filter resembles the inverse filter. We will soon see that noise sets a lower limit on useful C values.

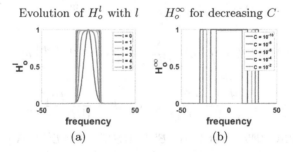

Evolution of H_o^l with l H_o^∞ for decreasing C

(a)   (b)

Fig. 2. (a) Evolution of the oracle-provided PSF H_o^l with PROBE iteration number l, starting with a known Gaussian kernel at $l = 0$ and converging to an ideal low-pass filter. (b) H_o^∞ approaches a flat spectrum as $C \to 0$, corresponding to an impulsive blur kernel h_o^∞ and implying that all blur has been removed.

The convergence process is visualized for the one-dimensional case in Fig. 2. Figure 2a shows the evolution of H_o^l, corresponding to the PSF estimated by the oracle, over several iterations. Starting with a Gaussian blur kernel represented by H_o^0, and $C = 10^{-2}$, H_o^l converges to an ideal low-pass filter as predicted by the analysis. Figure 2b shows that as C approaches 0, H_o^∞ approaches a flat spectrum, meaning the residual blur kernel approaches an impulse function and all blur has been removed.

4   PROBE Residual Error Dynamics

The convergence rate of the oracle kernel coefficients H_o^l to their stable solutions 0 or $1 - C$ varies with the regularization parameter C. Specifically, it can be shown that PROBE's convergence rate increases as we decrease C. This section analyzes the dynamics of PROBE's residual blur removal process, elucidating additional aspects of its convergence.

In PROBE, we refer to the difference between the restoration outcome in the current iteration and true latent image as *residual* error. The residual error evolves along the iterative process, ideally decreasing until convergence. Following iteration l, combining the oracle model (5) with the non-blind deblurring process, the residual blur can be expressed as

$$\epsilon^l = u - \hat{u}^l = u * (\delta - h_o^l * h_{\mathrm{RI}}^l) - n^l * h_{\mathrm{RI}}^l. \tag{10}$$

Viewing u, n^l as random processes, the current mean squared residual error is

$$\mathrm{MSE}^l \triangleq \mathrm{MSE}(\epsilon^l) = R_\epsilon(\mathbf{0}) = \int \left[S_u |1 - H_o^l H_{\mathrm{RI}}^l|^2 + S_n^l |H_{\mathrm{RI}}^l|^2 \right] d\mathbf{f}, \tag{11}$$

where $R_\epsilon(\boldsymbol{\tau})$ is the residual error autocorrelation, and S_u and S_n^l are the power spectral densities (PSD) of the latent image u and noise n^l respectively.

Substituting H_{RI} at the respective iteration, the MSE at the current and next iterations can be expressed as

$$\mathrm{MSE}^l = \int \left[\frac{S_u C^2}{(|H_o^l|^2 + C)^2} + \beta^l S_n^l \right] d\mathbf{f} \tag{12}$$

$$\mathrm{MSE}^{l+1} = \int \left[\alpha^l \frac{S_u C^2}{(|H_o^l|^2 + C)^2} + \beta^{l+1} \beta^l S_n^l \right] d\mathbf{f} \tag{13}$$

where

$$\alpha^l = \frac{(|H_o^l|^2 + C)^2}{(|H_o^{l+1}|^2 + C)^2} \quad \text{and} \quad \beta^l = \frac{|H_o^l|^2}{(|H_o^l|^2 + C)^2}. \tag{14}$$

We define PROBE's boost factor $\mathcal{B}^l \triangleq MSE^l - MSE^{l+1}$. We wish to ensure $\mathcal{B}^l > 0$, implying reduction of the residual error between consecutive iterations (boosting). Substituting Eqs. (12) and (13) we obtain

$$\mathcal{B}^l = \int S_u \cdot (1 - \alpha^l) \frac{C^2}{(|H_o^l|^2 + C)^2} d\mathbf{f} - \int S_n^l \cdot (\beta^{l+1} - 1)\beta^l d\mathbf{f}, \tag{15}$$

where the first (left) integral is a signal term and the second (right) integral is a noise term. Since Eq. (15) expresses $\mathcal{B}^l$ as the difference between the signal and noise term, we wish to increase the signal term and reduce the noise term.

Consider the noise term. Since β^l is non-negative, the noise term contributes to reduction of the residual error when $\beta^{l+1} < 1$. From Eqs. (8), (9) and (14), we derive conditions on C and H_o^0 to ensure $\beta^l < 1$:

$$H_o^0 \in (0, C) \vee (1 - C, 1) \;\rightarrow\; H_o^{l+1} < H_o^l \;\rightarrow\; \frac{|H_o^l|}{|H_o^l|^2 + C} < 1 \;\rightarrow\; \beta^l < 1 \quad (16)$$

In the signal term, since the signal power S_u is multiplied by $(1 - \alpha^l)$, the signal term is positive when $\alpha^l < 1$, contributing to residual noise reduction. From Eqs. (9) and (14), we derive conditions on C and H_o^0 to ensure $\alpha^l < 1$:

$$H_o^0 \in (C, 1 - C) \;\rightarrow\; H_o^{l+1} > H_o^l \;\rightarrow\; \alpha^l < 1 \quad (17)$$

Taken together, the constraints (16) and (17) on C are sufficient but not necessary. In fact, they are mutually exclusive. At high SNR cases, where $S_u \gg S_n$, we prefer to increase the (large) signal term (by decreasing C) even if it increases the (small) noise term. However, the noise term increases with PROBE iterations motivating a higher C. Nevertheless in this work our analysis assumes a constant C throughout the iterative PROBE process.

Once convergence has been reached, $H^{l+1} = H^l$. Substituting in (14), $\lim_{l \to \infty} \alpha^l = 1$ and and $\lim_{l \to \infty} \beta^l = 1$. Thus, after PROBE convergence, once α and β equal 1, both the signal and noise terms fall out, leaving $\mathcal{B}^\infty = 0$, i.e., further boosting is not possible.

5   From Theory to Practice

The analysis in Sects. 3 and 4 relies on modelling the blur-kernel estimation model by an oracle. However, in an actual PROBE system the oracle model no longer holds. In this respect, our PROBE system follows [5] and employs an ad-hoc $\mathrm{MAP}_{u_c, h}$ PSF estimation algorithm extracted from the implementation of Kotera et al. [7]. Compared to the oracle, the $\mathrm{MAP}_{u_c, h}$ PSF estimator adds an extra layer of error, appearing in the PROBE iterations as

$$g^l = (h_{\mathrm{MAP}_{u_c, h}}^l + n_{\mathrm{MAP}_{u_c, h}}^l) * u + n^l = h_{\mathrm{MAP}_{u_c, h}}^l * u + \tilde{n}^l, \quad (18)$$

where $\tilde{n}^l \triangleq n_{\mathrm{MAP}_{u_c, h}}^l * u + n^l$. Therefore, using an ad-hoc kernel estimation $h_{\mathrm{MAP}_{u_c, h}}$ rather than an oracle leads to higher effective noise $\tilde{n}$.

To corroborate our analysis, we compare the predicted, simulated and experimental PSNR obtained along PROBE iterations. $\mathrm{PSNR}^l = -10 \log_{10} \mathrm{MSE}^l$, since all gray levels are scaled to $[0, 1]$.

– Given a test image with synthetic blur and known additive noise, and assuming oracle PSF estimation in PROBE, we predicted MSE^l (hence PSNR^l) using Eq. (12), where we estimated S_u from the image, and evolved S_n^l from the spectral density S_n of the additive noise.

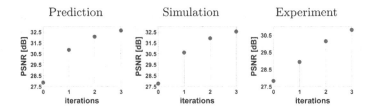

Fig. 3. PSNR for the first three PROBE iterations, based on analytic prediction (left), simulation (middle) and experimental results (right). Successful elimination of residual blur, and PSNR improvement, are consistently observed. The prediction and simulation results, based on the oracle model, are slighty superior to the experimental results.

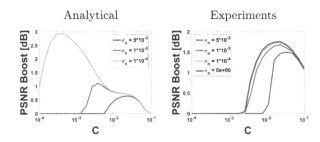

Fig. 4. PSNR boost in the 2nd PROBE iteration, breaking the limit of the sequential approach. Noise (σ_n) effects the PSNR boost and the range of possible C values. The real experiment is subject to an effective noise floor, due to imperfect kernel estimation.

- With the same blurred and noisy image, again assuming the oracle model, we *simulated* PROBE using its iterative equation, computing $PSNR^l$ by comparison to the latent image.
- We obtained $PSNR^l$ *experimentally* using an actual PROBE system with an ad-hoc $MAP_{u_c,h}$ kernel estimator.

The predicted, simulated and experimental results of three PROBE iterations are compared in Fig. 3. In all three evaluation approaches, PSNR improves in the iterative process. The prediction and simulation results, based on the oracle model, are very similar. They are both about 2dB better than the experimental results, obtained using a real PSF estimator rather than the oracle model.

A sequential algorithm for non-blind deblurring, consisting of blur-kernel estimation followed by non-blind deblurring, typically improves PSNR. The sequential approach coincides with the first PROBE iteration. With PROBE, additional iterations further boost the PSNR. The actual boost depends on the level of additive noise. Figure 4(left) shows the analytically predicted PSNR boost, using the oracle blur-kernel estimation model, as a function of C, for several values of σ_n, where $S_n^0 = \sigma_n^2$. Figure 4(right) shows the corresponding graphs for an actual experimental setup. The imperfect PSF estimation in the experimental case creates an effective noise floor, as described by Eq. (18), and that noise poses an effective lower limit on useful values of the regularization parameter C.

6   Experimental Evaluation

PROBE's building-blocks are an ad-hoc $\text{MAP}_{u_C,h}$ kernel estimator extracted from the implementation of Kotera *et al.* [7] and the low cost modified inverse filter (Eq. 6). Typical PROBE operation is shown in Figs. 5 (blurred Lena) and 6 (out-of-focus image acquired using a smartphone). We show the visual improvement by iteration, and the shrinkage of the residual blur kernels.

Figure 7 compares PROBE's deblurring result to highly sophisticated state of the art methods [3, 7, 9, 15]. Starting with the same Gaussian blurred Lena image, PROBE, using a modified inverse filter as its non-blind deblurring module, yields the best visual outcome (compare pupils) and the highest PSNR.

We provide systematic quantitative performance evaluation using the dataset of Levin *et al.* [10]. Figure 8 shows one blurred image from the dataset and its restoration using PROBE and alternative state of the art methods.

For quantitative comparison accross the dataset we employ two error ratio metrics, see Fig. 9. The error ratio is defined as $\frac{\text{SSD}(u,\hat{u})}{\text{SSD}(u,\hat{u}|h)}$ where the sum of squared differences (SSD) is equal to MSE up to a scalar factor. The numerator is the difference between the latent image and its blind restoration using the method under test, while the denominator differs between the absolute error ratio and the relative error ratio. It is customary to present the error ratio statistic over the whole dataset as the cumulative distribution of individual image error ratios.

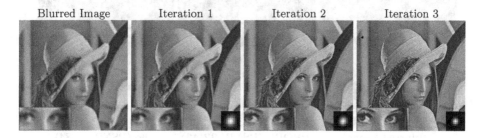

Fig. 5. PROBE deblurs Lena. Displayed: the blurred image and PROBE results by iteration with the corresponding shrinking residual blur kernels superimposed.

Fig. 6. PROBE deblurs an out-of-focus image obtained with a smartphone. Displayed: The blurred image and PROBE results by iteration with the corresponding shrinking residual blur kernels superimposed.

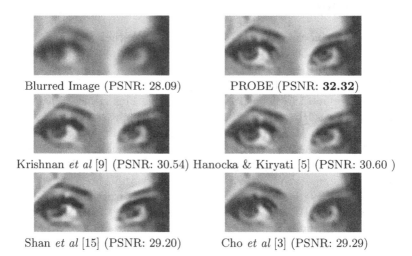

Fig. 7. Gaussian blurred Lena and the restoration results obtained using PROBE and five state-of-the-art blind deblurring algorithms. Only the eye region is shown. Note the pupils.

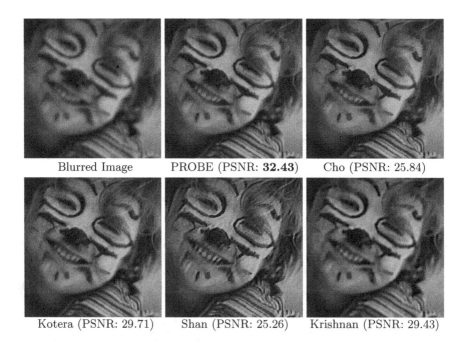

Fig. 8. Featured example from database [10]

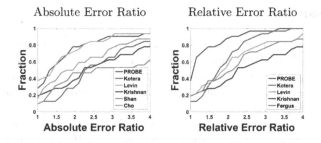

Fig. 9. Absolute error ratio and relative error ratio on dataset [10]

Table 1. PSNR and time computation per frame for modern deblurring methods over the dataset of [10]

Method	Mean PSNR	Time [sec]
Kotera [7]	26.3	3
Krishnan [9]	26.5	9
Shan [15]	25.4	NA
Cho [3]	27.1	NA
Levin [11]	28.4	36
PROBE	28.4	5

Figure 9(left) is the absolute error ratio (reported in several state-of-the-art methods [9,10,19]), whose denominator is the difference between the latent image and a reference non-blind restoration (in this case [8]) given the true blur kernel. PROBE and [11] are similarly superior to other methods, but note that PROBE is 7 times faster than [11].

A caveat in the error ratio measure is the non-blind reference algorithm used to compute the denominator. In straightforward sequential algorithms, it makes sense to use the non-blind deblurring module of the respective blind algorithm. We therefore propose an additional dataset-wide statistic, the relative error ratio, whose denominator is the difference between the latent image and its respective non-blind restoration given the true blur kernel. Figure 9(right) demonstrates PROBE's state of the art performance.

In Table 1 we present the mean PSNR for each method and the computation time for Matlab-implemented methods[1]. PSNR differences between PROBE and Levin *et al.* [11] are negligible. However, using unoptimized Matlab implementation, our method is 7 times faster than [11].

---

[1] Using a single-threaded Matlab on a 3.4Ghz CPU.

7   Discussion

PROBE is a novel recursive blind deblurring framework. Though inspired by the algorithm of [5], PROBE is cheaper and faster, with the simple modified inverse filter at its foundation. Quantitative comparative performance analysis on the standard database of Levin et al. [10] reveals that PROBE is second to none at the time of writing.

Using a novel oracle model for PROBE's PSF estimation module, we provide analytic convergence and error analysis for PROBE and demonstrate their validity and practical predictive value.

We are currently implementing a resource-limited PROBE application for cellphones. PROBE's computational bottleneck is in its PSF estimation module which, following [5], is still borrowed from an ad-hoc $MAP_{u_c,h}$ algorithm in the implementation of Kotera *et al.* [7]. Recent literature hints that PSF estimation should not be too difficult by itself [10,12]. Studying the use of a simpler PSF estimation module within PROBE is an interesting direction for future research.

References

1. Bar, L., Sochen, N., Kiryati, N.: Semi-blind image restoration via mumford-shah regularization. IEEE Trans. Image Process. **15**(2), 483–493 (2006)
2. Chan, T.F., Wong, C.: Total variation blind deconvolution. IEEE Trans. Image Process. **7**, 370–375 (1998)
3. Cho, S., Lee, S.: Fast motion deblurring. ACM Trans. Graph. (SIGGRAPH ASIA 2009) 28(5), Article no. 145 (2009)
4. Fergus, R., Singh, B., Hertzmann, A., Roweis, S.T., Freeman, W.T.: Removing camera shake from a single photograph. ACM Trans. Graph. **25**, 787–794 (2006). SIGGRAPH 2006 Conference Proceedings, Boston, MA
5. Hanocka, R., Kiryati, N.: Progressive blind deconvolution. In: Azzopardi, G., Petkov, N. (eds.) CAIP 2015. LNCS, vol. 9257, pp. 313–325. Springer, Cham (2015). doi:10.1007/978-3-319-23117-4_27
6. Komodakis, N., Paragios, N.: MRF-based blind image deconvolution. In: Lee, K.M., Matsushita, Y., Rehg, J.M., Hu, Z. (eds.) ACCV 2012. LNCS, vol. 7726, pp. 361–374. Springer, Heidelberg (2013). doi:10.1007/978-3-642-37431-9_28
7. Kotera, J., Šroubek, F., Milanfar, P.: Blind deconvolution using alternating maximum a posteriori estimation with heavy-tailed priors. In: Wilson, R., Hancock, E., Bors, A., Smith, W. (eds.) CAIP 2013. LNCS, vol. 8048, pp. 59–66. Springer, Heidelberg (2013). doi:10.1007/978-3-642-40246-3_8
8. Krishnan, D., Fergus, R.: Fast image deconvolution using hyper-laplacian priors. In: Neural Information Processing Systems NIPS (2009)
9. Krishnan, D., Tay, T., Fergus, R.: Blind deconvolution using a normalized sparsity measure. In: Computer Vision and Pattern Recognition (CVPR) (2011)
10. Levin, A., Weiss, Y., Durand, F., Freeman, W.T.: Understanding and evaluating blind deconvolution algorithms. In: Computer Vision and Pattern Recognition (CVPR), pp. 1964–1971 (2009)
11. Levin, A., Weiss, Y., Durand, F., Freeman, W.T.: Efficient marginal likelihood optimization in blind deconvolution. In: IEEE Conference on Computer Vision and Pattern Recognition (CVPR), pp. 2657–2664 (2011)

12. Michaeli, T., Irani, M.: Nonparametric blind super-resolution. In: IEEE International Conference on Computer Vision (ICCV), pp. 945–952 (2013)
13. Osher, S., Rudin, L.I.: Feature-oriented image enhancement using shock filters. SIAM J. Numer. Anal. **27**(4), 919–940 (1990)
14. Perrone, D., Favaro, P.: Total variation blind deconvolution: the devil is in the details. In: IEEE Conference on Computer Vision and Pattern Recognition (CVPR) (2014)
15. Shan, Q., Jia, J., Agarwala, A.: High-quality motion deblurring from a single image. ACM Trans. Graph. (SIGGRAPH) **27**, 73 (2008)
16. Shao, W.Z., Li, H.B., Elad, M.: Bi-l_0-l_2-norm regularization for blind motion deblurring. J. Vis. Commun. Image Represent. **33**, 42–59 (2015)
17. Xu, L., Jia, J.: Two-phase kernel estimation for robust motion deblurring. In: Daniilidis, K., Maragos, P., Paragios, N. (eds.) ECCV 2010. LNCS, vol. 6311, pp. 157–170. Springer, Heidelberg (2010). doi:10.1007/978-3-642-15549-9_12
18. You, Y.L., Kaveh, M.: A regularization approach to joint blur identification and image restoration. IEEE Trans. Image Process. **5**, 416–428 (1996)
19. Zhou, Y., Komodakis, N.: A MAP-estimation framework for blind deblurring using high-level edge priors. In: Fleet, D., Pajdla, T., Schiele, B., Tuytelaars, T. (eds.) ECCV 2014. LNCS, vol. 8690, pp. 142–157. Springer, Cham (2014). doi:10.1007/978-3-319-10605-2_10

Poster Session III

Feature Selection on Affine Moment Invariants in Relation to Known Dependencies

Aleš Zita[1,2($\boxtimes$)], Jan Flusser[1], Tomáš Suk[1], and Jan Kotera[1,2]

[1] Institute of Information Theory and Automation,
Czech Academy of Sciences, Prague, Czech Republic
`zita@utia.cas.cz`
[2] Faculty of Mathematics and Physics,
Charles University in Prague, Prague, Czech Republic

Abstract. Moment invariants are one of the techniques of feature extraction frequently used for pattern recognition algorithms. A moment is a projection of function into polynomial basis and an invariant is a function returning the same value for an input with and without particular class of degradation. Several techniques of moment invariant creation exist often generating over-complete set of invariants. Dependencies in these sets are commonly in a form of complicated polynomials, furthermore they can contain dependencies of higher orders. These theoretical dependencies are valid in the continuous domain but it is well known that in discrete cases are often invalidated by discretization. Therefore, it would be feasible to begin classification with such an over-complete set and adaptively find the pseudo-independent set of invariants by the means of feature selection techniques. This study focuses on testing of the influence of theoretical invariant dependencies in discrete pattern recognition applications.

Keywords: Affine invariants · Image moments · Feature selection · Machine learning · Pattern recognition

1 Introduction

One of the difficult tasks in image processing is a recognition of shapes degraded by some transformation. Several approaches to the invariant recognition exist. Those are methods based on brute force, i.e. methods which are focused on training data alteration in such way that they add artificial training data deformed with all possible transformations in question. This approach is adopted for example by deep convolutional networks. It has several disadvantages such as impossibility to generate all of the possible transformations or to cover some of the

This work was supported by Czech Science Foundation (GA ČR) under grant GA15-16928S.

This work was supported by The Charles University Grant Agency (GA UK) under grant 1094216.

transformation classes. Approach that overcomes these disadvantages is using computed features which are mathematically invariant to a certain family of transformations.

General moment is defined as a projection of a function to polynomial basis

$$M_{pq} = \iint P_{pq}(x,y)f(x,y)\mathrm{d}x\mathrm{d}y.$$

Moment invariant is then a function of moments satisfying invariance to particular class of deformations. As a simple example of invariant's creation we can demonstrate construction of invariants to translation from geometric moments. We define geometric moment as

$$m_{pq} = \iint x^p y^q f(x,y)\mathrm{d}x\mathrm{d}y,$$

with only m_{00} being invariant to translation. We can construct other invariants using m_{00} as follows

$$\mu_{pq} = \iint (x - x_t)^p (y - y_t)^q f(x,y)\mathrm{d}x\mathrm{d}y,$$

where $x_t = m_{10}/m_{00}$ and $y_t = m_{01}/m_{00}$.

This technique is equivalent to shifting the center of gravity to the origin. Moments μ_{pq} are called central moments. Similar approaches can be used to create invariants to rigid or affine transform. For more examples please refer to [1–3]. Several techniques of affine moment invariants generation exist. Most of them produce over-complete sets containing dependencies.

In our experiments, we use the graph method of generating affine invariants. The core of this approach is in generating invariants using undirected multigraphs [3]. If we define 'cross product' of two image points (x_1, y_1) and (x_2, y_2) as

$$C_{12} = x_1 y_2 - x_2 y_1,$$

then after the image affine transform it holds

$$C'_{12} = J \cdot C_{12},$$

where J denotes the Jacobian. Therefore, C_{12} is relative affine invariant. Basic idea of invariants creation is integration of the cross product as a moment. After Jacobian elimination by normalization we get the affine invariants. This means that for $N \geq 2$ (degree of the invariant – the number of moments multiplied in one term), we can define

$$I(f) = \int_{-\infty}^{+\infty} \cdots \int_{-\infty}^{+\infty} \prod_{k,j=1}^{N} C_{k,j}^{n_{kj}} \cdot \prod_{i=1}^{N} f(x_i, y_i)\mathrm{d}x_i\mathrm{d}y_i, \tag{1}$$

where $n_{k,j}$ are nonnegative integers. After affine transform, we get

$$I' = J^w |J|^N \cdot I,$$ (2)

where $w = \sum_{jk}^{N} n_{jk}$ is the invariant's weight. Normalizing by μ_{00}^{w+N} gives an invariant

$$\left(\frac{I}{\mu_{00}^{w+N}} \right)' = (\text{sign}(J))^w \left(\frac{I}{\mu_{00}^{w+N}} \right).$$ (3)

For example $N = 2; n_{12} = 2$ gives

$$I(f) = \int\limits_{-\infty}^{+\infty} \cdots \int\limits_{-\infty}^{+\infty} (x_1 y_2 - x_2 y_1)^2 f(x_1, y_1) f(x_2, y_2) \mathrm{d}x_1 \mathrm{d}y_1 \mathrm{d}x_2 \mathrm{d}y_2$$

$$= 2(m_{20}m_{02} - m_{11}^2).$$

Every affine invariant created by this method can be represented by a multigraph where each point (x_k, y_k) corresponds to a vertex, and the cross product $C_{jk}^{n_{jk}}$ corresponds to a multiedge with multiplication factor n_{jk} connecting vertices k and j. Then the problem of affine invariant generation is equivalent to problem of generation single connected component multigraphs with w edges and number of vertices grater or equal to 2.

1.1 Dependencies and Completeness

In the present time, the search for the dependencies among moment affine invariants generated by the graph method is based on brute force approaches. First, the duplicities in generated invariants are found by reduction of polynomials to their irreducible forms. Next, the trivial dependencies are eliminated (zero invariants), after that the linear and polynomial dependencies are searched using brute force algorithm. This is a time costly process, because the whole invariant space needs to be searched. Therefore, finding higher order dependencies is practically impossible.

For example, from all 2 533 942 752 generated invariants of order ≤ 12 there are 2 532 349 394 zero invariants and 1 575 126 identical invariants, 14 538 linear combinations and 2 105 products. After this first reduction, there are still 1 589 irreducible invariants from which we know that only 80 are independent.

The cardinality of complete and independent invariant set can be calculated using following formula [4]

$$c = \binom{r + d}{r} - DOF(T),$$ (4)

where r denotes the order of invariant; d is number of image dimensions and $DOF(T)$ denotes the degrees of freedom of the degradation operator T (in our case the affine transform). In this way we can easily calculate, that for moment invariants of 4th order and affine transform of 2D image the complete independent set has cardinality equal to 9. In other words, there are independent sets of cardinality 9 within $I_1 - I_{32}$ (affine invariants of order 4 generated by the graph method). Some of these sets are known and had been proven to form a complete set, e.g. $\{I_1, I_2, I_3, I_4, I_6, I_7, I_8, I_9, I_{22}\}$ [5].

In our study, we want to investigate the relations between theoretical properties of affine invariants in combination with practical methods of feature selection. Therefore, we designed our experiments to test the strength of known dependencies against discriminative powers of individual invariants.

From the known set of dependencies of 4th order, 5 can be produced by set of 9 invariants. Let d1–d5 denote the following known dependencies (corresponding to dependencies no. 1, 2, 6, 8 and 12 in [5]):

$$d1 : -4I_1^3I_2^2 + 12I_1^2I_2I_3^2 - 12I_1I_3^4 - I_2I_4^2 + 4I_3^3I_4 - I_5^2 = 0$$
$$d2 : -16I_1^3I_7^2 - 8I_1^2I_6I_7I_8 - I_1I_6^2I_8^2 + 4I_1I_6I_9^2$$
$$+ 12I_1I_7I_8I_9 + I_6I_8^2I_9 - I_7I_8^3 - 4I_9^3 - I_{10}^2 = 0$$
$$d3 : -4I_1I_2I_9 + 4I_1I_{16}^2 + I_2I_8^2 + 4I_3^2I_9 - 4I_3I_8I_{16} + I_{18}^2 = 0 \qquad (5)$$
$$d4 : -I_1I_2I_{15} - I_1I_2I_{16} + 2I_1I_3I_11 + I_2I_{22}$$
$$+ I_3^2I_{15} + I_3^2I_{16} - 2I_3I_{32} - I_4I_{11} = 0$$
$$d5 : 2I_1I_3I_{24} + I_1I_{15}I_{17} - I_4I_{24} - I_{15}I_{28} - I_{17}I_{22} + I_{18}I_{22} = 0$$

2   Method

We proposed several experiments to test whether the theoretical relations between individual image affine moment invariants are reflected in discrete world of pattern recognition tasks. For this purpose, we utilize well known feature selection algorithms, classifiers and datasets.

2.1   Feature Selection Algorithms

Sequential Forward Selection (SFS). A method which starts with an empty set and then sequentially selects features with best possible classification outcome. It is a basic method of selecting relatively good subset of features for low dimensional problems. Its main advantage is its computational speed. But the most prominent disadvantage is that the algorithm does not allow to remove any feature previously selected.

Sequential Forward Floating Search (SFFS). An algorithm, which in each turn adds the most significant feature and then repeatedly tries to remove features by comparing the performance with the best performance achieved so far for the same-sized subset. This way, it tries to deal with fore-mentioned disadvantage of greedy approach of SFS [6].

2.2   Classifiers

Support Vector Machine (SVM). Support Vector Machine [7] is one of the widely used and well performing classification algorithm. In our experiments, we used SVM with RBF kernel. Parameters which were used were tuned to give best possible classification performance on full feature set for given problems.

Neural Network (NNET). As a second classification algorithm, we used fully connected classification neural network with two hidden layers with 50 neurons both. The network was finely tuned to give best classification performance on both the problems. The reason behind using neural network for this task is its theoretical ability to discover complex dependencies.

2.3   Datasets

MNIST. A well-known database with handwritten digits [8]. The dataset consists of 60 000 training and 10 000 testing digits images (see Fig. 1 for illustration). For the purpose of this study, we calculated first 32 normalized affine moment invariants for all images in the dataset.

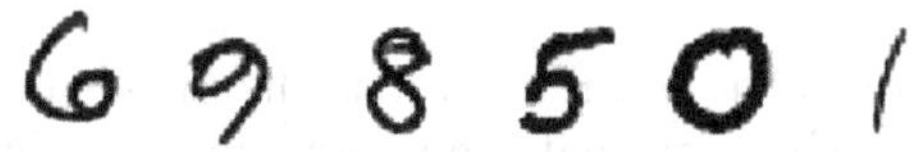

Fig. 1. Examples of MNIST database with handwritten digits. White corresponds to zero and black to one.

MEW 2014. The next database we used for our experiments is a database of segmented tree leaves [9, 10]. The affine moment invariants were calculated on the segmented images directly (see Fig. 2 for illustration). The database contains 15 074 images distributed to 201 classes. In each of the experiments we used subset of 100 classes to reduce computational time and complexity of the classification task.

3   Experiments

3.1   SFS

The experiments were designed to investigate probability with which the feature selection algorithm selects the set of invariants from all 4th order invariants that does not produce any of the known polynomial dependency. There is a known set of 32 irreducible affine invariants and the set of known dependencies [5]. All invariants were calculated from the original binary images from MNIST

Fig. 2. Examples of MEW 2014 database with segmented tree leaves (white = 0, black = 1).

and MEW 2014 databases. Because of the magnitude differences of individual invariants, it is usual to normalize them prior to classification. The normalization technique used in this work is based on two phase technique

$$\mu'_{pq} = \mu_{pq} \cdot \pi^{\frac{p+q}{2}} \cdot \left(\frac{p+q}{2} + 1 \right) \tag{6}$$

$$I' = \text{sign}(I) \cdot \sqrt[d]{|I|}, \tag{7}$$

where d denotes degree of the invariant. First, the moments within each invariant are normalized in such a way that their corresponding complex moments are equal to 1 when calculated on unitary circle (6). Next, the invariants are normalized to degree (7). This covers the cases in which the products of many moments within a invariant can result in very large numbers. Furthermore, each invariant was scaled by a learned factor to produce the best classification performance on both the databases independently.

We started with the SFS method to progressively select most discriminative feature (invariant), one at a time. The feature selection process was forced to continue after peak classification performance was reached, until set of 9 features was selected. To introduce diversity to the experiment, each feature selection process was executed on random subset of classes.

Subsequently, we performed a search for defined dependencies (Sect. 1.1) on the resulting sets. Our goal was to estimate the influence of theoretical dependencies of the invariants in continuous domain to discrete world of machine learning. The statistics of feature selection process can be viewed as the indicators of discriminative powers of individual moment invariants.

3.2 SFFS

Our next effort was to improve the feature selection performance by using SFFS method, again on both datasets. But, in this case we omitted the usage of the NNET classifier. The reason being, that the outcome of neural network classifiers depends on random initialization and the optimization function of neural network has typically many local minimums, and cannot provide the level of classification consistency SFFS process requires to run efficiently. To successfully utilize neural networks for SFFS would mean to run the classification many times over to produce meaningful statistics of current classification performance. This would be impractical and time consuming.

The outputs of the experiments are the same statistics as in case of SFS.

3.3 Adding Dependent Feature

Our next task was focused on studying the strength of particular invariants discriminability vs. invariant dependency. We took the histogram of all selected features in previous experiments as a measure of each invariant's discriminative power. Furthermore, we performed uncorrelated estimation of each invariant's discriminative power by running classification statistics on sets represented by single invariants at a time.

For all known dependencies of invariants of order 4, we started the feature selection process with all the invariants from given dependency except the one with highest discriminability. This experiment's goal is to study the level of particular dependency when in contradiction to strong discriminative ability.

4 Results

4.1 SFS

Starting from empty sets we have run 200 sequential feature selection trials on MNIST datasets with both classifiers. One of these runs (what is 0.5% cases) ended up with invariant set producing dependency d1, see Eq. (5). Histogram of selected invariants can be seen in Fig. 3 left.

The second part of this experiment was to run the same task for MEW 2014 database. The dependent set was again generated in 1 case out of 200 (0.5%) with dependency d2 in Eq. (5). See Fig. 3 right for resulting feature histogram.

Note that the sets of selected invariants differ, because discriminative powers of individual invariants changes with the classification task.

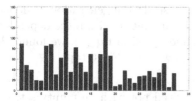 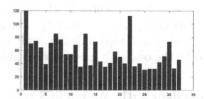

Fig. 3. Histogram of affine invariants selected by the SFS process. The images indicate relative discriminative powers of invariants $I_1 - I_{32}$ (x-axis) when used in classification task of MNIST (left) and MEW 2014 (right) datasets.

4.2 SFFS

In first batch of all 200 feature selections on MNIST database, one (0.5%) resulted in dependent set being generated. In this case dependency d2 in Eq. (5) emerged.

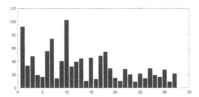 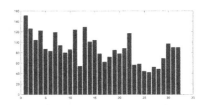

Fig. 4. Histogram of affine invariants selected by the SFFS process. The images indicate relative discriminative powers of invariants $I_1 - I_{32}$(x-axis) when used in classification task of MNIST (left) and MEW 2014 (right) datasets.

The next experiment was the same configuration run on MEW 2014, resulting in 4 dependent sets being generated out of 200 trial runs, all of them having dependency d2 in (5). Histogram of both experiments can be seen in Fig. 4, resp.

4.3 Adding Dependent Feature

Because the feature selection processes in our experiments produced only dependencies d1 and d2, see Eq. (5), we will focus in this experiment on generating those two. We estimated discriminability for the individual affine invariant by running classification on each of them separately. Because the discriminative strength of invariants is data-related, we performed the calculation for each dataset independently. We assumed, that the most discriminative invariant overall for both the dependencies is the invariant I_1, as it represents image reference ellipse and due to relatively small polynomial exponents is producing smallest numerical computational instabilities.

Our experiments showed that invariants with greatest discriminative power are I_{18} and I_{22} respectively. However, when we study the generation of dependency d1 in Eq. (5), we found I_4 to be most discriminative within the dependent group (see Fig. 5 top). For the study of dependency d2 in Eq. (5), we found the I_1 to be the one with greatest discriminative power (see Fig. 6 top).

Dependency d1. We ran SFS process initialized not with an empty set, but with set of $\{I_1, I_2, I_3, I_5\}$ (i.e. removing the strongest I_4 from dependent set) for both dataset and both the classifiers to test the strength of the dependency d1 in (5).

In the result, dependent feature sets were selected in 12/200 (6%) cases on MNIST database and 51/200 (25.5%) dependent feature set selected on MEW

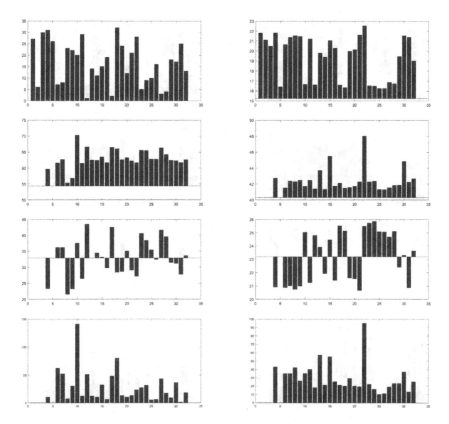

Fig. 5. Statistics on invariants discriminative powers when invoking d1 (5) dependency. Left: MNIST dataset, right: MEW 2014; top: the mean classification accuracies for individual invariants $I_1 - I_{32}$; middle top: the mean classification accuracies for invariants $I_4, I_6 - I_{32}$, when added to starting set (I_1, I_2, I_3 and I_5). Bad performance of individual invariants suggests strong correlation with the I_1, I_2, I_3, I_5 set. Middle bottom: the mean difference graph showing relative performance gain/loss for individual invariants. Note the relative decrease in performance for I_4 which completes the dependent set. Bottom: histogram of invariants selected in the process.

2014. See Fig. 5 bottom for histogram of both datasets. This corresponds to I_4 having greater relative discriminability in MEW 2014, or more correlated in MNIST dataset. The middle bottom images of Fig. 5 show the relative changes in individual invariants performance and indicate the strong correlation of invariants in relation to the starting set.

Dependency d2. The setup for testing the strength of d2 in (5) was the same, only we initiated Feature Selection with $\{I_5, I_6, I_7, I_8, I_9\}$.

Invoking dependency d2 resulted in 60/200 (30%) cases of selecting I_1 for MNIST dataset and 52/200 (26%) for MEW 2014, showing significant

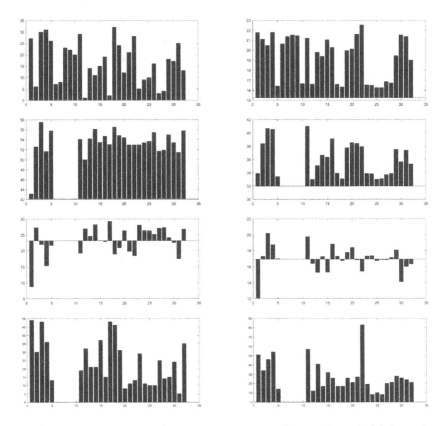

Fig. 6. Statistics on invariants discriminative powers when invoking d2 (5) dependency. Left: MNIST dataset, right: MEW 2014; top: the mean classification accuracies for individual invariants $I_1 - I_{32}$; middle top: the mean classification accuracies for invariants $I_1 - I_5$, $I_{11} - I_{32}$ when added to starting set $(I_6 - I_{10})$. Bad performance of individual invariants suggests strong correlation with the starting set. Middle bottom: the mean difference graph showing relative performance gain/loss for individual invariants. Note the decrease of I_1 performance due to the dependency to the starting set $I_6 - I_{10}$. Bottom: histogram of invariants selected in the process.

discriminative power of this invariant. The clear I_1 significance declination is depicted in middle bottom images in Fig. 6.

5   Conclusion

In our work, we have shown the importance of studying the theoretical dependencies between affine moment invariants and their direct impact on the classification performance. Our experiments show that the chance of generating set of affine moment invariants with polynomial dependency by the means of feature selection processes is less or equal to 2% in all cases. Some of these cases were positively identified as a situation where, during feature selection process, there were more invariants with the same classification performance, so the one with the smaller index was chosen, hence producing the dependent set. Those cases can be considered random noise.

In the second part of our experiments, we focused on the strength of dependencies d1 and d2 by removing most significant invariant from the dependent sets. We have confirmed that although it was possible to forcefully invoke independent set of affine moment invariants, this occurred in less than 30% of the cases, again showing power of the invariant dependencies in accordance to theory prediction.

This signifies that studying of the affine invariant dependencies have its purpose and is to be considered when using invariants as image features in pattern recognition tasks.

Such a study was never done before and up until now it was not clear, whether the theoretical properties of affine moment invariants have any noticeable relations to practical classification tasks in discrete computer domain, where continuous domain relations does not have to necessarily hold, due to the discretization process, numerical instabilities or even precision limitations of the computation.

In our ongoing work, we would like to broaden this study to include invariants of higher orders and confirm, that the same holds also for those more complex features. We would also like to be able to further study the numerical properties of discrete affine moment invariants in relation to calculation and usage in classification optimization processes.

References

1. Flusser, J., Suk, T.: Pattern recognition by affine moment invariants. Pattern Recogn. **26**(1), 167–174 (1993)
2. Suk, T., Flusser, J.: Affine moment invariants generated by graph method. Pattern Recogn. **44**(9), 2047–2056 (2011)
3. Flusser, J., Suk, T., Zitová, B.: 2D and 3D Image Analysis by Moments. Wiley, New York (2016)
4. Flusser, J., Zitová, B., Suk, T.: Moments and Moment Invariants in Pattern Recognition. Wiley, New York (2009)
5. Suk, T., Flusser, J.: Tables of affine moment invariants generated by the graph method. Technical report, Research Report 2156, Institute of Information Theory and Automation (2005)
6. Pudil, P., Novovičová, J., Kittler, J.: Floating search methods in feature selection. Pattern Recogn. Lett. **15**(11), 1119–1125 (1994)
7. Vapnik, V.: The Nature of Statistical Learning Theory. Springer, Heidelberg (2013)
8. LeCun, Y., Cortes, C., Burges, C.J.: The MNIST database of handwritten digits (1998). http://yann.lecun.com/exdb/mnist. Accessed 7 June 2017
9. Novotnỳ, P., Suk, T.: Leaf recognition of woody species in Central Europe. Biosyst. Eng. **115**(4), 444–452 (2013)
10. Suk, T., Novotny, P.: Middle European Woods (MEW 2012, 2014) (2014). http://zoi.utia.cas.cz/node/662. Accessed 7 June 2017

GMM Supervectors for Limited Training Data in Hyperspectral Remote Sensing Image Classification

AmirAbbas Davari$^{(\boxtimes)}$, Vincent Christlein, Sulaiman Vesal, Andreas Maier, and Christian Riess

Pattern Recognition Lab, Computer Science Department, Friedrich-Alexander-University Erlangen-Nuremberg, Erlangen, Germany
amir.davari@fau.de

Abstract. Severely limited training data is one of the major and most common challenges in the field of hyperspectral remote sensing image classification. Supervised learning on limited training data requires either (a) designing a highly capable classifier that can handle such information scarcity, or (b) designing a highly informative and easily separable feature set. In this paper, we adapt GMM supervectors to hyperspectral remote sensing image features. We evaluate the proposed method on two datasets. In our experiments, inclusion of GMM supervectors leads to a mean classification improvement of about 4.6%.

Keywords: Hyperspectral image classification · Remote sensing · Limited training data · GMM supervector

1 Introduction

Remote sensing plays an important role for various applications, including environmental monitoring, urban planning, ecosystem-oriented natural resources management, urban change detection and agricultural region monitoring [29]. Most of these monitoring and detection applications require the construction of a label map from the remotely sensed images. In these label maps, individual pixels are marked as members of specific classes like for example water, asphalt, or grass. The assignment of observations to labels is done via classification, which explains the importance of classification for remote sensing applications.

The availability of very high resolution hyperspectral remote sensing images (VHR-RSI) has drawn researchers' attention over the past decade. It has been shown that jointly exploiting spectral and spatial information improves classification performance compared to using spectral features alone [19]. To this end, extended multi-attribute profiles (EMAP) [8] are one of the most popular and powerful feature descriptors for such spectral-spatial pixel representations. By operating directly on connected components rather than pixels, EMAP allows to employ arbitrary region descriptors (e.g. shape, color, texture, etc.) and to

© Springer International Publishing AG 2017
M. Felsberg et al. (Eds.): CAIP 2017, Part II, LNCS 10425, pp. 296–306, 2017.
DOI: 10.1007/978-3-319-64698-5_25

support object-based image analysis. In addition, EMAP can be implemented efficiently via various tree-based hyperspectral (HS) image representation techniques like max- and min-tree [25] or alpha tree [26]. EMAPs unique characteristics make it a popular tool in the hyperspectral remote sensing image analysis community [14, 21, 32].

A long-standing problem in hyperspectral remote sensing is image classification based on only a limited number of labeled pixels, as the process of labeling pixels for training data is a manual, time-consuming and expensive procedure. Additionally, popular descriptors like EMAP typically provide high dimensional features. These two factors together lead to a relatively small ratio of labeled data compared to the overall feature dimensionality, and can potentially cause a problem known as Hughes phenomenon [15]. Researchers have put considerable effort into developing different algorithms to address this problem. These approaches can be categorized into two groups: (1) developing new classifiers or reformulating existing ones in order to work well with the limited training data [3, 6, 12, 16, 28, 30], and (2) using dimensionality reduction. In particular, supervised dimensionality reduction techniques like NWFE [18], DAFE [11], or DBFE [20] has been shown to oftentimes outperform unsupervised reduction techniques [4] like PCA. However, both approaches, i.e. designing classifiers capable of handling limited amount of training data and feature vector dimensionality reduction are challenged in extreme cases when training data is severely limited.

Gaussian Mixture Model supervectors have been successfully used to handle limited training data in several other applications, such as online signature verification [33], writer identification [7], or speech analysis [1, 5, 17, 27]. In this paper, we adopt GMM supervectors with a universal background model (GMM-UBM) to hyperspectral remote sensing image classification in the presence of limited training data. We apply the GMM-UBM approach to feature extraction. To deal with small size training sets, We concatenating all the means of the GMM components to form the supervectors. We show that such supervectors are highly effective in addressing the limited data problem.

Section 2 introduces the proposed workflow and tools used for feature extraction in the context of our framework for hyperspectral remote sensing images. Section 3 describes the experimental setup, datasets, feature extraction mechanism and classification algorithm and illustrates the results. Finally, our work will be concluded in Sect. 4.

2 Methodology

We first introduce Gaussian supervectors, then present the proposed workflow to limited data classification.

2.1 GMM Supervector

The computation of GMM supervectors consists of three main components, namely universal background model computation, adaptation to the data and normalization [24].

Universal Background Model. The universal background model (UBM) essentially is a GMM fitted to the labeled training data. In detail, let $\lambda = \{w_k, \boldsymbol{\mu}_k, \boldsymbol{\Sigma}_k \,|\, k = 1, \ldots, K\}$ denote the parameters of a GMM with K mixture components, where w_k, $\boldsymbol{\mu}_k$, $\boldsymbol{\Sigma}_k$ denote the k-th mixture weight, mean vector and covariance matrix, respectively.

Given a feature vector $\boldsymbol{x} \in \mathcal{R}^D$, its likelihood function is defined as

$$p(\boldsymbol{x} \,|\, \lambda) = \sum_{k=1}^{K} w_k g_k(\boldsymbol{x}), \tag{1}$$

where $g_k(\boldsymbol{x})$ is a function to evaluate the k-th Gaussian at position $\boldsymbol{x}$, i.e.,

$$g_k(\boldsymbol{x}) = g(\boldsymbol{x}; \boldsymbol{\mu}_k, \boldsymbol{\Sigma}_k) = \frac{1}{\sqrt{(2\pi)^D |\boldsymbol{\Sigma}_k|}} e^{-\frac{1}{2}(\boldsymbol{x}-\boldsymbol{\mu}_k)^\top \boldsymbol{\Sigma}_k^{-1}(\boldsymbol{x}-\boldsymbol{\mu}_k)}. \tag{2}$$

The mixture weights are positive real numbers, i.e. $w_k \in \mathbb{R}_+$, and satisfy the constraint $\sum_{k=1}^{K} w_k = 1$.

Finally, the posterior probability of a feature vector $\boldsymbol{x}_j$ to be generated by the Gaussian mixture k is

$$\gamma_k(\boldsymbol{x}_j) = p(k \,|\, \boldsymbol{x}_j) = \frac{w_k g_k(\boldsymbol{x}_j)}{\sum_{l=1}^{K} w_l g_l(\boldsymbol{x}_j)}. \tag{3}$$

For estimating the GMM parameters, expectation maximization is used for a maximum likelihood estimation [10]. The parameters λ are iteratively refined to increase the log-likelihood $\log p(\boldsymbol{X} \,|\, \lambda) = \sum_{m=1}^{M} \log p(\boldsymbol{x}_m \,|\, \lambda)$ of the model for the set of training samples $\boldsymbol{X} = \{\boldsymbol{x}_1, \ldots, \boldsymbol{x}_M\}$.

In the original formulation of Gaussian supervectors, all parameters, i.e., weights w_k, means $\boldsymbol{\mu}_k$ and covariances $\boldsymbol{\Sigma}_k$ are used. However, in our case of severely limited training data, we found that supervectors consisting only of the means are much more robust. Thus, we will eventually only use these means for supervector formation.

GMM Adaptation and Mixing. A key idea of the Gaussian supervectors is to adapt the GMM components to the distribution of the test set. Since we will only use the means, we only describe the adaptation of the GMM mean vectors.

Thus, let $\boldsymbol{X}_f = \{\boldsymbol{x}_1, \ldots, \boldsymbol{x}_T\}$ denote the D-dimensional feature representations of the T pixels in the test set. Let furthermore $n_k = \sum_{t=1}^{T} \gamma_k(\boldsymbol{x}_t)$. Then, the first order statistic for adaptation of the UBM to the test data is

$$E_k^1 = \frac{1}{n_k} \sum_{t=1}^{T} \gamma_k(\boldsymbol{x}_t)\boldsymbol{x}_t, \tag{4}$$

where $E_k^1 \in \mathbb{R}^D$. The adapted mean vectors $\hat{\boldsymbol{\mu}}_k$ can then be computed as

$$\hat{\boldsymbol{\mu}}_k = \alpha_k E_k^1 + (1 - \alpha_k)\boldsymbol{\mu}_k, \tag{5}$$

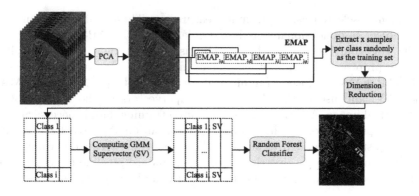

Fig. 1. Proposed workflow.

where

$$\alpha_k = \frac{n_k}{n_k + r}, \qquad (6)$$

and r denotes a fixed relevance factor that controls the strength of the adaptation.

Finally, the supervector s is formed by concatenating the adapted GMM parameters. As stated above, we use only the adapted mean components, leading to

$$s = \left(\hat{\boldsymbol{\mu}}_1^\top, \ldots, \hat{\boldsymbol{\mu}}_K^\top \right)^\top. \qquad (7)$$

Normalization. The purpose of normalization is to transform the s in Eq. 7 into a common range. Several groups used for this task a feature mapping inspired by the symmetrized Kullback-Leibler divergence [7,31]. This mapping is referred to as KL-normalization and is computed for the mean vectors as

$$\tilde{\boldsymbol{\mu}}_k = \sqrt{w_k} \boldsymbol{\sigma}_k^{-\frac{1}{2}} \odot \hat{\boldsymbol{\mu}}_k, \qquad (8)$$

where $\boldsymbol{\sigma}_k$ represents the GMM's k-th component standard deviation, $\tilde{\boldsymbol{\mu}}_k$ denotes the normalized adapted mean vector, and $\odot$ denotes the Hadamard product. In the case of mean adaptation, the normalized supervector $\tilde{s}_{\mathrm{m}}$ is represented as

$$\tilde{s}_{\mathrm{m}} = \left(\tilde{\boldsymbol{\mu}}_1^\top, \ldots, \tilde{\boldsymbol{\mu}}_K^\top \right)^\top. \qquad (9)$$

2.2 Proposed Workflow

A high-level overview of the proposed method is shown in Fig. 1. For hyperspectral remote sensing image classification, we use a standard dimensionality-reduction workflow. Here, spectral bands are first reduced via principle component analysis (PCA). Then, extended multi-attribute profile (EMAP) [8] is computed as the feature vector. From the computed EMAP, c samples per class are randomly selected as the training feature set. These training features are again subject to dimensionality reduction, denoted by EMAP-reduced.

The key contribution of the method is injected right before classification: we propose to compute Gaussian mixture model supervectors from the EMAP/EMAP-reduced training features, denoted by EMAP-SV/EMAP-reduced-SV, and use in the classifier the supervectors as the features of the hyperspectral image. A GMM from such limited training data is necessarily only a coarse approximation of the underlying distribution. Nevertheless, we show that it is just good enough to support the classifier in better determining the class boundaries. The parametrization of the standard pipeline follows dataset-dependent recommendations from the literature, and is reported in Sect. 3.2.

3   Experiments

We first introduce the datasets that were used for the evaluation in Sect. 3.1. Feature extraction and classification are presented in Sects. 3.2 and 3.3, respectively. Quantitative and qualitative results are presented in Sects. 3.4 and 3.5, respectively.

3.1   Data Sets

In order to evaluate our method, we use two popular datasets that were acquired by two different sensors. First, Pavia Centre dataset has been acquired by the ROSIS sensor in 115 spectral bands during a flight campaign over Pavia, northern Italy. 13 of these bands are removed due to noise. Therefore, 102 bands are used in this work. The scene image is 1096×715 pixels with geometrical resolution of 1.3 m. For the computation of EMAP, we use the first three principle components of this dataset, containing 99.12% of the total spectral variance. Second, the Salinas dataset was acquired by AVIRIS sensor in 224 spectral bands over Salinas Valley, California. 20 of the water absorption bands were discarded and therefore 204 bands are used in this work. The scene image is 512×217 pixels with high spatial resolution of 3.7 m pixels. We use the first three principle components of this dataset, containing 99.14% of the total spectral variance, for the computation of EMAP.

3.2   Feature Extraction

The host feature vector used in this work is the extended multi-attribute profile (EMAP) with four attributes and four thresholds, λ, per each attribute. For the Pavia Centre dataset we use the same threshold values as in [8]. For the Salinas dataset, we use the same threshold values as in [21]. The attributes and their corresponding threshold values for the Pavia Center are

– Area of the connected components: $\lambda_a = [100, 500, 1000, 5000]$;
– Length of the bounding box diagonal fit over the of the connected components: $\lambda_d = [10, 25, 50, 100]$;

- Standard deviation of the gray values of the connected components: $\lambda_s = [20, 30, 40, 50]$;
- Moment of inertia [13]: $\lambda_i = [0.2, 0.3, 0.4, 0.5]$;

The attributes and their corresponding threshold values for the Salinas dataset are

- Area of the connected components: $\lambda_a = [100, 500, 1000, 5000]$;
- Length of the bounding box diagonal fit over the of the connected components: $\lambda_d = [10, 25, 50, 100]$;
- Standard deviation of the gray values of the connected components: $\lambda_s = [15, 20, 25, 30]$;
- Moment of inertia: $\lambda_i = [0.1, 0.15, 0.2, 0.25]$.

For calculating the EMAPs, the Max-tree hierarchical image representation is used. We use max-filtering [9] for filtering the Max-tree with each value in λ.

We implemented two variants of the second dimensionality reduction (DR) to investigate its impact on our proposed approach. Specifically, we reduce the EMAP dimensionality in one variant with principle component analysis (PCA), and in another variant with non-parametric weighted feature extraction (NWFE) [18]. PCA is a popular unsupervised DR method. NWFE is supervised and very strong performance has been reported for this method [4]. For the Pavia Centre dataset, 7 PCA dimensions and 6 NWFE dimensions were used to preserve 99% of the variance of the input EMAP features. For the Salinas dataset, 4 PCA dimensions and 7 NWFE dimensions were used to preserve 99% of the variance of the input EMAP features.

The supervectors (SV) are computed over the aforementioned raw EMAP/ EMAP-reduced feature vectors. The number of GMM components is set to 3. However, in preliminary experiments, we found that the choice of this parameter is not critical to this work. The relevance factor is set to be $r = 16$ as it is commonly used in the literature [17,24]. Kullback-Leibler divergence is used for normalization of the supervectors. The supervectors are computed over EMAP results (denoted as EMAP-SV) and EMAP-reduced results (denoted as EMAP-reduced-SV), respectively.

3.3   Classification

We use the random forest classifier with 100 trees. The tree depth and the bagging number is set to be square root of the number of input variables by default as suggested in [2]. For training, we randomly select c pixels per class from the image as the training set. All the remaining pixels are used for testing. In order to simulate the severely limited training data case, we choose c to be 13 and 20 as two different training set sizes. For each experiment, this procedure was repeated 25 times. The performance metrics are overall accuracy, average accuracy and kappa, abbreviated as OA, AA, and Kappa:

- OA: The overall accuracy is the number of correctly classified instances divided by the total number of data points (pixels).

- AA: The average accuracy is the average of class-based accuracies.
- Kappa: The kappa statistic is a measure of how closely the instances classified by the classifier matched the ground truth. By measuring the expected accuracy, it gives a statistic for the accuracy of a random classifier.

3.4 Quantitative Results

For quantitative evaluation, we compare a total of twelve combinations of the proposed algorithm: we use PCA and NWFE as dimensionality reduction of the EMAP features, and we apply these two variants on both datasets. We compare the classification performance on raw EMAP, EMAP-PCA and EMAP-NWFE to the supervectors computed over each of them. For each of these sets, we choose two selections of training sets, namely 12 and 20 pixels per class. Tables 1 and 2 show the classification results of the aforementioned feature sets computed over Pavia Centre dataset and Salinas dataset, respectively.

Both Tables 1 and 2 show that the smaller the training size, the higher the performance gain achieved by using GMM supervectors. For example, consider the case of EMAP feature computed over Pavia Centre dataset in Table 1. In the case of 20 pixels per class, using supervectors results in a Kappa improvement of 0.0312. With a training data size of only 13 pixels per class, the Kappa improvement is even 0.0598. Thus, the proposed method has a bigger impact in applications with severely limited training data.

To study the effect of the second level dimensionality reduction algorithms on our idea, we reduced the dimensionality of EMAP by means of PCA and NWFE to a number that preserves 99% of the data variance. It turns out that the performance boost achieved by the supervectors over raw EMAP variants is consistent over variants of dimensionality reduction algorithms, i.e. PCA and NWFE. All metrics show improvement for both EMAP-PCA supervector and EMAP-NWFE supervector comparing to raw EMAP classification. This shows the consistency of the method over different dimensionality reduction techniques. We also note that the performance gained by NWFE is higher than for PCA. Furthermore, the standard deviation of the proposed method is consistently low. Thus, by using supervectors, robustness of the training set is increased with respect to the class-wise structure. By extension, the classifier becomes more robust and consistent on different training samples.

Finally, supervectors computed over EMAP variants using less training samples oftentimes lead to a comparable or sometimes even higher performance than raw EMAP variants using more training samples without synthetic samples. For example in Table 1, EMAP-PCA-SV on 13 training samples achieves a Kappa of 0.9198. This is higher than the Kappa of 0.8974 that is achieved by EMAP-PCA-raw on 20 training samples. Similarly, in Table 2, EMAP-NWFE-SV on 13 training samples obtain higher Kappa values than EMAP-NWFE-raw on 20 training samples.

Table 1. Classification performances of raw EMAP, EMAP-PCA and EMAP-NWFE vs. their supervector (SV) correspondences, computed over Pavia Centre dataset. This table shows the results for two training data sizes namely 13 and 20 pixels per class.

Algorithm	Feature	AA% (±SD)	OA% (±SD)	Kappa (±SD)
13 Pix/Class				
EMAP	raw	77.87 (±2.97)	90.01 (±3.78)	0.8600 (±0.0495)
	SV	**88.73** (±1.30)	**94.28** (±0.94)	**0.9198** (±0.0129)
EMAP-PCA	raw	73.51 (±3.00)	86.38 (±3.61)	0.8089 (±0.0493)
	SV	**82.07** (±1.96)	**91.70** (±1.67)	**0.8838** (±0.0225)
EMAP-NWFE	raw	80.06 (±3.56)	91.37 (±2.67)	0.8787 (±0.0365)
	SV	**88.02** (±1.17)	**95.39** (±0.42)	**0.9349** (±0.0059)
20 Pix/Class				
EMAP	raw	81.80 (±2.07)	92.73 (±1.23)	0.8974 (±0.0171)
	SV	**90.43** (±1.22)	**94.92** (±0.77)	**0.9286** (±0.0106)
EMAP-PCA	raw	79.07 (±1.69)	90.89 (±1.22)	0.8717 (±0.0169)
	SV	**83.12** (±2.07)	**92.37** (±1.09)	**0.8928** (±0.0151)
EMAP-NWFE	raw	83.32 (±2.24)	93.28 (±1.30)	0.9053 (±0.0181)
	SV	**88.91** (±0.86)	**95.67** (±0.55)	**0.9389** (±0.0077)

Table 2. Classification performances of raw EMAP, EMAP-PCA and EMAP-NWFE vs. their supervector (SV) correspondences, computed over Salinas dataset. This table shows the results for two training data sizes namely 13 and 20 pixels per class.

Algorithm	Feature	AA% (± SD)	OA% (±SD)	Kappa (±SD)
13 Pix/Class				
EMAP	raw	83.84 (±2.06)	76.30 (±2.74)	0.7380 (±0.0292)
	SV	**90.90** (±0.98)	**85.37** (±1.34)	**0.8378** (±0.0147)
EMAP-PCA	raw	82.50 (±2.06)	74.96 (±3.63)	0.7230 (±0.0378)
	SV	**86.84** (±1.77)	**78.30** (±2.90)	**0.7606** (±0.0311)
EMAP-NWFE	raw	88.68 (±1.20)	80.42 (±2.34)	0.7838 (±0.0247)
	SV	**91.43** (±1.00)	**83.09** (±1.72)	**0.8132** (±0.0186)
20 Pix/Class				
EMAP	raw	86.81 (±1.63)	79.74 (±2.56)	0.7756 (±0.0269)
	SV	**92.88** (±0.71)	**88.09** (±1.29)	**0.8680** (±0.0142)
EMAP-PCA	raw	86.59 (±1.06)	78.70 (±2.33)	0.7643 (±0.0249)
	SV	**86.99** (±1.28)	**79.40** (±1.88)	**0.7725** (±0.0205)
EMAP-NWFE	raw	90.56 (±1.26)	82.26 (±2.62)	0.8038 (±0.0280)
	SV	**91.91** (±0.81)	**83.40** (±1.94)	**0.8168** (±0.0208)

3.5   Qualitative Results

Figure 2 shows example label maps corresponding to the classification results for training size of 13 pixels per class for the Salinas dataset. In this Figure, (a) shows the ground truth labeling, (b) shows the output of raw EMAP, (c) shows the EMAP supervector results. Analogously, (d) shows the output of raw EMAP-PCA, and (e) shows EMAP-PCA supervector, (f) shows the corresponding results for raw EMAP-NWFE, and (g) shows the output of EMAP-NWFE supervector, i.e., identical processing pipelines using NWFE dimensionality reduction, with raw and supervector feature sets. Comparing (e) and (g) confirms the superiority of EMAP-NWFE supervectors over the EMAP-PCA supervector. Comparing (b) with (c), (d) with (e) and (f) with (g), specially in the large homogeneous areas, clearly shows that using the supervectors avoids a number of misclassification that are present in the raw EMAP, EMAP-PCA and EMAP-NWFE.

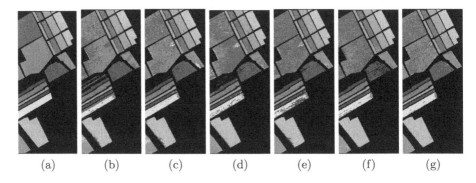

    (a)   (b)   (c)   (d)   (e)   (f)   (g)

Fig. 2. Example label maps on Salinas using 13 training samples per class. (a) ground truth (b) EMAP, (c) EMAP-SV (d) EMAP-PCA, (e) EMAP-PCA-SV, (f) EMAP-NWFE, (g) EMAP-NWFE-SV.

4   Conclusion

Limited training data is a common issue in hyperspectral remote sensing image classification. This limitation severely challenges classifiers, particularly when using high dimensional feature vectors. We propose to use GMM supervectors with a universal background model to address the limited data problem. In our results on real data, we show the performance gain on the Pavia Centre and the Salinas datasets. It turns out that supervectors consistently increase the overall accuracy, average accuracy, and kappa coefficient. Furthermore, the performance boost using supervectors is consistent over different dimensionality reduction algorithms and different training data sizes. It can also be observed that using supervectors decreased the standard deviations of the error metrics.

Quantitatively, the exact performance improvement depends on the details of the processing chain and on the dataset. The mean improvement in our experiments is almost 4.6%, with variations between one percent and almost ten percent. These results are encouraging, as the approach itself is quite straightforward, and can be smoothly integrated into any classification pipeline.

References

1. Bahari, M.H., Saeidi, R., Hamme, H.V., Leeuwen, D.V.: Accent recognition using i-vector, Gaussian mean supervector and Gaussian posterior probability supervector for spontaneous telephone speech. In: 2013 IEEE International Conference on Acoustics, Speech and Signal Processing, Institute of Electrical and Electronics Engineers. IEEE, May 2013
2. Breiman, L.: Random forests. Mach. Learn. **45**(1), 5–32 (2001)
3. Bruzzone, L., Chi, M., Marconcini, M.: A novel transductive svm for semisupervised classification of remote-sensing images. IEEE Trans. Geosci. Remote Sens. **44**(11), 3363–3373 (2006)
4. Castaings, T., Waske, B., Atli Benediktsson, J., Chanussot, J.: On the influence of feature reduction for the classification of hyperspectral images based on the extended morphological profile. Int. J. Remote Sens. **31**(22), 5921–5939 (2010)
5. Cerva, P., Silovsky, J., Zdansky, J.: Comparison of generative and discriminative approaches for speaker recognition with limited data. Radioengineering **18**(3), 307–316 (2009)
6. Chi, M., Feng, R., Bruzzone, L.: Classification of hyperspectral remote-sensing data with primal SVM for small-sized training dataset problem. Adv. Space Res. **41**(11), 1793–1799 (2008)
7. Christlein, V., Bernecker, D., Hönig, F., Maier, A., Angelopoulou, E.: Writer identification using GMM supervectors and exemplar-SVMs. Pattern Recogn. **63**, 258–267 (2017)
8. Dalla Mura, M., Atli Benediktsson, J., Waske, B., Bruzzone, L.: Extended profiles with morphological attribute filters for the analysis of hyperspectral data. Int. J. Remote Sens. **31**(22), 5975–5991 (2010)
9. Dalla Mura, M., Benediktsson, J.A., Waske, B., Bruzzone, L.: Morphological attribute profiles for the analysis of very high resolution images. IEEE Trans. Geosci. Remote Sens. **48**(10), 3747–3762 (2010)
10. Dempster, A., Laird, N., Rubin, D.: Maximum likelihood from incomplete data via the EM algorithm. J. Roy. Stat. Soc.: Ser. B (Methodol.) **39**(1), 1–38 (1977)
11. Fukunaga, K.: Introduction to Statistical Pattern Recognition. Academic Press, San Diego (2013)
12. Hoffbeck, J.P., Landgrebe, D.A.: Covariance matrix estimation and classification with limited training data. IEEE Trans. Pattern Anal. Mach. Intell. **18**(7), 763–767 (1996)
13. Hu, M.-K.: Visual pattern recognition by moment invariants. IRE Trans. Inf. Theory **8**(2), 179–187 (1962)
14. Huang, X., Guan, X., Benediktsson, J.A., Zhang, L., Li, J., Plaza, A., Dalla Mura, M.: Multiple morphological profiles from multicomponent-base images for hyperspectral image classification. IEEE J. Sel. Top. Appl. Earth Obs. Remote Sens. **7**(12), 4653–4669 (2014)

15. Hughes, G.: On the mean accuracy of statistical pattern recognizers. IEEE Trans. Inf. Theory **14**(1), 55–63 (1968)
16. Jackson, Q., Landgrebe, D.A.: An adaptive classifier design for high-dimensional data analysis with a limited training data set. IEEE Trans. Geosci. Remote Sens. **39**(12), 2664–2679 (2001)
17. Kelly, F.: Automatic recognition of ageing speakers. Ph.D. thesis, Trinity College Dublin (2014)
18. Kuo, B.-C., Landgrebe, D.A.: Nonparametric weighted feature extraction for classification. IEEE Trans. Geosci. Remote Sens. **42**(5), 1096–1105 (2004)
19. Landgrebe, D.A.: Signal Theory Methods in Multispectral Remote Sensing, vol. 29. Wiley, Hoboken (2005)
20. Lee, C., Landgrebe, D.A.: Feature extraction based on decision boundaries. IEEE Trans. Pattern Anal. Mach. Intell. **15**(4), 388–400 (1993)
21. Liu, T., Gu, Y., Jia, X., Benediktsson, J.A., Chanussot, J.: Class-specific sparse multiple kernel learning for spectral-spatial hyperspectral image classification. IEEE Trans. Geosci. Remote Sens. **54**(12), 7351 (2016)
22. McLachlan, G., Peel, D.: Finite Mixture Models. Wiley, Hoboken (2004)
23. Oliveira-Brochado, A., Martins, F.V.: Assessing the number of components in mixture models: a review. Technical report, Universidade do Porto, Faculdade de Economia do Porto (2005)
24. Reynolds, D.A., Quatieri, T.F., Dunn, R.B.: Speaker verification using adapted Gaussian mixture models. Digit. Signal Proc. **10**(1–3), 19–41 (2000)
25. Salembier, P., Oliveras, A., Garrido, L.: Antiextensive connected operators for image and sequence processing. IEEE Trans. Image Process. **7**(4), 555–570 (1998)
26. Soille, P.: Constrained connectivity for hierarchical image partitioning and simplification. IEEE Trans. Pattern Anal. Mach. Intell. **30**(7), 1132–1145 (2008)
27. Srinivasan, B.V., Zotkin, D.N., Duraiswami, R.: A partial least squares framework for speaker recognition. In: 2011 IEEE International Conference on Acoustics, Speech and Signal Processing, ICASSP. Institute of Electrical and Electronics Engineers (IEEE), May 2011
28. Tadjudin, S., Landgrebe, D.A.: Covariance estimation for limited training samples. In: 1998 Proceedings of the IEEE International Geoscience and Remote Sensing Symposium, IGARSS 1998, vol. 5, pp. 2688–2690. IEEE (1998)
29. Valero, S., Salembier, P., Chanussot, J.: Hyperspectral image representation and processing with binary partition trees. IEEE Trans. Image Process. **22**(4), 1430–1443 (2013)
30. Vatsavai, R.R., Shekhar, S., Burk, T.E.: A semi-supervised learning method for remote sensing data mining. In: 2005 Proceedings of the 17th IEEE International Conference on Tools with Artificial Intelligence, ICTAI 2005, IEEE (2005). 5 pp
31. Xu, M., Zhou, X., Li, Z., Dai, B., Huang, T.S.: Extended hierarchical Gaussianization for scene classification. In: 2010 17th IEEE International Conference on Image Processing (ICIP), Hong Kong, pp. 1837–1840, September 2010
32. Xu, X., Li, J., Huang, X., Dalla Mura, M., Plaza, A.: Multiple morphological component analysis based decomposition for remote sensing image classification. IEEE Trans. Geosci. Remote Sens. **54**(5), 3083–3102 (2016)
33. Zapata-Zapata, G.J., Arias-Londoño, J.D., Vargas-Bonilla, J.F., Orozco-Arroyave, J.R.: On-line signature verification using gaussian mixture models and small-sample learning strategies. Revista Facultad de Ingeniería Universidad de Antioquia **79**, 86–97 (2016)

A New Shadow Removal Method Using Color-Lines

Xiaoming Yu[1,2], Ge Li[1](✉), Zhenqiang Ying[1], and Xiaoqiang Guo[3]

[1] School of Electronic and Computer Engineering, Shenzhen Graduate School,
Peking University, Shenzhen, China
`geli@ece.pku.edu.cn`
[2] Computer Science and Technology, Dalian University of Technology, Dalian, China
[3] Academy of Broadcasting Science, SAPPRFT, Peking, China

Abstract. In this paper, we present a novel method for single-image shadow removal. From the observation of images with shadow, we find that the pixels from the object with same material will form a line in the RGB color space as illumination changes. Besides, we find these lines do not cross with the origin due to the effect of ambient light. Thus, we establish an offset correction relationship to remove the effect of ambient light. Then we derive a linear shadow image model to perform color-line identification. With the linear model, our shadow removal method is proposed as following. First, perform color-line clustering and illumination estimation. Second, use an on-the-fly learning method to detect umbra and penumbra. Third, estimate the shadow scale by the statistics of shadow-free regions. Finally, refine the shadow scale by illumination optimization. Our method is simple and effective for producing high-quality shadow-free images and has the ability for processing scenes with rich texture types and non-uniform shadows.

Keywords: Shadow removal · Enhancement · Color line · Offset correction

1 Introduction

Shadows are prevalent in natural images and provide important cues for understanding shape of the occlusion, contact between objects and illumination position etc. But shadows will affect the performance of many computer vision tasks such as image segmentation, object recognition etc [21–24]. Thus, shadow removal is an important preprocess for these tasks. The main difficulty for shadow removal is the lack of illumination information in shadow regions. Many approaches are proposed for estimating the shadow illumination by matching regions [8,16,19,25] or shadow boundary detection [4,5,7,12,20]. Those methods do not perform well when mismatching regions or failing to recognize shadow boundary.

From observation of images with shadow, we find that pixels from an object with same material will form a color-line [13] in the RGB color space as illumination changes. Inspired by [3], which use the haze-lines to perform image dehazing,

M. Felsberg et al. (Eds.): CAIP 2017, Part II, LNCS 10425, pp. 307–319, 2017.
DOI: 10.1007/978-3-319-64698-5_26

we may use the color-lines to cluster different materials in shadow images. But unlike the haze-lines which have a uniform intersection point, the color-lines are not easy for identification since the uncertain offsets. Thus, an offset correction method is proposed to derive a linear shadow image model. Through the linear model, color-lines clustering and illumination estimation can be effectively implemented. For performing shadow removal, an on-the-fly learning approach is used to do shadow detection. Then, an initial shadow-free result is presented based on the statistics of shadow-free regions. Finally, we use the illumination optimization to refine the results.

2   Related Work

Shadow removal generally involves two subtasks: shadow detection and shadow relighting. Many approaches have been proposed including automatic methods [5,8,10,25] and user-assisted methods [1,7,16,18,19,25] by a single image, multiple images and video sequences [9,14]. A complete review of existing works is beyond the scope of this paper, we refer readers to [1,25] for excellent overviews on these methods. In this paper, we mainly focus on single-image shadow removal.

Shadow removal based on the gradient domain was suggested by Finlayson et al. [4,5]. They construct a shadow-free image based on zeroing the gradients on the shadow boundaries. The similar idea is used in [12,20] for shadow removal and modification.

Inspired by color transfer technique [15], several shadow removal methods have been proposed [16,25]. Shor and Lischinski [16] perform shadow removal by estimating an affine shadow formation model. However, this method cannot handle multi-texture shadows well since it assumes that the textures are similar in the shadow area. Thus, Xiao et al. propose a multi-scale illumination transfer technique [25] to remove multi-texture shadows. But these methods are not suitable for soft shadow cases since the penumbra regions are not considered in their model.

In addition, many approaches use the image matting to guide their shadow removal. Wu et al. [17,18] consider an image as a linear combination of a shadow-free image and a shadow matte image. Guo et al. present a region-based approach [8] for shadow detection. By using the image matting technique to get the illumination attenuation factor, they recover the shadow illumination from paired regions. Since this method does not take into account the reflectance variation, it can't recover the texture details well. Zhang et al. propose a method [25] by matching the corresponding texture patches between the shadow and lit regions. With the coherent optimization processing among the neighboring patches, they produce shadow-free results with consistent illumination. However, this method is sensitive to user initial setting when they performing image matting for illumination factor estimation.

Arbel and Hel-Or [1] use intensity surfaces and texture anchor points to remove shadows lying on curved surfaces. Gong and Cosker [7] estimate the shadow scale of some single-material strips and then interpolate the scales in other regions using image inpainting technique. But this method may get poor results when shadow lies on the boundary of different materials since it cannot get enough strips to perform reliable interpolation.

3 Proposed Method

3.1 Shadow Image Model

Following the image formation equation in [2], an image $I(x)$ can be considered to be composed of an illumination $L(x)$ and a reflectance $R(x)$:

$$I_c(x) = L_c(x)R_c(x) \quad c \in \{\text{R,G,B}\}. \tag{1}$$

The illumination $L_c(x)$ in lit region consists of the direct illumination L_c^d and the ambient illumination L_c^a. Assuming shadows in the scene are cast since the direct illumination L_c^d is blocked, the illumination can be expressed as:

$$L_c(x) = \hat{S}(x)L_c^d + L_c^a \quad \hat{S}(x) \in [0,1], \tag{2}$$

where $\hat{S}(x)$ is the attenuation factor of the direct illumination. $\hat{S}(x) = 0$ for umbra pixels, $\hat{S}(x) = 1$ for lit pixels, and others for penumbra pixels. Equation (2) is an affine function with the offset L_c^a.

Mathematically, $L_c(x)$ can be rewritten as the following affine function:

$$L_c(x) = (\hat{S}(x) + \alpha)L_c^d + (L_c^a - \alpha L_c^d) \quad \alpha \in \mathbb{R}, \tag{3}$$

where α is an offset factor. The first term $(\hat{S}(x)+\alpha)L_c^d$ describes the illumination changes in the direction of L^d, and the second term $(L_c^a - \alpha L_c^d)$ denotes the offset with a given α. Consider the reflectance of the offset in Eq. (3),

$$D_c^\alpha(x) = (L_c^a - \alpha L_c^d)R_c(x), \tag{4}$$

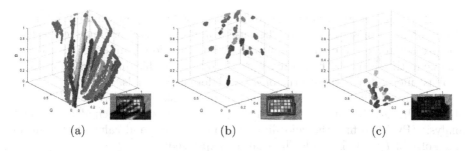

(a)   (b)   (c)

Fig. 1. (a) Shadow image and its RGB space distribution. (b) Shadow-free image and its RGB space distribution. (c) Umbra image and its RGB space distribution. (Color figure online)

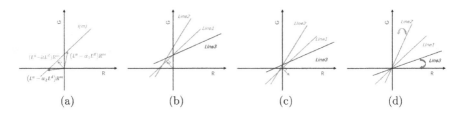

Fig. 2. (a) Color-line m with different offsets. (b) Different color-lines in RGB space. (c) Eliminate uniform offset of Line1. (d) Correct offset in different color-lines. (Color figure online)

the shadow image can be expressed based on Eqs. (1), (2) and (4):

$$I_c(x) = (1 + \alpha)S(x)L_c^d R_c(x) + D_c^\alpha(x) \quad S(x) \in [\frac{\alpha}{1 + \alpha}, 1], \tag{5}$$

where $S(x)$ is the new illumination attenuation factor.

From Eq. (5), there are three key observations in the RGB color space. First, if fix the $R(x)$, Eq. (5) is an affine function and pixels will form a color-line according to changes of direct light (like Fig. 1a). Second, when infinite extension of the attenuation factor ($S \in [-\infty, +\infty]$), the formation of color-lines does not intersect the origin unless the direction of L^d and L^a are identical. Third, for a given α, there is no single offset that will cause all the color-lines to intersect the origin since the color-line is reflectance dependent. Some similar conclusions are presented in [11].

From the above observations, we learn that the color-lines implies some important cues about the changes in illumination for objects of same materials. Furthermore, if the offset of different color-lines can be eliminated, we can distinguish different objects by different color-lines, and recover the shadow scale to achieve shadow removal.

3.2   Offset Correction

In [23, 24], Ying et al. assume that different color-lines approximately intersect at one point in most instances. They use a uniform offset for image correction and present some interesting applications, but the global uniform offset is not reliable in the scenes with rich textures. To overcome this difficulty, we propose a robust method to estimate the offset in different materials.

First, manually select an area m_u where has similar reflectance R^{m_u} and obvious changes in illumination (Fig. 3a). Then use the principal component analysis (PCA) to find the color-line direction v of m_u and calculate the mean value of p of $I(m_u)$, The color-line can be expressed:

$$I_c(m_u) \approx t(m_u)v_c + p_c, \tag{6}$$

where $t(m_u)$ is the independent variable for the line.

As discussed above, different α will cause different offsets in Eq. (5). Intuitively, if the smallest offset is eliminated, the pixel value will change smallest and we may cause fewer errors. So we choose the smallest offset $D_c^{\hat{\alpha}}(m_u)$ which is perpendicular to the color-line (Fig. 2a):

$$D_c^{\hat{\alpha}}(m_u) = (L_c^a - \hat{\alpha}L_c^d)R_c^{m_u} \approx p_c - \frac{<p,c>}{||v||}v_c, \tag{7}$$

where $< .,. >$ is the dot product operation.

Next, we can get a rough offsets correction image $I_c'(x)$ (Fig. 2c) by eliminating uniform offset $D_c^{\hat{\alpha}}(m_u)$:

$$I_c'(x) = I_c(x) - D_c^{\hat{\alpha}}(m_u). \tag{8}$$

As we discussed earlier, a uniform offset is inaccurate. But with this rough offset correction, the offset in different color-lines is reduced to some extent, and the pixel vector direction approximates the product of the direct illumination L^d and the reflectance R. Since the L^d is globally fixed, the ration between different pixel directions is determined by the pixel reflectance. So the offset $D_c^{\hat{\alpha}}(x)$ can be estimated as:

$$D_c^{\hat{\alpha}}(x) = D_c^{\hat{\alpha}}(m_u)\frac{R_c(x)}{R_c^{m_u}} \approx D_c^{\hat{\alpha}}(m_u)\frac{||v||}{v_c}\frac{I_c'(x)}{||I'(x)||}. \tag{9}$$

The final offset correction image $\hat{I}_c(x)$ (Figs. 2d and 3c) can be expressed as:

$$\hat{I}_c(x) = I_c(x) - D_c^{\hat{\alpha}}(x) = (1 + \hat{\alpha})S(x)L_c^d R_c(x). \tag{10}$$

3.3   Color-Lines Clustering and Illumination Estimation

After offset correction, all the color-lines will intersect the origin when we infinite extension them. $\hat{I}_c(x)$ can be expressed in spherical coordinates as:

$$\hat{I}_c(x) = [d(x), \theta(x), \phi(x)], \tag{11}$$

(a)   (b)   (c)   (d)

Fig. 3. (a) Shadow image and user selected region (blue line). (b) Original image clustering. (c) Offset correction image. (d) Correction image clustering. (Color figure online)

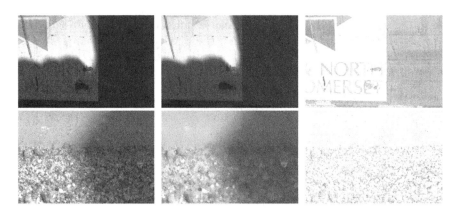

Fig. 4. First column: Shadow images. Second column: Illumination attenuation factors that estimated by color-lines. Third column: Shadow-free images by Eq. (15).

where $d(x)$ is the distance to the origin ($||\hat{I}_c(x)||$), $\theta(x)$ and $\phi(x)$ are the longitude and latitude, respectively.

For the same reflectance, changes in S affect only $d(x)$ without changing either $\theta(x)$ and $\phi(x)$. It means that we can find different color-lines by clustering the similar $[\theta(x), \phi(x)]$.

Similarly to [3], we use a uniform sampling of a sphere to build a KD-Tree for color-lines clustering. The different color clusters represent different materials (see Fig. 3d).

For a given color-line m, $d(x)$ depends on the pixel illumination:

$$d(x) = (1 + \hat{\alpha})S(x)||L^d \circ R^m||, \tag{12}$$

where $\circ$ denotes the pixel-wise multiplication.

Assume that the furthest pixel of different color-lines is shadow-free (this assumption will be relaxed in Sect. 3.4), the distance $d_{\max}$ of a color-line is:

$$d_{\max} = (1 + \hat{\alpha})||L^d \circ R||. \tag{13}$$

According to (12) and (13), the illumination attenuation factor $S(x)$ can be calculated as:

$$S(x) = \frac{d(x)}{d_{\max}(x)}. \tag{14}$$

A simple method for shadow removal is to perform global illumination uniformity:

$$I_c^{\hat{f}}(x) = \frac{\hat{I}_c(x)}{S(x)} + D_c^{\hat{\alpha}}(x), \tag{15}$$

where $I_c^{\hat{f}}(x)$ is the estimated shadow-free image. This method work well in the simple flat scenes (first row in Fig. 4). But the results may look overexposed and flat (second row in Fig. 4). In natural shadow-free scenes, the illumination will change upon different distance between the scene and camera, or the existence of self-shadow etc. Thus, the shadow-free pixels are some compact clusters (Fig. 1b) at the top of the color-lines rather than the points. According to this fact, we refine our shadow removal method in the following sections.

3.4 Refined Shadow Removal Method

Shadow Detection. In order to perform shadow removal, we need to detect the shadows firstly based on [7].

First, two user-supplied rough inputs that indicates sample lit and shadow pixels are required to construct a KNN classifier (K = 3) in the Log-RGB color space.

Then, a label (zero and one) image will be received by classifying the pixels. Next, a spatial filtering with a Gaussian kernel is applied to the label image of posterior probability, and we can get the rough shadow mask by binarizing the filtered image using a threshold of 0.5.

After getting the mask, the sampling lines perpendicular to the shadow boundary are sampled by the gradient of Illumination factor $S(x)$. Instead of processing aligned and selected samples [7] which are time consuming, our sampling work is just to distinguish and get the rough penumbra regions as in [1,10].

Finally, the scene is classified as lit region (N), penumbra (P) and umbra (U) like Fig. 5d.

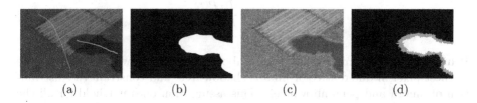

| (a) | (b) | (c) | (d) |

Fig. 5. (a) Shadow image with user inputs. (b) Rough shadow mask. (c) Penumbra sampling lines (red lines) in illumination factor $S(x)$. (d) Shadow mask: white for umbra U, gray for penumbra P, and black for lit region N. (Color figure online)

Umbra Removal. Similar to the distribution of lit pixels in RGB space, the umbra pixels are a more compact cluster (Fig. 1c). We can remove the umbra by making the distribution of it to be same as that of N as following:

$$\hat{I}_c^f(U^m) = (\hat{I}_c(U^m) - E(\hat{I}_c(U^m)))\frac{SD(\hat{I}_c(N^m))}{SD(\hat{I}_c(U^m))} + E(\hat{I}_c(N^m)), \quad (16)$$

where m is the color-line index, $E(.)$ is the mean operator and $SD(.)$ is the standard deviation operator.

The operation in Eq. (16) keeps similar statistical distribution between U^m and N^m. Similar ideas are also found in [16, 19] to estimate parameters of a shadow affine model. However, they ignore the penumbra. Besides, these methods require the single material in shadow regions [16] or similar material distribution between shadow and non-shadow regions [19].

Penumbra Removal. As intensity of illumination changes gradually, penumbra cannot be removed like the method used in umbra. Through the umbra-free regions $\hat{I}_c^f(U^m)$ and the illumination factor S, we relight the penumbra by pixel shadow scale estimation. First, we calculate the mean umbra scale k_c^U:

$$k_c^U = E\left(\frac{\hat{I}_c(U^m)}{\hat{I}_c^f(U^m)}\right). \tag{17}$$

Then, amplify scale k^U through the illumination factor S to estimate penumbra scale $K(P^m)$:

$$K_c(P^m) = k_c^U + (1 - k_c^U)\frac{S(P^m) - E(S(U^m))}{E(S(N^m)) - E(S(U^m))}. \tag{18}$$

In Eq. (18), we assume the scale in penumbra $K(P^m)$ is gradually changed from umbra scale k^U to non-shadow scale 1. The penumbra removal image can be express as:

$$\hat{I}_c^f(P^m) = \frac{\hat{I}_c(P^m)}{K_c(P^m)}. \tag{19}$$

Illumination Optimization. In the above assumption, we assume that the furthest pixel in different color-lines is shadow-free, and constrain the distribution of umbra and penumbra pixels. This assumption does not hold for all the scenes (like Fig. 6b). In order to improve the robustness, we modify the previous assumption to the following:

1. Most pixels under the shadow regions corresponded to similar material pixels in the lit regions.
2. The illumination changes are locally smooth;

Assumption 1 means that we can use the proposed method to remove the shadow of most pixels. Assumption 2 implies that we can estimate the illumination in unknown shadow regions through the pixels that have been relighted.

Using the above assumptions, we refine the shadow scale $K_c(x) = \frac{\hat{I}_c(x)}{\hat{I}_c^f(x)}$ by minimizing the following energy function:

$$\operatorname*{argmin}_{K_c'} \sum_x \left(\frac{K_c(x) - K_c'(x)}{\sigma^S(x)}\right)^2 + \lambda \sum_x \sum_{y \in N_x} \left(\frac{K_c'(x) - K_c'(y)}{||I(x) - I(y)||}\right)^2, \tag{20}$$

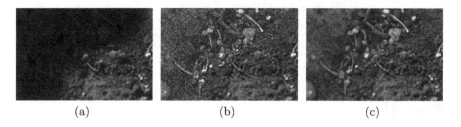

(a)   (b)   (c)

Fig. 6. (a) A shadow image that has many unknown materials. (b) Initial shadow-free image. (c) Optimized shadow-free image.

where $\sigma^S(x)$ is the standard deviation of $S(x)$ which is calculated per color-line, and allows us to apply our estimate only to pixels where the assumptions hold, λ is a smoothing factor for controlling the balance in the data term (left) and the smoothness term (right), and N_x denotes the set of four nearest neighbors of each pixel x in the image. In our experiments, we set $\lambda = 0.001$.

The final shadow-free image $I^f(x)$ is:

$$I_c^f(x) = \frac{\hat{I}_c(x)}{K_c'(x)} + D_c^{\hat{\alpha}}(x). \tag{21}$$

The main framework is summarized in Algorithm 1. The user scribbles include a region with obvious illumination changes (a color line) and part of the regions which indicate the scene illumination feature. Since our method focuses on illumination recovery, we simplify the operation with the user assistance which can be replaced by some automatic methods.

Algorithm 1. Shadow Removal

Input   : $I(x)$, user scribbles
Output: Shadow-free image $I^f(x)$

1 *Estimate the color-line offsets $D^{\hat{\alpha}}(x)$ and perform offset correction;*
2 *Perform Color-line clustering and illumination estimation $S(x)$;*
3 *Shadow detection;*
4 **foreach** *color-line* **do**
5   *Umbra removal: let umbra pixels subject to similar statistical distribution with lit pixels;*
6   *Penumbra removal: estimate the penumbra scale by umbra scale and illumination;*
7 **end**
8 *Optimize shadow scale to get shadow-free image $I^f(x)$.*

Table 1. Shadow removal errors by online benchmark site [6], standard derivations are shown in brackets.

Degree	All pixels errors				Shadow pixels errors			
	Zhang et al. [25]	Guo et al. [8]	Gong and Cosker [7]	Ours	Zhang et al. [25]	Guo et al. [8]	Gong and Cosker [7]	Ours
Texturness								
Weak	0.35(0.17)	0.53(0.50)	0.26(0.16)	**0.25(0.16)**	0.16(0.20)	0.42(0.57)	**0.10(0.09)**	**0.09(0.11)**
Medium	0.39(0.25)	0.59(1.09)	0.26(0.11)	**0.24(0.10)**	0.28(0.25)	0.47(1.15)	0.12(0.09)	**0.10(0.07)**
Strong	0.58(0.38)	0.71(0.60)	0.49(0.40)	**0.36(0.25)**	0.39(0.50)	0.64(1.03)	0.36(0.44)	**0.20(0.18)**
Mean	0.44(0.27)	0.61(0.73)	0.34(0.22)	**0.29(0.17)**	0.27(0.38)	0.51(0.92)	0.19(0.21)	**0.13(0.12)**
Softness								
Weak	0.37(0.24)	0.52(1.08)	**0.23(0.10)**	**0.23(0.10)**	0.24(0.42)	0.39(1.13)	0.10(0.09)	**0.09(0.07)**
Medium	0.40(0.20)	0.70(0.36)	0.34(0.15)	**0.30(0.15)**	0.25(0.26)	0.64(0.43)	0.15(0.10)	**0.12(0.11)**
Strong	0.69(0.49)	1.09(0.75)	0.60(0.27)	**0.44(0.24)**	0.49(0.62)	1.01(0.97)	0.40(0.25)	**0.24(0.19)**
Mean	0.48(0.31)	0.77(0.73)	0.39(0.18)	**0.32(0.16)**	0.33(0.43)	0.68(0.84)	0.22(0.15)	**0.15(0.13)**
Brokenness								
Weak	0.37(0.23)	0.59(0.98)	0.25(0.13)	**0.24(0.11)**	0.24(0.40)	0.48(1.04)	0.11(0.09)	**0.10(0.08)**
Medium	0.43(0.22)	0.42(0.29)	0.29(0.14)	**0.28(0.16)**	0.27(0.27)	0.27(0.35)	0.14(0.11)	**0.12(0.11)**
Strong	1.07(0.47)	1.42(1.06)	0.69(0.30)	**0.59(0.23)**	0.88(0.72)	1.55(1.84)	0.52(0.32)	**0.39(0.17)**
Mean	0.63(0.31)	0.81(0.78)	0.41(0.19)	**0.37(0.17)**	0.46(0.46)	0.76(1.08)	0.26(0.17)	**0.20(0.12)**
Colorfulness								
Weak	0.36(0.18)	0.48(0.64)	0.24(0.11)	**0.23(0.10)**	0.21(0.24)	0.36(0.78)	0.10(0.08)	**0.09(0.07)**
Medium	0.60(0.50)	1.67(2.29)	0.48(0.18)	**0.43(0.18)**	0.57(1.06)	1.56(2.07)	0.24(0.14)	**0.22(0.15)**
Strong	0.78(0.57)	1.20(0.99)	0.56(0.31)	**0.55(0.27)**	0.72(1.00)	1.34(2.33)	0.46(0.48)	**0.40(0.30)**
Mean	0.58(0.41)	1.12(1.31)	0.43(0.20)	**0.40(0.18)**	0.50(0.77)	1.09(1.73)	0.27(0.23)	**0.24(0.17)**
Others								
	0.35(0.16)	0.38(0.52)	**0.19(0.06)**	0.20(0.07)	0.16(0.22)	0.25(0.58)	**0.06(0.02)**	0.07(0.05)

4   Experiment

4.1   Quantitative Results

We evaluate our method on the online benchmark site and dataset [6] provided by Gong and Cosker, which is for open comparison of single image shadow removal. As shown in Table 1, our method outperforms the previous methods [7,8,25] in most cases.

4.2   Visual Results

In Fig. 7, we compare our results to that of state-of-the-art methods [7,8,25]. Our result is more natural than that of the other methods in most cases. The methods of [8,25] are region-based, and reliant on image matting for estimating illumination attenuation factor, but their estimation is unreliable when the shadow regions are rich in texture. The results of [7] exist high-light artifacts since it estimates the image relighting scale by sparse sampling in the penumbra.

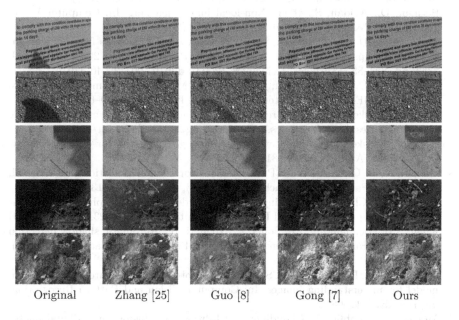

| Original | Zhang [25] | Guo [8] | Gong [7] | Ours |

Fig. 7. Comparisons using images from dataset [6].

5   Conclusion

We present a novel single image shadow removal method based on the color-line regularity in natural images. We demonstrate that the offsets of different color-lines are different and derive a linear shadow image model by our offset correction method. Then we use this model for color-line clustering and shadow illumination recovery. Since existing shadow removal methods only use the local regions for illumination estimation, they are sensitive to the choice of regions. By exploiting umbra and penumbra regions in detail, we propose a refine shadow removal method using color-lines for global illumination estimation and local optimization, which is more robust and effective.

Acknowledgments. This work was supported by the grant of National Science Foundation of China (No. U1611461), Shenzhen Peacock Plan (20130408-183003656), Guangdong Province Projects (2014B010117007), and Science and Technology Planning Project of Guangdong Province, China (No. 2014B090910001).

References

1. Arbel, E., Hel-Or, H.: Shadow removal using intensity surfaces and texture anchor points. IEEE Trans. Pattern Anal. Mach. Intell. **33**(6), 1202–1216 (2011)
2. Barrow, H., Tenenbaum, J.: Recovering intrinsic scene characteristics from images. In: Hanson, A., Risenman, E. (eds.) Computer Vision Systems (1978)
3. Berman, D., Avidan, S., et al.: Non-local image dehazing. In: Computer Vision and Pattern Recognition, pp. 1674–1682. IEEE (2016)
4. Finlayson, G.D., Hordley, S.D., Drew, M.S.: Removing shadows from images. In: Heyden, A., Sparr, G., Nielsen, M., Johansen, P. (eds.) ECCV 2002. LNCS, vol. 2353, pp. 823–836. Springer, Heidelberg (2002). doi:10.1007/3-540-47979-1_55
5. Finlayson, G.D., Hordley, S.D., Lu, C., Drew, M.S.: On the removal of shadows from images. IEEE Trans. Pattern Anal. Mach. Intell. **28**(1), 59–68 (2006)
6. Gong, H., Cosker, D.: Shadow removal dataset and online benchmark for variable scene categories. http://www.cs.bath.ac.uk/~hg299/shadow_eval/eval.php
7. Gong, H., Cosker, D.: Interactive removal and ground truth for difficult shadow scenes. JOSA A **33**(9), 1798–1811 (2016)
8. Guo, R., Dai, Q., Hoiem, D.: Paired regions for shadow detection and removal. IEEE Trans. Pattern Anal. Mach. Intell. **35**(12), 2956–2967 (2013)
9. Huang, J.B., Chen, C.S.: Moving cast shadow detection using physics-based features. In: Computer Vision and Pattern Recognition, pp. 2310–2317. IEEE (2009)
10. Khan, S.H., Bennamoun, M., Sohel, F., Togneri, R.: Automatic shadow detection and removal from a single image. IEEE Trans. Pattern Anal. Mach. Intell. **38**(3), 431–446 (2016)
11. Maxwell, B.A., Friedhoff, R.M., Smith, C.A.: A bi-illuminant dichromatic reflection model for understanding images. In: Computer Vision and Pattern Recognition, pp. 1–8. IEEE (2008)
12. Mohan, A., Tumblin, J., Choudhury, P.: Editing soft shadows in a digital photograph. IEEE Comput. Graph. Appl. **27**(2), 23–31 (2007)
13. Omer, I., Werman, M.: Color lines: Image specific color representation. In: Computer Vision and Pattern Recognition, vol. 2, p. II. IEEE (2004)
14. Prati, A., Mikic, I., Trivedi, M.M., Cucchiara, R.: Detecting moving shadows: algorithms and evaluation. IEEE Trans. Pattern Anal. Mach. Intell. **25**(7), 918–923 (2003)
15. Reinhard, E., Adhikhmin, M., Gooch, B., Shirley, P.: Color transfer between images. IEEE Comput. Graph. Appl. **21**(5), 34–41 (2001)

16. Shor, Y., Lischinski, D.: The shadow meets the mask: pyramid-based shadow removal. In: Computer Graphics Forum, vol. 27, pp. 577–586 (2008)
17. Wu, T.P., Tang, C.K.: A Bayesian approach for shadow extraction from a single image. In: Computer Vision-ICCV, vol. 1, pp. 480–487. IEEE (2005)
18. Wu, T.P., Tang, C.K., Brown, M.S., Shum, H.Y.: Natural shadow matting. ACM Trans. Graph. **26**(2), 8 (2007)
19. Xiao, C., She, R., Xiao, D., Ma, K.L.: Fast shadow removal using adaptive multi-scale illumination transfer. Computer Graphics Forum, vol. 32, pp. 207–218 (2013)
20. Xu, L., Qi, F., Jiang, R.: Shadow removal from a single image. In: Intelligent Systems Design and Applications, vol. 2, pp. 1049–1054. IEEE (2006)
21. Ying, Z., Li, G.: Robust lane marking detection using boundary-based inverse perspective mapping. In: 2016 IEEE International Conference on Acoustics, Speech and Signal Processing (ICASSP), pp. 1921–1925. IEEE (2016)
22. Ying, Z., Li, G., Tan, G.: An illumination-robust approach for feature-based road detection. In: 2015 IEEE International Symposium on Multimedia (ISM), pp. 278–281. IEEE (2015)
23. Ying, Z., Li, G., Wen, S., Tan, G.: ORGB: Offset correction in RGB color space for illumination-robust image processing. In: International Conference on Acoustics, Speech and Signal Processing. IEEE (2017, in press)
24. Ying, Z., Li, G., Zang, X., Wang, R., Wang, W.: A novel shadow-free feature extractor for real-time road detection. In: Proceedings of the 2016 ACM on Multimedia Conference, pp. 611–615. ACM (2016)
25. Zhang, L., Zhang, Q., Xiao, C.: Shadow remover: image shadow removal based on illumination recovering optimization. IEEE Trans. Image Process. **24**(11), 4623–4636 (2015)

Improving Semantic Segmentation with Generalized Models of Local Context

Hasan F. Ates$^{(\boxtimes)}$ and Sercan Sunetci

Department of Electrical and Electronics Engineering,
Isik University, Istanbul, Turkey
hasan.ates@isikun.edu.tr, sercan.sunetci2@isik.edu.tr

Abstract. Semantic segmentation (i.e. image parsing) aims to annotate each image pixel with its corresponding semantic class label. Spatially consistent labeling of the image requires an accurate description and modeling of the local contextual information. Superpixel image parsing methods provide this consistency by carrying out labeling at the superpixel-level based on superpixel features and neighborhood information. In this paper, we develop generalized and flexible contextual models for superpixel neighborhoods in order to improve parsing accuracy. Instead of using a fixed segmentation and neighborhood definition, we explore various contextual models to combine complementary information available in alternative superpixel segmentations of the same image. Simulation results on two datasets demonstrate significant improvement in parsing accuracy over the baseline approach.

Keywords: Image parsing · Segmentation · Superpixel · MRF

1 Introduction

Semantic segmentation (i.e. image parsing) is a fundamental problem in computer vision. The goal is to segment the image accurately and annotate each segment with its true semantic class. Recently superpixel-based methods have shown superior performance in image parsing/labeling problems [1–3]. In superpixel-based segmentation, the image is segmented into visually meaningful atomic regions that agree with object boundaries. Then the parsing algorithm assigns the same semantic label to all the pixels of a superpixel, and thus achieves a spatially consistent labeling of the whole image.

Recent literature have seen a surge of image parsing methods that use superpixels. Some of these methods use nonparametric, data-driven approaches that do not require classifier training [4,5]; while others are based on advanced learning techniques, such as deep convolutional neural networks [3,6]. Nonparametric methods, such as SuperParsing [4], try to match the superpixels of a test image with the superpixels of most similar training images. Then the label of the test superpixel is determined based on the class conditional likelihoods computed from the labels of matching superpixels. Modifications to SuperParsing include methods that improve the matching set [5], that adapt the likelihood estimation

© Springer International Publishing AG 2017
M. Felsberg et al. (Eds.): CAIP 2017, Part II, LNCS 10425, pp. 320–330, 2017.
DOI: 10.1007/978-3-319-64698-5_27

[7,8] and that use advanced descriptors [1,9]. Most existing methods also incorporate contextual inference in superpixel neighborhoods, typically in the form of Markov Random Field (MRF) optimization, to improve parsing consistency and accuracy [5,9].

Despite the ability of superpixels to adhere to object boundaries, committing to a single segmentation for parsing is rather restrictive for describing the rich contextual information in an image. Superpixel segments are determined based on color homogeneity of the pixels, which is not semantically informative. In a recent work [10] we have shown that the use of multiple alternative segmentations could improve the parsing accuracy. In this paper we extend our previous work and investigate various adaptive contextual models to combine complementary information available in alternative superpixel segmentations. These models incorporate not only the spatial neighborhood of adjacent superpixels within the same segmentation but also the neighborhood of intersecting superpixels from different segmentations. We claim that these segmentations provide complementary information about the underlying object classes in the image. The proposed approach provides a unified framework to fuse the information obtained from multiple segmentations. In other words, MRF optimization is used to code contextual constraints both in the spatial and inter-segmentation superpixel neighborhoods for more consistent labeling. As a result, a more flexible description of neighborhood context is achieved, when compared to the fixed context of a single segmentation. In particular, this paper provides contributions in the following two aspects:

– Alternative labelings of the same image are produced by using multiple superpixel segmentations and different encodings of the superpixel features.
– A generalized model of local context is proposed to encode contextual constraints between alternatives and to fuse complementary information available in different segmentations.

The proposed MRF models are tested and compared on SIFT Flow [11] and 19-class subset of LabelMe [12] datasets. Simulation results demonstrate the significant improvement in parsing accuracy over the baseline approach. Results are also competitive with those of state-of-the-art algorithms that use advanced features and classifiers.

Section 2 develops the general framework of the MRF contextual model. Section 3 gives the details of the model and the parsing algorithm. Section 4 provides and discusses the simulations results. Section 5 concludes the paper with ideas for future work.

2   Contextual Modeling of Alternative Segmentations

In superpixel image parsing, a fixed segmentation is typically used to derive local features, estimate class likelihoods, label each segment and perform MRF smoothing of labels. Hence parsing performance heavily depends on how well the size, shape, boundary and content of superpixels represent the underlying object classes. Farabet [3] uses a multiscale set of segmentations to learn hier-archical features and tries to find an optimal cover of the image out of many segmentations. In the end, this method also commits to a final fixed segmenta-tion, which is claimed to be optimal, but does not consider a joint optimization of alternative representations.

In this paper we develop "Multi-hypothesis MRF model" and show that labeling accuracy could benefit from the joint use of multiple segmentations. In this approach the local context incorporates not only the spatial neighborhood of adjacent superpixels within the same segmentation but also the neighborhood of intersecting superpixels from different segmentations. These inter-segmentation neighborhoods help fuse alternative representations coming from multiple seg-mentations. Hence, the proposed MRF model describes both intra-segmentation and inter-segmentation contextual information. Intra- neighborhood contains adjacent superpixels of a given segmentation. Inter- neighborhood contains inter-secting superpixels from different segmentations. The MRF model is used to code contextual constraints in both intra- and inter- neighborhoods for more consis-tent labeling. We explore different data and neighborhood models for a more generalized contextual framework within the MRF formalization.

In the following, the generalized MRF model is described for two alternative segmentations; but it could be easily generalized to any number of alternatives. Let the set of superpixels be defined as $SP_m = \{s_i^m\}$ $(m = 1, 2)$. We define a third segmentation $SP_3 = \{s_k^3\}$ based on the intersection of the superpixels of the two alternatives (see Fig. 1):

$$s_k^3 = s_i^1 \cap s_j^2 \neq \emptyset, \quad \forall\, s_i^1 \in SP_1,\, s_j^2 \in SP_2 \tag{1}$$

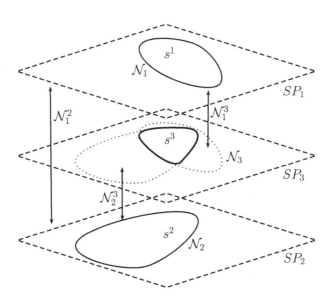

Fig. 1. Multi-hypothesis MRF model

For each segmentation, let $\mathcal{N}_m$ represent the contextual neighborhood that contains pairs of adjacent superpixels. In addition to these intra- neighborhoods, we define inter-segmentation context $\mathcal{N}_n^m(n, m = 1, 2, 3)$ as follows:

$$(s_i^n, s_j^m) \in \mathcal{N}_n^m \quad \text{if} \quad s_i^n \cap s_j^m \neq \emptyset \tag{2}$$

The image is parsed by assigning to each superpixel s_i^m a class label c_i^m. Each segmentation produces an alternative parsing result, $\mathbf{c}^m = \{c_i^m\}$. We formulate the labeling problem as an MRF energy minimization over the whole set of superpixel labels $\mathbf{c} = \{\mathbf{c}^1, \mathbf{c}^2, \mathbf{c}^3\}$ as follows:

$$J(\mathbf{c}) = \sum_{m=1}^{3} \left(\sum_{s_i^m \in SP_m} D(s_i^m, c_i^m) + \lambda_m \sum_{(s_i^m, s_j^m) \in \mathcal{N}_m} E(c_i^m, c_j^m) \right)$$

$$+ \sum_{(n,m) \in IC} \lambda_n^m \sum_{(s_i^n, s_j^m) \in \mathcal{N}_n^m} E(c_i^n, c_j^m) \tag{3}$$

where D and E are appropriate data cost and smoothness cost terms, respectively, of related label assignments; IC is the set of inter-segmentation neighborhoods ($IC = \{(1, 2), (1, 3), (2, 3)\}$), and λ_m, λ_n^m are smoothness constants for each corresponding context. In the next section we discuss the details of the MRF model, in particular how to define data and smoothness costs and how to select the smoothness constants.

3   Image Parsing with Generalized MRF Model

In the MRF formulation given above, the data cost $D(s_i, c)$ represents the confidence with which a superpixel s_i is assigned to a class c; the smoothness cost $E(c_i, c_j)$, on the other hand, is a measure of likelihood that two neighboring superpixels are assigned to distinct class labels c_i, c_j.

In this paper the data costs for SP_1 and SP_2 are computed using the nonparametric approach of [4]. The test superpixel is matched with the superpixels of a suitable subset of the training images, i.e. "retrieval set", based on a rich set of superpixel features. A scene-level matching is performed to find a good retrieval set that contains images similar to the tested image. For each superpixel feature, the nearest R superpixels are determined from the retrieval set. The labels of these matching superpixels are used to compute class-conditional log-likelihood ratio scores for each class. Assuming independent features, the overall log-likelihood ratio $L(s_i, c)$ is equal to the sum of log-likelihood ratios of all features (see [4] for details). Then the data term is defined as $D(s_i, c) = w_i \sigma(L(s_i, c))$, where w_i is the superpixel weight, and $\sigma(\cdot)$ is the sigmoid function. Here $w_i = |s_i|/\mu_s$, where $|s_i|$ is the number of pixels in s_i, μ_s is the mean of $|s_i|$.

As in [4], the smoothness costs for SP_1 and SP_2 are based on probabilities of label co-occurrence: $E(c_i, c_j) = -\log[(P(c_i|c_j) + P(c_j|c_i))/2] \times \delta(c_i \neq c_j)$, where $P(c|c_j)$ is the conditional probability of one superpixel having label c given that its neighbor has label c_j, estimated by counts from the training set. The delta function $\delta(c_i \neq c_j)$ is used to assign zero cost when $c_i = c_j$.

Since SP_3 is generated from the other two segmentations, its data and smoothness costs are defined as functions of the corresponding costs in SP_1 and SP_2. In order to fuse the complementary information coming from the two segmentations, we differentiate the set of classes in segment SP_1 and SP_2 and select from the union of those two sets for segment SP_3. In other words, a semantic class c in SP_1 and SP_2 is shown as $c^{(1)}$ and $c^{(2)}$, respectively, and these two are treated as separate classes in SP_3 (Note that, c_i^m represents the class label assigned to superpixel s_i^m, while $c^{(m)}$ represents any given class of segmentation SP_m). Then the data cost for $s_k^3 \in SP_3$ is given by $(m = 1, 2)$:

$$D(s_k^3, c^{(m)}) = f_m(D(s_i^1, c), D(s_j^2, c)) \quad \text{if} \quad (s_i^1, s_k^3) \in \mathcal{N}_1^3 \ \& \ (s_j^2, s_k^3) \in \mathcal{N}_2^3 \quad (4)$$

Likewise, the smoothness costs in inter- neighborhoods $\mathcal{N}_n^m$ are based on the costs in $\mathcal{N}_1$ and $\mathcal{N}_2$ as follows:

$$E(c_i^{(1)}, c_j^{(2)}) = g(E(c_i^{(1)}, c_j^{(1)}), E(c_i^{(2)}, c_j^{(2)})) \quad (5)$$

The specifics of f_m and g will be discussed in the next Section.

The smoothness constants λ_m, λ_n^m control the level of contextual dependency in different neighborhoods of the MRF model in Eq. (3). These values are determined from the training set by a leave-one-out strategy: each training image is removed from the training set, and then parsed by the proposed algorithm to obtain its labeling accuracy under different parameter settings. Then, the set of parameters that maximizes the mean accuracy in the training set is chosen.

The MRF energy function is minimized by the α-expansion method of [13]. The outcome leads to three alternative labelings $\mathbf{c} = \{\mathbf{c}^1, \mathbf{c}^2, \mathbf{c}^3\}$. The set of labels $\mathbf{c}^3$ of the segmentation SP_3 is selected as the final labeling of the image.

4 Simulations and Discussions

4.1 Implementation Details

For parsing a test image, retrieval set is obtained and then graph-based segmentation [14] is applied to obtain the superpixels. In [14], K controls superpixel color consistency and S determines the smallest superpixel size. We test two alternative segmentations: $SP_1 = \{K = 200, S = 100, R = 30\}$, $SP_2 = \{K = 400, S = 200, R = 15\}$. Here R is the nearest neighbor set size that is used to define superpixel feature neighborhoods.

20 different superpixel features are used to find matching superpixels from the retrieval set. These features include shape (e.g. superpixel area), location (e.g. superpixel mask), texture (e.g. texton/SIFT), color (e.g. color histogram) descriptors [4]. There are six SIFT features that are defined over different subregions of the superpixel. These SIFT features are encoded using either LLC (Locality-constrained Linear Coding) [15] (256-word dictionary) or KCB (Kernel Codebook Encoding) [16] (512-word dictionary) algorithms.

The smoothness constants of the MRF model are selected from the set $\{l\lambda | l \in \{0, 1, 2\}; \ 5 \leq \lambda \leq 25, \lambda \in \mathbb{Z}\}$. The function g is set as $g(x, y) = 0.5x + 0.5y$. For data costs of SP_3, we test three alternatives ($w_k = 0.5|s_k^3|(1/|s_i^1| + 1/|s_j^2|)$):

- **DC1**: $D(s_k^3, c^{(m)}) = 0.5(w_k/w_i^1)D(s_i^1, c) + 0.5(w_k/w_j^2)D(s_j^2, c), \ m = 1, 2$

- **DC2**: $D(s_k^3, c^{(1)}) = (w_k/w_i^1)D(s_i^1, c); \ D(s_k^3, c^{(2)}) = (w_k/w_j^2)D(s_j^2, c)$

- **DC3**: A feature vector is obtained for s_k^3 by concatenating the data cost vectors: $[(w_k/w_i^1)D(s_i^1, \mathbf{c})^T, (w_k/w_j^2)D(s_j^2, \mathbf{c})^T]^T$. Boosted decision tree (BDT) classifiers [17] are trained using this feature vector. This classifier outputs a likelihood ratio score, $L_{BDT}(s_k^3, c)$, and we set

$$D(s_k^3, c^{(m)}) = 5w_k\sigma(L_{BDT}(s_k^3, c)), \ m = 1, 2.$$

The first model uses an average of two data costs from SP_1 and SP_2 for superpixels of SP_3. The second model assigns two different data costs to the same class c under two different hypotheses of SP_1 and SP_2. The third model treats the data costs of the second model as superpixel features to train BDT classifiers; the likelihood score output by the classifier is used to define the data cost of s_k^3. The superpixel weight w_k is set proportional to the relative size of s_k^3 with respect to s_i^1 and s_j^2; if the intersection of the two superpixels is small, then the data cost for s_k^3 is assigned lower weight since it is deemed unreliable for labeling s_k^3.

Different realizations are obtained by using either LLC or KCB coded SIFT features in the above data models, and separately for the data costs of SP_1 and SP_2. Simulation results for some of the best performing alternatives are tabulated and discussed in the next section.

4.2   Results and Comparisons

The proposed models are evaluated on two well-known datasets, namely SIFT Flow [11] and 19-class subset of LabelMe [12]. In experiments overall pixel-level classification accuracy (i.e. correctly classified pixel percentage) and average per-class accuracies are compared.

SIFT Flow dataset contains 2,688 images and 33 labels. This dataset includes outdoor scenery such as mountain view, streets, etc. There are objects from 33 different semantic classes, such as *sky, sea, tree, building, cars*, at various densities. There are also 3 geometric classes, which are not considered in our simulations. Dataset is separated into 2 subsets; 2.488 training images and 200 test images. The retrieval set size is set at 200 images, as in original SuperParsing.

19-class LabelMe dataset contains 350 images with 19 classes (such as *tree, field, building, rock*, etc.). The dataset is split into 250 training images and 100 test images. The retrieval set size is set at 100 images, due to the smaller size of training set.

Table 1. Per-pixel and average per-class labeling accuracies for SIFT Flow dataset

Method	Percentage Accuracy	
	Per-pixel (%)	Per-class (%)
Model-1	78.5	30.0
Model-2	79.5	32.1
Model-3	80.6	31.8
Seg-1	77.8	29.1
Seg-2	78.1	27.9
SuperParsing [4]	76.2	29.1
Tighe [2]	78.6	39.2
Nguyen [5]	78.9	34.0
George [1]	81.7	50.1
Kernel+Multi-hyp. [8]	77.8	29.4

Table 1 reports the pixel-level and average per-class labeling accuracies of the tested methods for SIFT Flow. For each data model, the following parameter settings are reported:

- **Model-1**: DC1; KCB encoding; $\lambda_3 = \lambda_1^2 = 0$; $\lambda = 18$ for all the other $\mathcal{N}$.
- **Model-2**: DC2; KCB encoding; $\lambda_3 = \lambda_1^2 = 0$; $\lambda_2 = \lambda_1^3 = \lambda_2^3 = 18$; $\lambda_1 = 36$.
- **Model-3**: DC3; LLC encoding for SP_1, SP_2; KCB encoding for SP_3; $\lambda_3 = \lambda_1^2 = 0$; $\lambda = 18$ for all the other $\mathcal{N}$.

For comparison, results are provided for original SuperParsing [4], with per-exemplar detectors (Tighe [2]), adaptive nonparametric parsing (Nguyen [5]), our Kernel based Multi-hypothesis SuperParsing [8], George [1], and for SP_1 and SP_2 segmentations without the inter- neighborhoods:

- **Seg-1**: $\mathbf{c}^1$ labeling; KCB encoding; $\lambda_n^m = 0$, $\forall (n, m) \in IC$; $\lambda_1 = 10$.
- **Seg-2**: $\mathbf{c}^2$ labeling; LLC encoding; $\lambda_n^m = 0$, $\forall (n, m) \in IC$; $\lambda_2 = 2.5$.

Model-1 provides an improvement of around 0.5% in pixel accuracy over Seg-1,2 with the use of inter- neighborhoods. Model-2 brings an additional 1% increase, showing that it is more effective to keep data costs separate when fusing the complementary information of the two hypotheses. In other words, Model-2 allows for a more flexible representation to select from alternatives with different data costs and contextual constraints. Model-3 improves pixel accuracy of Model-2 by about 1%. Simulations show that 0.5% of this improvement is due to the use of BDT classifiers; the other 0.5% comes from the joint use of LLC and KCB encoding for superpixel features. Note that LLC encoding is used for SP_1 and SP_2, while KCB encoding is used to define data costs of SP_3. When both encoding methods are jointly used to define all the data costs, there is little gain over using a single encoding method. This shows once again that our MRF framework is able to combine complementary information coming from alternative encodings. The proposed models provide up to 3% increase in average per-class accuracy as well. c^3 defines a labeling on regions at a finer scale, which are created by the intersection of SP_1 and SP_2 segments. Therefore our MRF model avoids over-smoothing the labeling and improves the accuracies of rare classes in the dataset.

Overall, Model-3 is 4.4% and 2.7% higher, respectively, in per-pixel and per-class accuracies over SuperParsing. Labeling accuracy is better than or comparable to the other listed methods from literature. George [1] uses classifier fusion for BDT classifiers trained on high-dimensional feature vectors that include Fisher Vector descriptors. Tighe [2] combines likelihood scores from SuperParsing with detection scores from per-exemplar SVM detectors trained for all classes in the dataset. Compared to these methods, our MRF framework provides a much simpler and more intuitive framework for the fusion of likelihood scores coming from alternative representations.

In general, state-of-art results on SIFT Flow [18–20] use advanced classifiers, such as deep convolutional networks, for feature learning and/or likelihood estimation. Our model does not require any classifier training, except for Model-3 which trains BDTs on low-dimensional data cost feature vectors. Therefore the proposed method has better scalability than most of these state-of-art techniques; when a new class is added to the set, a BDT will be trained for the new class and no re-training will be necessary.

The average per-class accuracies of our models are lower than Tighe [2], George [1] and other methods in literature that use balanced training sets to improve the accuracies of rare classes. We believe that our contextual framework could also provide significantly better per-class accuracy with the help of complementary information coming from a carefully selected balanced training set. However this is left as a future work.

Table 2 reports the pixel-level and average per-class labeling accuracies of the tested methods for LabelMe. Results from literature are provided for Nguyen [5] and Myeong [21]. The following parameter settings are reported:

- **Model-2(LM):** DC2; LLC encoding; $\lambda_3 = \lambda_1^2 = 0$; $\lambda_2 = \lambda_1^3 = \lambda_2^3 = 18$; $\lambda_1 = 36$.

Table 2. Per-pixel and average per-class labeling accuracies for 19-class LabelMe dataset

Method	Percentage Accuracy	
	Per-pixel (%)	Per-class (%)
Model-2(LM)	83.0	52.9
Model-3(LM)	83.8	55.6
Seg-1(LM)	81.9	54.7
Seg-2(LM)	79.5	49.9
Myeong [21]	81.8	54.4
Nguyen [5]	82.7	55.1

- **Model-3(LM)**: DC3; KCB encoding for SP_1, SP_2; LLC encoding for SP_3; $\lambda_3 = \lambda_1^2 = 0$; $\lambda = 38$ for all the other $\mathcal{N}$.
- **Seg-1(LM)**: $\mathbf{c}^1$ labeling; LLC encoding; $\lambda_n^m = 0$, $\forall (n, m) \in IC$; $\lambda_1 = 7.5$.
- **Seg-2(LM)**: $\mathbf{c}^2$ labeling; LLC encoding; $\lambda_n^m = 0$, $\forall (n, m) \in IC$; $\lambda_2 = 5$.

Model-3(LM) achieves state-of-art results on this dataset, both in per-pixel and mean per-class accuracies. Note that Seg-2(LM) performance is significantly worse than Seg-1(LM), which also affects $\mathbf{c}^3$ labeling accuracy. A better selection of superpixel segmentation parameters for SP_1 and SP_2 could further boost the performance of Model-3(LM).

At Fig. 2, parsing results of SuperParsing and Model-3 are compared visually for some selected test images from SIFT Flow. Model-3 labelings are generally more consistent and accurate than those of SuperParsing. In the top figure, *mountain* is correctly identified, even though it is hard to notice even with visual

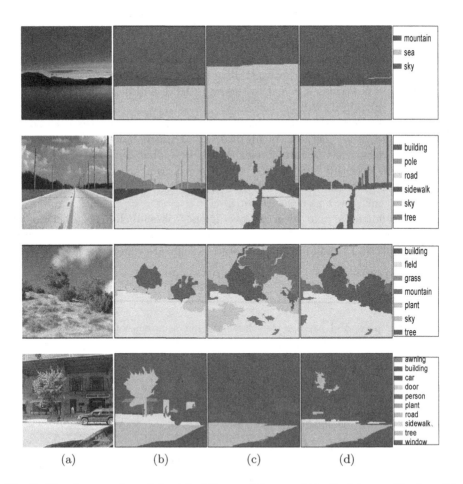

(a)   (b)   (c)   (d)

Fig. 2. Visual comparison: (a) original Image; (b) ground Truth; (c) superParsing; (d) Model-3

inspection. The other figures also indicate finer and spatially consistent results, even though there is still room for improvement.

5   Conclusion

In this paper a novel contextual modeling framework is introduced for superpixel image parsing. This framework defines contextual constraints over inter- and intra- neighborhoods for multiple segmentations of the same image. In addition to producing spatially more consistent parsing results, the proposed approach carries out labeling at a finer scale over the intersecting regions of alternative segmentations.

As future work, we plan to advance our contextual models to improve pixel-level and class average parsing accuracies. A more balanced training set will be used to further improve labeling accuracy of rare classes in the dataset. More advanced classifiers and feature learning will also be tested within this MRF framework. Optimization of superpixel segmentation parameters could also boost the performance of the algorithm.

Acknowledgements. This work is supported in part by TUBITAK project no: 115E307 and by Isik University BAP project no: 14A205.

References

1. George, M.: Image parsing with a wide range of classes and scene-level context. In: CVPR, pp. 3622–3630 (2015)
2. Tighe, J., Niethammer, M., Lazebnik, S.: Scene parsing with object instance inference using regions and per-exemplar detectors. Int. J. Comput. Vision **112**, 150–171 (2015)
3. Farabet, C., Couprie, C., Najman, L., LeCun, Y.: Learning hierarchical features for scene labeling. PAMI **35**, 1915–1929 (2013)
4. Tighe, J., Lazebnik, S.: Superparsing: scalable nonparametric image parsing with superpixels. Int. J. Comp. Vision **101**, 329–349 (2013)
5. Nguyen, T., Lu, C., Sepulveda, J., Yan, S.: Adaptive nonparametric image parsing. IEEE Trans. Circuits Syst. Video Tech. **25**, 1565–1575 (2015)
6. Sharma, A., Tuzel, O., Liu, M.: Recursive context propagation network for semantic scene labeling. Adv. Neural Inf. Process. Syst. **27**, 2447–2455 (2014)
7. Eigen, D., Fergus, R.: Nonparametric image parsing using adaptive neighbor sets. In: CVPR, pp. 2799–2806 (2012)
8. Ates, H.F., Sunetci, S., Ak, K.E.: Kernel likelihood estimation for superpixel image parsing. In: Campilho, A., Karray, F. (eds.) ICIAR 2016. LNCS, vol. 9730, pp. 234–242. Springer, Cham (2016). doi:10.1007/978-3-319-41501-7_27
9. Yang, J., Price, B., Cohen, S., Yang, M.H.: Context driven scene parsing with attention to rare classes. In: CVPR, pp. 3294–3301 (2014)
10. Ak, K., Ates, H.: Scene segmentation and labeling using multi-hypothesis super-pixels. In: Signal Processing and Communications Applications Conference (SIU), pp. 847–850 (2015)

11. Liu, C., et al.: SIFT flow: dense correspondence across difference scenes. In: ECCV (2008)
12. Jain, A., Gupta, A., Davis, L.S.: Learning what and how of contextual models for scene labeling. In: Daniilidis, K., Maragos, P., Paragios, N. (eds.) ECCV 2010. LNCS, vol. 6314, pp. 199–212. Springer, Heidelberg (2010). doi:10.1007/978-3-642-15561-1_15
13. Boykov, Y., Kolmogorov, V.: An experimental comparison of min-cut/max-flow algorithms for energy minimization in vision. IEEE PAMI **26**, 1124–1137 (2004)
14. Felzenszwalb, P., Huttenlocher, D.: Efficient graph-based image segmentation. Int. J. Comp. Vision **59**, 167–181 (2004)
15. Wang, J., et al.: Locality-constrained linear coding for image classification. In: CVPR (2010)
16. van Gemert, J.C., Geusebroek, J.-M., Veenman, C.J., Smeulders, A.W.M.: Kernel codebooks for scene categorization. In: Forsyth, D., Torr, P., Zisserman, A. (eds.) ECCV 2008. LNCS, vol. 5304, pp. 696–709. Springer, Heidelberg (2008). doi:10.1007/978-3-540-88690-7_52
17. Hoiem, D., Efros, A., Hebert, M.: Recovering surface layout from an image. Int. J. Comput. Vision **75**, 151–172 (2007)
18. Shelhamer, E., Long, J., Darrell, T.: Fully convolutional networks for semantic segmentation. IEEE PAMI **39**, 640–651 (2016)
19. Liu, W., Rabinovich, A., Berg., A.: Parsenet: Looking wider to see better. In: ICLR Workshop (2016)
20. Liang, M., Hu, X., Zhang, B.: Convolutional neural networks with intra-layer recurrent connections for scene labeling. In: NIPS, pp. 937–945 (2015)
21. Myeong, H., Lee, K.: Tensor-based high-order semantic relation transfer for semantic scene segmentation. In: CVPR, pp. 3073–3080 (2013)

An Image-Matching Method Using Template Updating Based on Statistical Prediction of Visual Noise

Nobuyuki Shinohara[✉] and Manabu Hashimoto

Chukyo University, 101-2 Yagoto-Honmachi, Showa-ku, Nagoya, Aichi 466-8666, Japan
shinohara@isl.sist.chukyo-u.ac.jp

Abstract. An image-matching method that can continuously recognize images precisely over a long period of time is proposed. On a production line, although a multitude of the same kind of components can be recognized, the appearance of a target object changes over time. Usually, to accommodate that change in appearance, the template used for image recognition is periodically updated by using past recognition results. At that time, information other than that concerning the target object might be included in the template and cause false recognition. In this research, we define the pixels which become those factors as "noisy-pixel". With the proposed method, noisy pixels in past recognition results are extracted, and they are excluded from the processing to update the template. Accordingly, the template can be updated in a stable manner. To evaluate the performance of the proposed method, 5000 images in which the appearance of the target object changes (due to variation of lighting and adhesion of dirt) were used. According to the results of the evaluation, the proposed method achieves recognition rate of 99.5%, which is higher than that of a conventional update-type template-matching method.

Keywords: Template matching · Template update · Noisy-pixel prediction

1 Introduction

At manufacturing sites, although object detection by image sensors proving useful, handling changes to the appearance of objects over time is a well-known problem. As for template matching, an object imaged at a certain time, is used as a template; consequently, if the appearance of the object changes for some reason (such as lighting variations, shifting of focus, adherence of dirt, and individual differences between objects), recognition precision decreases, and that is a serious problem.

Conventionally, such changes are handled by one of three main approaches. As for the first approach, a method that updates the template is applied. Generally referred to as "update-type template matching" [1–4], this method uses past recognition results and periodically updates the template. Accompanying this method, however, is the risk that in the case the past recognition results available for the update are scarce, the template might be wrongly updated. The memory that can save past recognition results is constrained in terms of capacity by the equipment used in a plant; consequently, updating the template with a small amount of data poses a problem. For the second approach,

M. Felsberg et al. (Eds.): CAIP 2017, Part II, LNCS 10425, pp. 331–341, 2017.
DOI: 10.1007/978-3-319-64698-5_28

methods that abstract images are adopted. With these methods, namely, "increment sign correlation" (ISC) [5] and "orientation code matching" (OCM) [6], information about pixel intensity is abstracted in coded form as a slope of pixel intensity and a magnitude relation of pixel intensity, and the influence of changes to appearance is thereby reduced. However, these methods cannot handle changes to object appearance that invert the code. As for the third approach, a "keypoint-based matching method" is adopted [7, 8]. With this method, however, it is necessary to process not only the template but also input images. As a result of this increased need for processing, keypoint-based matching is not often used on production lines, where processing costs are a major concern.

The aim of the present study is to handle changes to the appearance of an object over a long period of time. In this study, a method called "updating-type template matching" is proposed, and the template is updated in a manner that responds correctly to the code even if the appearance of an object changes as an inversion of the code. Moreover, two research targets, namely, stabilizing update of a template when little data is available and improving trackability of changes to the appearance of objects, were attained. Also in this research, we define the pixels which become factors of wrongly updated as "noisy-pixel".

2   Basic Idea

The basic idea behind stabilizing updating of a template with little data is explained as follows: noisy pixels are predicted from past recognition results, and the noisy pixels are excluded from the processing for updating the template. An example of predicting noisy pixels is shown Fig. 1.

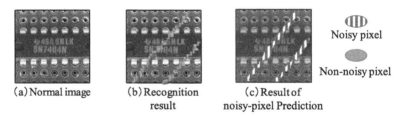

(a) Normal image   (b) Recognition   (c) Result of
                           result   noisy-pixel Prediction

Noisy pixel

Non-noisy pixel

Fig. 1. Example of predicting noisy pixels

In the recognition result shown in Fig. 1(b), there are regions in which objects other than the target object appear in the form of lead wires. Pixels included within such regions are predicted as "noisy pixels." The result of this inference of noisy pixels is shown in Fig. 1(c). For predicting the noisy pixels, a model that analyzes pixel intensity of the target object and plots its distribution is used. This model expresses occurrence probability of pixel intensity for each pixel composing an object, and pixels having pixel intensity with high occurrence probability are considered to be statistically stable and observable. Such pixels are predicted as "non-noisy pixels." On the contrary, pixels having pixel intensity with low occurrence probability are considered to be noisy pixels.

In this study, the model used for predicting noisy pixels is referred to as the "no-noise probability model."

2.1 Template Update Using Non-noisy Pixels

The difference between templates after updating by update-type template matching by the conventional method and the proposed method when noise (which does not exist on the object in the image per se) occurs in the past recognition results is shown in Fig. 2.

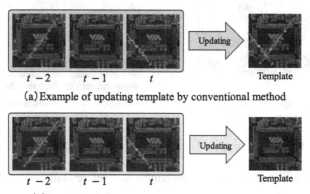

(a) Example of updating template by conventional method

(b) Example of updating template by proposed method

Fig. 2. Difference between templates after updating

As shown in the figure, by applying the basic idea described above, even in the case that noise occurs in the past recognition results, the proposed method can update the template while excluding the noise.

2.2 Selection of Reference Pixels Using Non-noisy Pixels

As for the conventional method, distinguishing pixels are extracted from an image, and those images are selected for use as reference ("reference images" hereafter) [9]. The advantage of template matching by selecting pixels in this manner is shown in Fig. 3.

By selecting a small proportion of pixels in a template image, it is possible to decrease the number of pixels used for reference; consequently, as shown in Fig. 3(a), processing is speeded up in comparison with the case that all pixels in the template are used. Moreover, as shown in Fig. 3(b), even in the case that some regions of the target object are obscured, by selecting pixels in a manner that avoids those obstructions, it is possible to successfully recognize the target object. In this manner, by selecting pixels appropriately, high recognition rate and robustness against noise can be enhanced.

In this study, which aims to acquire high recognition rate and robustness against noise, an algorithm for selecting reference pixels used in template matching is adopted. The results of predicting noisy pixels are used, and pixels with high frequency of being predicted as non-noisy pixels are selected as reference pixels. Moreover, in the same way as the method described in [9], in consideration of the results of pixel co-occurrence

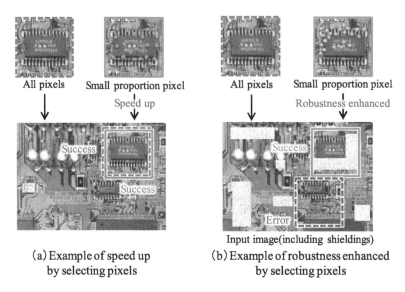

(a) Example of speed up by selecting pixels

(b) Example of robustness enhanced by selecting pixels

Fig. 3. Advantages of template matching by selecting pixels

analysis, images with high statistical distinctness are preferentially selected as reference images.

3 Proposed Method

The proposed method is based on update-type template matching and pixel-selection-type template matching. In particular, when a template is updated, the images in a template (pixel intensity) and the location (coordinates) of pixels for referring to in that

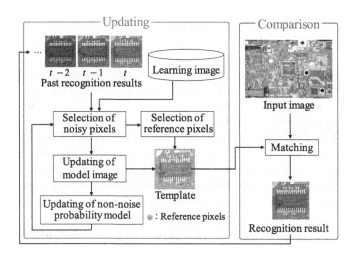

Fig. 4. Flow of proposed method

image is updated, so the template is redefined as a "model image" with pixel intensity data and as reference pixels with coordinate data. The flow of the proposed method is shown schematically in Fig. 4.

The method is composed of a comparison module and an update module. As for the comparison module, degree of similarity of each point in an input image is calculated by using a template, and the points with the highest degree of similarity are retrieved. At those points, a template-sized image is extracted from the input image and taken as a recognition result. For the scale of degree of similarity, a normalized cross-correlation is used.

As for the comparison module, first, the no-noise probability model is used to estimate noisy pixels in the N frames of past recognition results used for updating. Then, only images predicted to be non-noisy ones are used to create "average images," which are updated as "model images". After that, in respect to the N frames of past recognition results, the results of estimating noisy pixels and the results of the pixel co-occurrence analysis are utilized, and M pixels are selected as reference pixels from pixels with high frequency of being predicted as statistically non-noisy and pixels with high individuality.

3.1 Selection of Noisy Pixels

As for the no-noise probability model (used for predicting noisy pixels), an image reflecting the target object is generated as a "learning image" by analyzing the distribution of pixel intensity from the learning image.

A number (N_m) of learning images $L_n (n = 0, 1, 2, \dots, N_m-1)$ are used, and statistical occurrence probability of the pixel intensity of each pixel, assumed to be given by a statistical pixel intensity variation model (M_v), is calculated from Eqs. (1) and (2) given as

$$M_v(i,j,v) = \frac{1}{N_m} \sum_{n=0}^{N_m-1} \delta(L_n(i,j), v) \tag{1}$$

$$\begin{cases} \delta = 1 \ if & L_n(i,j) = v \\ \delta = 0 \ otherwise \end{cases} \tag{2}$$

$M_v(i,j,v)$ expresses the probability that pixel intensity v will be generated at coordinates (i, j). In the current study, M_v is approximated as a mixed Gaussian distribution (i.e., a Gaussian mixture model, GMM), and variation of pixel intensity in the learning image. A schematic diagram of prediction of noisy pixels using the no-noise probability model is shown in Fig. 5.

As for all pixels of the recognition results, the no-noise probability model is referenced, and no-noise probability $M_v(i,j,v)$ is calculated. And noisy pixels are predicted on the basis of the magnitude of $M_v(i,j,v)$ at that time. Moreover, in this study, images of the target object captured under different lighting conditions are used as initial learning images, and after method execution, the recognition results are added to the learning image. The memory capacity required to maintain the no-noise probability model is equivalent to that for $3 \times K$ images, where K is number of mixed Gaussian distributions in the GMM.

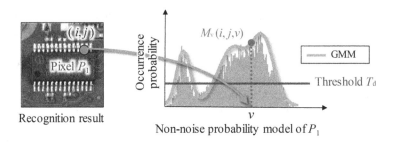

Fig. 5. Schematic diagram of noisy-pixel selection using no-noise probability model

3.2 Updating of Model Image

"Average" images generated by using only non-noisy pixels from the recognition results of the past N frames are updated as "model image." A schematic diagram of the updating of the model image is shown in Fig. 6.

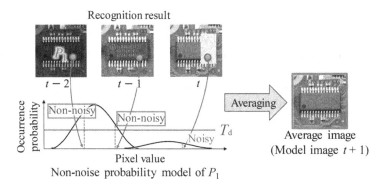

Fig. 6. Schematic diagram of updating model image

When the template is updated, in regard to all pixels in the past N frames of recognition results $f_n(n = 0, 1, 2, \ldots, N - 1)$, no-noise probability is calculated, and if it equals or exceeds a threshold value (T_d), it is assumed that the recognition results are composed of a non-noisy pixels. An average image, $a(i, j)$, is generated from Eq. (3) as

$$a(i,j) = \frac{V(i,j)}{C(i,j)} \tag{3}$$

where $C(i, j)$ is the number of pixels assumed to be non-noisy ones in regard to the (i, j)-th pixel in the past N frames of recognition results, and $V(i, j)$ is the sum total of the pixel intensity possessed by those pixels. Average image a acquired in this manner is updated as a model image to be used from the next frame.

3.3 Selection of Reference Pixels

In regard to the past N frames, pixels with high frequency of being predicted as non-noisy ones and high individuality are selected as new reference pixels. A schematic diagram of the selection of reference pixels is shown in Fig. 7.

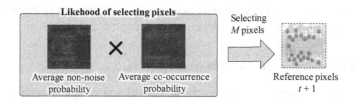

Fig. 7. Schematic diagram of selection of reference pixels

First, from the past N frames of recognition results, no-noise probability for each pixel is obtained as an "average no-noise probability" (P_{ud}). Next, pixel co-occurrence analysis [9] is performed on the past N frames of recognition results, and average individuality (namely, pixel co-occurrence probability) for each pixel is obtained as O_s. With the product of average no-noise probability Pud and average individuality O_s taken as the likelihood of selecting an image (P_l), M pixels are selected as reference pixels in the order of highest P_l.

4   Experiment and Considerations

The effectiveness of the proposed method was evaluated by the experiment described in the following section.

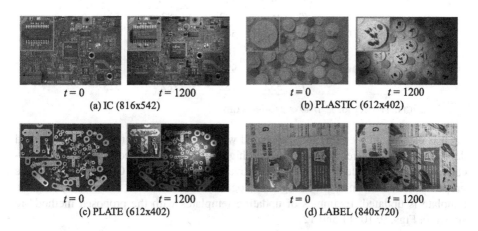

Fig. 8. Examples of input images and target objects (Color figure online)

4.1   Experimental Conditions

In the experiment, time series of images in which the target object varies dramatically as a result of variations of lighting conditions and adhesion of dirt were used. Four kinds of objects, namely, an IC chip ("IC" hereafter), a plastic product ("PLASTIC"), a metal product ("PLATE"), and glossy labels ("LABEL"), were used as recognition targets, and 1250 images of each kind were used as input images. Examples of the input images and target objects (indicated on the upper left by blue squares) are shown in Fig. 8.

4.2   Evaluation of Recognition Performance

The recognition performance of the proposed method was compared with that of other methods as explained as follows. In the comparison, as methods for abstracting images, ISC [5] and OCM [6] were used, and as methods for updating templates, SCPTM [3] and adaptive STC [4] were used. Recognition success rate Pr [%] and verification time [ms] for each method are listed in Table 1. Note that verification time is the time taken for verification, and it does not include the time taken by updating the template. As for the number of reference pixels, ISC and OCM use all the pixels in the template are used, but the proposed method used 2% of the pixels in the template. As for the methods used for template matching (namely, SCPTM, adaptive STC, and the proposed method), the template is updated once every 50 input images. The past five frames of input images were taken as the recognition results used for updating the template.

Table 1. Recognition success rate P_r [%] and verification time [ms] for each dataset

Method			IC	PLASTIC	PLATE	LABEL
ISC	P_r		100.0	96.9	94.1	70.3
	T		17338	7268	7268	21077
OCM	P_r		99.2	100.0	92.2	99.8
	T		17338	7268	7268	21077
SCPTM	P_r		74.0	72.6	99.5	99.8
	T		352	182	182	523
Adaptive STC	P_r		64.1	54.1	99.8	60.2
	T		352	182	182	523
Proposed method	P_r		98.9	99.3	99.9	99.9
	T		352	182	182	523

(CPU: Intel ®CORE™   i7-4600U,Memory: 8 GB)

It is clear from the table that compared with ISC and OCM, the proposed method demonstrates higher recognition rate and higher processing speed (i.e., shorter verification time) when selecting pixels. Moreover, compared with SCPTM and STC, the proposed method demonstrates higher recognition rate and uses fewer data when the template is updated. Examples of updating templates with the proposed method are shown in Figs. 9, 10, 11 and 12.

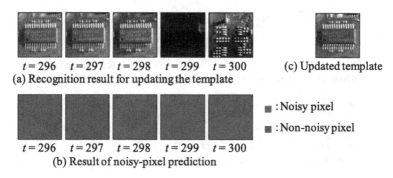

$t=296$ $t=297$ $t=298$ $t=299$ $t=300$ (c) Updated template
(a) Recognition result for updating the template

■ : Noisy pixel

■ : Non-noisy pixel

$t=296$ $t=297$ $t=298$ $t=299$ $t=300$
(b) Result of noisy-pixel prediction

Fig. 9. Example of updating a template (IC)

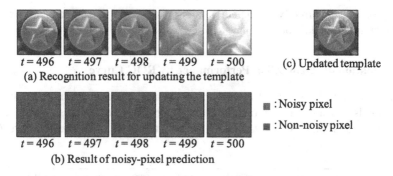

$t=496$ $t=497$ $t=498$ $t=499$ $t=500$ (c) Updated template
(a) Recognition result for updating the template

■ : Noisy pixel

■ : Non-noisy pixel

$t=496$ $t=497$ $t=498$ $t=499$ $t=500$
(b) Result of noisy-pixel prediction

Fig. 10. Example of updating a template (PLASTIC)

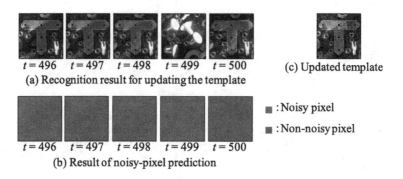

$t=496$ $t=497$ $t=498$ $t=499$ $t=500$ (c) Updated template
(a) Recognition result for updating the template

■ : Noisy pixel

■ : Non-noisy pixel

$t=496$ $t=497$ $t=498$ $t=499$ $t=500$
(b) Result of noisy-pixel prediction

Fig. 11. Example of updating a template (PLATE)

In Figs. 9, 10, 11 and 12, past recognition results used for updating the template are the five (a) images, prediction results for noisy pixels corresponding to those results are the five (b) images, and the template updated by using the (a) images is shown in image (c). From the figures, it is clear that even in the case that recognition errors occurred in the past recognition results, the template is still updated correctly. This is because the mistaken-recognition results and, as shown in Fig. 12(b), the region over which shadow

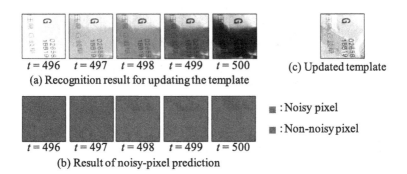

$t = 496$ $t = 497$ $t = 498$ $t = 499$ $t = 500$ (c) Updated template

(a) Recognition result for updating the template

■ : Noisy pixel

■ : Non-noisy pixel

$t = 496$ $t = 497$ $t = 498$ $t = 499$ $t = 500$

(b) Result of noisy-pixel prediction

Fig. 12. Example of updating a template (LABEL)

suddenly occurs is correctly predicted as noisy pixels, and those pixels are excluded from the template-update processing.

4.3 Evaluation of Recognition Performance by ROC Curve

The recognition performance of the proposed method was compared with that of the other methods by using receiver operating characteristic (ROC) curves. With threshold (T_h) of degree of similarity as a variable, objects with degree of similarity above T_h are

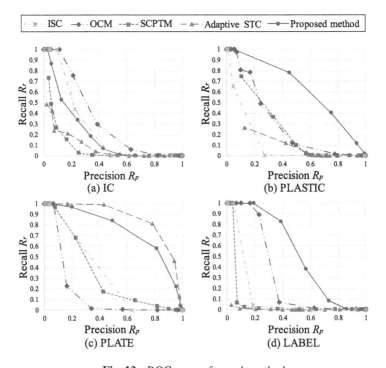

Fig. 13. ROC curves for each method

judged as targets, and objects with degree of similarity below T_h are judged as non-targets. Compliance rate R_p and recall rate R_r are calculated as time T_h is varied, and ROC curves are obtained. The ROC curves for each method are plotted in Fig. 13.

As shown in Fig. 13, although the proposed method demonstrates slightly lower recognition performance for part of the dataset than that of the other methods, it demonstrates a relatively stable high recognition performance compared to that of the other methods.

5 Concluding Remarks

A method for stably updating a template in the case of sparse data is proposed. Stability of pixels is predicted from statistical occurrence probability of pixel intensity, and unstable pixels are predicted as noisy ones. With the proposed method, the results of the noisy-pixel prediction are utilized in updating the model image in the template and the reference pixels. As a result, the template can be updated in a stable manner. On evaluating the recognition performance of the proposed method in relation to four types of datasets in which the appearance of a target object changes (due to lighting variation and adhesion of dirt), it was found that in comparison with conventional update-type template matching methods, the proposed method demonstrated a 13% higher recognition rate. It also demonstrated stable recognition in regard to four datasets concerning different materials. Accordingly, it was demonstrated that the proposed method has superior versatility in comparison to the conventional methods.

References

1. Amit, Y., Grenander, U., Piccioni, M.: Structural image restoration through deformable template. J. Am. Stat. Assoc. **86**(414), 376–387 (1991)
2. Jain, A.K., Zhong, Y., Lakshmanan, S.: Object matching using deformable templates. IEEE Trans. Pattern Anal. Mach. Intell. **18**(3), 267–278 (1996)
3. Saito, M., Hashimoto, M.: A fast and robust image matching for illumination variation using stable pixel template based on co-occurrence analysis. Trans. Inst. Electr. Eng. Jpn. C Publ. Electron. Inf. Syst. Soc. **133**(5), 1010–1016 (2013)
4. Taguchi, T., Noguchi, K., Syakunaga, K.: Object tracking by adaptive sparse template. Trans. Inst. Electr. Eng. Jpn. D Publ. Electron. Inf. Syst. Soc. **93**(8), 1502–1511 (2010)
5. Kaneko, S., Murase, I., Igarashi, S.: Robust image registration by increment sign correlation. Pattern Recogn. **35**(10), 2223–2234 (2002)
6. Ullah, F., Kanek, S., Igarashi, S.: Orientation code matching for robust object search. IEICE Trans. Inf. Syst. **84-D**(8), 999–1006 (2001)
7. Lowe, D.G.: Distinctive image features from scale-invariant keypoints. Int. J. Comput. Vis. (IJCV) **60**(2), 91–110 (2004)
8. Rublee, E., Rabaud, V., Konolige, K., Bradski, G.: ORB: an efficient alternative to SIFT or SURF. In: Proceeding of International Conference on Computer Vision (ICCV) (2011)
9. Hashimoto, M., Okuda, H., Sumi, K., Fujiwara, T., Koshimizu, H.: High-speed image matching using unique reference pixels selected on the basis of co-occurrence probability. Trans. Inst. Electr. Eng. Jpn. D Publ. Ind. Appl. Soc. **131**(4), 531–538 (2011)

Robust Accurate Extrinsic Calibration
of Static Non-overlapping Cameras

Andreas Robinson[✉], Mikael Persson, and Michael Felsberg

Linköping University, 581 83 Linköping, Sweden
{andreas.robinson,mikael.persson,michael.felsberg}@liu.se
https://liu.se/organisation/liu/isy/cvl

Abstract. An increasing number of robots and autonomous vehicles are equipped with multiple cameras to achieve surround-view sensing. The estimation of their relative poses, also known as extrinsic parameter calibration, is a challenging problem, particularly in the non-overlapping case. We present a simple and novel extrinsic calibration method based on standard components that performs favorably to existing approaches. We further propose a framework for predicting the performance of different calibration configurations and intuitive error metrics. This makes selecting a good camera configuration straightforward. We evaluate on rendered synthetic images and show good results as measured by angular and absolute pose differences, as well as the reprojection error distributions.

1 Introduction

Autonomous vehicles and robots benefit greatly from visual surround view sensing for both spatial and semantic understanding [15]. Surround view in turn requires multiple cameras with highly accurate extrinsics. Achieving the latter is an issue in camera configurations with limited or non-overlapping fields-of-view, as the standard methods for overlapping cameras cannot be applied. Since calibration errors cause biased estimates in crucial tasks such as visual odometry and spatial reconstruction, highly accurate estimates are required. Unfortunately non-overlapping configurations are common due to monetary and computational cost considerations and existing applicable methods rely on complicated tools and software.

Two approaches to minimal camera surround view sensing exist: either maximize the field-of-view per camera, with fisheye lenses or a catadioptric mirror, or allow more cameras with narrower field-of-view but rectilinear wide-angle lenses. While both fisheye lenses and catadioptric mirrors have excellent field-of-view, they can be costly and suffer from distortion that is difficult to calibrate well, compared to rectilinear lenses. In addition, with very few cameras it is hard avoid the body of the vehicle obstructing the view. Hence, this work will focus on the rectilinear type.

The intrinsic parameters of each lens in a camera configuration is straightforward to estimate independently [20]. We address the extrinsic estimation

M. Felsberg et al. (Eds.): CAIP 2017, Part II, LNCS 10425, pp. 342–353, 2017.
DOI: 10.1007/978-3-319-64698-5_29

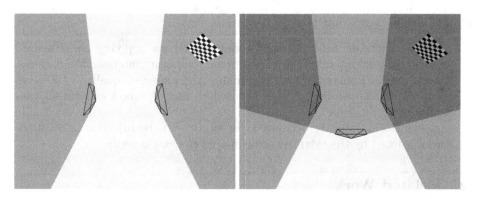

Fig. 1. Left: Camera pair without overlap seen from above. Extrinsic calibration not possible in a static setting. Right: A temporarily placed additional camera enables extrinsic calibration.

Fig. 2. Calibrating two marginally overlapping cameras on a truck, with the addition of a third camera.

problem, under the assumption that intrinsic calibration has been achieved to sufficient accuracy.

We propose adding intermediate cameras during the extrinsic calibration process, as exemplified by the schematic in Fig. 1, for the purpose of calibrating cameras on vehicles such as the one shown in Fig. 2. This strategy creates or increases the overlap between fields-of-view and allows the use of any standard stereo calibration toolbox, with minimal development effort. Since the extra cameras are temporary, the cost, weight and performance limitations of the permanently installed cameras do not apply. Thus, better-performing cameras can be used to further improve the accuracy, if required. We show that greater overlap is beneficial and analyze the trade-off between additional cameras and propagated calibration errors.

In this paper we contribute

- a simple extrinsic calibration procedure based on applying the standard method through the novel use of additional temporary intermediate cameras.
- an evaluation framework for determining the expected quality of a given setup. This framework can also be used to identify good calibration configurations.
- a geometrically intuitive extrinsic error metric - the reprojection error distribution caused by the extrinsic error in a given camera setup.

2   Related Work

Approaches to calibrating non-overlapping cameras can be sorted into two broad categories, static and dynamic: Static methods either use mirrors to allow all cameras to see the same calibration object or large calibration objects that are only partially visible in each camera. Dynamic methods use structure-from-motion to capture the structure of a scene and to subsequently infer rigid relationships between cameras.

In the first category, a fairly recent method by Sturm et al. [18] recovers the poses of virtual cameras reflected in three or more planar mirrors, in order to find the pose of a "true" camera. Later work by several other authors [9–11,14,17,19] present extensions and alternative approaches that also use planar mirrors. In contrast, Agrawal et al. [1] show that calibration with a spherical mirror is advantageous; it lacks degeneracies and only one mirror is required, although the accuracy is comparably low. Additionally, methods using calibration objects that are only partially visible in each camera have been suggested although this seems less common. Liu et al. [13] performs extrinsic calibration with a long stick, with the ends visible in each of the two cameras. In [12], this stick is substituted by lasers.

In the second category, calibration with structure-from-motion or photogrammetry, Esquivel et al. [6] and Carrera et al. [4] capture sequences of the surrounding environment on vehicles with rigidly coupled cameras, finding the relative camera poses after 3D reconstruction. Pagel et al. [16] implements online calibration based on a system for visual odometry and more recently Dong et al. [5] places easily identifiable calibration objects in a scene, uses photogrammetry to pinpoint their locations and then relate the cameras to be calibrated to these objects. Note that structure-from-motion approaches require additional information to determine an absolute scale. This is typically achieved with a calibration object of known size.

There are several ways to verify and compare camera intrinsic/extrinsic calibration quality. One choice is to verify that the reprojection errors on the calibration features is sensible, i.e. less than one pixel on average with no severe outliers. Another reasonable metric for comparison is the difference between true and estimated pose, as they map to intuitive physical properties. However, as our target application is structure-from-motion, the reprojection error distribution provides an intuitive understanding of both error severity and acceptable limits. In this paper, we propose a new way to estimate this extrinsic error metric.

Both static and dynamic methods can yield good results but have their downsides. Mirrors must have the reflective surface on the front to avoid chromatic aberrations, which adds to their fragility. Placing them to ensure the calibration object is reflected into the camera can be complicated. Large calibration objects warp under their own weight, while lasers have little pixel coverage and are hard to spot in cameras equipped with wide-angle lenses. Structure-from-motion approaches can be labor intensive to execute well in uncontrolled environments. We have also found that the estimated metric scale is very sensitive to noise. In addition, the perhaps most important aspect of these methods is that they require significant software development efforts to get up and running.

In contrast, with our method a few additional cameras makes the stereo calibration tools in the OpenCV software library [3] sufficient. The performance of the proposed hardware extension based approach has to our knowledge not been investigated previously.

3   Method

This section presents a general model of the extrinsic camera calibration problem, the proposed method, the imaging model and concludes with the error metrics.

3.1   The Extrinsic Graph

Consider the extrinsic camera configuration problem for N cameras as an acyclic, un-weighted and un-directed graph. Each node corresponds to a camera pose and each edge to an observed relative pose. A connected graph implies that all poses are constrained and can be determined. Since the graph is acyclic, the relative poses can be found separately per constraint. General camera configurations need not result in a connected graph, but by adding cameras we can always create one. Each extra camera adds two new edges to the graph. Since acyclic connected graphs are sufficient, there is no need to add any cycle. Adding cycles and appropriate weights could improve accuracy, but would require joint optimization.

3.2   Calibration Procedure

Given the relative poses, any camera can be selected as origin and the poses of each other camera in relation to it are trivially computed by chaining the transformations. We use unit quaternion based poses to avoid compounding numerical errors. Thus we create an acyclic connected graph and estimate the relative poses in an interleaved manner as follows:

Estimating Relative Transforms. The metric relative transform between two overlapping cameras with known intrinsics can be estimated using a chessboard pattern of known proportions and a small number, (here $N = 5$), of images. The chessboard pattern is found by OpenCV findChessBoardCorners() and the relative pose is computed using stereoCalibrate() which implements a homography based solver. Note that the signal-to-noise ratio of the estimation is related to the detection noise over the pixel distance between corner projections [8]. This suggests it is a good idea to have a large overlap between cameras and for the chessboard to cover most of it. Thus we require a lower bound of overlap to consider two cameras connected, how the overlap relates to the error is explored in Sect. 4.

Adding Cameras. The goal is to create a fully connected acyclic graph by adding cameras. Finding the optimal configuration is both difficult and unnecessary. This motivates the following simple greedy scheme:
    Iterate until the graph is connected:

- Find the two closest, in terms of angular distance, unconnected, cameras.
- Place an evenly distributed sequence of cameras with sufficient overlap along the shortest path.

The suitable number of cameras can be found by simulation similar to the one in Sect. 4. Note that regardless of the number of intermediate camera positions, at most two extra physical cameras are required since they can be moved.

3.3   Camera Noise Model

The image capture and feature extraction processes contain three distinct sources of noise:
    In the optical stage, light enters the camera and is focused by the lens before hitting the sensor. Imperfections due to both lens design and manufacturing variability, attenuates (vignetting), diffuses (blurring) and distorts the light. We model the first two effects directly as

$$I_1(\mathbf{x}) = \cos^4(k_1|\mathbf{x}|) \circ (H_\sigma * I_0)(\mathbf{x}) \tag{1}$$

where $|\mathbf{x}|$ is the distance from the camera's optical axis, $H_\sigma(\mathbf{x})$ is a small point spread (blur) Gaussian function of width σ, and k_1 controls the falloff rate of the vignetting function. In addition $(\cdot \circ \cdot)$ and $(\cdot * \cdot)$ denote point-wise multiplication and convolution, respectively. In the experiments, the blur kernel is fixed to 5×5 pixels, with $\sigma = 0.5$. k_1 is chosen so that the maximum intensity at the left and right image edges are 25% of the maximum in the center. The third effect, distortion, is modeled below.
    In the electronic stage, light is converted into electric charge in the sensor, is read out, amplified and digitized to form a digital image. Here, a significant

source of error is the shot noise. We model this with a Poisson process approximation, specifically Gaussian noise with variance proportional to the light intensity:

$$I_2(\mathbf{x}) = \mathcal{N}(I_1(\mathbf{x}), b + k_2\sqrt{I_1(\mathbf{x})}) \tag{2}$$

where b is the minimum noise variance and k_2 a scale factor. We found that with $I(x) \in [0, 255]$, setting the parameters to $b = 2$ and $k_2 = 0.1$ produced noise that looked "reasonably realistic" as is shown in Fig. 4. Estimating the parameters of a specific camera sensor would further improve the model accuracy.

In addition, pixel intensities are clamped to the range $[0, 255]$, i.e.

$$I_3(\mathbf{x}) = \min(\max(I_2(\mathbf{x}), 0), 255) \tag{3}$$

In the digital stage, a detector locates the calibration pattern landmarks with subpixel-precision. This operation does not have to be modeled explicitly since either the OpenCV function findChessboardCorners() or cornerSubPix() can be applied directly. The output from the detector is a set of 2D points $\mathbf{y}_k$. $k \in [1, K]$ where K is the number of calibration pattern landmarks.

In a real-world setting, these coordinates will have rather large errors due to lens distortion. However, for the purpose of the simulation we assume that distortion is mostly corrected by a function $U(\mathbf{y})$ and model the residual error with a Gaussian noise process. The final coordinates are thus

$$\mathbf{z}_k = U(\mathbf{y}_k) + \epsilon_d \tag{4}$$

with $\epsilon_d \sim \mathcal{N}(\mathbf{0}, \sigma_d\mathbf{I})$, where $\mathbf{I}$ denotes the 2×2 identity matrix.

3.4   Error Metrics

As mentioned in Sect. 2, we consider two quality measures: the pose (rotational and translational) error and a novel estimate of the distribution of reprojection errors. The latter provides a intuitive understanding of what impact the former is expected to have, on a structure-from-motion system.

Rotation and Translation Errors. Let the 4×4 matrix $\mathbf{P}$ represent a 3D rigid transform, composed of a rotation matrix $\mathbf{R}$ and translation vector $\mathbf{t}$. Let $\mathbf{P}$ denote the ground truth and $\hat{\mathbf{P}}$ the associated estimate. $\mathbf{P}_i$ is the transform from world coordinates to the local coordinate system of camera i and $\mathbf{P}_{ji}$ is the relative transform from the local coordinate system of camera i to that of camera j.

We define the translation error of $\hat{\mathbf{P}}$ relative to $\mathbf{P}$ as $\epsilon_t = \|\hat{\mathbf{R}}^{-1}\hat{\mathbf{t}} - \mathbf{R}^{-1}\mathbf{t}\|_2$. Similarly, the rotation error is defined as $\epsilon_R = |\mathbf{R}\hat{\mathbf{R}}^{-1}|$, where $|\cdot|$ is the magnitude of the angle around the axis of rotation in $\mathbf{R}\hat{\mathbf{R}}^{-1}$.

Expected Reprojection Error. The distribution of reprojection errors is approximated by sampling points $\mathbf{y}$ and computing the error as follows. Let $\mathbf{y} = (x, y, z, 1)^T$ be a homogeneous 3D point in the local coordinate system of camera i and that $\mathbf{x} = \mathbf{P}_i^{-1}\mathbf{y}$ is the same point in the world coordinate system. With the camera projection operator defined as $\mathrm{Proj} : (x, y, z, 1)^T \mapsto (x/z, y/z, 1)^T$, the reprojection error (in pixels) of $\mathbf{x}$ in camera i is $\epsilon_p = f\|\mathrm{Proj}(\hat{\mathbf{P}}_i\mathbf{x}) - \mathrm{Proj}(\mathbf{P}_i\mathbf{x})\|_2$, where f is the focal length.

4   Experiments

The following experiments on synthetic imagery will show how the calibration error as measured by our chosen metrics varies with the camera configuration and the noise. Care has been taken to simulate realistic noise, as simpler models may hide interesting behaviors.

Our use-case, shown in Fig. 2, is a mobile platform (truck) with two 1600×1200 pixel cameras at $90°$ relative yaw angle but with very wide-angle lenses (focal length 1.67 mm). Due to the wide fields-of-view, there is an overlap of approximately $10°$ between the cameras. However, this image area is small and calibration patterns must be placed a significant distance away to be simultaneously visible in both cameras. The procedure has been performed on the mobile setup but the accuracy cannot be determined without ground truth or additional sensors. Thus, the experiments are performed on synthetic data.

4.1   Setup

To evaluate the impact of additional cameras providing significant overlap, a synthetic calibration pipeline is constructed. The pipeline generates noise-free images, applies noise, and finds the extrinsic camera parameters. We experiment with two cameras with a $90°$ interval, three cameras at $45°$, and five cameras at $22.5°$.

First, images are generated with the Blender 3D modeling software [7]. Pairwise camera configurations and associated pattern placements are shown in Fig. 3. The cameras are outward facing on a circle with 1.5 m radius and relative rotations of either 22.5, 45, or $90°$. In each case, five checkerboard calibration patterns (7-by-9 squares, 84×108 cm) are manually placed to maximally cover the joint field of view. As is shown in the figure, larger relative rotations provide less overlap and require the patterns to be placed farther away from the cameras. Note that cases such as 45-to-$90°$ or 45-to-$67.5°$ rotations are omitted, as they are equivalent to simultaneously rotating all objects in any of the existing setups.

In each configuration one calibration pattern at a time is placed inside the joint field-of-view of the camera pair. Images are rendered and the associated ground-truth image 2D coordinates of all pattern landmarks are stored.

Second, errors caused by the camera are simulated and added to the images. An example image with blur, vignetting and shot noise applied, is shown in Fig. 4.

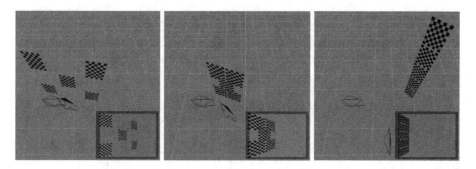

Fig. 3. Camera and pattern setups with 22.5, 45 and 90° rotations between cameras. Insets show the pattern placements in the viewport of the right camera in the pair. Their appearance in the left camera are mirror-symmetric. The ground-plane grid lines are 1 m apart.

Fig. 4. Example of a rendered image with added vignetting, blur and shot noise. Inset: Closeup of camera shot noise - best viewed on-screen.

Third, the calibration pattern landmarks are located with sub-pixel precision (findChessboardCorners() in OpenCV). Since a coarse alignment can and should be verified manually, we only perform the subpixel refinement (cornerSubPix()) with ground-truth landmarks as input. Gaussian noise is then added, simulating the effects of the residual camera distortion error. This is in line with other work, such as [1] and is in effect the main source of noise in the calibration pipeline.

Finally, these noisy landmark coordinates are used to estimate the relative transform between the two cameras in a pair (stereoCalibrate()).

For each of the camera configurations, we generate $M = 1000$ calibration runs, with different noise added each time. This provides a reasonable sampling of the error distribution. N-view camera setups are simulated by linking the relative poses of pairs of cameras $\hat{\mathbf{P}}_{N,1} = \prod_{k=1}^{N-1} \hat{\mathbf{P}}_{k+1,k}$. For example, in a three-view setup, $\hat{\mathbf{P}}_{31} = \hat{\mathbf{P}}_{32}\hat{\mathbf{P}}_{21}$. In addition we compute the reprojection errors of 10,000 random points $\mathbf{y} = (x, y, z, 1)^T$, uniformly distributed across the image plane at uniformly random depths $z \in [0.1, 100]$.

4.2   Results

The mean errors ϵ_R and ϵ_t in $\hat{\mathbf{P}}_{N,1}$ are shown in Fig. 5. The distribution of reprojection errors in the presence of varying levels σ_d of distortion noise ϵ_d, are shown in Fig. 6.

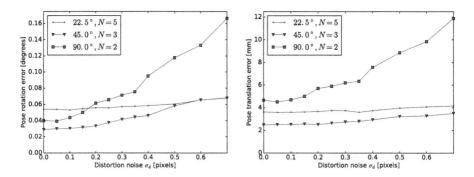

Fig. 5. Expected pose rotation errors and expected pose translation errors as functions of added ϵ_d noise of magnitude σ_d.

5   Discussion

It is clear from Fig. 5 that the calibration is most sensitive to noise when no cameras have been added $(N = 2)$ and the overlap is small. In contrast, additional cameras $(N = 3$ or $N = 5)$ and greater overlap significantly improves the resilience to noise and is likely to compensate for the error accumulation along the chain of transforms. The difference between $N = 3$ and $N = 5$ is small. Depending on the assumed error, it is better to use one rather than three additional cameras. This trade-off may change if we expect significantly higher noise but that could also indicate poor intrinsic estimation quality. Presumably this generalizes to similar configurations, but should be confirmed using the provided toolbox to analyze the configuration on a case by case basis.

While it is difficult to compare the expected accuracy under the radically different setups and scenarios, we compare our results with those from other papers in Table 1. When possible, the values corresponding to a noise level of $\sigma_d = 0.5$ are used.

Subject to differences in noise models, setup, cameras, and evaluation, we conclude that the performance of the proposed system is favorable or at least comparable to the baselines. Note that the rotation error represents a lower bound on projection accuracy since it will affect distant points which are otherwise unaffected by translation error.

The accuracy of the calibration in terms of the absolute pose error is good and is likely to exceed the precision of tools like tape measures and protractors. However, the reprojection error distributions in Fig. 6 indicate that the calibration

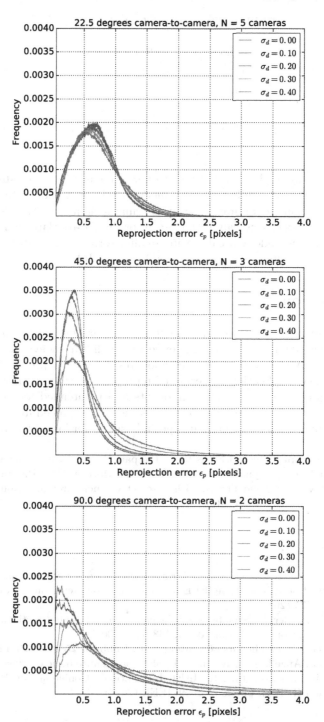

Fig. 6. Distribution of reprojection errors with varying noise magnitude σ_d, computed for the relative pose between outermost cameras at $90°$, with $N-2$ cameras in between.

Table 1. Comparison of rotation errors.

Method	Rotation error
Ours	0.04°
Esquivel et al. [6]	0.5°
Antone et al. [2]	0.1°
Agrawal et al. [1]	1°

error is not negligible but should be accounted for in a structure-from-motion system. In particular, while the error mode is relatively moderate, the heavy tail of the two-camera calibration setup is remarkable and motivates the use of additional cameras during calibration, even for limited-overlap configurations.

The calibration code and calibration quality estimation toolbox, along with the dataset and evaluation code, is available on GitHub[1].

6   Conclusion

We have presented a simple and novel extrinsic calibration method and an evaluation framework for determining the expected quality using synthetic data. We have shown that the system achieves good results with a predictable impact on the intended application in terms of reprojection error distributions. The method is based on off-the-shelf components and is intended for multi-camera setups with small or no overlapping fields-of-view. The procedure is robust and simple to implement which makes it especially suited for industrial applications.

Acknowledgement. This work was funded in part by Vinnova, Sweden's innovation agency, through grant iQmatic, Daimler AG, EC's Horizon 2020 Programme, grant agreement CENTAURO and The Swedish Research Council through a framework grant for the project Energy Minimization for Computational Cameras (2014-6227).

References

1. Agrawal, A.: Extrinsic camera calibration without a direct view using spherical mirror. In: Proceedings of the IEEE International Conference on Computer Vision, pp. 2368–2375 (2013)
2. Antone, M., Teller, S.: Scalable extrinsic calibration of omni-directional image networks. Int. J. Comput. Vision **49**(2), 143–174 (2002)
3. Bradski, G., et al.: The OpenCV library. Dr. Dobbs J. **25**(11), 120–126 (2000)
4. Carrera, G., Angeli, A., Davison, A.J.: Slam-based automatic extrinsic calibration of a multi-camera rig. In: 2011 IEEE International Conference on Robotics and Automation (ICRA), pp. 2652–2659. IEEE (2011)

---

[1] https://github.com/midjji/non-overlapping-extrinsic-cameracalibration.git.

5. Dong, S., Shao, X., Kang, X., Yang, F., He, X.: Extrinsic calibration of a non-overlapping camera network based on close-range photogrammetry. Appl. Opt. **55**(23), 6363–6370 (2016)
6. Esquivel, S., Woelk, F., Koch, R.: Calibration of a multi-camera rig from non-overlapping views. In: Hamprecht, F.A., Schnörr, C., Jähne, B. (eds.) DAGM 2007. LNCS, vol. 4713, pp. 82–91. Springer, Heidelberg (2007). doi:10.1007/978-3-540-74936-3_9
7. Foundation, B.: Blender version 2.78c (2017). https://www.blender.org/
8. Hedborg, J., Forssén, P.-E., Felsberg, M.: Fast and accurate structure and motion estimation. In: Bebis, G., et al. (eds.) ISVC 2009. LNCS, vol. 5875, pp. 211–222. Springer, Heidelberg (2009). doi:10.1007/978-3-642-10331-5_20
9. Hesch, J.A., Mourikis, A.I., Roumeliotis, S.I.: Extrinsic camera calibration using multiple reflections. In: Daniilidis, K., Maragos, P., Paragios, N. (eds.) ECCV 2010. LNCS, vol. 6314, pp. 311–325. Springer, Heidelberg (2010). doi:10.1007/978-3-642-15561-1_23
10. Kumar, R.K., Ilie, A., Frahm, J.M., Pollefeys, M.: Simple calibration of non-overlapping cameras with a mirror. In: IEEE Conference on Computer Vision and Pattern Recognition, CVPR 2008, pp. 1–7. IEEE (2008)
11. Lébraly, P., Deymier, C., Ait-Aider, O., Royer, E., Dhome, M.: Flexible extrinsic calibration of non-overlapping cameras using a planar mirror: application to vision-based robotics. In: 2010 IEEE/RSJ International Conference on Intelligent Robots and Systems (IROS), pp. 5640–5647. IEEE (2010)
12. Liu, Q., Sun, J., Liu, Z., Zhang, G.: Global calibration method of multi-sensor vision system using skew laser lines. Chin. J. Mech. Eng. **25**(2), 405–410 (2012)
13. Liu, Z., Zhang, G., Wei, Z., Sun, J.: Novel calibration method for non-overlapping multiple vision sensors based on 1d target. Opt. Lasers Eng. **49**(4), 570–577 (2011)
14. Long, G., Kneip, L., Li, X., Zhang, X., Yu, Q.: Simplified mirror-based camera pose computation via rotation averaging. In: Proceedings of the IEEE Conference on Computer Vision and Pattern Recognition, pp. 1247–1255 (2015)
15. Maddern, W.: Industry solutions: autonomous vehicle driven by vision. Vis. Syst. Des. **22**(2), February 2017. http://www.vision-systems.com/articles/print/volume-22/issue-2/features/industry-solutions-autonomous-vehicle-driven-by-vision.html
16. Pagel, F.: Extrinsic self-calibration of multiple cameras with non-overlapping views in vehicles. In: IS&T/SPIE Electronic Imaging, p. 902606. International Society for Optics and Photonics (2014)
17. Rodrigues, R., Barreto, J.P., Nunes, U.: Camera pose estimation using images of planar mirror reflections. In: Daniilidis, K., Maragos, P., Paragios, N. (eds.) ECCV 2010. LNCS, vol. 6314, pp. 382–395. Springer, Heidelberg (2010). doi:10.1007/978-3-642-15561-1_28
18. Sturm, P., Bonfort, T.: How to compute the pose of an object without a direct view? In: Narayanan, P.J., Nayar, S.K., Shum, H.-Y. (eds.) ACCV 2006. LNCS, vol. 3852, pp. 21–31. Springer, Heidelberg (2006). doi:10.1007/11612704_3
19. Takahashi, K., Nobuhara, S., Matsuyama, T.: A new mirror-based extrinsic camera calibration using an orthogonality constraint. In: 2012 IEEE Conference on Computer Vision and Pattern Recognition (CVPR), pp. 1051–1058. IEEE (2012)
20. Zhang, Z.: A flexible new technique for camera calibration. IEEE Trans. Pattern Anal. Mach. Intell. **22**(11), 1330–1334 (2000)

An Integrated Multi-scale Model for Breast Cancer Histopathological Image Classification with Joint Colour-Texture Features

Vibha Gupta[(✉)] and Arnav Bhavsar

School of Computer and Electrical Engineering,
Indian Institute of Technology Mandi, Mandi, India
vibha_gupta@students.iitmandi.ac.in, arnav@iitmandi.ac.in

Abstract. Breast cancer is one of the most commonly diagnosed cancer in women worldwide, and is commonly diagnosed via histopathological microscopy imaging. Image analysis techniques aid physicians by automating some tasks involved in the diagnostic workflow. In this paper, we propose an integrated model that considers images at different magnifications, for classification of breast cancer histopathological images. Unlike some existing methods which employ a small set of features and classifiers, the present work explores various joint colour-texture features and classifiers to compute scores for the input data. The scores at different magnifications are then integrated. The approach thus highlights suitable features and classifiers for each magnification. Furthermore, the overall performance is also evaluated using the area under the ROC curve (AUC) that can determine the system quality based on patient-level scores. We demonstrate that suitable feature-classifier combinations can largely outperform the state-of-the-art methods, and the integrated model achieves a more reliable performance in terms of AUC over those at individual magnifications.

Keywords: Histopathological images · Joint colour-texture features · Receiver operating characteristics (ROC) · Area under the ROC curve (AUC)

1 Introduction

Breast cancer (BC) is the most common type of cancer and the fifth most common cause of cancer mortality among women globally [1]. While, different types of imaging technologies, have been employed for diagnosis of BC, histopathology biopsy imaging has been a 'gold standard' in diagnosing breast cancer because it captures a comprehensive view of effect of the disease on the tissues [2].

However, image examination by pathologists is often subjective and may not be easily quantified. Thus, computer-aided diagnosis (CADx) systems provide valuable assistance for physicians and specialists. These help in overcoming the subjective interpretation and relieve some workload of pathologists. An important part

© Springer International Publishing AG 2017
M. Felsberg et al. (Eds.): CAIP 2017, Part II, LNCS 10425, pp. 354–366, 2017.
DOI: 10.1007/978-3-319-64698-5_30

of such CADx systems is the automation of image analysis to determine whether a tissue sample is malignant or a benign. Due to automated image analysis some tasks involved in diagnostic workflow can be made more efficient and precise.

However, automated image analysis can be challenging as inconsistency in histopathology slide preparation such as differences in fixation, staining protocol, non-standard imaging condition, etc. can cause variability in tissue appearance (colour and texture). The texture variation is typically captured by classifiers employing traditional texture features. To mitigate the effect of colour variability, a straightforward approach is to use gray-scale images [3,4]. On the other hand, a stain (or colour) normalization preprocessing can be performed, which is typically a more sophisticated process involving methods such as histogram matching, colour transfer, colour map quantile matching approach and spectral matching etc. [5,6].

However, it is observed that some inter-image colour variation might be informative [5]. Similarly, recent research in digital histopathology has indicated significance of colour information in quantitative analysis on histopathology [7,8]. As can be seen from Fig. 1, along with texture, colour information is also available in images which can be utilized to get a more discriminating representation.

From a machine learning perspective, various methods which do not employ normalization have also been proposed [9–11]. Our proposed method falls in this category where features are directly extracted from image (without normalization). This follows the philosophy that instead of reducing the colour variation, we learn the colour variation (along with the texture variation) as a part of the classification process.

We believe that the colour-texture variability can be better captured with joint colour-texture features [12]. Such features consider the mutual dependency between colour channels and texture information. These features can be defined with individual colour channels, or with correlated pairs of colour channels. Such jointly defined colour-texture features can locally adapt to the variation in the image content [13].

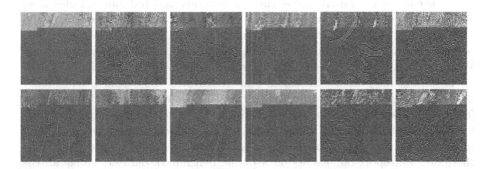

Fig. 1. Sample of histopathological images (first row: benign tumor, second row: malignant tumor) from BreakHis dataset at magnification factor of 40X.

In addition, different from existing works where a small set of classifier was utilized, here a total of 22 classification frameworks experimented with. These classification frameworks include Quadratic Discriminant, Subspace Discriminant, RUSBoosted Trees, Boosted Tree, Coarse Gaussian SVM, Weighted KNN etc. We argue that such an exploration of joint colour-texture features and various classifiers leads to the selection of better suited features and classifiers to this specific problem. Due to space-constraints, we report the features and classifiers which correspond to top five results for each image magnification.

1.1 Related Work

In recent years, a number of methods have been investigated for BC histopathology classification. However, most of these method use traditional morphology and texture features. Kowal et al. [14] utilized four different clustering algorithms for nuclei segmentation and extracted 21-dimensional feature vector. In [14], three different classifiers are reported for each clustering algorithm separately. This s carried out on a dataset which contained 500 images of cytological samples that were extracted from 50 patients. Filipczuk et al. [15] presented a diagnosis system where nuclei were estimated by the Hough transform. Four different classifiers trained on 25-dimensional feature vector was used for classification using 737 images of cytological sample which had drawn from the same place as [14]. Based on above discussed methods, it is realized that for accurate system nuclei should be segmented properly as subsequent analysis is based on segmentation. However, segmentation of histological images is not a trivial task and is prone to mistakes. Instead of relying on the accurate segmentation, [16] investigated multiple image descriptors along with random subspace ensembles and proposed two-stage cascade framework with a rejection option using a dataset composed of 361 images. In another work [17], an ensembles of one-class classifiers were assessed by the same authors on the same dataset.

The works in [9,10] also propose the use colour information in addition to texture. Milagro et al. [9] combinations of traditional texture features and colour spaces is considered. Furthermore, they have also considered different classifiers such as Adaboost learning, bagging trees, random forest, Fisher discriminant and SVM. In [10], authors utilized colour and differential invariants to assign class posterior probabilities to pixels and then performs a probabilistic classification. While our intuition of using colour information and a set of classifiers is similar to [9], our integrated joint colour-texture features also consider dependency between colour channels and texture, rather than extracting traditional texture features independently from colour channels. Moreover, unlike ours, and similar to the above discussed works, [9,10] do not consider experimentation with respect to different optical magnifications, which is an important aspect [4]. Furthermore, we report our results on a public benchmark dataset, unlike all the above approaches.

With regards to the concern of benchmarking, it has been observed that the dataset used in the above works are not publicly available to the scientific community, and such datasets contain rather small number of images.

Table 1. Detailed description of BreaKHis dataset

	Magnifications				Total	Patient
	40x	100x	200x	400x		
Benign	625	644	623	588	2480	24
Malignant	1370	1437	1390	1232	5429	58
Total	**1995**	**2081**	**2013**	**1820**	**7909**	**82**

Spanhol et al. [4] introduced the BreakHis dataset which intended to take away the impediment of publicly available data set. The BreakHis dataset contains fairly large amount of microscopic biopsy images (7909) that were collected from 82 patients in four different magnifications (40x, 100x, 200x, 400x). The details of dataset are provided in Table 1. Figure 1 shows the images of benign and malignant tumor given in different magnifications.

In the same study, a series of experiments utilizing six different texture descriptors and four different classifiers were evaluated and showed the accuracy at patient-level. In [18], Alexnet [19] was used for extracting features, and classification was reported on image-level as well as patient-level, using this dataset. Bayramoglu at el. [20] proposed a magnification independent model utilizing deep learning and reported accuracies for both multi-task network which predicts magnification factor and malignancy (benign/malignant) simultaneously, and single task network which predicts malignancy.

1.2 Salient Aspects of This Work

Considering that the area of BC histopathology image analysis is still an emerging one, as new approaches are developed, the evaluation and comparison among such frameworks is of increasing importance from a clinical perspective. In this context we consider the following aspects about methodology and evaluation which drives our work.

(a) As implied above, there is scope for further exploration of suitable features and classifiers for this problem, which can better capture the discriminative

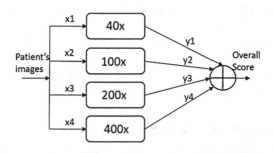

Fig. 2. An Integrated model for BC image classification.

information to address the classification task. Thus, in this work we look into employing joint colour-texture features for this task. Motivated from [4], where conventional texture features (GLCM, CLBP, PFTAS etc.) along with small set of well known classifiers were utilized, in this work we explore a relatively larger set of classifiers for the joint-color texture features. This provides their comparative performance under one roof, and indeed, for some classifiers we demonstrate an improved performance over the state-of-the-art.

(b) The above discussed methods yield a continuous value for a scoring, rather than single value for making decisions. In the discussed methods [4,18], patient and image level score were used as performance measures. However, it is also important to convert a patient score to a decision (benign or malignant), using a decision threshold on the patient-level scores, and finally comment on the quality of the diagnostic test in the context of the accuracy of such patient-level decisions. For such a quality check, the receiver operating characteristics (ROC) curve that includes all the decision threshold, offers a more compressive assessment. In diagnostic test assessment, area under the ROC curve (AUC) can be used to judge the quality of approaches. A value of AUC that lies in range 0 to 1, where 0 and 1 correspond to inaccurate and accurate test respectively. A value of 0.5 for AUC indicates no discrimination, 0.7 to 0.8 is considered acceptable, 0.8 to 0.9 taken as fair or good or some time excellent test, and more than 0.9 is considered as outstanding [21]. In light of this, we suggest an integrated model using all magnifications (as elaborated below), that uses the AUC as a performance measure.

(c) In some previous work [4,18] a model corresponding to each magnification was built independently based on different combinations of features and classifiers. We believe that instead of just relying on the individual scores correspond to each magnifications, assessment of overall score calculated as the ratio of total images classified to total images of patient, can also yield useful information with respect to a beneficial decision. For instance, for a patient who has large variation in scores, the decision cannot be made reliably by just looking at the highest score. In this work, while we report results on individual magnifications, we also suggest an integrated model that makes use of all magnifications, and can yield a more reliable system in terms of the AUC. Figure 2 depicts the proposed integrated model, wherein, x1, x2, x3, and x4 are the total number of input images of four different magnifications, and y1, y2, y3, and y4 are the corresponding classified images.

2   Methodology

In this section, we briefly discuss about the images descriptors and classifiers we have utilized for this study.

2.1   Joint Colour-texture Features

In order to find suitable feature for each magnification various features are utilized. Due to space constraints, we provide only a short introduction of features which are included in combinations that yields the top results. For more details please refer [12].

1. **Normalized colour space representation** [22]: The matrix of complex numbers (C1+iC2), where C1 and C2 are the normalized colour channel chosen based on the range and average values of the colour channels is used to extract textural (Gabor filter) features.
2. **Multilayer coordinate clusters representation** [23]: To describe the textural and colour content of an image, it splits the original colour image into a bundle of binary images, where each binary image represents a colour code based on a predefined palette (quantized colour space). Patches of such binary patters are then clustered and the method computes the histograms of occurrence of the binary patches based on the cluster centres. This process is repeated for each layer, and the resulting histograms are concatenated. Depending on, how many samples (n) are taken on each axis of the colour space, resulting palettes ($N=n^3$) will be 8, 27 and 64 levels.
3. **Gabor features on Gaussian colour model** [24]: The following two stages are used to extract color-texture: (1) Measurement of color in transformed space (based on a Gaussian colour model), (2) Utilization of Gabor filter bank for texture measurement.
4. **Complex wavelet features and chromatic features** [25]: Dual Tree Complex Wavelet Transform (DT-CWT) is applied to each color channel separately. The final feature vector is a concatenation of all DT-CWTS from different channels.
5. **Opponent colour local binary pattern (OCLBP)** [26]: This is an extension of standard Local Binary Pattern (LBP) which developed as the joint colour-texture operator for colour images. It is a concatenation of all LBPs extracted from different channels including colour channels separately (intra channel) and opponent colour channel ((c_1, c_2), (c_1, c_3) and (c_2, c_3)) jointly.

2.2 Classifiers

We explore various supervised classifiers, for which we provide a short description below [27].

1. **Support Vector Machine (SVM)**: It is a supervised machine learning algorithm that learns a hyperplane which separates a samples of one class from samples of other class with maximum margin. Depending on the type of the kernel and, its scale that used to make the distinction between classes, a variety of SVMs exists.
   (a) Linear SVM
   (b) Quadratic SVM (Quadratic kernel)
   (c) Cubic SVM (Cubic kernel)
   (d) Fine Gaussian SVM (Radial Basis Function (RBF) kernel, kernel scale set to $\sqrt{P}/4$)

(e) Medium Gaussian SVM (RBF kernel, kernel scale set to $\sqrt{P}$)
(f) Coarse Gaussian SVM (RBF kernel, kernel scale set to $2\sqrt{P}$)
where P is the number of predictors.

2. **Decision Tree**: It is a top-down approach that uses a tree-like graph of possible solutions including resource costs, and utility. Several variations of tress are exist based on maximum number of splits utilized in the tree.
(a) Simple Tree (maximum number of splits is 4)
(b) Medium Tree (maximum number of splits is 20)
(c) Complex Tree (maximum number of splits is 100)

3. **Nearest Neighbors Classifier**: It does not make any underlying assumptions about the distribution of data. It locates the data into some clusters, or groups and classified an unclassified point into the cluster for which it has a higher probability of getting classified based on distance metrics. Depending on number of neighbors and metric used, a variety of k-NN exists.
(a) Fine KNN (number of neighbors is set to 1, euclidean metric)
(b) Medium KNN (number of neighbors is set to 10, euclidean metric)
(c) Coarse KNN (number of neighbors is set to 100, euclidean metric)
(d) Cosine KNN (number of neighbors is set to 10, Cosine distance metric)
(e) Cubic KNN (number of neighbors is set to 10, cubic distance metric)
(f) Weighted KNN (number of neighbors is set to 10, distance based weight)

4. **Discriminant Analysis**: It assumes that different classes generate data based on different gaussian distributions and predicts membership in a group or category based on observed values. We consider two types of discriminant analysis based on boundary type formed between classes.
(a) Linear Discriminant (linear boundaries)
(b) Quadratic Discriminant (non-linear boundaries such as ellipse, parabola)

5. **Ensemble Classifier** [28]: It is a set of classifiers trained to solve same problem and, their output are combined to classify a new sample. The employment of logistics to make different schemes (combination) leads to different ensemble methods:
(a) Boosted Tree
(b) Bagged Tree
(c) RUSBoosted Trees

3 Experimental Results and Discussion

For fair comparison with existing approach [4,18,20], we have randomly chosen 58 patients (70%) for training and remaining 25 for testing (30%). We train the above mentioned classifiers using image representations of chosen 58 patients, and also used five trials of random training-testing data selection. These trained models are tested using remaining image representations of 25 patients. Due to the disproportionate ratio of normal and abnormal cases, the same procedure is repeated for five trails (each time different patients for training and testing are chosen) and average results are reported. The discussed protocol is followed for all magnification, i.e. same patients are used for training for all magnifications. In subsequent subsections, we will discuss the evaluation metrics used to discuss the present work, performance evaluation of each magnification as well as for integrated model, AUC performance evaluation and performance comparison.

3.1   Evaluation Metric

There can be various ways to evaluate the model when the observed variable lies in continuous range (discussed in introduction section). In some previous work [4,18], patient recognition rate (PRR) that further depends on patient score (PS), and image recognition rate (IRR) were used to report the results. The first measure takes the decision patient-level while second at image-level (i.e. without using patient information) The definition of these measures are given as follows:

$$PRR = \frac{\sum_{i=1}^{N} PS_i}{N} \tag{1}$$

where N is the total number of patients (available for testing). The patient score is define as follows,

$$PS = \frac{N_{rec}}{N_P} \tag{2}$$

N_{rec} and N_P are the correctly classify and total cancer image of patient P respectively.

$$IRR = \frac{TCCI}{TI} \tag{3}$$

where, $TCCI$ and TI are the total correctly classified image and total images respectively.

In addition, we also employ the ROC curve and AUC computation [29] to grade quality of the framework as a system for patient-level diagnosis.

3.2   Performance Evaluation

Tables 2, 3, 4 and 5 illustrate the performance of the models corresponding to each magnification. For each magnification, results are reported for five best combinations which are ranked based on the obtained patient score.

In proposed study, we compute the AUC based on the ROC obtained using the patients scores. Hence, in Tables 2, 3, 4 and 5, we give more prominence to the patient score. In each table, fourth and the fifth row shows the patient and the image score obtained for top combinations and the corresponding features and classifiers are given in the second and third row.

It is observed from the tables that, all the features are not appropriate for same classifier. Hence, suitable combinations of features and classifiers are more advantageous to quantify the images of different magnification.

We also suggest an integrated model, where we consider best feature-classifier combination (based on the patient score) for each magnification. The integrated model yields a patient-level score of 88.40% and image-level score of 88.09%. We note that the integrated model is performing similar to the individual magnifications in terms of score. However, as we demonstrate next, the integration can be considered more reliable, based on the AUC analysis.

Table 2. Top 5 features and classifiers combination for 40x magnification.

Top 5 combination	Combination 1	Combination 2	Combination 3	Combination 4	Combination 5
Feature used	Opponent colour LBP	Normalized colour space representation	Multiple CCR at level 8	Gabor feat. on Gauss. col. mod.	Multiple CCR at level 27
Classifier used	Linear SVM	Subspace Discriminant	Bagged Tree	Linear SVM	Simple Tree
Patient score	**86.74 ± 2.37**	86.01 ± 1.62	86.23 ± 1.75	85.85 ± 1.41	84.32 ± 0.68
Image score	**86.40 ± 2.77**	86.03 ± 1.82	85.39 ± 2.96	85.69 ± 2.04	84.50 ± 2.62

Table 3. Top 5 features and classifiers combination for 100x magnification.

Top 5 combination	Combination 1	Combination 2	Combination 3	Combination 4	Combination 5
Feature used	Gabor feat. on Gauss. col. mod.	Multiple CCR at level 8	Multiple CCR at level 27	Multiple CCR at level 64	Normalized colour space representation
Classifier used	Simple Tree	Bagged Tree	Boosted Tree	Bagged Tree	Simple Tree
Patient score	**88.56 ± 2.73**	88.41 ± 1.68	87.34 ± 1.91	87.05 ± 3.79	87.13 ± 2.42
Image score	**86.70 ± 3.23**	88.31 ± 2.99	87.81 ± 3.17	87.32 ± 4.49	85.60 ± 2.67

Table 4. Top 5 features and classifiers combination for 200x magnification.

Top 5 combination	Combination 1	Combination 2	Combination 3	Combination 4	Combination 5
Feature used	Multiple CCR at level 8	Multiple CCR at level 64	Gabor feat. on Gauss. col. mod	Normalized colour space representation	Multiple CCR at level 27
Classifier used	Bagged Tree	Simple Tree	Subspace Discriminant	Medium Gaussian SVM	Simple Tree
Patient score	**90.31 ± 3.76**	89.57 ± 4.62	88.51 ± 3.52	88.55 ± 3.61	88.04 ± 9.74
Image score	**91.86 ± 3.21**	89.84 ± 3.85	88.82 ± 4.39	88.89 ± 3.58	88.75 ± 8.54

Table 5. Top 5 features and classifiers combination for 400x magnification.

Top 5 combination	Combination 1	Combination 2	Combination 3	Combination 4	Combination 5
Feature used	Normalized colour space representation	DT-CWT	Gabor feat. on Gauss. col. mod.	Gabor chromatic features	Multiple CCR at level 8
Classifier used	Linear SVM	Coarse Gaussian SVM	Coarse Gaussian SVM	Coarse Gaussian SVM	Medium Gaussian SVM
Patient score	**88.31 ± 3.01**	87.98 ± 3.99	86.83 ± 4.21	86.59 ± 6.38	85.87 ± 3.38
Image score	**87.06 ± 3.49**	86.37 ± 3.85	85.07 ± 4.29	85.13 ± 5.67	85.41 ± 2.52

Table 6. AUC comparison for different magnification.

	Magnifications				Integrated model
	40x	100x	200x	400x	
Area under the curve	69.97	74.15	71.93	74.24	**81.96**
Optimum threshold range	64.71–83.87	66.0–97.22	91.67–100	55.56–95.85	81.25–93.06

3.3   AUC Evaluation

As discussed in Sect. 1.2, it is important to take decisions on patients (rather than images), and that ROC and the related AUC is an effective way to rate such diagnostic systems. Here, we consider the same in context of reliability of the test for patient-level decisions, by thresholding patient-level scores. Note that this ROC computation on patient-level scores is different from the traditional ROC analysis for in pattern classifiers (e.g. for image-level classification).

Table 6 details the value of AUC obtained for all magnification levels as well as for integrated model. A threshold on the real-valued scores determines a final label (benign or malignant). The ROC curve is computed using different values of threshold. We also compute the optimal threshold for the ROC curve [30]. Table 6 illustrates the range of this optimal threshold estimated using five trials.

From the reported results in Table 6, it is clear that the AUC for models corresponding to single magnification, is lower than that for the integrated model, thus ascertaining the good quality of inference of the integrated model. The value of 81.92 for AUC for the integrated model signifies a good quality test [21]. We also note that the variation of the optimum threshold among the five trials is one of the lowest. This suggests that the integrated model yields a stable value of the optimum threshold.

3.4   Performance Comparison

Table 7 compares the proposed method with state-of-the-art methods which use same dataset and also the same protocol. We can observe from the table that, except for the 40x magnification case, the proposed framework outperforms the others approaches. Furthermore, one can also observe that the proposed work

Table 7. Performance comparison.

Methods & Score		Spanhol et al. [4]	Spanhol et al. [18]	Bayramoglu et al. [20]	Proposed
40x	Patient level	83.8 ± 4.1	**90.6±6.7**	83.08 ± 2.08	86.74 ± 2.37
	Image level	NA	**89.6 ± 6.5**	NA	86.40 ± 2.77
100x	Patient level	82.1 ± 4.9	88.4 ± 4.8	83.17 ± 3.51	**88.56 ± 2.73**
	Image level	NA	85.0 ± 4.8	NA	**86.70 ± 3.23**
200x	Patient level	85.1 ± 3.1	85.3 ± 3.8	84.63 ± 2.72	**90.31 ± 3.76**
	Image level	NA	84.0 ± 3.2	NA	**91.86 ± 3.21**
400x	Patient level	82.3 ± 3.8	86.1 ± 6.2	82.10 ± 4.42	**88.31 ± 3.01**
	Image level	NA	80.8 ± 3.1	NA	**87.06 ± 3.49**

yields the least variance in scores. Thus, we demonstrate that suitable joint colour-texture features and classifier combination are effective for BC histopathology image classification.

4 Conclusion

This study proposes an integrated model over multiple magnifications for breast cancer histopathological image classification. In this work, we employ a wide range of joint colour-texture features and classifiers. We demonstrate that some of these features and classifiers are indeed effective for a superior classification performance. In addition, the present study also focuses on measuring the performance of the integrated model based on the AUC criteria, and deduce that the this yields better results than the classification at individual magnifications.

References

1. American Cancer Society: Breast cancer facts & figures 2011–2012. American Cancer Society INC., vol. 1, no. 34 (2011)
2. Gurcan, M.N., Boucheron, L.E., Can, A., Madabhushi, A., Rajpoot, N.M., Yener, B.: Histopathological image analysis: A review. IEEE Rev. Biomed. Eng. **2**, 147–171 (2009)
3. Basavanhally, A.N., Ganesan, S., Agner, S., Monaco, J.P., Feldman, M.D., Tomaszewski, J.E., Bhanot, G., Madabhushi, A.: Computerized image-based detection and grading of lymphocytic infiltration in HER2+ breast cancer histopathology. IEEE Trans. Biomed. Eng. **57**(3), 642–653 (2010)
4. Spanhol, F.A., Oliveira, L.S., Petitjean, C., Heutte, L.: A dataset for breast cancer histopathological image classification. IEEE Trans. Biomed. Eng. **63**(7), 1455–1462 (2016)
5. Sethi, A., Sha, L., Vahadane, A.R., Deaton, R.J., Kumar, N., Macias, V., Gann, P.H.: Empirical comparison of color normalization methods for epithelial-stromal classification in H and E images. J. Pathol. Inform. **7**, 17 (2016). doi:10.4103/2153-3539.179984
6. Li, X., Plataniotis, K.N.: A complete color normalization approach to histopathology images using color cues computed from saturation-weighted statistics. IEEE Trans. Biomed. Eng. **62**(7), 1862–1873 (2015)
7. Gorelick, L., Veksler, O., Gaed, M., Gómez, J.A., Moussa, M., Bauman, G., Fenster, A., Ward, A.D.: Prostate histopathology: Learning tissue component histograms for cancer detection and classification. IEEE Trans. Med. Imaging **32**(10), 1804–1818 (2013)
8. Nguyen, K., Sarkar, A., Jain, A.K.: Structure and context in prostatic gland segmentation and classification. In: Ayache, N., Delingette, H., Golland, P., Mori, K. (eds.) MICCAI 2012. LNCS, vol. 7510, pp. 115–123. Springer, Heidelberg (2012). doi:10.1007/978-3-642-33415-3_15
9. Fernández-Carrobles, M.M., Bueno, G., Déniz, O., Salido, J., García-Rojo, M., González-López, L.: Influence of texture and colour in breast TMA classification. PloS one 10(10), e0141556 (2015)

10. Amaral, T., McKenna, S., Robertson, K., Thompson, A.: Classification of breast-tissue microarray spots using colour and local invariants. In: 2008 5th IEEE International Symposium on Biomedical Imaging: From Nano to Macro, ISBI 2008, pp. 999–1002. IEEE (2008)

11. Tabesh, A., Teverovskiy, M.: Tumor classification in histological images of prostate using color texture. In: 2006 Fortieth Asilomar Conference on Signals, Systems and Computers: ACSSC 2006, pp. 841–845. IEEE (2006)

12. Bianconi, F., Harvey, R., Southam, P., Fernández, A.: Theoretical and experimental comparison of different approaches for color texture classification. J. Electron. Imaging 20(4), 043006 (2011)

13. Ilea, D.E., Whelan, P.F.: Image segmentation based on the integration of colour-texture descriptorsa review. Pattern Recogn. 44(10), 2479–2501 (2011)

14. Kowal, M., Filipczuk, P., Obuchowicz, A., Korbicz, J., Monczak, R.: Computer-aided diagnosis of breast cancer based on fine needle biopsy microscopic images. Comput. Biol. Med. 43(10), 1563–1572 (2013)

15. Filipczuk, P., Fevens, T., Krzyzak, A., Monczak, R.: Computer-aided breast cancer diagnosis based on the analysis of cytological images of fine needle biopsies. IEEE Trans. Med. Imaging 32(12), 2169–2178 (2013)

16. Zhang, Y., Zhang, B., Coenen, F., Wenjin, L.: Breast cancer diagnosis from biopsy images with highly reliable random subspace classifier ensembles. Mach. Vis. Appl. 24(7), 1405–1420 (2013)

17. Zhang, Y., Zhang, B., Coenen, F., Xiao, J., Lu, W.: One-class kernel subspace ensemble for medical image classification. EURASIP J. Adv. Signal Process. 2014(1), 17 (2014)

18. Spanhol, F.A., Oliveira, L.S., Petitjean, C., Heutte, L.: Breast cancer histopathological image classification using convolutional neural networks. In: 2016 International Joint Conference on Neural Networks (IJCNN), pp. 2560–2567 IEEE (2016)

19. Krizhevsky, A., Sutskever, I., Hinton, G.E.: ImageNet classification with deep convolutional neural networks. In: Advances in Neural Information Processing Systems, pp. 1097–1105 (2012)

20. Bayramoglu, N., Kannala, J., Heikkilä, J.: Deep learning for magnification independent breast cancer histopathology image classification, 2440–2445 (2016)

21. Hanley, J.A., McNeil, B.J.: The meaning and use of the area under a receiver operating characteristic (ROC) curve. Radiology 143(1), 29–36 (1982)

22. Vertan, C., Boujemaa, N.: Color texture classification by normalized color space representation. In: 2000 Proceedings of the 15th International Conference on Pattern Recognition, vol. 3, pp. 580–583. IEEE (2000)

23. Bianconi, F., Fernández, A., González, E., Caride, D., Calviño, A.: Rotation-invariant colour texture classification through multilayer CCR. Pattern Recogn. Lett. 30(8), 765–773 (2009)

24. Hoang, M.A., Geusebroek, J.-M., Smeulders, A.W.M.: Color texture measurement and segmentation. Signal Process. 85(2), 265–275 (2005)

25. Barilla, M.E., Spann, M.: Colour-based texture image classification using the complex wavelet transform. In: 2008 5th International Conference on Electrical Engineering, Computing Science and Automatic Control, CCE 2008, pp. 358–363. IEEE (2008)

26. Mäenpää, T., Pietikäinen, M.: Texture analysis with local binary patterns. In: Handbook of Pattern Recognition and Computer Vision, vol. 3, pp. 197–216 (2005)

27. Classification-learner-app. https://in.mathworks.com/help/stats/classification-learner-app.html

28. Rokach, L.: Ensemble-based classifiers. Artif. Intell. Rev. **33**(1–2), 1–39 (2010)
29. Rosner, B.: Fundamentals of Biostatistics. 6th ed. Duxbury (2005). Chapter 3
30. Briggs, W.M., Zaretzki, R.: The skill plot: a graphical technique for evaluating continuous diagnostic tests. Biometrics **64**(1), 250–256 (2008)

Multi-label Poster Classification into Genres Using Different Problem Transformation Methods

Miran Pobar and Marina Ivasic-Kos[(✉)]

Department of Informatics, University of Rijeka, Rijeka, Croatia
{mpobar,marinai}@uniri.hr

Abstract. Classification of movies into genres from the accompanying promotional materials such as posters is a typical multi-label classification problem. Posters usually highlight a movie scene or characters, and at the same time should inform about the genre or the plot of the movie to attract the potential audience, so our assumption was that the relevant information can be captured in visual features.

We have used three typical methods for transforming the multi-label problem into a number of single-label problems that can be solved with standard classifiers. We have used the binary relevance, random k-labelsets (RAKEL), and classifier chains with Naïve Bayes classifier as a base classifier. We wanted to compare the classification performance using structural features descriptor extracted from poster images, with the performance obtained using the Classeme feature descriptors that are trained on general images datasets. The classification performance of used transformation methods is evaluated on a poster dataset containing 6000 posters classified into 18 and 11 genres.

Keywords: Multi-label classification · RAKEL ensemble method · Binary relevance · Classifier chains · Movie poster · Classemes

1 Introduction

The purpose of a movie poster is to attract the potential audience to go and see the movie and to promote a movie by emphasizing the key information on actors, plot, genre, mood, etc. To achieve the marketing goals, the poster should present in an attractive manner a focused and obvious message that is tailored to the plot of the movie.

Therefore, the design of movie posters usually follows a set of conventions on the layout, colors, fonts and composition. The conventions can be dependent upon genre and target audience. For example, the title can say a lot about a film, but even the title colors can suggest the genre since some colors are considered to be more suitable for certain genres, e.g. dark and red colors for horror, light blue, and pink for romance.

The challenge in the poster design is even greater for many movies that belong to several genres. For example, according to the data from The movie database (TMDB) [1], "Fist fight" only belongs to the Comedy genre, while "Logan" belongs to Action, Drama and Science Fiction genres. In fact, a movie can be classified into a large number of genres, with no constraints.

© Springer International Publishing AG 2017
M. Felsberg et al. (Eds.): CAIP 2017, Part II, LNCS 10425, pp. 367–378, 2017.
DOI: 10.1007/978-3-319-64698-5_31

The issue of classifying movies into genres using the accompanying promotional materials (trailers or posters) is thus a typical multi-label classification problem that is recently attracting increasing attention in the research community. Multi-label classification problems occur in different domains, ranging from text classification (news, web pages, e-mails etc.), scene classification, video annotations, poster classification to the functional genomics classification (gene and protein function). Many different approaches have been developed and their comparison is given in [2].

The early attempts of classifying posters into genres took into account only one genre per movie, in order to simplify the problem into a classical single-label classification. In [3], low-level features are extracted from movie trailers and are used to classify 100 movies into four genres (drama, action, comedy, horror). In [4], a visual vocabulary is formed from low-level features obtained from a collection of temporally ordered static key frames. This visual vocabulary is used for genre classification on 1239 movie trailers. In [5] the same visual features were used as in [3] but only three kinds of the genre were considered: action, drama, and thriller.

Lately, suitable methods to deal with the multi-label problem were applied to poster classification. In [6], algorithm independent transformation methods were used, but at most two labels (genres) per poster were considered. The classification performance was tested for two out of two correct labels and for at least one of two correctly detected labels with Naive Bayes, distance-ranking classifiers and with random k-labelsets (RAKEL) and all methods have performed uniformly well. The experiment was conducted on a set of 1500 posters classified into six genres. The features used in the classification were low-level features based on color and edge combined with the number of faces detected on posters.

In [7], a much larger poster dataset was used containing 6000 movie posters. The authors have stated that the best results were obtained using Naïve Bayes (NB) with 740-dimensional feature vector comprising GIST [8] and color-based features. In [9] movies are classified into genres using posters and synopsis. The posters are represented with multiple visual features and the synopsis is represented with vector space model. A test film is classified based on the 'OR' operation on the outputs of support vector machine (SVM) classifiers trained on posters and the synopsis.

Our goal was to determine the movie genres automatically from movie posters using both the global structural GIST features extracted from poster images and the classifier-based image descriptor Classemes [10], that are pre-trained on "abstract categories" of 2659 real world object classes.

To solve the multi-label poster classification problem, we have used the binary relevance, RAKEL method, and classifier chains with Naïve Bayes classifier as a base classifier.

Apart from examining how is the classification performance related to different problem transformation methods, we want to compare the effect of different feature sets.

The rest of this article is organized as follows. Section 2 gives a brief overview of typical methods of transforming the multi-label classification problem. Section 3 provides details about the experimental setup, along with benchmark metrics for algorithm evaluation. In Sect. 4, the experimental results are presented and discussed, and in Sect. 5, the contribution of the paper is given.

2   Multi-label Problem Transformation Methods

Our aim was to develop a method that would automatically provide a list of relevant movie genres (labels) for a given, previously unseen poster. The assumption was that a movie could simultaneously belong to several genres, so we had to deal with the problem of multi-label classification.

A multi-label classification for an example e_j can be formally defined as:

$$\exists e_j \in E : \varphi(e_j) = Z_j, Z_j \subseteq C, |Z_j| \geq 2 \, and \, \varphi : E \rightarrow \mathcal{P}(C).$$

where E is a set of examples, C is a set of class labels and the function φ is a classifier, so that exists at least one example e_j that is mapped into two or more classes from C.

The generality of multi-label classification makes it more difficult for implementation and training than the single-label classification.

Methods that can cope with the multi-label classification problem can be roughly divided into algorithm adaptation (AA) methods and problem transformation (PT) methods [11, 12].

The algorithm adaptation (AA) methods are those that adapt, extend or customize an existing classification algorithm to solve multi-label problems [10]. In recent years, several of such multi-label classifiers methods have been developed, such as ML-kNN derived from kNN, Adaboost.MH and Adaboost.MR, derived from the Adaboost method [12]. Since the PT methods performed significantly better than AA methods on the poster dataset in [7], we will here focus on PT methods.

In the PT multi-label transformation approach, the multi-label classification problem is transformed into more than one single-label classification problem, each focusing on one label of the multi-label case [12]. The aim is to transform the data so that any classification method, designed for single-label classification, can be applied. Typically boosting, kNN, decision trees, neural networks, and SVM [12 and references within] are used it this case. The most common transformation methods are label power set (LP) and binary relevance (BR) methods. Ensemble methods transform the data and fuse the results from multiple ordinary classifiers, thus can be considered as problem transformation methods. Some of the widely known ensemble methods are Random k label sets (RAKEL), classifier chains (CC), random forest based predictive clustering trees (RF-PCT), random decision tree (RDT) etc.

Binary relevance
In the BR method, for each class label, a binary classifier is independently trained. Each classifier then decides whether its corresponding class label should be assigned to an example or not. The overall classification result is the set of all assigned labels.

The original data can be transformed similarly as when a binary classifier is used to deal with multi-class problems. The original dataset of examples and corresponding class labels is split into $|C|$ datasets D_i so that each dataset D_i contains all examples belonging to the

C_i class and all the others are treated as negative examples and are labelled as $\neg C_i$. Considering for example the movies "Logan" that belongs to Action, Drama and Science Fiction genres, and "Fist fight" that only belongs to the Comedy genre, the corresponding data sets are: $D_{\text{Action}} = \{(Logan, Action), (First Fight, \neg Action), \ldots\}, D_{\text{Drama}}, D_{\text{SF}} = \{(Logan, Science Fiction), (First Fight, \neg Science Fiction), \ldots\} \ldots D_{\text{Comedy}} = \{(Logan, \neg Comedy), (First Fight, Comedy), \ldots\}$. Then, a single-label classifier is trained for each label $C_i \in C$. The overall classification result consists of labels that are outputs of binary classifiers: $\varphi(e_j) = \bigcup_i C_i : \varphi_i''(e_j) = C_i; \varphi_i'' : E \to \{C_i, \neg C_i\}$.

Rakel

RAKEL method randomly breaks up a set of labels into a number of smaller label sets and trains a multi-label classifier for each of them. Each classifier φ_j solves the single-label classification problem for classes $C_i \in C$ that are the subsets of one of m k-sized label sets $R_j, j = 1 \ldots m$, sampled randomly from all distinct k-label sets of C.

The label set $Y_i = \{C_1, C_m, \ldots, C_r\}$ of each example e_i in the training set is replaced with a new label that corresponds to the intersection of labels in Y_i and in R_j. For example, if a label set R_j is defined as $R_j = \{action, comedy, drama\}$ an example-label set pair in the training set $(e_i, \{action, comedy, crime\})$ will be replaced with the pair $(e_i, action\&comedy)$. For classifying an unknown example e, the results of all classifiers φ_j are joined into final decision using a voting process.

Classifier chains

The classifier chain (CC) involves $|C|$ binary classifiers as in BR [13]. Classifiers are arranged as a chain $\varphi_1'', \ldots, \varphi_{|C|}''$ where each classifier handles the binary relevance problem associated with a label $C_i \in C$. The feature set of each classifier φ_i'' in the chain is extended with the label predictions of all previous classifiers $\varphi_1'', \ldots, \varphi_{i-1}''$. For instance, an example/single-label pair in the training set (e_i, C_i) will be transformed into the $((e_i\varphi_1''(e_i), \ldots, \varphi_{i-1}''(e_i)), C_i)$. The overall classification result of all classifiers φ_j are joined into final decision using a voting process.

The advantage of the BR method is that can predict new label combinations, unseen in the training data. Also, BR is computationally simple, because for $|C|$ labels, only $|C|$ classifiers must be trained. However, BR ignores the dependence between labels and cannot profit from the information about the mutual occurrence of labels in the training data.

RAKEL and CC methods overcome the assumption of BR approach that all labels are mutually independent. The classifier chain method passes label information between classifiers, allowing CC to take into account label correlations. RAKEL solves the problem of power set label explosion since it takes into account only those combinations that exist in the data set.

3 Experimental Setup

The classification experiments are performed on a poster dataset obtained from the TMDB. The aim was to compare the influence of different problem transformation methods that deal with the multi-label classification problem, therefore in all cases the same base classifier is used. The methods have been tested with the GIST structural descriptor and with the Classemes classifier-based descriptor.

3.1 Data and Preprocessing Step

We have used poster dataset containing movie posters from 24 years since 1990 [7]. The database includes 6739 movie posters belonging to 18 genres: Action, Adventure, Animation, Comedy, Crime, Disaster, Documentary, Drama, Fantasy, History, Horror, Mystery, Romance, Science Fiction, Suspense, Thriller, War, and Western.

In order to get class balance in the dataset, a set of poster images for each of the selected genres was gathered by taking the top 20 most popular movies in each year in the range. Despite the efforts to collect the same number of posters for every genre, the collected data was not entirely balanced. The distribution of movies for each selected genre is shown in Table 1.

Table 1. Distribution of movies per selected 18 genres.

C_i	Action	Adventure	Animation	Comedy	Crime	Disaster	Documentary	Drama	Fantasy	History	Horror	Mystery	Romance	SF	Suspense	Thriller	War	Western
$|C_i|$	1815	1081	734	1524	874	64	924	2610	762	675	888	745	934	908	390	1947	527	273

In addition, as every movie could belong to other genres besides the selected genres, some additional genre labels outside the set of 18 selected genres have appeared in the data (e.g. Noir, Musical and Sport). Hence, the total number of genres was 34, and a problem occurred that some genres did not have enough data to define a model.

We solve the problem of lack of data and of unbalanced data in two ways. In the first, we have considered only the data belonging to the 18 selected genres and discarded the additional genres, further referred to as 18G classification task. In the second, we have merged the genres for which there was insufficient data with similar genres per our judgment, such as Neo-Noir and Crime with Thriller, Music with Documentary. After the merge, the number of genres was reduced to 11, forming our 11G classification task. The way the genres are merged is detailed in Table 2.

Table 2. List of originally obtained, 11 merged and 18 selected genres

Merged genre (11G)	Selected genres (18G)	Additional genres outside the list of selected genres
Action/Adventure	**Action**	
	Adventure	
	Disaster	
	War	
Animation	**Animation**	
Comedy	**Comedy**	
Drama	**Drama**	Musical
Crime/Thriller	**Crime**	
	Thriller	
	Mystery	Film Noir, Neo-noir
	Suspense	
Family/Romance	**Romance**	Holiday, Family, Kids
Horror	**Horror**	
Western	**Western**	Road Movie
Science fiction	**Science Fiction**	
	Fantasy	
Documentary	**History**	
	Documentary	Music
Misc.		Eastern, Erotic, Sport, Short, Indie, Sports Film, Foreign, TV movie

3.2   Features

We wanted to examine if features extracted from a poster image, such as spatial structure, or higher-level descriptors like classemes, have the discriminative ability in terms of automatic genre classification.

To capture the information on the poster spatial layout, the GIST descriptor [8] was used. This descriptor refers to the dominant spatial structure of the image characterized by properties of its boundaries (e.g., the size, the degree of openness, perspective) and its content (e.g., naturalness, roughness) [8]. The spatial properties are estimated using global features computed as a weighted combination of Gabor-like multi scale-oriented filters. The dimension of GIST descriptor is $n \times n \times k$ where $n \times n$ is the number of samples used for encoding and k is the number of different orientation and scales of image components. GIST descriptor of each genre is implemented with 8×8 encoding samples obtained by projecting the averaged output filter frequency within 8 orientations per 8 scales. The size of GIST feature vector is 500.

Classemes [10] are classifier-based features that are pre-trained on "abstract categories" of real-world object classes. The abstract categories are constrained to be a super-category obtained by grouping a set of object classes that can be distinguished from other sets of categories and could be used as features for training linear models.

For the extraction of the classeme features, the LP-beta classifier [14] is used, which is defined as a linear combination of M nonlinear SVMs, each trained on a different low-level representation of the image in a 1-vs-rest manner. The low-level representation of the image used for the classemes classifiers contains a set of 15 low-level features concatenated into 22860-dimensional vector capturing different visual cues concerning the color distribution, the spatial layout in the image as GIST, oriented and unoriented HOGs, SIFT and SSIM [15].

The classifiers were pre-trained on the first 150 images retrieved with bing.com image search per each of 2659 classes from the LSCOM ontology [16], so the size of the classemes feature vector is 2659.

The authors stated that classemes provide a general image representation, which can be used to describe and recognize arbitrary categories and novel classes not present in the training set that was used to learn the descriptor [10]. Even though the posters may differ from general images, the object categories of classemes may still provide a high-level representation of layout suitable for use with linear classifiers.

3.3 Classification Methods

For classifying unknown posters into movie genres, we have used the binary relevance, RAKEL method, and classifier chains with Naïve Bayes classifier as base single-label classifier of each method. All three methods are tested on both 11G and 18G classification tasks, with GIST and Classeme features.

In the BR and CC cases, 11 and 18 binary classifiers were trained, one for each genre in the 11G and 18G tasks. To classify an unknown poster, all binary NB classifiers are applied. Each classifier gives a decision whether a movie belongs to its genre or not, and the overall results are obtained by gathering the decisions of all classifiers. In the CC case, we performed 20 runs of training and testing, with the randomly picked order of classifiers in the chain in the each run. We report the average result of all runs.

We have applied the variant of the RAKEL with the Naïve Bayes classifier. RAKEL randomly selects m distinct label sets R_j. The number of label sets m was twice the number of labels in the set C in order to achieve a high level of predictive performance, so for the 11G task it was $m_{11G} = 22$ and for the 18G task it was $m_{18G} = 36$. The size k of label sets R_j is small to avoid problems of LP (in our case $k = 3$).

3.4 Evaluation Measures

Multi-label classification problem requires appropriate evaluation measures, that differs from those used in single-label classification, example-based and label based. Example based measures are calculated on the average differences between the ground truth and the predicted sets of labels in the test set.

We used the following example-based measures. Hamming loss is the symmetric difference of the predicted label set and the true label set, i.e. the fraction of the wrong labels to the total number of labels:

$$HammingLoss = \frac{1}{N}\sum_{i=1}^{N}\frac{|(Y_i\backslash Z_i)\cup(Z_i\backslash Y_i)|}{|C|} \qquad (1)$$

In the best case, the value of Hamming loss is zero.

To define the evaluation measures, we assume that an example $e_j \in E, j = 1..N$ has the set of ground truth labels $Y_j = \{C_l, C_m, \ldots, C_r\}$, $Y_j \subseteq C$ where E is a set of feature vectors, and C is a set of all labels. The set of labels for the example e_j that are predicted by a classifier is Z_j.

Accuracy is defined as the average ratio of correctly predicted labels and all predicted and true labels for each example:

$$Accuracy = \frac{1}{N}\sum_{i=1}^{N}\frac{|Y_i\cap Z_i|}{|Y_i\cup Z_i|}. \qquad (2)$$

Precision is defined as the average ratio of correctly predicted labels and all predicted label for each example.

$$Precision = \frac{1}{N}\sum_{i=1}^{N}\frac{|Y_i\cap Z_i|}{|Z_i|}. \qquad (3)$$

Recall is defined as the average ratio of correctly predicted labels classifier and the ground truth labels for each example.

$$Recall = \frac{1}{N}\sum_{i=1}^{N}\frac{|Y_i\cap Z_i|}{|Y_i|}. \qquad (4)$$

F-Measure can be interpreted as the harmonic mean of precision and recall.

$$F1 = \frac{1}{N}\sum_{i=1}^{N}\frac{2|Y_i\cap Z_i|}{|Z_i|+|Y_i|} \qquad (5)$$

These measures reach their best value at 1 and the worst at 0.

Label based recall, precision, and F1 measures are calculated for each label separately similarly as in binary classification. For average based results, two kinds of averaging operations called macro-averaging and micro-averaging can be used.

A micro-average evaluation measure gives equal weight to each example and is an average over all the example and label pairs. A macro-average evaluation measure gives equal weight to each label, regardless of its frequency, so is a per-label average [17].

4   Results and Discussion

We have tested the classification performance on 11 genres (11G) and 18 genres (18G) classification tasks. All experiments were run using 5-fold cross-validation.

In the following tables, example-based and label-based results are presented. The micro-averaged label-based measures were very similar to example-based measures, so here the macro-averaged measures are shown.

For the macro-averaged evaluation measures, all tested transformation methods have performed similarly on both tasks (Table 3). All transformation methods have achieved better results for the 11G task, which was expected because it is a simpler task. Classeme features have generally proved to be better poster data representatives, although the obtained results are only slightly better than those achieved with 5 times shorter GIST vector.

Table 3. Label-based macro-averaged evaluation results for 11G and 18G tasks.

	Measure	Classifier					
		BR+NB	CC+NB	RAKEL+NB	BR+NB	CC+NB	RAKEL+NB
		Gist features			Classeme features		
11G	Precision	0.29	0.30	0.30	0.30	**0.31**	0.30
	Recall	0.61	0.51	0.55	**0.62**	0.60	0.61
	F1 score	0.38	0.36	0.38	**0.39**	**0.39**	**0.39**
18G	Precision	0.21	0.21	0.21`	0.21	**0.22**	**0.22**
	Recall	0.60	0.38	0.49	**0.62**	0.55	0.58
	F1 score	**0.30**	0.25	**0.30**	**0.30**	**0.30**	**0.30**

CC+NB has the same or worse results concerning label-based evaluation measures than BR+NB, Fig. 1. Passing the genre information between the classifiers in CC+NB chain has not been proved useful in case of label-based evaluations since the scores have not been improved in comparison with BR+NB. This may be explained by error propagation between classifiers in the chain, where a misclassification in one step may impact the classification in all further steps. BR+NB has in all cases achieved the best recall, while precision was very similar for all methods.

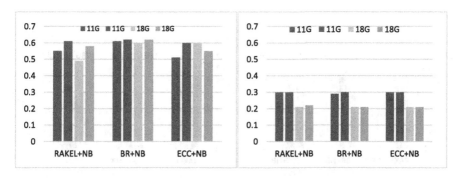

Fig. 1. Macro-averaged recall (left) and precision (right) scores.

Detailed example-based classification results are presented in Table 4 for both 11G and 18G tasks and for GIST and Classeme feature sets. All classifiers except RAKEL have obtained better F1 scores on the label-based than on example-based evaluation measures. RAKEL has achieved slightly better F1 scores for example-based measures in all cases. Simultaneously, RAKEL has significantly better results than other methods for all example-based measures except for Hamming loss.

Table 4. Example based evaluation results for 11 genres (11G) and 18 genres (18G) tasks.

	Measure	Classifier					
		BR+NB	CC+NB	RAKEL+NB	BR+NB	CC+NB	RAKEL+NB
		Gist features			Classeme features		
11G	Hamming loss	0.38	**0.35**	0.37	0.39	0.37	0.37
	Accuracy	0.21	0.21	0.27	0.20	0.20	**0.29**
	Precision	0.25	0.26	0.32	0.24	0.25	**0.33**
	Recall	0.47	0.41	0.56	0.46	0.44	**0.63**
	F1 score	0.31	0.31	0.39	0.30	0.30	**0.41**
18G	Hamming loss	0.38	**0.29**	0.32	0.39	0.34	0.35
	Accuracy	0.15	0.15	**0.21**	0.14	0.14	**0.21**
	Precision	0.17	0.18	**0.24**	0.15	0.16	0.23
	Recall	0.44	0.33	0.54	0.46	0.40	**0.59**
	F1 score	0.23	0.22	**0.32**	0.30	0.22	**0.32**

The highest example-based accuracy, precision, and recall are achieved also with RAKEL, Figs. 2 and 3. Considering that all methods performed similarly according to label-based measures and that RAKEL achieved significantly better example-based measures, we can conclude that RAKEL transformation method performs the best for both our tasks. It seems that information on the mutual occurrence of genres in the training data was significant in this case, so results using RAKEL were improved in comparison to BR. However, the results with CC are not improved, possibly due to propagation of errors through the chain.

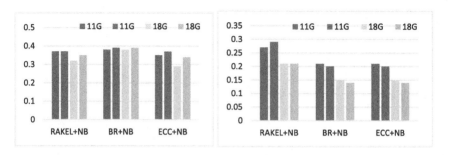

Fig. 2. Hamming loss (left) and accuracy (right) scores.

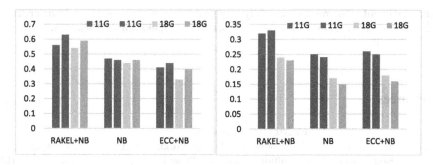

Fig. 3. Example-based recall (left) and precision (right) scores.

In comparison with published results in [7] for the same classification tasks, we have shown that introduction of RAKEL+NB and classeme features contribute the improvement of the classification performance regarding F1 for more than 20%.

5 Conclusion

In this paper, we compared the impact of different problem transformation methods on multi-label classification performance of poster images. The comparison was based on standard label-based and example-based multi-label evaluation measures. The considered problem transformation methods were binary relevance, RAKEL, and classifier chains. Naïve Bayes classifier was used as the base classifier in all cases. We have used a poster dataset of 6739 movie posters classified into 11 or 18 genres. The classification task was multi-label because each poster can be classified into more than one genre. The posters were represented with GIST and classeme features. The GIST feature vector is 5 times shorter than Classemes, but the achieved classification results are similar for both features.

For the macro-averaged evaluation measures, all tested transformation methods have performed similarly on both tasks and achieved better results for the 11G task. BR and CC have achieved better F1 scores on the label-based than on example-based evaluation measures, while the opposite was true for RAKEL. However, RAKEL has achieved significantly better example-based measures than other methods.

Considering that all methods performed similarly according to label-based measures and that RAKEL achieved significantly better example-based measures, we can conclude that RAKEL transformation method performs the best for both our tasks. It seems that information on the mutual occurrence of genres in the training data was significant in this case, so results using RAKEL were improved in comparison to BR. However, the results achieved with CC are not improved, possibly due to the propagation of errors through the chain.

In the future work, we plan to test a deep learning approach with a much larger poster dataset and different features.

Acknowledgment. This research was fully supported by Croatian Science Foundation under the project Automatic recognition of actions and activities in multimedia content from the sports domain (RAASS).

References

1. The movie database, March 2014. http://www.themoviedb.org/
2. Madjarov, G., Kocev, D., Gjorgjevikj, D., Džeroski, S.: An extensive experimental comparison of methods for multi-label learning. Pattern Recogn. **45**(9), 3084–3104 (2012)
3. Rasheed, Z., Sheikh, Y., Shah, M.: On the use of computable features for film classification. IEEE Trans. Circ. Syst. Video Technol. **15**(1), 52–64 (2005)
4. Zhou, H., Hermans, T., Karandikar, A.V., Rehg, J.M.: Movie genre classification via scene categorization. In: Proceedings of the International Conference on Multimedia, pp. 747–750. ACM (2010)
5. Huang, H.-Y., Shih, W.-S., Hsu, W.-H.: A film classifier based on low-level visual features. In: 9th IEEE Workshop on Multimedia Signal Processing, MMSP 2007, pp. 465–468 (2007)
6. Ivašić-Kos, M., Pobar, M., Mikec, L.: Movie posters classification into genres based on low-level features. In: Proceedings of International Conference MIPRO, pp. 1448–1453, Opatija (2014)
7. Ivasic-Kos, M., Pobar, M., Ipsic, I.: Automatic movie posters classification into genres. In: Bogdanova, A.M., Gjorgjevikj, D. (eds.) ICT Innovations 2014. AISC, vol. 311, pp. 319–328. Springer, Cham (2015). doi:10.1007/978-3-319-09879-1_32
8. Fu, Z., Li, B., Li, J., Wei, S.: Fast film genres classification combining poster and synopsis. In: He, X., Gao, X., Zhang, Y., Zhou, Z.-H., Liu, Z.-Y., Fu, B., Hu, F., Zhang, Z. (eds.) IScIDE 2015. LNCS, vol. 9242, pp. 72–81. Springer, Cham (2015). doi:10.1007/978-3-319-23989-7_8
9. Torresani, L., Szummer, M., Fitzgibbon, A.: Efficient object category recognition using classemes. In: Daniilidis, K., Maragos, P., Paragios, N. (eds.) ECCV 2010. LNCS, vol. 6311, pp. 776–789. Springer, Heidelberg (2010). doi:10.1007/978-3-642-15549-9_56
10. Tsoumakas, G., Vlahavas, I.: Random k-label sets: an ensemble method for multi-label classification. In: Machine Learning: ECML 2007, pp. 406–417 (2007). Springer
11. Tsoumakas, G., Katakis, I.: Multi-label classification: an overview. Int. J. Data Warehouse. Min. **3**(3), 1–13 (2007)
12. Read, J., Pfahringer, B., Holmes, G., Frank, E.: Classifier chains for multi-label classification. Mach. Learn. J. **85**(3), 254–269 (2011). Springer
13. Oliva, A., Torralba, A.: Modeling the shape of the scene: a holistic representation of the spatial envelope. Int. J. Comput. Vis. **42**(3), 145–175 (2001)
14. Gehler, P., Nowozin, S.: On feature combination for multiclass object classification. In: 2009 IEEE 12th International Conference on Computer Vision, pp. 221–228. IEEE, September 2009
15. Shechtman, E., Irani, M.: Matching local self-similarities across images and videos. In: CVPR 2007, pp. 1–8. IEEE, June 2007
16. Naphade, M., Smith, J.R., Tesic, J., Chang, S.F., Hsu, W., Kennedy, L., Curtis, J.: Large-scale concept ontology for multimedia. IEEE Multimed. **13**(3), 86–91 (2006)
17. Yang, Y.: An evaluation of statistical approaches to text categorization. Inf. Retrieval **1**(1–2), 69–90 (1999)

Hybrid Cascade Model for Face Detection in the Wild Based on Normalized Pixel Difference and a Deep Convolutional Neural Network

Darijan Marčetić[(✉)] [iD], Martin Soldić [iD], and Slobodan Ribarić [iD]

Faculty of Electrical Engineering and Computing, University of Zagreb, Zagreb, Croatia
{darijan.marcetic,martin.soldic,slobodan.ribaric}@fer.hr

Abstract. The main precondition for applications such as face recognition and face de-identification for privacy protection is efficient face detection in real scenes. In this paper, we propose a hybrid cascade model for face detection in the wild. The cascaded two-stage model is based on the fast normalized pixel difference (NPD) detector at the first stage, and a deep convolutional neural network (CNN) at the second stage. The outputs of the NPD detector are characterized by a very small number of false negative (FN) and a much higher number of false positive face (FP) detections. The FP detections are typically an order of magnitude higher than the FN ones. This very high number of FPs has a negative impact on recognition and/or de-identification processing time and on the naturalness of the de-identified images. To reduce the large number of FP face detections, a CNN is used at the second stage. The CNN is applied only on vague face region candidates obtained by the NPD detector that have an NPD score in the interval between two experimentally determined thresholds. The experimental results on the Annotated Faces in the Wild (AFW) test set and the Face Detection Dataset and Benchmark (FDDB) show that the hybrid cascade model significantly reduces the number of FP detections while the number of FN detections are only slightly increased.

Keywords: Face detection in the wild · Normalized pixel difference model · Deep convolutional neural networks

1 Introduction

The main precondition for successful face-based authentication (verification or identification) and face de-identification for privacy protection [1] is efficient face detection in real scenes. The detection of faces in the wild is a challenging and very hard computer vision task. Some of the main constraining factors are the variability and diversity of the face poses, occlusions, expression variability, different illumination conditions, scale variations, and the richness of colour and texture.

Recently, many methods have been proposed for face detection: the unified model for face detection, pose estimation and landmark localization called the Deformable Parts Model (DPM) [2], a detector based on multiple registered integral image channels [3], a face detector based on Normalized Pixel Difference (NPD) [4], a deep neural

© Springer International Publishing AG 2017
M. Felsberg et al. (Eds.): CAIP 2017, Part II, LNCS 10425, pp. 379–390, 2017.
DOI: 10.1007/978-3-319-64698-5_32

network detector [5], and a very deep convolutional network detector [6]. An alternative approach to robust face detection is based on cascades consisting of multilevel homogenous or hybrid stages. In general, the first stages are very fast but less accurate in the sense of FP detections, and the following stages are used for reducing FP detections with minimal impact on FN detections. One of the earliest homogenous cascade models for face detection is described in [7]. The model was used for fast face detection and localization in images using nonlinear Support Vector Machine (SVM) cascades of a reduced set of support vectors. The cascaded model achieved a thirty-fold speed-up compared to using the single level of a SVM. In [8], a face detection algorithm, called DP2MFD, capable of detecting faces of various sizes and poses in unconstrained conditions is proposed. It consists of deep pyramid convolutional neural networks (CNNs) at the first stage, and a deformable part model (DPM) at the second stage. The inputs of the detector are a colour image resolution pyramid with seven levels. Different CNNs are used for each level. The detector output is based on a root-filter DPM and a DPM score pyramid. Extensive experiments on AFW, FDDB, MALF, IJB-A unconstrained face detection test sets have demonstrated state-of-the-art detection performance of the cascade model. A joint cascade face detection and alignment is described in [9]. It combines the Viola-Jones detector with a low threshold to ensure high recall at the first stage, and pose indexed features with boosted regression [10] are used for face detection at the second stage. A two-stage cascade model for robust head-shoulder detection is introduced in [11]. It combines several methods as follows: a histogram of gradients (HOG) and a local binary patterns (LBP) feature-based classifier at the first stage, and a Region Covariance Matrix (RCM) at the second stage. In [12], cascade architecture built on CNNs with high discriminative capability and performance is proposed. The CNN cascade operates at multiple resolutions and quickly rejects the background regions at fast low-resolution stages, and carefully evaluates a small number of challenging candidates at the last high-resolution stage. To improve localization effectiveness and reduce the number of candidates at later stages, a so-called CNN-based calibration stage is introduced. The proposed cascade model achieves state-of-the-art performance and near real time performance for VGA resolution (14 FPS on a single CPU). In [13], we proposed a two-stage cascade model for unconstrained face detection called 2SCM. The first stage is based on the NPD detector, and the second stage uses the DPM. The experimental results on the Annotated Faces in the Wild (AFW) [14] and the Face Detection Dataset and Benchmark (FDDB) [15] showed that the two-stage model significantly reduces false positive detections while simultaneously the number of false negative detections is increased by only a few. These recent papers have shown that a multi-stage organization of several detectors significantly improves face detection results compared to "classical" one-stage approaches.

In this paper, we present a modification of the two-stage cascade model 2SCM described in [13] in such a way that the second stage is implemented by a CNN instead of a DPM. This modified hybrid two-cascade model is called the HCM.

2 Theoretical Background

In the proposed hybrid two-stage cascade HCM for face detection in the wild, the NPD-based detector [4] and the CNN [16] are used. A short description of both stages follows.

2.1 Normalized Pixel Difference

Authors [4], inspired by Weber's Fraction [17] in experimental psychology, devised an unconstrained face detector based on the normalized pixel difference. The NPD features are defined as: $f(v_{i,j}, v_{k,l}) = (v_{i,j} - v_{k,l})/(v_{i,j} + v_{k,l})$, where $v_{i,j}, v_{k,l} \geq 0$ are intensity values of two pixels at the positions (i, j) and (k, l). By definition, $f(0,0)$ is 0. The NPD feature has the following properties [4]: (i) the NPD is antisymmetric; (ii) the sign of NPD is an indicator of the ordinal relationship; (iii) the NPD is a scale invariant; (iv) the NPD is bounded in $[-1, 1]$. These properties are useful because they reduce feature space, encode the intrinsic structure of an object image, increase robustness to illumination changes, and the bounded property makes an NPD feature amenable for threshold learning in tree-based classifiers. Due to limitations with a stump, as a basic tree classifier with only one threshold that splits a node in two leaves, the authors [4] used for learning a quadratic splitting strategy for deep tree, where the depth was eight.

For practical reasons, the values of the NPD features are quantized into $L = 256$ bins. In order to reduce redundant information contained in the NPD features, the AdaBoost algorithm is used to select the most discriminative features. Based on these features, a strong classifier is constructed. Pre-computed multiscale detector templates, based on the basic 20×20 learned face detector, are applied to detect faces at various scales. The score $ScoreNPD(I, S_i)$ is obtained based on the result of the 1226 deep quadratic trees and the 46401 NPD features, but the average number of feature evaluations per detection window is only 114.5.

The output of the NPD detector is represented by the square regions S_i, $i = 1, 2, \ldots$ n, where n is the number of the detected regions of interest in an image I. For each region S_i in an image I, the score $ScoreNPD(I, S_i)$ is calculated [4]. The regions S_i, $i = 1, 2, \ldots,$ $j \leq n$, with scores $ScoreNPD(I, S_i)$ greater than some predefined threshold θ_{NPD}, are classified as faces. Originally, to achieve a minimum FN, a threshold $\theta_{NPD} = 0$ was used [4].

2.2 Convolution Neural Network

In general, a deep CNN is composed of a series of stacked stages. Each stage can be further decomposed into several stacked layers, including a convolutional layer, a recti-fying unit or a non-linear activation function, a pooling layer and sometimes a normal-ization layer [18]. In our model, we used the deep CNN architecture proposed in [19], which is supported by the Dlib library [19]. The CNN architecture (see Table 1) is designed for face detection and localization. The localization is based on combination of the sliding window and max-margin object detection (MMOD) approaches [16]. This specific CNN consists of seven stages implemented as convolutional layers with a different number of convolutional kernels all having the same size of the first two dimensions, i.e. (5 × 5). Note that the CNN does not have pooling layers. The last stage is implemented with max-margin object detection (MMOD) [16]. The output of the CNN is specific in such a way that it specifies both the size and location of faces.

Table 1. Architecture of the CNN for face detection and localization

Stage	Layer		Kernels			Number of parameters
No.	No.	Type	Number	Size	Stride	
1	1	con	16	5 × 5×3	2 × 2	16 × 5×5 × 3=1,200
	2	relu	–	–	–	0
2	3	con	32	5 × 5×16	2 × 2	32 × 5×5 × 16 = 12,800
	4	relu	–	–	–	0
3	5	con	32	5 × 5×32	2 × 2	32 × 5×5 × 32 = 25,600
	6	relu	–	–	–	0
4	7	con	45	5 × 5×32	1 × 1	45 × 5×5 × 32 = 36,000
	8	relu	–	–	–	0
5	9	con	45	5 × 5×45	1 × 1	45 × 5×5 × 45 = 50,625
	10	relu	–	–	–	0
6	11	con	45	5 × 5×45	1 × 1	45 × 5×5 × 45 = 50,625
	12	relu	–	–	–	0
7	13	con	1	9 × 9×45	1 × 1	1 × 9×9 × 45 = 3,645
	14	MMOD	–	–	–	0
Total number of parameters						180,495

An input to the CNN is an original image $M \times N$ pixels. From this image a resolution image pyramid consisting of six levels is created. The first six stages, layers 1 to 12, of the CNN have a convolutional layer followed by a rectifying unit with an activation function defined as $f(x) = max(0, x)$, where the first three and the last three stages have a 2×2 and 1×1 stride, respectively. The seventh stage consists of only the convolutional layer with one $9 \times 9 \times 45$ kernel with a 1×1 stride (see Table 1). This last stage of the CNN is specific due to the characteristics of the face localization problem (where both the size and location of a face are required) and thus it is implemented with max-margin object detection (MMOD). Specific details of the CNN architecture and number of parameters are given in Table 1. The 180,495 parameters of the CNN were learned on a dataset of 6,975 faces [20]. This dataset is a collection of face images selected from ImageNet, AFLW, Pascal VOC, the VGG dataset, WIDER, and FaceScrub (excluding the AFW and FDDB dataset).

3   Hybrid Cascade Face Detector

The first stage of the HCM is based on the NPD, and the second stage is the CNN. The NPD is very fast and it achieves low FN face detections for unconstrained scenes. The drawback of the NPD is a high number of FPs (typically an order of magnitude higher than FN face detections) when the operating point is set to achieve minimal FN detections. The originally proposed threshold θ_{NPD} for NDP detection is zero [4]. Figure 1 illustrates a typical result of the NPD detector for an image with a rich texture in which the number of FP face detections is relatively high.

Fig. 1. Example of the results of an NPD detector (applied on the AFW test set) for the threshold $\theta_{NPD} = 0$. The green squares denote ground truth, the red true positive, and the blue squares denote false positive. The values *NS* corresponding to *ScoreNPD(I, S_i)* are given at the bottom of each square. (Color figure online)

The number of FP face detections for the NPD detector can be reduced by increasing the θ_{NPD}, but this has negative effects on FNs.

The output of the NPD detector is represented by square regions $S_i = (x_i, y_i, s_i)$, $i = 1$, $2, \ldots, j$, where x_i and y_i are the coordinates of a square region centre, s_i is the size of the region ($s_i \times s_i$), and j is the number of detected faces in an image I. Note that for all S_i, $i = 1, 2, \ldots, j$, the score *ScoreNPD* (I, S_i) is greater than zero [4].

In order to reduce FP face detections, but to keep the FNs as low as possible, the outputs of the NPD detector that have a *ScoreNPD(I, S_i)* in the interval corresponding to vague face region candidates are forwarded to the CNN detector to classify the regions S_i as face or non-face.

The procedure of the HCM is described as follows:

NPD decision stage

For every output square region S_i of the NPD in an image I

(i) **IF** *ScoreNPD(I,S_i)* $\in [0, \theta_1]$, **THEN** the square region S_i is classified as non-face region and it is labelled as a non-face. The square region S_i is not forwarded to the CNN stage;

(ii) **IF** *ScoreNPD(I,S$_i$)* $\in$ [θ_2, ∞], **THEN** the square region S_i is classified as a face region and it is labelled as a face. The square region S_i is not forwarded to the CNN stage;

(iii) **IF** *ScoreNPD(I,S$_i$)* $\in$ < θ_1, θ_2 > , i.e. *ScoreNPD* falls in an interval corresponding to the vague face region candidates, **THEN** the square region S_i is forwarded to the CNN detector;

CNN decision stage

(iv) expand the square region S_i and resize S_i to uniform size;

(v)   **IF** the output of the CNN, called the confidence value *ConValCNN(I,S$_i$)*, is higher than θ_3, **THEN** the vague face region candidate S_i with the original dimensions is labelled as a face;

(vi) otherwise the vague face region candidate S_i is labelled as a non-face.

Note that *ConValCNN(I,S$_i$)* expresses the confidence that a face is detected in an image I at a region S_i and is defined in [16, 19].

The first two thresholds θ_1 and θ_2 define three intervals for the NPD score *ScoreNPD(I,S$_i$)*. The third threshold $\theta3$ defines two intervals for the CNN confidence value *ConValCNN(I,S$_i$)*. These thresholds are determined experimentally on a subset of AFW as $\theta1 = 5$, $\theta2 = 67$, $\theta3 = 0.4$. They define the operating point of the HCM face detector, and are selected to maximize a sum of Precision and Recall, where Precision = TP/(TP + FP) and Recall = TP/(TP + FN), where TP is the number of correctly detected faces. All S_i which are inputs to the CNN stage are expanded by 75% of the original size in each direction and then resized to 225×225. In general, the CNN detector implemented on a single CPU (for high resolution images, e.g. 10 M pixels or more) is typically about an order of magnitude slower than the NPD detector.

This shortcoming of the CNN is circumvented in the HCM in such a way that the CNN is applied *only* to vague face candidate regions S_i (all scaled to small resolution ~ 50 K pixels). These characteristics justify using the CNN detector at the second stage only on *a relatively small number of scaled regions S_i*, selected based on the criterion in (iii) (see the HCM procedure), and these regions are a small fraction of the whole area of an image I. For the implementation of the NPD and CNN we used program implementations [4, 19], respectively.

4   Experimental Results

In all experiments, we selected the same two test sets that were used in our previous work to compare the results obtained by the HCM with 2SCM [13]. The test sets are the AFW [14], consisting of 205 images that contain in total 468 unconstrained faces, and the subset of the FDDB [15] which contains 2,845 annotated images, with 5,171 unconstrained faces in total. On these two test sets, we performed five experiments as follows:

(i)   The NPD detector (*ScoreNPD(I,S$_i$)* > θ_{NPD}, where the threshold $\theta_{NPD} = 0$ as originally proposed in [4]) achieved the results given in Table 2 and in Figs. 5(a) and (e).

(ii)   The DPM detector (threshold $\theta_{DPM} = -0.65$ [2]) achieved the results given in Table 2 and in Figs. 5(b) and (f).

Based on the above results of the first two experiments, we can see that the NPD detector, in general, achieved relatively low FN but high FP face detections in comparison with the DPM.

(iii) The CNN detector ($ConValCNN(I,S_i) > \theta_{CNN}$, where the threshold $\theta_{CNN} = 0.4$ [16]) achieved the results given in Table 2 and in Figs. 5(c) and (g). The results showed that the CNN outperforms both NPD and DPM detectors in the sense of FNs and FPs. Note that all 23 FPs (for the AFW test set) are due to superficial manual annotation. Figure 2 depicts all 23 FPs for the AFW (first row) and all 29 FPs for FDDB (second row) which are not manually annotated as faces thus true FPs is 0 for both test sets.

Fig. 2. Faces detected by the CNN in the AFW (first row) and FDDB test sets (second row) which are not manually annotated as faces.

Note that the face detection time of the CNN is greater than that of the NPD and the DPM when a single CPU is used. Using a GPU, the face detection time of the CNN is comparable with that of the NPD (on a single CPU). For example, the single CPU processing time for the NPD detector is 12.59 s for an image resolution 2138×2811, while for the CNN detector, the processing time for a single CPU and the GPU (384 cores) is 143.54 s and 3.55 s, respectively. The single CPU processing time for the NPD detector is 0.94 s for an image resolution 768×1024, while for the CNN detector, the processing time for a single CPU and the GPU (384 cores) is 17.05 s and 2.10 s, respectively.

(iv)   The two-stage cascade model 2SCM described in [13] achieved the results given in Table 2 and in Figs. 5(d) and (h).

(v)   The proposed HCM described in Sect. 3, achieved the following results (Table 2, Figs. 5(i) and (j)): on the AFW test set 28 faces were not detected (FNs) among 468 faces in 205 images. Note that the 783 FPs from the first stage of the HCM are reduced to only 4 true FPs at the second stage. The remaining FPs, i.e. $19 = 8 + 11$, where all FPs that are the result of poor manual annotation (see Fig. 3) - 8 FPs are resulted in the first stage (which are not forwarded to the CNN stage), and 11 FPs are obtained at the second stage of the HCM (from vague face candidate regions). Similar arguments are valid for the FDDB where from 54 FPs there are 14 (i.e. 13 +1) true FPs while the remaining 40 (i.e. 28 + 12) FPs are the result of poor manual annotation.

Fig. 3. Images, detected by the HCM, which are not manually annotated as faces (ground truth). Faces detected at the first stage (first 8 images) and at the second stage (11 images) in the AFW test set (first row); Faces detected by the HCM at the first stage (28 images in the second row) and 12 images in the third row for FDDB test set.

When HCM is compared with 2SCM on the AFW test set, the FNs at the second stage are reduced by more than 1.71 times (from 48 to 28 FPs), while the FPs are the same (23 FPs). Note that most FP face detections (19 from 23 FPs for the second stage of HCM) are due to the poor annotation of the AFW test set by humans as was noted in [21]. The CNN can detect "difficult faces" that are missed by most human annotators, which increases FPs. The results obtained for the FDDB test set are given in Table 3. When the HCM is compared with 2SCM on the FDDB test set, the FNs are reduced by more than 1.19 times (from 1081 to 901 FNs), while the FPs are decreased by 2.25 times (from 122 to 54 FNs). The results in terms of precision and recall are given in Table 3.

Our proposed HCM face detector achieves face detection results comparable to the results of the state-of-the-art CNN detector. The main advantage of the HCM is that it has CPU face detection time that is about 15 times faster than the CNN run on a single CPU (image resolution 768×1024). CPU face detection time for the HCM is comparable with the CNN detection time when using a GPU (384 cores). An illustration of results of the HCM applied on image from Fig. 1 is depicted in Fig. 4.

Table 2. Results of the experiments performed on AFW test set.

Test set			AFW [14]			
Number of images			205			
Number of faces			468			
Method		Thresholds	FN	FP	Precision	Recall
1	NPD	$\theta_{NPD} = 0$	28	783	0.360	0.940
2	DPM	$\theta_{DPM} = -0.65$	161	72	0.810	0.656
3	CNN	$\theta_{CNN} = 0.4$	7	23 (0[a])	0.952 (1[a])	0.985
4	2SCM	$\theta_1 = 44, \theta_2 = 5,$ $\theta_3 = -0.65$	48 (28 + 20)	23	0.948	0.897
5	**HCM**	$\theta_1 = 5, \theta_2 = 67, \theta_3 = 0.4$	**28**	**23** **12(4[b]) + 11(0[b])**	**0.952** **(0.991[b])**	**0.940**

[a] Please see (iii) Sect. 4,

[b] See (v) Sect. 4.

Table 3. Results of the experiments performed on FDDB test set

Test set			FDDB [15]			
Number of images			2845			
Number of faces			5171			
Method		Thresholds	FN	FP	Precision	Recall
1	NPD	$\theta_{NPD} = 0$	857	1902	0.694	0.834
2	DPM	$\theta_{DPM} = -0.65$	1250	63	0.984	0.758
3	CNN	$\theta_{CNN} = 0.4$	778	29 (0[a])	0.993 (1[a])	0.850
4	2SCM	$\theta_1 = 44, \theta_2 = 5,$ $\theta_3 = -0.65$	1081 (857 + 224)	122	0.971	0.791
5	**HCM**	$\theta_1 = 5, \theta_2 = 67,$ $\theta_3 = 0.4$	**901**	**54** **41(13[b]) + 13(1[b])**	**0.988** **(0.997[b])**	**0.826**

[a] Please see (iii) in Sect. 4,
[b] See (v) Sect. 4.

Fig. 4. Example of the results of the HCM (applied on the AFW test set). The green squares denote ground truth, the red true positive, and the blue squares denote false positive. The values NS and CS correspond to *ScoreNPD(I, S_i)* and *ConValCNN(I,S_i)*, respectively. (Color figure online)

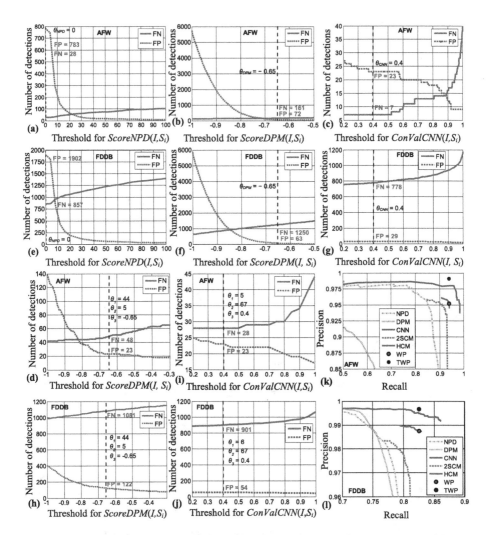

Fig. 5. The results of experiments for the AFW and FDDB test sets. The results of experiments: (a) and (e) NPD detector; (b) and (f) DPM detector; (c) and (g) CNN detector; (d) and (h) 2SCM detector; (i) and (j) the proposed HCM detector; (k) precision and recall of experiments for (a)–(d), (i) for AFW test set; (l) precision and recall of experiments for (e)–(h), (j) for FDDB test set. Working point WP and true working point TWP correspond to results obtained with HCM parameters (thresholds: $\theta_1 = 5, \theta_2 = 67, \theta_3 = 0.4$) for ground truth and manually verified annotation, respectively.

5   Conclusion

In this paper, we have proposed the hybrid cascade model HCM for unconstrained face detection. The first stage of the model is based on the NPD detector, and the second on the deep CNN-based detector. The model is introduced to reduce FP face detections, but to keep FNs as low as possible. This is achieved by conditionally forwarding the outputs of the NPD detector that represent vague face candidate regions to the CNN stage. The condition for forwarding is based on a value of the NPD detector score. The arguments for using the proposed model are: (1) the NPD detector is used at the first stage of the HCM because it is much faster (about 15 times) than the CNN for face detection and localization on a single CPU; (2) the CNN detector is used conditionally as *a post classier* and operates *only on a small number* of rescaled vague face candidate regions which are the outputs of the NPD detector. This enables effective implementation of a second stage of the HCM. The achieved time performance is comparable to the NPD which is considered one of the fastest state-of-the-art detectors [4]. The HCM is suitable for mobile and embedded platforms. (3) NPD and CNN detectors use features which are uncorrelated (normalized pixel differences vs. convolution-based features) what makes the HCM more robust then the NPD.

Experiments performed on the AFW test set showed that FPs were reduced by more than 34 times (from 783 to 23 FPs), while FNs were not increased compared with the NPD detector for the originally proposed threshold [4]. For the FDDB test set, FPs were reduced by more than 35 times (from 1902 to 54 FPs), while FNs were increased 1.05 times (from 857 to 901 FNs) for the same NPD threshold. We are aware that the CNN-based detector applied on the whole image outperforms (in terms of the Recall and Precision) the HCM, but the CNN computational load is about 15 times higher than the computational load of the HCM. Our future work will aim to improve the cascade model to decrease the number of false negative face detections and introduce the stage for face pose estimation.

Acknowledgments. This work has been supported by the Croatian Science Foundation under project 6733 De-identification for Privacy Protection in Surveillance Systems (DePPSS).

References

1. Ribarić, S., Ariyaeeinia, A., Pavešić, N.: De-identification for privacy protection in multimedia content: A survey. Sig. Process. Image Commun. **47**, 131–151 (2016)
2. Zhu, X., Ramanan, D.: Face detection, pose estimation, and landmark localization in the wild. In: Proceedings of IEEE Conference on Computer Vision and Pattern Recognition, pp. 2879–2886 (2012)
3. Dollár, P., Tu, Z., Perona, P., Belongie, S.: Integral channel features. In: Proceedings of British Machine Vision Conference, pp. 1–11 (2009)
4. Liao, S., Jain, A.K., Li, S.Z.: A fast and accurate unconstrained face detector. IEEE TPAMI **38**(2), 211–223 (2016)
5. Taigman, Y., Yang, M., Ranzato, M., Wolf, L.: DeepFace: closing the gap to human-level performance in face verification. In: IEEE Conference on Computer Vision and Pattern Recognition (CVPR), pp. 1701–1708 (2014)
6. Simonyan, K., Zisserman, A.: Very deep convolutional networks for large-scale image recognition. In: Proceedings of International Conference on Learning Representations (2014). http://arxiv.org/abs/1409.1556

7. Romdhani, S., Torr, P., Schölkopf, B., Blake, A.: Efficient face detection by a cascaded support–vector machine expansion. Proc. Roy. Soc. London A Math. Phys. Eng. Sci. **460**(2051), 3283–3297 (2004)

8. Ranjan, R., Patel, V.M., Chellappa, R.: A deep pyramid deformable part model for face detection. In: IEEE 7th International Conference on Biometrics Theory, Applications and Systems (BTAS), pp. 1–8 (2015)

9. Chen, D., Ren, S., Wei, Y., Cao, X., Sun, J.: Joint cascade face detection and alignment. In: Fleet, D., Pajdla, T., Schiele, B., Tuytelaars, T. (eds.) ECCV 2014. LNCS, vol. 8694, pp. 109–122. Springer, Cham (2014). doi:10.1007/978-3-319-10599-4_8

10. Dollár, P., Welinder P., Perona, P.: Cascaded pose regression. In: IEEE Conference on Computer Vision and Pattern Recognition (CVPR), pp. 1078–1085 (2010)

11. Ronghang, H., Ruiping, W., Shiguang, S., Xilin, C.: Robust head-shoulder detection using a two-stage cascade framework. In: 22nd International Conference on Pattern Recognition (ICPR), pp. 2796–2801 (2014)

12. Li, H., Lin, Z., Shen, X., Brandt J., Hua, G.: A convolutional neural network cascade for face detection. In: Proceedings of the IEEE Conference on Computer Vision and Pattern Recognition, pp. 5325–5334 (2015)

13. Marčetić, D., Hrkać, T., Ribarić, S.: Two-stage cascade model for unconstrained face detection. In: IEEE International Workshop on Sensing, Processing and Learning for Intelligent Machines (SPLINE), pp. 1–4 (2016)

14. The Annotated Faces in the Wild (AFW) testset. https://www.ics.uci.edu/~xzhu/face/. Accessed 21 Mar 2017

15. Jain, V., Learned-Miller, E.: FDDB: a benchmark for face detection in unconstrained settings. In: Technical report UM-CS-2010-009, Dept. of Computer Science, University of Massachusetts, Amherst (2010)

16. King, D.E.: Max-margin object detection. In: arXiv preprint arXiv:1502.00046 (2015)

17. Weber, E.H.: Tastsinn und Gemeingefühl. In: Wagner, R. (ed.) Hand-wörterbuch der Physiologie, vol. III, pp. 481–588. Vieweg, Braunschweig (1846)

18. Wu, Z., Huang, Y., Wang, L., Wang, X., Tan, T.: A comprehensive study on cross-view gait based human identification with deep cnns. IEEE TPAMI **39**(2), 209–226 (2017)

19. King, D.E.: Dlib-ml: a machine learning toolkit. J. Mach. Learn. Res. **10**, 1755–1758 (2009)

20. http://dlib.net/files/data/dlib_face_detection_dataset-2016-09-30.tar.gz. Accessed 21 Mar 2017

21. Mathias, M., Benenson, R., Pedersoli, M., Gool, L.: Face detection without bells and whistles. In: Fleet, D., Pajdla, T., Schiele, B., Tuytelaars, T. (eds.) ECCV 2014. LNCS, vol. 8692, pp. 720–735. Springer, Cham (2014). doi:10.1007/978-3-319-10593-2_47

Labeling Color 2D Digital Images in Theoretical Near Logarithmic Time

F. Díaz-del-Río[1], P. Real[1(✉)], and D. Onchis[2,3]

[1] H.T.S. Informatics' Engineering, University of Seville, Seville, Spain
fdiaz@atc.us.es, real@us.es
[2] Faculty of Mathematics, University of Vienna, Vienna, Austria
darian.onchis@univie.ac.at
[3] Faculty of Mathematics and Computer Science, West University of Timisoara, Timisoara, Romania

Abstract. A design of a parallel algorithm for labeling color flat zones (precisely, 4-connected components) of a gray-level or color 2D digital image is given. The technique is based in the construction of a particular Homological Spanning Forest (HSF) structure for encoding topological information of any image. HSF is a pair of rooted trees connecting the image elements at inter-pixel level without redundancy. In order to achieve a correct color zone labeling, our proposal here is to correctly building a sub-HSF structure for each image connected component, modifying an initial HSF of the whole image. For validating the correctness of our algorithm, an implementation in OCTAVE/MATLAB is written and its results are checked. Several kinds of images are tested to compute the number of iterations in which the theoretical computing time differs from the logarithm of the width plus the height of an image. Finally, real images are to be computed faster than random images using our approach.

Keywords: Digital image · Gray-level · Color · Adjacency · Flat zone · Contour · Connected component · Hole · Parallel algorithm · Homological spanning forest

1 Introduction

In general, n-xel's values of biomedical digital images have a relevant physical meaning. In this context, to find a semantically correct segmentation of a gray-level or color nD digital image based on merging original regions (or flat zones) of constant color represents an important processing problem in image understanding. For undertaking this task, we can start with an initial decomposition of the domain of the image into a set $\{R_i\}$ given by the color flat zones of I (connected components where color is constant). This is called here *color-constant region pre-segmentation* and the process in order to construct it, *connected component labeling* (or *CCL*, for short) of the original image. Within the digital context, this pre-segmentation strongly depends on the adjacency relationship we choose for connecting n-xels. We can interpret this problem as well as the

© Springer International Publishing AG 2017
M. Felsberg et al. (Eds.): CAIP 2017, Part II, LNCS 10425, pp. 391–402, 2017.
DOI: 10.1007/978-3-319-64698-5_33

generation of subsequent high level segmentations in terms of topological reductions (in number of cells) of abstract cell complexes (or ACC, for short) [18] representing the segmentations. Depending of the dimensionality (0 or n) of the cells of the different labeled connected components (CCs) of the image, we get a Region-Incidency-ACC or a Contour-Incidency-ACC. Classically, this issue has been treated in the literature with ACCs of dimension one (that is, graphs), like the Region-Adjacency-Graph and its dual (see, for instance, [4,8,17] for a 2D treatment). Problems yet unsolved of different nature and difficulty appears in the process of building a model for digital images capable to be efficiently used in topological recognition tasks. Recently, an important advance in this sense has been the development of the HSF (Homological Spanning Forest) framework for topological parallel computing of 2D digital images [7,20]. Roughly speaking, an HSF of a digital image I is a set of trees living at interpixel level within an abstract cell complex version of I and appropriately connecting all cells without redundancy. This notion is in principle independent of the pixel's values of the image. If there is an interest to analyze a concrete digital object D within it, it is possible to construct an admissible sub-HSF structure for D, "adapting" an original HSF of the whole image to the interpixel frontier or membrane between the object and the background. In this way, such HSF structure can be a useful tool for topologically classifying digital closed curves inside the object.

In this paper, working with the connectivity criterion of 4-adjacency for all the pixels, we develop a parallel CCL algorithm for a gray-level or color 2D digital image based on the previous computation of a particular HSF structure of this image called *region-contour HSF*. This HSF is "adapted" to the image's contour (at interpixel level, the set of digital curves, formed by crack vertices and edges, between neighbor color flat zones). Again, this adaptation process is done in parallel via a combinatorial optimization technique called *crack transport* applied to an initial HSF of the image (called Morse Spaning Forest). Figure 1 gives an idea of how this technique works.

Distinguishing some cells in the region-contour HSF, it is possible to generate a labeling procedure of 4-CCs, including their corresponding area and perimeter measurements.

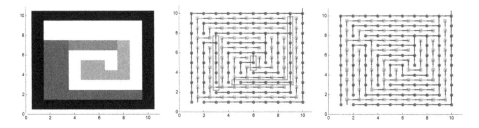

Fig. 1. (Left) A gray level 2D digital image. (Middle) HSF structure of the gray-level image consisting in two trees that is not adapted to the image's contour; (Right) Resulting Region-contour HSF after three crack transports

A sledgehammer to crack nuts? The advantages of this new CCL approach based on HSF structure are:

- It can be considered as one of the more parallel procedures for labeling CCs up to know. Its theoretical time complexity is near the logarithm of the width plus height of the image and the experimental results obtained certify that the method correctly labels even large images. In this sense, we think that an implementation over a many-core processor could significantly improve the speed of the algorithm.
- It is susceptible to be extended to any dimension, due to the fact that the topological HSF structures can be defined in nD space.
- Its versatility regarding the type (binary, gray-level, color, ...) of the analyzed image. Most of the CCL algorithms existing in the literature exclusively works for binary images.
- It is susceptible to be promoted to obtain a useful and efficient HSF-based model representation for recognition tasks or other high level computer vision applications. A possible idea for properly extending the above CCL algorithm is to distinguish in a region-contour HSF extra critical cells of dimension 1 and 2 that allow to find relations of the kind "to be surrounded by" between sets of neighbor CCs.

There is a plethora of CCL algorithms available in the literature. Regarding the technique, there are mainly two classes of algorithms: *raster-scan algorithms*, such as one-pass ([1, 15]) and two-pass ([11, 12, 24, 26, 30]), and *label propagation algorithms* ([3, 5, 14, 27]). Regarding the processing, we have sequential ([25, 28]) and parallel ([2, 10, 13, 16, 21]) algorithms. Regarding the type of image, there are also two classes: binary CCL algorithms (most of the previous references enter into this category) and gray-level and color ([6, 19, 22, 23, 29]) CCL algorithms. A historical overview of this fundamental low-level image processing operation is given in [9]. Most of the previous references are valid in 2D digital context and the criterion of pixel connectivity relies on 4-adjacency or 8-adjacency.

2   Generation of HSF Trees

CCL is one of the fundamental operations in real time applications. The labeling operation transforms an image into a symbolic matrix in which all elements (pixels) belonging to a CC are assigned to a unique label. One of the possible applications of a region-contour HSF is obtaining a CCL. The information of CCs and their contours is registered in some special cells, which are called critical cells. For example, in Fig. 2, we show a region-contour HSF of a gray-level 2D digital image and the same HSF in which we have distinguished on it some critical cells of dimension 0 (square), 1 (circle) and 2 (hollow square), attending to some criteria. Precisely, the 0-critical cells we have choose are representative pixels of the different 4-CCs. The 1-critical cells named by A1 and B1 specify 1-holes of the regions A and B, respectively. For instance, the hole B1 surrounds the 8-connected set formed by the 4-connected regions C, D, E and F. The 1-critical

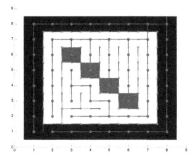 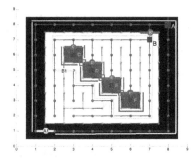

Fig. 2. (left) Region-contour HSF structure of a gray-level 2D digital image; (right) Distinguishing some critical cells on the HSF.

cells named by C', D', E', and F' determine 1-holes of the contour of the image. The three 2-critical cells (hollow squares) are the crossing points of the image's contour. Let us limit ourselves to say that good choice of extra critical cells of dimension 1 and 2 on a region-contour HSF would allow to discover efficient topological recognition solvers.

From now on, we focus on strategies allowing the implementation of an efficient and parallel computation of a region-contour HSF. In this respect, we keep the mathematical background to a minimum and we avoid the use of a too much technical and not informative pseudo-code style. The main three stages for constructing an HSF structure of a digital image are in order:

1. **Input data.** Input data are 2-dimensional positive integer-valued matrices of size $m \times n$ associated to a color or gray-level 2D digital image I. The outermost border of I is filled with an inexistent color (e.g. -1).
2. **Generation of initial HSF trees.** The topological interpretation of the image is condensed at inter-pixel level into two trees of an HSF of I. One of them is composed by 0 and 1-cells (called 0–1 tree) whereas the rest of 1-cells and 2-cells lives on the 1–2 tree. In principle, an HSF is a notion that is independent of the pixel intensities in the image. In our algorithm, the initial constructed HSF is called Morse Spanning Forest (MrSF) and we limit ourselves to say that it can be built in parallel [7], with an architecture of one processing unit element (PE) for each image's pixel.
   This process is explained here using a 8×8-pixel image in Fig. 3. The 0-cells are drawn with small solid red circles, 1-cells with crosses and 2-cells with squares. Edge-vectors of the HSF trees (called links in [7]) having a 0, 1 or 2-cell as its heads are respectively shown with red, blue and green lines. The final result is that the 0–1 tree has a root at the most Northeast corner of the image, while the 1–2 tree is rooted at the most southwest 1-cell. Some special cells (called critical cells) are registered in this stage: a) The 0-cells c (which play a similar role to the sinks in [7]) which are the heads of a link (c, c') whose tail is a 1-cell c' surrounded by two 0-cells of different colors. In Fig. 3 (Left and Middle), sinks are marked with dotted circles; b) Those 1-cells c

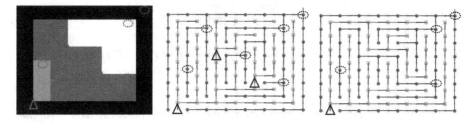

Fig. 3. (Left) A 2D digital image containing a stair-like shape. (Middle) the MrSF of the image. 0-cells are drawn with small solid red circles, 1-cells with little crosses and 2-cells with squares Links coming from 0, 1 and 2-cell are shown with red, blue and green lines, resp. (Right) a resulting HSF after optimization based on crack transport. (Color figure online)

(which play a similar role to the sources in [7]) that, being surrounded by two 0-cells of same color, are the head of a link (c, c') whose tail is a 2-cell c' of a contour of the CC. In Fig. 3 (Left and Middle), sources are marked with triangles. In fact, associated to each sink and source c, there is a pair of neighbor links: (c, c') and (c', c''), such that the dimension of c'' is the same than that of c. This pair of links is called a *crack of the MrSF*.

3. **Crack transport.** It is a combinatorial optimization process (in terms of "transports" of cracks) in order to get two new trees (an HSF as in Fig. 3, right) from the original MrSF. In the resulting region-contour HSF, there is only one sink for each CC and one source for each hole. Finally in Fig. 3 (right), there are only four sinks (one for each region) and one source (the dummy contour of the image surrounds the rest of regions). This means that the number of critical cells must be reduced to its minimum. The correct and parallel implementation of these transports is detailed in the next section.

3   A Parallel Algorithm for Building the HSF of Color Images

In this paper, we propose a variation of our previous works [7,20] so that two important achievements are attained: (1) The parallel algorithm presented here is extended to color images and; (2) No crack transport is done in a sequential manner. The first goal is accomplished thanks to the proper definition of sinks and sources for color images. This yields to the construction of an MrSF which is valid for the following parts of the processing. The nature and degree of the parallel computing in the MrSF construction remains the same than in [7]. In consequence, working with a processing element per pixel, the time complexity is preserved to be the logarithm of the width plus height of the image. We refer the reader to these works for specific details of the algorithm, or directly to our implementation (http://es.mathworks.com/matlabcentral/fileexchange/62644--labeling-color-2d-digital-images-in-theoretical-near-logarithmic-time-).

B	*		A	B
A	C		A	A

Fig. 4. Pixel patterns for detecting sinks (left) and sources (right). (*) means any color. Letters A, B, C are specific colors, being A different from B and C.

The second aspect comes from certain properties of the MrSF trees. In this case, the number of parallel crack's transport to get to the final HSF is considerably reduced in relation to the previous work. In fact, the biggest number of iterations for images up to 4 Mpixels has been found to be 5 for only some random images (and usually inferior for real images). To sum up, the time order complexity results to be very near to that of the MrSF building (that is, logarithmic). In the rest of this section several issues to construct the resulting HSF are fully detailed.

MrSF is built in a similar manner to [7], that is, each 0-cell link must travel only to its North or to its East. The next priority rules are followed in order to compute the link direction of each 0-cell. First, connection between contour pixels of the same color is preferred. If both (North and East) neighbors are in the CC contour, the North direction is (arbitrarily) chosen. Second (if previous rule is not satisfied), North direction is chosen if both neighbors have the same color. Finally, if both neighbors have a different color, the 0-cell is marked as a sink. Note that the case in which only one (North and East) neighbor has the same color that of the current 0-cell is included in the first rule.

Besides, when the two North and East neighbors of a pixel have the same color but the North-East one is of different color, one of the 1-cells that are between these three identical color pixels is marked as critical. To sum up, in order to build an MrSF for color images using a 4-adjacency criterion, the definitions of sink and source are that given by Fig. 4. Previous link direction rules and the search of the critical cells can be done fully in parallel, which supposes a time order complexity of $O(1)$.

Once the rules for the 0-cell links are defined, the rules for the rest of links is done in the same manner than in [7]. The next step is the building of a provisional labeling over the MrSF trees. As our aim is finding topological magnitudes of an image, it is necessary the extraction of global information. This force unavoidably to insert a global searching across the whole image. In this step, parallelization remains the same than in [7], which yields to a time order complexity equal to $O(log(m + n))$. This is the most time consuming part of the parallel labeling.

After the previous MrSF provisional labeling, we proceed to transform the MrSF into an HSF, which must contain the minimum possible number of critical 0-cells and 1-cells. This conversion can be understood from several points of view:

1. If a CC held several 0–1 trees (each one had a sink as a root), those trees must be fused into one. This fusion supposes a crack transport that also changed some links in the 1–2 tree. An example of a region with several 0–1 trees is the biggest CC in Figure 3 (left). This region contains three sinks (marked

with dotted circles), thus it holds three 0–1 trees. After the fusion process, the region will contain one tree (Fig. 3, right)

2. If a CC contained several sinks and several sources, they must be paired so that finally only one sink (and one source per hole) would remain. This cancellation process involves crack transports in both the 0–1 and the 1–2 trees. Using the same example, the biggest region in Fig. 3 (left) has three sinks and two sources, which implies that two sink/source pairs must be canceled.

3. Previous viewpoints are useful to understand our goal but they do not give any insight about how to proceed with the crack transports or cancellations in a parallel manner. In order to achieve parallelism, we must find a procedure to detect the maximum number of sink-source pairs that can be simultaneously canceled. In this regard, the two MrSF tree structures provide uniqueness conditions that help us finding a suitable parallel cancellation method.

In fact, the method proposed here develop the third point of view. In Fig. 5 (Left) some possible indexations of three sinks (named 1, 2, 3) along the 1–2 tree are shown. Indexation of sources A, B and C (along the 0–1 tree) are also drawn. The indexation along a tree is unique. However, two nodes can arrive to the same node as depicted in this figure. When a sink points to a source (through the 1–2 tree), and this source points to the same sink (through the 0–1 tree), this implies that this sink/source pair can be canceled. Moreover, because indexations are unique, all the pair cancellations can be done in parallel pairs. For example, in Fig. 5 (Left) the continuous arrows determine that the pairs 1/A and 3/C can be canceled in parallel.

Furthermore, in the case of 2D images, two possible indexations of each sink and each source must be taken into consideration. Each sink splits the 1–2 tree into two parts: one of them would travel to the South and the other to the West. Likewise, each source divides the 0–1 tree into two parts: one of them would travel to the East, and the other to the North. This yields to two different indexations for each critical cell and, thus, the possibility of considering two possible cancellations for a same sink (with two different sources). This case is

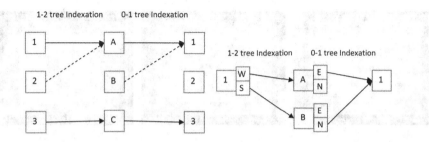

Fig. 5. (Left) An example of tree indexations from two sinks and two sources. (Right) an example of two possible pair cancellations for the same sink through two different sources.

depicted in Fig. 5 (right), and can frequently occur for stair-like shapes as it happens in the center sink of the biggest region of Fig. 3.

In conclusion, in order to proceed with parallel cancellations, in a first iteration sink/source pairs must be searched, for instance, using the South indexation of every sink and the North indexation of every source (which can be called South-then-North cancellation). After that, in a second iteration, different pairs can be found using the West indexation of every sink and the East indexation of every source (which can be called West-then-East cancellation). Note that the selected order in our method is arbitrary: a similar procedure can be done in the opposite order, or, even, using first the indexations from sources and then from sinks. Evidently the selected order would produce different results in terms of numbers of iterations. These numbers depend on the shapes of the regions that are unknown when trying to label them.

To understand the computational difficulties involved in the crack transport, a more complex example to transform the MrSF into HSF through two iterations is given by the following Fig. 6. Figure 6 (left) shows an image that contains a spiral, an L-shape and a reflected L-shape. The superfluous sink and source (4 and A) of the L-shape region can be promptly canceled using a first iteration (South-then-North cancellation). In consequence, this implies that cracks of the trees are transported and those sinks that pointed to A now points to B, and likewise, those sources that pointed to 4 must be now redirected to 5. The resulting trees after the first South-then-North cancellation is displayed at Fig. 6 (middle). Nevertheless, it must be noted that the superfluous sink and source of the spiral cannot be canceled in the first iteration. Thus, the spiral must wait until a second iteration (West-then-East cancellation), resulting in the final Fig. 6 (right).

Previous redirections in the 0–1 and 1–2 trees suppose a searching of the corresponding root that must jump across several hops. If the number of hops is high, it may imply an increment of the total time processing. However, the number of cells involved in these redirections is not very high, which means that the total number of operations is very much lower than that of initial MrSF building. One question remains: When should these iterations be stopped? This

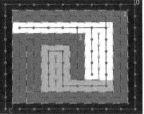

Fig. 6. from left to right: The MrSF of an image containing a spiral, an L-shape and a reflected L-shape; Its MrSF after a first iteration (South-then-North cancellation); Its HSF, obtained after a second iteration (West-then-East cancellation). Letters indicate the sources and numbers the sinks.

question can be answered if the number of false sources are counted. A true source, which denotes the presence of a hole in a region, is that one whose East and North indexations along the 0–1 tree points to the same sink (e.g. the source C in Fig. 6, right). Conversely, a false source is that one whose East and North indexations along the 0–1 tree points to two different sinks. After each iteration, the number of false sources can be computed: if this number is exactly zero, iterations must be stopped. In conclusion, the number of iterations for the parallel crack transport process cannot be a priori known and depends on the shapes of the regions of an image. As explained in the next section, our tests show that this number is very low.

4   Experimental Results

In order to corroborate the correctness of our algorithm, an implementation in OCTAVE/MATLAB is written. Results are checked against those values returned by functions like bweuler() and bwlabel(). All the figures along this work are generated with these codes. Several kinds of images are tested to compute the number of iterations and the number of remaining sinks after each iteration. In general, real images are computed faster than random images. The number of iterations is lower for the first ones due to the fact that they usually present fewer spiral-like shapes. In Table 1 results for random images of different gray levels and sizes are presented. These images are generated by multiplying the number of gray levels by the function rand(). From left to right, the rate in which the number of sinks and false sources is reduced is shown (until no false source remains). In general the less the number of levels, the more difficult to process the image is (it needs more parallel iterations). Nevertheless, we have not found any image requiring more than 5 iterations (for sizes until 2048x2048 pixels). Table 2 shows the corresponding result from some very viewed medical images (taken from http://goldminer.arrs.org/top-40.php).

Another interesting parameter is the number of hops along the redirections when searching for the root of each sink and source in the cancellation process. This gives an idea of the total time spent on this stage. The mean results after four tests using big random images are shown in Table 3. This implies that for the image of 2-gray levels and 4 MiPixels only an amount of $(1263 + 311 + 35) = 1609$ of access operations will be necessary; this number is reduced to 58 when processing the 8-gray level image of the same size. Evidently this quantities are very much lower than the total number of operations for obtaining the initial MrSF building, which means that this stage does not have a considerable impact on the processing times. It can be seen that for random images the number of hops decreases considerably along with the number of gray levels. Due to this, real images present usually a much smaller number of hops: the maximum number is 386 in the third iteration for the larger medical tested image.

Table 1. Results for random images of different gray levels.

# gray levels	height	width	#iter.	# initial sinks	# sinks after iter. 1	# sinks after iter. 2	# sinks after iter. 3	# sinks after iter. 4	# sinks after iter. 5	# initial false sources	# false sources after iter. 1	# false sources after iter. 2	# false sources after iter. 3	# false sources after iter. 4	# false sources after iter. 5
2	128	128	3	3976	2532	2169	2153			1905	406	16	0		
2	256	256	4	16175	10222	8594	8520	8518		7889	1776	87	2	0	
2	512	512	4	65434	41871	35162	34876	34868		31615	7347	329	8	0	
2	1024	1024	5	261152	165643	138842	137611	137564	137563	127924	29566	1428	56	2	0
2	2048	2048	5	1049780	667224	559273	554159	553986	553985	513098	119079	5857	193	2	0
8	128	128	2	12225	12035	12034				191	1	0			
8	256	256	2	49548	48703	48692				856	11	0			
8	512	512	2	199045	195433	195384				3661	49	0			
8	1024	1024	2	799253	785023	784823				14430	200	0			
8	2048	2048	2	3207902	3151400	3150598				57304	802	0			
32	128	128	2	14897	14883	14883				14	0	0			
32	256	256	2	60503	60452	60452				51	0	0			
32	512	512	2	244186	243963	243963				223	0	0			
32	1024	1024	2	980744	979775	979775				969	0	0			
32	2048	2048	2	3928115	3924217	3924214				3901	3	0			

Table 2. Results for several medical images of different sizes.

Image Name	# gray levels	height	width	# iter.	# initial Sinks	# sinks after iter. 1	# sinks after iter. 2	# sinks after iter. 3	# sinks after iter. 4	# initial false sources	# false sources after iter. 1	# false sources after iter. 2	# false sources after iter. 3	# false sources after iter. 4
EURORAD2015	256	600	498	4	202861	196423	195659	195645	195644	7323	808	19	1	0
AJR2004	256	295	166	3	13417	11286	10457	10432		4006	1378	39	0	
AJNR2013	256	1196	1800	4	581353	471635	460343	460275	460274	122025	11479	71	2	0
AJR2009	256	295	260	2	57208	55608	55492			1729	117	0		
Radiology2007	256	526	1274	2	264588	262253	262127			2503	131	0		
PartFibreToxicol2014	256	1000	1200	4	391612	353944	348131	347876	347850	45881	6664	422	48	0

Table 3. number of hops along the redirections when searching for link transports for random images of different sizes and gray levels.

# gray levels	height	width	# iter.	# hops during iter. 1	# hops during iter. 2	# hops during iter. 3	# hops during iter. 4
2	512	512	4	0	32	166	375
2	1024	1024	4	0	34	414	674
2	2048	2048	4	0	35	311	1263
8	512	512	2	0	18		
8	1024	1024	2	0	27		
8	2048	2048	2	0	58		

5   Conclusions

In this paper, we design a parallel 4-adjacency CCL algorithm based on region-contour HSF of a 2D digital image. In a near future, we intend to progress in several directions: (a) to develop a fully functional implementation of the HSF framework in a language (like C++ or python) which allows us to efficiently exploit the parallelism over multicore architectures; (b) to extend the CCL algorithm based on HSF structure to 3D and 4D.

At long term, we are interested in developing a topologically consistent and robust nD digital image analysis and recognition, trying to establish meaningful and efficient topological representation models of images and objects based on HSF structures. For future high level computer vision applications, CCs would be the fundamental bricks whose adjacency relationships would be established through an image HSF structure.

Acknowledgments. This work has been supported by the Spanish research projects (AEI/FEDER,UE) TEC2016-77785-P and MTM2016-81030-P. The last co-author gratefully acknowledges the support of the Austrian Science Fund FWF-P27516.

References

1. Abubaker, A., Qahwaji, R., Ipson, S., Saleh, M.: One scan connected component labeling technique. In: IEEE International Conference on Signal Processing and Communications, pp. 1283–1286. IEEE (2007)
2. Alnuweiri, H.M., Prasanna, V.K.: Parallel architectures and algorithms for image component labeling. IEEE T. Pattern Anal. **10**, 1014–1034 (1992)
3. Ballard, D.H., Brown, C.M.: Computer Vision. Prentice-Hall, Upper Saddle River (1982)
4. Braquelaire, J.P., Brun, L.: Image segmentation with topological maps and inter-pixel representation. J. Vis. Commun. Image R. **9**(1), 62–79 (1998)
5. Chang, F., Chen, C.J., Lu, C.J.: A linear-time component-labeling algorithm using contour tracing technique. Comput. Vis. Image Und. **93**(2), 206–220 (2004)
6. Crespo, J., Schafer, R.W.: The flat zone approach and color images. In: Serra, J., Soille, P. (eds.) Mathematical Morphology and Its Applications to Image Processing Computational Imaging and Vision, vol. 2, pp. 85–92. Springer, Dordrecht (1994)
7. Diaz-del-Rio, F., Real, P., Onchis, D.M.: A parallel homological spanning forest framework for 2D topological image analysis. Pattern Recogn. Lett. **83**, 49–58 (2016)
8. Felzenszwalb, P.F., Huttenlocher, D.P.: Efficient graph-based image segmentation. Internat. J. Comput. Vis. **59**(2), 167–181 (2004)
9. Grana, C., Borghesani, D., Cucchiara, R.: Optimized Block-based connected-component labeling with decision trees. IEEE T. Image Process. **19**(6), 1596–1609 (2010)
10. Han, Y., Wagner, R.A.: An efficient and fast parallel-connected component algorithm. J. ACM **37**(3), 626–642 (1990)
11. He, L., Chao, Y., Suzuki, K.: A run-based two-scan labeling algorithm. IEEE T. Image Process. **17**(5), 749–756 (2008)
12. He, L., Chao, Y., Yang, Y., Li, S., Zhao, X., Suzuki, K.: A novel two-scan connected-component labeling algorithm. In: Yang, G.-C., Ao, S.-L., Gelman, L. (eds.) IAENG Transactions on Engineering Technologies. Lecture Notes in Electrical Engineering, vol. 229, pp. 445–459. Springer, Dordrecht (2013)
13. Hesselink, H., Meijster, A., Bron, C.: Concurrent determination of connected components. Sci. Comp. Programm. **41**, 173–194 (2001)
14. Hu, Q., Qian, G., Nowinski, W.L.: Fast connected-component labeling in three-dimensional binary images based on iterative recursion. Comput. Vis. Image Und. **99**(3), 414–434 (2005)

15. Johnston, C.T., Bailey, D.G.: FPGA implementation of a single pass connected components algorithm. In: 4th IEEE International Symposium on Electronic Design, Test and Applications, pp. 228–231. IEEE(2008)
16. Kalentev, O., Rai, A., Kemnitz, S., Schneider, R.: Connected component labeling on a 2D grid using CUDA. J. Parallel Distrib. Comput. **71**(4), 615–620 (2011)
17. Kropatsch, W.G.: Building irregular pyramids by dual-graph contraction. IEEE Proc. Vis. Image Signal Process. **142**(6), 366–374 (1995)
18. Kovalevsky, V.: Geometry of Locally Finite Spaces. Publishing House Dr. Baerbel Kovalevski, Berlin (2008)
19. Meyer, F.: From connected operators to leveling. In: Mathematical Morphology and its Applications to Image and Signal Processing. Computational Imaging and Vision, vol. 12, pp. 191–198. Kluwer Academic Publishers (1998)
20. Molina-Abril, H., Real, P.: Homological spanning forest framework for 2D image analysis. Annals Math. Artificial Intell. **4**(64), 385–409 (2012)
21. Montanvert, A., Meer, P., Rosenfeld, A.: Hierarchical image analysis using irregular tessellations. IEEE Trans. Pattern Anal. Mach. Intell. **13**(4), 307–316 (1991)
22. Mandler, E., Oberlander, M.F.: One-pass encoding of connected components in multi-valued images. In: Proceedings of the IEEE International Conference on Pattern Recognition, vol. 2, pp. 65–69 (1990)
23. Niknam, M., Thulasiraman, P., Camorlinga, S.A.: A parallel algorithm for connected component labeling of gray-scale images on homogeneous multicore architectures. J. Phys: Conf. Ser. **256**(012010), 1–7 (2010)
24. Rosenfeld, A., Pfaltz, J.L.: Sequential operations in digital picture processing. J. ACM **13**(4), 471–494 (1966)
25. Samet, H.: Connected-component labeling using quadtrees. J. ACM **28**(3), 487–501 (1981)
26. Schwenk, K., Huber, F.: Connected-component labeling algorithm for very complex and high-resolution images on an FPGA platform. In: SPIE Remote Sensing, ISOP (2015)
27. Shima, Y., Murakami, T., Koga, M., Yashiro, H., Fujisawa, H.: A high-speed algorithm for propagation-type labeling based on block sorting of runs in binary images. In: Proceedings the 10th International Conference Pattern Recognition, pp. 655–658, June 1990
28. Suzuki, K., Horiba, I., Sugie, N.: Linear-time connected-component labeling based on sequential local operations. Comput. Vis. Image Und. **89**(1), 1–23 (2003)
29. Sang, H., Zhang, J., Zhang, T.: Efficient multi-value connected component labeling algorithm and its ASIC design. In: Proceedings of the SPIE Medical Imaging Conference (2007). 67892I
30. Wu, K., Otoo, E., Suzuki, K.: Optimizing two-pass connected-component labeling algorithms. Pattern Anal. Appl. **12**(2), 117–135 (2009)

What Is the Best Depth-Map Compression for Depth Image Based Rendering?

Jens Ogniewski$^{(\boxtimes)}$ and Per-Erik Forssén

Department of Electrical Engineering, Linköping University,
581 83 Linköping, Sweden
{jenso,perfo}@isy.liu.se
http://www.isy.liu.se/

Abstract. Many of the latest smart phones and tablets come with integrated depth sensors, that make depth-maps freely available, thus enabling new forms of applications like rendering from different view points. However, efficient compression exploiting the characteristics of depth-maps as well as the requirements of these new applications is still an open issue. In this paper, we evaluate different depth-map compression algorithms, with a focus on tree-based methods and view projection as application.

The contributions of this paper are the following: 1. extensions of existing geometric compression trees, 2. a comparison of a number of different trees, 3. a comparison of them to a state-of-the-art video coder, 4. an evaluation using ground-truth data that considers both depth-maps and predicted frames with arbitrary camera translation and rotation.

Despite our best efforts, and contrary to earlier results, current video depth-map compression outperforms tree-based methods in most cases. The reason for this is likely that previous evaluations focused on low-quality, low-resolution depth maps, while high-resolution depth (as needed in the DIBR setting) has been ignored up until now. We also demonstrate that PSNR on depth-maps is not always a good measure of their utility.

Keywords: Depth map compression · Quadtree · Triangle tree · 3DVC · View projection

1 Introduction

In recent years depth-map compression has become an important research issue, with the advent of commonly available depth sensors, like the Microsoft Kinect, which even are included in some modern mobile phones. The data generated by these sensors becomes more and more accurate, thus enabling new applications, like efficient algorithms to create astonishingly accurate geometric models from this type of data, e.g. [1]. Alas, video cameras with included depth sensors (so called RGB+D sensors) generate tremendous amounts of data, which only gets worse due to ever increasing image resolutions. Thus, efficient compression algorithms are increasingly important.

© Springer International Publishing AG 2017
M. Felsberg et al. (Eds.): CAIP 2017, Part II, LNCS 10425, pp. 403–415, 2017.
DOI: 10.1007/978-3-319-64698-5_34

While compression of RGB images and videos are well understood, depth images have different statistics and applications, and how to compress them is still an open research topic. Some of the first depth-map compression techniques used mesh based approaches, for example [2], which suggested a hierarchical decomposition of the depth-map to generate the mesh. Current methods include e.g. [3] which adapts a current video coder for depth-map compression, or geometric primitives [4], which uses a block based plane model. In this paper however we concentrate mainly on treebased methods.

Tree-based approaches have earlier be proven to be beneficial for depth-map compression. They allow alignment of the borders of the nodes with the edges of the depth-map, thus creating sharp edges in the compressed depth-map, a property that we try to enforce further with adaptive trees (see Sect. 3). In contrast, most texture encoding techniques cause fuzzy edges in depth maps, and these will cause stray points when rendering from a different view. Also, tree-based methods usually encode plane equations and can thus recreate slanted areas with higher fidelity, less effected by quantization step effects typically found in texture encoding. Furthermore, tree-based methods describe the depth-map by a number of geometric primitives, which can be rendered very quickly, compared to point wise projection as is necessary when using image based methods to compress depth-maps. For all these reasons, trees are a natural choice for depth-map compression for *Depth based image rendering* (DBIR). Finally, tree-based approaches have previously shown superior compression performance, albeit only compared to still-coder or (now outdated) video compression schemes in low quality scenarios. However, high quality scenarios, which are more interesting for projection purposes, have not been examined further. Nor have comparisons been done with a current state-of-the-art video encoder, neither have the different tree-based approaches been compared to each other. Here, we compare quad trees [5–8], triangle trees [9–11], as well as own enhancements aimed at improving coding efficiency. We also compare with the 3DVC module [14] from the latest video standard (HEVC [17]). All this is done in a high-quality scenario, which is a requirement for good performance in DIBR.

There are different protocols for evaluation of depth-map compression quality. [12,13] suggest a specific distortion metric for depth-maps, based on the quality of warped views rather than the PSNR of the encoded depth-map itself. While the former considers a stereo scenario, the latter includes a scenario with a number of parallel camera positions, translated along the x axis. We find the basic idea promising, but consider a more flexible framework that allows for arbitrary camera translation and rotation. [18] examines the influence of different warping techniques on the visual quality. Here, we present the PSNR values of the compressed depth-maps, as well as PSNR and Multi-Scale SSIM [16] results of projections using the compressed depth-maps, compared to projections using the ground-truth values. We use our own, state-of-the-art projection algorithm [15] for that. For our evaluation, we use the depth extension of the Sintel datasets [19] (see also Fig. 1), which provide ground-truth for RGB, depth and camera poses. Thus, we are certain that any errors and artifacts are introduced by the algorithms themselves rather than noise or inaccurate input data. While the results

sleeping2   alley2   temple2   bamboo1   mountain1

Fig. 1. The first frames of the Sintel sequences used in this evaluation.

for the different projections are interesting by themselves, it should be pointed out they could easily be extended to similar applications (like e.g. construction of models from depth data).

The rest of the paper is organized as follows: Sect. 2 describes the Sintel dataset as well as some small modifications we did to increase the accuracy of our evaluation. Section 3 describes the different tree-based methods, Sect. 4 the used encoding methods. In Sect. 5 we present a short introduction to our projection algorithm (for more detail the reader is referred to our earlier paper [15]). Section 6 presents the evaluation and the results, while Sect. 7 closes with some concluding remarks.

2   The Sintel Dataset

For the evaluation, we use the 2015 depth extension of the Sintel datasets [19]. This provides ground-truth data for both depth and camera poses. This is necessary to ensure that all errors or artifacts are introduced by the algorithms themselves rather than by noisy or erroneous input data. The Sintel data set provides two different texture streams for each sequence: *clean* without any after-effects (like lightning and blur) as well as *final* with the after-effects included. Due to the very nature of these effects, they remove fine detail from the sequences, and thus the differences between the different projections (using ground-truth depth or any of the compressed depth maps) are more pronounced in the clean sequences. Therefore we only used the clean sequences in our evaluation.

The sequences we chose were *sleeping2, alley2, temple2, bamboo1* and *mountain1* (see Fig. 1), since they contain only low to moderate amounts of moving objects (which are currently not handled by our projection algorithm), but moderate to high camera movement/rotation. Thus, they represent cases where the inaccuracies introduced by depth-map coding are noticeable.

To adapt the sequences to our evaluation, and further increase the accuracy of the results, we did the following changes to the sequences:

1. We removed the last 4 rows of all images and depth-data, to make the height divisible by 16. This is done to increase the efficiency of the reference video coder, which would otherwise have to resort to use very small blocks, which would lead to an increased bitrate.
2. The depth map of the first frame of the sleeping 2 sequence contains some pixels with enormously high values. These are probably introduced by pixels which were never written to during the rendering of the scene. Since these pixels would dominate both the encoding and the evaluation, we clamp them to the second highest value found in this depth-map.

3   Tree-Based Methods

The trees are divided into inner nodes and leaves, where only the inner nodes have children (which might either be leaves or inner nodes, too), while only the leaves contain information describing the actual depth values, normally in the form of one plane equation per leaf. The geometric primitives which are typically chosen for tree-based coding are triangles and quads.

Triangles are typically split in one of their sides, thus each inner node has 2 children (and therefore triangular trees are also called binary trees). In the case of quads, they are split in each of their four sides, thus leading to four children. Therefore, given the same number of leaves, triangle trees will have more inner nodes and thus need more bits to encode the actual tree structure. On the other hand, quads are only able to model vertical and horizontal edges correctly; for depth maps containing edges with arbitrary direction a high number of quads is needed to describe them with adequate accuracy. Triangles on the other hand can also model slanted edges correctly (if the angle coincides with the angle found in the triangles). They thus offer slightly more flexibility.

Normally only triangles are used that are isosceles and right triangles, and they are always split in the middle of their long side, thus creating two children that are isosceles and right triangles as well. This has however two disadvantages:

1. This requires that the depth-map can be divided into a few, comparably big triangles that are both isosceles and right triangles. If the biggest possible triangles are too small, the reached compression will be quite small as well since it is not possible to merge neighboring triangles that can be approximated by the same (or nearly the same) plane equation.
2. We can only approximate edges in the depth-map that are horizontal, vertical or diagonal. Although this is better than in the case of quad trees, we would like a more flexible method.

Thus, we introduce here the notion of *adaptive trees*. In an adaptive quad tree, both the split-positions in x and in y direction can be selected freely, i.e. they do not need to split the sides exactly in half as is normally done in quad trees. It would of course be possible to split all four sides of the quad freely, however this would mean that we have to encode double the amount of data, which therefore does not seem to be beneficial.

For triangle trees, we allow them to be split in any of their sides, as well as an arbitrary split position. Thus, we end up with 5 different trees:

1. *Quadtrees*, which are always split in the middle of their sides.
2. *Adaptive quadtrees*, where the splits are on an arbitrary position of each side. However, the split position of the two horizontal sides has to be the same, as well as the split position of the two vertical sides has to be the same.
3. *Isoscele Trees*, which are triangle trees that are always split in the middle of their longest side.
4. *Triangle Trees*, which are triangle trees that are split in the middle of one of their sides.

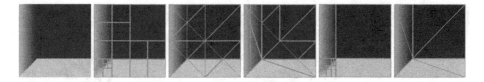

Fig. 2. Example for tree-based encoding, from left to right: Input, quadtree, isoscele trees, triangle trees, adaptive quadtree, adaptive triangle trees

5. *Adaptive Triangle Trees*, which are triangle trees split in an arbitrary position in any of their sides.

Examples for the different trees can be found in Fig. 2.

In most tree-based methods the image is divided into a number of trees first. However, here we omit this and encode the whole image directly in one tree in case of the quadtrees, and 4 in case of the triangle trees. This enables a more flexible division of the depth map, and thus a smaller bitrate. Using only one (resp. four) tree(s) means that we might have to transmit a couple of more bits to describe additional splits. On the other hand, the data needed to describe these splits is very small, compared to the data used for the plane equation. Thus, depending on the input data we can actually save data with our approach. The standard approach might lead to a situation where several neighboring trees contain only one leaf with the same plane equation (or at least nearly the same). Using only one tree, these could be merged to one leaf instead, thus avoiding sending the same plane equation several times and therefore saving data rate, even if using a prediction scheme for the encoding of the plane equations.

For the actual modeling of the depth-values, we use one plane equation per leaf, as is usually done. We tried different ways to express this plane equation, and finally settled for a differential form (i.e. describing the plane by the depth in the middle point as well as how the depth changes in both x and y directions), since no other form offered a higher accuracy or encoding gains, but this form simplifies the computations.

4   Depth-Map Encoding

The search for an efficient encoding is done using the following algorithm:

We calculate all possible splits available in the tree. That even includes evaluating splitting each leaf in each of its sides (in case of the triangle trees) and each split position (in case of the adaptive trees). We then choose the split that leads to the highest increase in PSNR. For efficiency, we save the calculated PSNR improvements, to avoid recalculating them in later steps. This step is repeated until a given bitrate is reached.

In case of the triangle and isoscele trees, we first have to divide the image in a number of triangles. We choose 4, with the exact middle point as the common point of all 4 triangles. For the adaptive triangle tree, this point was allowed to

Fig. 3. Depth-map of the first image from the Sintel Bamboo 1 sequence, as used for the warping. (a) (top left) input depth-map (b) (top right) result using HEVC/3DVC compression, (c) (middle-up, left) using isoscele trees, (d) (middle-up, right) using triangle trees, (e) (middle-down, left) using quad tree-compression, (f) (middle-down, right) using triangle tree-compression and (g) (bottom) using adaptive triangle tree-compression.

be placed on an arbitrary position. We only split one of the trees in each step of our algorithm.

In some cases large leaves will not be split since this only lead to a comparably small change. To counter this, we introduced a *look-ahead* functionality: rather than calculating the PSNR after just one split, we calculate the PSNR after the leaf is split n number of times. For adaptive trees this could be done in two different ways:

1. Calculate each split for itself.
2. Choose one certain split, than calculate all possible consecutive splits which could result from this certain split, before evaluating the next split.

While it would of course be preferably to evaluate all possible combinations, this would increase the runtime exponentially and is therefore not feasible. On the other hand, this means that the look-ahead functionality improved the results for the adaptive trees only marginally. However, it improved the reached quality significantly for the non-adaptive trees.

After the tree division was done, the final tree was encoded using the following approach: Different streams were created for the split information and the plane equations. In the case of adaptive trees, an additional stream was added containing information where exactly the splits occur. This was done to separate the different data, which have different stochastic properties. In a real world application, the arithmetic encoder needs decoding trees to be able to decode these streams, which could either be included directly in the encoded-video-sequence, or general ones might be applied, based on the statistics derived from a high number of different input sequences. Here, we emulated this step instead.

Most of the compression gain of modern coding schemes is reached by applying arithmetic coding, and thus for an accurate comparison, an arithmetic coder needs to be added to the compression of the trees. However, developing such an encoder is unfortunately time consuming and therefore out of scope for this project. Instead, we approximate the outcome of the arithmetic coding by calculating the minimum number of bits that are required to encode the sequences. The possible symbols that were not included in our actual tree were added with a very low probability for this calculation. This emulation gives quite accurate results since modern arithmetic coder can actually reach (or at least come very close to) this minimum. Also, since prediction did not improve our results, it is highly unlikely that context-sensitive encoding could improve on this either. Finally, if the trees do not perform better than other approaches using this theoretical optimal compression, they will never perform better, and thus there is no need to actually develop an arithmetic coder for them.

Firstly, the tree structure was encoded. For the isoscele and the quadtrees, this was done by using a 1 to describe an inner node, a 0 to describe a leaf. For the (adaptive) triangle trees another symbol was added which describes in which side it was split. It was assumed that each side has the same probability to be split.

Secondly, for the adaptive trees the split position were encoded, and for the adaptive triangle tree, we add the position of the common point of the 4 trees to this encoding stream.

We tried different value ranges for this, and settled to use the difference of the actual position to the middle of the side. For large nodes, it is more likely that they are split close to the middle than towards the edges. For small nodes on the other hand their sides will be very small and thus no split far away from the middle can occur. Thus, using the difference to the middle creates a distribution where values around zero are much more likely, and thus enables a more efficient encoding. We were even able to wittness this during our experiments.

Finally, the plane equations were encoded using 8 bits for each of the 3 values (depth in the center, depth change in x and depth change in y direction, both calculated at the edge of the segment the leaf is describing). We tried to use different schemes to predict these values from surrounding other leaves, which however did not lead to an improved bitrate. If two leaves have very similar plane equations, in all likelihood their parent node will not have been split in the first place. Also, if the plane equation of two neighboring leaves are very different, this might lead in an increase of the number of symbols needed to encode the tree (if the values range from -1 to 1, and both extremes are discovered in neighboring leaves, prediction might require increasing the symbol range to a range from -2 to 2).

5 View Projection

For the projection needed in our evaluation, we use our own, state-of-the-art projection algorithm, described in [15]. It consists of a flexible framework integrating a multitude of different methods, both state-of-the-art methods as well as own contributions, which were mainly introduced to counter artifacts we discovered during this work. This framework was then finely tuned using an exhaustive search for the optimal parameters and methods. In the end, it is a forward warping projection scheme, which employs an internal upscale (by both 3 in width and height; Gaussian filtering is used for the downscale), and agglomerative clustering to merge different candidate pixels and projections from different view points, among other things. Hole-filling is provided by a hierarchical approach using scale pyramids for a rough estimate and cross-bilateral filtering for refinement, which can be seen as an enhancement of *hierarchical hole-filling* (HFF) [20].

6 Evaluation

We choose the depth-map of the first and the last frame from each of the test sequences to be encoded. Thus, the projection evaluation could be done by using a combined projection from these two frames to each other frame in the sequence.

Fig. 4. Example of the projection process: Top row: input images, image 1 (left), image 50 (middle) and mask-image (right) Bottom row: combined projection using ground-truth depth (left), and ground-truth frame 34 (right)

Table 1. Depth map quality of the different approaches measured in PSNR for the different sequences. An average value of the two frames of each sequence is shown.

Method	Sequence				
	Alley	Bamboo	Mountain	Sleeping	Temple
HEVC	49.07	45.53	47.49	46.05	49.51
Quad tree	32.93	25.56	31.39	32.28	30.55
Isoscele tree	29.01	21.47	28.32	28.32	26.91
Triangle tree	29.60	21.39	28.71	29.25	27.29
Adpative triangle tree	29.17	19.06	27.27	26.69	27.09
Adaptive quad tree	32.7	23.86	31.96	31.38	31.63

Since the depth-maps are in a floating point format, they first need to be converted to an intermediate gray scale image to make it possible to encode them with HEVC/3DVC [14]. The lowest value in each depth-map was mapped to a luminance of -0.5 in the intermediate image, the highest value to 255.5 (rounding was performed always towards 0). This mapping was done linearly. During earlier experiments we also tested an inverse mapping, and found that it leads to similar or slightly worse results. We used the standard settings of the encoder to allow for easier comparison with our experiments. After encoding, the luminance of the resulting images was transferred back to the original floating point range.

Note that video encoder typically reach much higher compression even for still images than most still image coders like e.g. JPEG2000, since these normally do not take correlation between the different blocks into account, while video coders tend to predict blocks from other blocks contained in the same (or other) image(s).

We then encoded the original depth-maps with all different tree methods, aiming at a similar bit rate as reached by the HEVC/3DVC reference encoder. The resulting quality levels are given in Table 1 (an average of the two encoded frames from each sequence is given), and example depth-maps are shown in Fig. 3. The average bitrates per frame were 23480 for the alley sequence, 99924 for bamboo, 17208 for mountain, 29960 for sleeping and 36264 for the temple sequence.

For the projection evaluation, we did a combined projection from the first and the last image of each sequence to each other image of the same sequence. To reach higher accuracy, we mask part of the projected images that block regions with holes caused by disocclusion and regions containing moving objects. For that, we created mask images beforehand, by projecting every point of the input frames to the target frame, and rounding the obtained x- and y-positions both up and down. To exclude moving objects from the masks, the depth of the projected point is compared to the depth of the ground-truth image for each pixel of this 2×2 region, and only the pixels for which this difference was

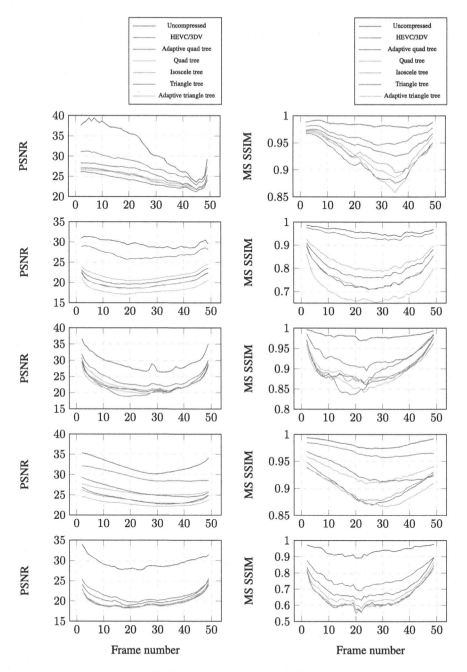

Fig. 5. Measured PSNR (a), (left) and MS SSIM (b), (right) between the warped images and the original images of the sequences: From top to bottom: alley, bamboo, mountain, sleeping and temple. Note that the curves are ordered according to their performance in the legend, the curve with the highest MS SSIM values is mentioned first.

below a predetermined threshold were then set in the mask. An example of the projection process is shown in Fig. 4. The resulting PSNR and MS SSIM values of the projected images compared to the groundtruth images are given in Fig. 5(a) and (b) respectively.

In terms of PSNR/Multiscale SSIM for the warped images, the depth maps encoded with the reference HEVC/3DVC performed best by far, with the exception of the mountain sequence. Simple quad-trees performed surprisingly well as well. On first glance, the PSNR values of the compressed depth-maps correspond well to the PSNR and MS SSIM results reached by the projected images. However, this is not the case for the mountain sequence, neither for the sleeping sequence if comparing quad- and adaptive quad-tree. This means that the PSNR of a depth-map is not in all cases a good indicator for the visual quality that can be reached by images that are results of image warping using its depth values, and is thus not reliable as a performance measure in this case.

The reason why the adaptive methods performed worse than their non-adaptive counterparts lies in the encoding algorithm. While the look-ahead functionality works very well in the non-adaptive cases, it did not perform much better for the adaptive cases, sometimes even worse. The reason is that a split with a certain split position might be beneficial for the exact lookahead-level, but not for a level that is larger. Using an extreme large look-ahead level, or evaluating all possible combinations might solve the problem, but is however not feasible from a computational viewpoint.

7   Conclusion and Future Work

We have presented an evaluation of the most important tree-based compression methods, and introduced modifications aimed at increasing their quality and coding performance. However, the latest standard video encoder for depth-maps (HEVC/3DVC) outperformed the tree-based methods in all but one of the tested sequences in our high-quality evaluation setting.

All tree-based approaches presented in this paper could potentially be improved if a better encoder was found (especially the adaptive ones). As exhaustive search is out of the question due to computational constraints this would require designing a better, novel parameter search strategy. Even if such a strategy could be found it is highly questionable if the performance could be raised to the level of HEVC/3DVC. Thus, while it still might be preferable to use tree-based approaches in narrow, very specialized cases, such as the low-quality setting previously explored, HEVC/3DVC will outperform them at least in the high quality setting that is needed for accurate DIBR. It might however be interesting to consider other recent compression schemes, e.g. [21].

While high PSNR values of compressed depth-maps often coincides with a good projection performance, we have found cases where they do not. Thus, the PSNR of a depth-map is not a good predictor for the quality of projection using this depth-map. To further improve depth-map compression for DIBR purposes a better error metric should therefore be developed. One solution is of course to directly measure the projection performance. However, using one or even several high quality projections might not always be possible due to the increased computational complexity.

References

1. Newcombe, R.A., Izadi, S., Hilliges, O., Molyneaux, D., Kim, D., Davison, A.J., Kohi, P., Shotton, J., Hodges, S., Fitzgibbon, A.: KinectFusion: real-time dense surface mapping and tracking. In: Proceedings of the 10th IEEE International Symposium on Mixed and Augmented Reality, pp. 127–136 (2011)
2. Kim, S.-Y., Ho, Y.-S.: Mesh-based depth coding for 3D video using hierarchical decomposition of depth maps. In: Proceedings of IEEE International Conference on Image Processing, pp. 117–120 (2007)
3. Jingjing, F., Miao, D., Weiren, Y., Wang, S., Yan, L., Li, S.: Kinect-like depth data compression. IEEE Trans. Multimed. **15**, 1340–1352 (2013)
4. Merkle, P., Múller, K., Marpe, D., Wiegand, T.: Depth intra coding for 3d video based on geometric primitives. IEEE Trans. Circ. Syst. Video Technol. **99**, 570–582 (2015)
5. Chai, B.B., Sethuraman, S., Sawhney, H.S., Hatrack, P.: Depth map compression for real-time view-based rendering. Pattern Recogn. Lett. **25**, 755–766 (2004)
6. Morvan, Y., Farin, D., de With, P.H.N.: Multiview video coding using depth based 3D warping. In: Proceedings of IEEE International Conference on Image Processing, vol. 5, pp. 105–108 (2007)
7. Colleu, T., Pateux, S., Morinc, L., Labit, C.: A polygon soup representation for multiview coding. J. Vis. Commun. Image Represent. **21**, 561–576 (2010)
8. Sandberg, D., Forssén, P.E., Ogniewski, J.: Model-based video coding using colour and depth cameras. In: Proceedings of International Conference on Digital Image Computing Techniques and Applications, pp. 158–163 (2011)
9. Chai, B.-B., Sethuraman, S., Sawhney, H.S.: A depth map representation for real-time transmission and view-based rendering of a dynamic 3D scene. In: Proceedings of First International Symposium on 3D Data Processing Visualization and Transmission, pp. 107–114 (2002)
10. Sarkis, M., Zia, W., Diepold, K.: Fast depth map compression and meshing with compressed tritree. In: Zha, H., Taniguchi, R., Maybank, S. (eds.) ACCV 2009. LNCS, vol. 5995, pp. 44–55. Springer, Heidelberg (2010). doi:10.1007/978-3-642-12304-7_5
11. Oh, B.T., Lee, J., Park, D.-S.: Binary tree decomposition depth coding for 3D video applications. In: Proceedings of IEEE International Conference on Multimedia and Expo, pp. 1–6 (2011)
12. Byung Tae, O., Lee, J., Park, D.-S.: Depth map coding based on synthesized view distortion function. IEEE J. Sel. Topics Signal Process. **5**, 1344–1352 (2011)
13. Wang, L., Yu, L.: Rate-distortion optimization for depth map coding with distortion estimation of synthesized view. In: Proceedings of IEEE International Symposium on Circuits and Systems, pp. 17–20 (2013)
14. Müller, K., Schwarz, H., Marpe, D., Bartnik, C., Bosse, S., Brust, H., Hinz, T., Lakshman, H., Merkle, P., Rhee, F.H., Tech, G., Winken, M., Wiegand, T.: 3D high-efficiency video coding for multi-view video and depth data. IEEE Trans. Image Process. **22**, 2266–2278 (2013)

15. Ogniewski, J., Forssén, P.E.: Pushing the limits for view prediction in video coding. In: 12th International Joint Conference on Computer Vision, Imaging and Computer Graphics Theory and Applications (2017)
16. Wang, Z., Simoncelli, E.P., Bovik, A.C.: Multi-scale structural similarity for image quality assessment. In: 37th IEEE Asilomar Conference on Signals, Systems and Computers (2003)
17. Sullivan, G.J., Ohm, J.R., Han, W.J., Wiegand, T.: Overview of the high efficiency video coding (HEVC) standard. IEEE Trans. Circ. Syst. Video Technol. **22**, 1649–1668 (2012)
18. Iyer, K.N., Maiti, K., Navathe, B., Kannan, H., Sharma, A.: Multiview video coding using depth based 3D warping. In: Proceedings of IEEE International Conference on Multimedia and Expo, pp. 1108–1113 (2010)
19. Butler, D.J., Wulff, J., Stanley, G.B., Black, M.J.: A naturalistic open source movie for optical flow evaluation. In: Fitzgibbon, A., Lazebnik, S., Perona, P., Sato, Y., Schmid, C. (eds.) ECCV 2012. LNCS, vol. 7577, pp. 611–625. Springer, Heidelberg (2012). doi:10.1007/978-3-642-33783-3_44
20. Solh, M., Al Regib, G.: Hierarchical Hole-Filling(HHF): depth image based rendering without depth map filtering for 3D-TV. In: IEEE International Workshop on Multimedia and Signal Processing (2010)
21. Chao, Y.-H., Ortega, A., Wei, H., Cheung, G.: Edge-adaptive depth map coding with lifting transform on graphs. In: Picture Coding Symposium (2015)

Nonlinear Mapping Based on Spectral Angle Preserving Principle for Hyperspectral Image Analysis

Evgeny Myasnikov[(⊠)]

Samara University, 34, Moskovskoye shosse, Samara 443086, Russia
mevg@geosamara.ru
http://www.ssau.ru

Abstract. The paper proposes three novel nonlinear dimensionality reduction methods for hyperspectral image analysis. The first two methods are based on the principle of preserving pairwise spectral angle mapper (SAM) measures for pixels in a hyperspectral image. The first method is derived in Cartesian coordinates, and the second one in hypersherical coordinates. The third method is based on the approximation of SAM measures by Euclidean distances. For the proposed methods, the paper provides both the theoretical background and fast numerical optimization algorithms based on the stochastic gradient descent technique. The experimental study of the proposed methods is conducted using publicly available hyperspectral images. The study compares the proposed nonlinear dimensionality reduction methods with the principal component analysis (PCA) technique that belongs to linear dimensionality reduction methods. The experimental results show that the proposed approaches provide higher classification accuracy compared to the linear technique when the nearest neighbor classifier using SAM measure is used for classification.

Keywords: Dimensionality reduction · Spectral angle mapper · Nonlinear mapping · Hyperspectral image · Classification · Stochastic gradient descent

1 Introduction

Nonlinear dimensionality reduction techniques attract more and more attention of researchers in the field of multi- and hyperspectral image processing in the last years. Unlike linear techniques, the nonlinear techniques allow, for example, to improve the accuracy of land cover classification and target detection. In hyperspectral image visualization, these techniques also allow to obtain false color (pseudo color) representations with desirable properties. All these advantages can be attributed to nonlinear techniques, which are sometimes called manifold learning techniques.

The nonlinear dimensionality reduction techniques that have been successfully applied in multi- and hyperspectral image processing are locally linear embedding (LLE) [1], laplacian eigenmaps (LE) [2], isometric embedding

© Springer International Publishing AG 2017
M. Felsberg et al. (Eds.): CAIP 2017, Part II, LNCS 10425, pp. 416–427, 2017.
DOI: 10.1007/978-3-319-64698-5_35

(ISOMAP) [3], curvilinear component analysis (CCA) [4,5], and nonlinear mapping (NLM) [6]. In all these techniques, it is required to measure dissimilarity between points of the hyperspectral image. The Euclidean distance is the most often used measure for this purpose. However, there are several other dissimilarity measures that have been successfully used in hyperspectral image analysis. The most extensively used among them is the spectral angle mapper (SAM) [7]. This measure proved to have some advantages over the Euclidean distance in the field of hyperspectral image analysis, but it has been rarely used in the field of nonlinear dimensionality reduction.

In particular, we can point to only a few studies [3,8,9] in which they used SAM measure with nonlinear dimensionality reduction techniques. In paper [3], SAM measure was used along with Euclidean distance at the first stage of ISOMAP method to determine the neighbor points in hyperspectral space. In paper [8], the authors study the effectiveness of SAM measure and Euclidean distance for dimensionality reduction using laplacian eigenmaps technique. In paper [9], SAM measure is used to improve laplacian eigenmaps technique for reducing the sensitivity to missing data in time series of multispectral satellite images. As can be seen, only two nonlinear dimensionality reduction techniques, namely ISOMAP and laplacian eigenmaps, were hitherto applied in combination with SAM measure for hyperspectral image analysis. However, one of the oldest and most well-known nonlinear techniques - nonlinear mapping- was not applied so far in combination with SAM measure for hyperspectral image analysis.

The purpose of this paper is to fill this gap for a nonlinear mapping method [11]. Specifically, we propose several novel nonlinear mapping methods based on spectral angle preserving principle. At present, the author is not aware of publications where nonlinear mapping method [11] was used in combination with SAM measure.

The paper is organized as follows. The next section consists of five subsections. Subsection 2.1 describes the basics of the nonlinear mapping technique. In Subsect. 2.2, we introduce the objective function that represents the spectral angle preserving error in Cartesian coordinates, and derive a novel nonlinear mapping method based on spectral angle preserving principle. In Subsect. 2.3, we introduce the analogous objective function in hyperspherical coordinates, and derive corresponding nonlinear mapping method in hyperspherical coordinates. Subsection 2.4 describes the alternative nonlinear mapping approach based on the approximation of spectral angles by Euclidean distances. In Subsect. 2.5, we develop the numerical optimization procedure that can be applied to hyperspectral images. This procedure is based on stochastic gradient descent approach. Section 3 describes the experimental studies. In this section, using publicly available hyperspectral images, we compare the proposed methods to the principal component analysis (PCA) technique that belongs to linear dimensionality reduction methods, and show the effectiveness of the proposed methods in terms of the classification accuracy. The paper ends with conclusions in Sect. 4.

2   Methods

Before starting the description of the proposed methods, let us give the basics of the nonlinear mapping method.

2.1   Nonlinear Mapping

A nonlinear mapping method belongs to a class of nonlinear dimensionality reduction techniques operating on the principle of preserving pairwise distances between data points. It minimizes the following data mapping error:

$$\varepsilon = \mu \cdot \sum_{i,j=1, i<j}^{N} (\rho_{i,j}(d(x_i, x_j) - d(y_i, y_j))^2), \tag{1}$$

where N is a number of data points, x_i are coordinates of the data points in a multidimensional space R^M, y_i are coordinates of the corresponding points in a lower dimensional space R^L, $d(\cdot)$ is a distance function, μ and ρ_{ij} are some constants determining the particular error function. For Eq. (1), the most common practice is to use $\mu = 1/\sum_{i<j} d^2(x_i, x_j)$, $\rho_{i,j} = 1$, or $\mu = 1/\sum_{i<j} d(x_i, x_j)$, $\rho_{i,j} = (d(x_i, x_j))^{-1}$. In the first case, Eq. (1) is known as a Kruskal stress [10], and in the second case, the same equation is known as a Sammon data mapping error [11]. The distance function $d()$ in Eq. (1) is usually the Euclidean distance:

$$d(x_i, x_j) = ||x_i - x_j|| = \sqrt{\sum_{k=1}^{M} (x_{ik} - x_{jk})^2}. \tag{2}$$

To minimize the error (1), a number of numerical optimization techniques can be used, but the methods based on the gradient descent algorithm [12] are used more often. In this article, we will also use similar methods.

Let us consider the basic gradient descent algorithm. Treating the coordinates $Y = (y_1, ...y_N)$ in the lower dimensional space as optimized parameters, the base version of the gradient descent algorithm sequentially narrows the initial configuration $Y(0)$ of data points in the output space using the following equation:

$$Y(t+1) = Y(t) - \alpha \nabla \varepsilon, \tag{3}$$

where t is a number of an iteration, $\nabla \varepsilon$ is a gradient of the objective function, α is a coefficient of the gradient descent. Thus, the iterative optimization process is defined by the following recurrent equation for the coordinates of data points in the output space:

$$y_{ik}(t+1) = y_{ik}(t) + 2\alpha\mu \sum_{j=1; i\neq j}^{N} \rho_{i,j} \cdot \frac{d(x_i, x_j) - d(y_i, y_j)}{d(y_i, y_j)} \cdot (y_{ik}(t) - y_{jk}(t)). \tag{4}$$

This equation allows us to find some suboptimal solution to the $\varepsilon \to_Y min$ problem by initializing the output coordinates $y_i(0)$, and following by the iterative optimization using (4) until coordinates $y_i(t)$ become stable.

2.2   Spectral Angle Preserving Mapping in Cartesian Coordinates

As it was outlined in the introduction, the spectral angle mapper [7] measure

$$\theta(x_i, x_j) = arccos\left(\frac{x_i \cdot x_j}{||x_i||||x_j||}\right), \tag{5}$$

is often used to measure the dissimilarity between points of the image in the hyperspectral space. Following the approach described in the previous section, we introduce the spectral angle mapping error in the similar form:

$$\varepsilon_{SAM_C} = \mu \sum_{i,j=1(i<j)}^{N} \left(\rho_{i,j}\left(\theta(x_i, x_j) - \theta(y_i, y_j)\right)^2\right). \tag{6}$$

Then we apply the gradient descent algorithm to obtain a suboptimal solution. The latter requires calculating the partial derivatives of the spectral angle mapping error:

$$\frac{\partial \varepsilon_{SAM_C}}{\partial y_{ik}} = \mu \sum_{j=1(j\neq i)}^{N} \left(-2\rho_{ij}\left(\theta(x_i, x_j) - \theta(y_i, y_j)\right)\frac{\partial\theta(y_i, y_j)}{\partial y_{ik}}\right). \tag{7}$$

By applying some simple transformations, we obtain the following expression for partial derivatives:

$$\frac{\partial \varepsilon_{SAM_C}}{\partial y_{ik}} = -2\mu \sum_{j=1(j\neq i)}^{N} \rho_{ij}(\theta(x_i, x_j) - \theta(y_i, y_j))\frac{y_{jk} - y_{ik}(y_i \cdot y_j)/||y_i||^2}{\sqrt{||y_i||^2||y_j||^2 - (y_i \cdot y_j)^2}}. \tag{8}$$

Finally, the iterative equation for the coordinates of data points in the output space takes the following form:

$$y_{ik}(t+1) = y_{ik}(t)$$
$$+ 2\alpha\mu\sum_{j=1(j\neq i)}^{N}\left(\rho_{ij}\left(\theta(x_i, x_j) - \theta(y_i, y_j)\right)\frac{y_{jk} - y_{ik}(y_i \cdot y_j)/||y_i||^2}{\sqrt{||y_i||^2||y_j||^2 - (y_i \cdot y_j)^2}}\right). \tag{9}$$

The latter equation defines the spectral angle preserving mapping (SAPM-C), an analog of the nonlinear mapping for the spectral angle measure (5) derived in Cartesian coordinates. As it can be seen, a correction vector $y_i(t+1) - y_i(t)$ is orthogonal to the corrected vector $y_i(t)$ that raises an unbounded growth of vectors with the number of iterations. To avoid this growth, one can control, for example, the average vector length or force all the vectors lengths to be equal to one by normalizing vectors at every step. The latter means that all data points are mapped onto the unit hypersphere. In this case Eq. (9) can be rewritten as

$$y_{ik}(t+1) = y_{ik}(t)$$
$$+ 2\alpha\mu \sum_{j=1(j\neq i)}^{N} \left(\rho_{ij}\left(\theta(x_i, x_j) - \theta(y_i, y_j)\right) \frac{y_{jk} - y_{ik}(y_i \cdot y_j)}{\sqrt{1 - (y_i \cdot y_j)^2}} \right). \quad (10)$$

It should be noted that normalization does not cause any loss of spectral angle information. Moreover, as vector lengths are not used in this method, we can reduce the dimensionality of the output space by transforming output data to hyperspherical coordinates (the details on the direct and inverse transformations are shown below) and truncating radiate coordinate containing information about lengths of vectors. Thus, mapping onto L dimensional unit hypersphere can be considered as mapping into $(L-1)$ dimensional output space for the considered method.

Transformations to Hyperspherical and Cartesian Coordinates. Let us define a spherical coordinate system in an L-dimensional Euclidean space. To do this, let us introduce a radiate coordinate r and angular coordinates $\phi_1, \phi_2, ...\phi_{L-1}$. A transformation to spherical coordinate system can be computed using the following equations [13]:

$$r = \sqrt{\sum_{i=1}^{L} x_i^2},$$
$$\phi_k = arccos(\frac{x_k}{\sqrt{\sum_{i=k}^{L} x_i^2}}), k = 1..(L-1),$$

$$\phi_{L-1} = \begin{cases} arccos(\frac{x_{L-1}}{\sqrt{x_{L-1}^2 + x_L^2}}), & x_L \geq 0, \\ 2\pi - arccos(\frac{x_{L-1}}{\sqrt{x_{L-1}^2 + x_L^2}}), & x_L < 0. \end{cases} \quad (11)$$

The inverse transformation can be done using the following equations [13]:

$$x_k = \begin{cases} r\prod_{i=1}^{k-1} sin(\phi_i)cos(\phi_k), & k = 1..(L-1), \\ r\prod_{i=1}^{L-1} sin(\phi_i), & k = L. \end{cases} \quad (12)$$

2.3 Spectral Angle Preserving Mapping in Hyperspherical Coordinates

As it was said above, the spectral angle preserving mapping can be considered as a mapping to a unit hypersphere. In this section, we derive an analogue of the above mapping in hyperspherical coordinates. To do this we define $r = 1$. By denoting

$$P_m^n(\phi_i, \phi_j) = \prod_{k=m}^{n} sin(\phi_{ik})sin(\phi_{jk}) \quad (13)$$

we obtain the spectral angle mapping measure (5) in the hyperspherical coordinate system:

$$\theta(\phi_i, \phi_j) = arccos \left(\sum_{k=1}^{L-1} P_1^{k-1}(\phi_i, \phi_j) cos(\phi_{ik}) cos(\phi_{jk}) + P_1^{L-1}(\phi_i, \phi_j) \right). \quad (14)$$

Following the basic approach, we introduce the nonlinear mapping error using SAM measure:

$$\varepsilon_{SAM_H} = \mu \sum_{i,j=1(i<j)}^{N} \rho_{i,j} \left(\theta(\psi_i, \psi_j) - \theta(\phi_i, \phi_j) \right)^2 \quad (15)$$

where ψ_i are hyperspherical coordinates of the data points in a multidimensional space, ϕ_i are (hyper-) spherical coordinates of the corresponding points in a lower dimensional space.

Then we apply the gradient descent algorithm to obtain a suboptimal solution. The latter requires calculating the partial derivatives of the spectral angle mapping error:

$$\frac{\partial \varepsilon_{SAM_H}}{\partial \phi_{ik}} = \mu \sum_{j=1(j\neq i)}^{N} -2\rho_{ij} \left(\theta(\psi_i, \psi_j) - \theta(\phi_i, \phi_j) \right) \frac{\partial \theta(\phi_i, \phi_j)}{\partial \phi_{ik}}. \quad (16)$$

Here

$$\frac{\partial \theta(\phi_i, \phi_j)}{\partial \phi_{ik}} = - \left(1 - arg^2(\phi_i, \phi_j) \right)^{-1/2} \frac{\partial arg(\phi_i, \phi_j)}{\partial \phi_{ik}}, \quad (17)$$

where arg is the argument of $arccos$ function in (14):

$$arg(\phi_i, \phi_j) = \sum_{k=1}^{L-1} P_1^{k-1}(\phi_i, \phi_j) cos(\phi_{ik}) cos(\phi_{jk}) + P_1^{L-1}(\phi_i, \phi_j). \quad (18)$$

The partial derivative of arg is expressed by the equation:

$$\frac{\partial arg(\phi_i, \phi_j)}{\partial \phi_{ik}} = P_1^{k-1}(\phi_i, \phi_j) \left(- sin(\phi_{ik}) cos(\phi_{jk}) + cos(\phi_{ik}) sin(\phi_{jk}) \right.$$
$$\left. \cdot \left(\sum_{m=k+1}^{L-1} P_{k+1}^{m-1}(\phi_i, \phi_j) cos(\phi_{im}) cos(\phi_{jm}) + P_{k+1}^{L-1}(\phi_i, \phi_j) \right) \right). \quad (19)$$

Finally, the iterative equation for the coordinates of data points in the output space takes the following form:

$$\phi_{ik}(t+1) = \phi_{ik}(t) + 2\alpha\mu \sum_{j=1(j\neq i)}^{N} \left(\frac{\theta(\psi_i, \psi_j) - \theta(\phi_i, \phi_j)}{\sqrt{1 - arg^2(\phi_i, \phi_j)}} \frac{\partial arg(\phi_i, \phi_j)}{\partial \phi_{ik}} \right) \quad (20)$$

The latter equation defines the spectral angle preserving mapping (SAPM-H) in the hyperspherical coordinate system.

Unfortunately, the described above optimization algorithm tends to get into local minima, especially in the case of a high output dimensionality. In particular, in experiments, described algorithm provided higher error values compared to other proposed methods when simple stochastic gradient descent was used as an optimization technique (see Sect. 2.5). For this reason, it can be useful to consider simplified approach in hyperspherical coordinate system. This approach is based on the modified angle dissimilarity measure:

$$\theta^*(\phi, \psi) = \sum_{k=1}^{L} sin^2\big((\phi_k - \psi_k)/2\big). \tag{21}$$

Corresponding optimization technique based on the above measure can be expressed by the equation:

$$\phi_{ik}(t+1) = \phi_{ik}(t)$$
$$+ \alpha\mu \sum_{j=1(j \neq i)}^{N} \left(\rho_{ij}(\theta(\phi_i, \phi_j) - \theta(\psi_i, \psi_j)) sin\frac{\phi_{ik} - \phi_{jk}}{2} cos\frac{\phi_{ik} - \phi_{jk}}{2} \right). \tag{22}$$

2.4 Nonlinear Mapping Based on the Approximation of Spectral Angles by Euclidean Distances

Yet another approach to reduce dimensionality, using spectral angle measure is based on the approximation of SAM values by Euclidean distances (SAED) between points in the low-dimensional space. The fact that the distance between two points on the unit hypersphere is equal to the angle between corresponding vectors allows us to hope that Euclidean distances can be good approximations of spectral angles. This is also conforming to the multidimensional scaling approach. According to this approach, we just treat SAM values as some dissimilarity measures taking values from the range [0; Pi]. In this case, Eq. (1) takes the form

$$\varepsilon_{SAM \to ED} = \mu \cdot \sum_{i,j=1(i<j)}^{N} \left(\rho_{ij}(\theta(x_i, x_j) - d(y_i, y_j))^2 \right) \tag{23}$$

and corresponding optimization technique can be directly derived from Eq. (4):

$$y_i(t+1) = y_i(t) + 2\alpha\mu \sum_{j=1(i \neq j)}^{N} \rho_{i,j} \cdot \frac{\theta(x_i, x_j) - d(y_i, y_j)}{d(y_i, y_j)} \cdot (y_i(t) - y_j(t)). \tag{24}$$

2.5 Numerical Optimization Based on the Stochastic Gradient Descent

Unfortunately, the described above basic gradient descent based methods cannot be directly applied to hyperspectral satellite images due to the high computational complexity and memory limitations of the base method [12]. All these methods require to execute $O(N^2)$ operations per one iteration of the optimization process, and to store the matrix of precomputed distances in the multidimensional space that takes $N(N-1)/2$ floating point values. For this reason, in this paper we adopted stochastic gradient descent algorithm based on minibatches. For this algorithm, the value of the gradient $\nabla \varepsilon$ in Eq. (3) is estimated using a random subsample:

$$\tilde{\nabla}\varepsilon = \sum_{j=1}^{R} \nabla \varepsilon_{r_j}, \tag{25}$$

where r is a random subsample (mini-batch) used to approximate the gradient at the iteration t of the optimization process, r_j is the j-th element of this subsample, R is the cardinality of subset r. Using this approach, the cardinality of mini-batch determines the computational complexity of the algorithm per one iteration. Thus for subsets of size R, the computational complexity is reduced to $O(RN)$.

3   Experimental Results

In this section, we present results of an experimental study of the proposed dimensionality reduction methods, namely spectral angle preserving mappings (SAPM-C and SAPM-H), described in the Sects. 2.2 and 2.3, and nonlinear mapping based on the approximation of spectral angles by Euclidean distances (SAED), described in the Sect. 2.4. The experiments were conducted on well-known hyperspectral images [14]. Here we present results for the Indian Pines Test Site 3 hyperspectral image (see Fig. 1).

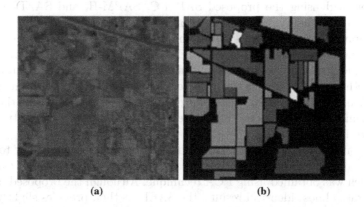

Fig. 1. Indian Pines Test Site 3 hyperspectral image: (a) false color representation of the image produced using nonlinear mapping technique; (b) ground truth image (classified pixels are shown with colors) (Color figure online).

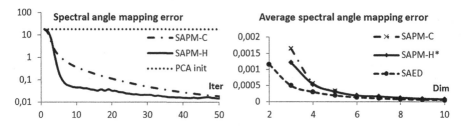

Fig. 2. The dependency of the spectral angle mapping error on the number of iterations (*Iter*) for the SAPM methods, and the error value obtained after initialization using PCA technique (left). The dependency of the average spectral angle mapping error on the dimensionality (*Dim*) of the reduced space for the proposed methods (right).

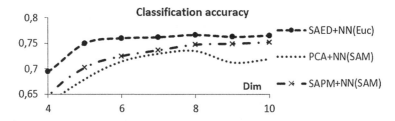

Fig. 3. The dependency of the classification accuracy on the dimensionality of the reduced space for the proposed nonlinear mapping methods and the PCA technique.

This image was acquired by the AVIRIS sensor in North-western Indiana. It contains 145×145 pixels containing 224 spectral reflectance bands.

The purpose of the first experiment was to estimate the quality of the mappings obtained using the proposed SAPM-C, SAPM-H, and SAED methods. To estimate the quality, we used the appropriate quality indicators: the spectral angle mapping errors (6) and (15) for the SAPM-C and SAPM-H, respectively, and the mapping error (23) for the SAED method. During the experiment, we changed the dimensionality of an output space from 2 to 10, and ran the proposed methods, starting with the initial configuration, which was obtained using the PCA method. The results of the first experiment are shown in the Fig. 2. Some examples of the spectral angle preserving mappings for 3D output space are shown in the Fig. 4.

As it can be seen from the results, all the proposed methods allow to reduce the spectral angle mapping error significantly compared to the initial configuration, which was obtained using PCA technique. Although the proposed methods demonstrate almost identical results, the SAED method provides slightly better error values. As it was outline in Sect. 2.2, a mapping into output L-dimensional space can be considered as a mapping onto $(L-1)$-dimensional manifold for both SAPM methods (SAPM-C and SAPM-H), so corresponding curves are shifted by 1 along the *Dim* axis compared to the curve for SAED method.

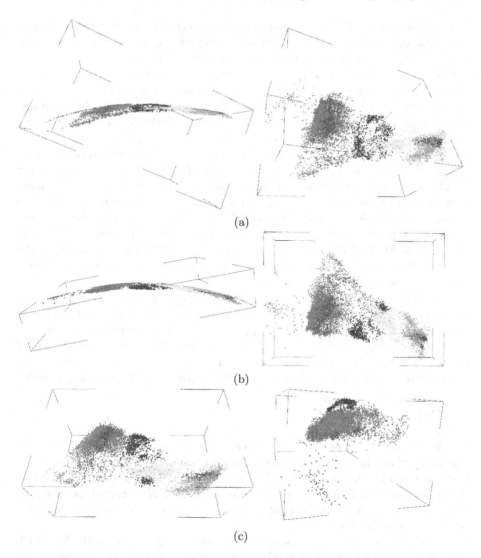

(a)

(b)

(c)

Fig. 4. Examples of spectral angle preserving mappings for 3D output space: SAPM-C (a), SAPM-H (b), SAED (c). Classified pixels of the groundtruth image are shown with corresponding colors (Color figure online).

It should be noted that for all the proposed methods, the corresponding errors decrease with the growth of the dimensionality. This observation is logical, as it is easier to preserve pairwise dissimilarities in higher dimensional spaces.

The purpose of the second experiment was to compare the proposed techniques and the linear PCA considering classification quality. To objectively estimate the classification quality, we used the ground truth classification of the test image shown in the Fig. 1b. To perform the experiment, we divided the whole set of the groundtruth samples into a training subset and a test subset, consisting of 60 and 40 percent of samples respectively. The dimensionality of the reduced spaces ranged from 2 to 10, as it was in the first experiment. To assess the classification quality, we used the classification accuracy, which is defined as a proportion of correctly classified pixels of a test subset. We used the nearest neighbor (NN-) classifier using SAM measures to perform a classification for SAPM and PCA based features, and the NN-classifier using Euclidean distances for SAED based features. The Fig. 3 shows the results of the experiment. We do not provide results for the dimensionalities from 2 to 4 here, as the classification quality was unsatisfactory for all the considered methods in such low dimensions. The Fig. 3 shows that both proposed methods provide better classification accuracy than PCA technique. Here SAED method provided best results again. It led to 4.6 percent higher classification accuracy on average for $Dim = 5..10$ compared to PCA+SAM technique (3.1 percent for PCA+Euc, not shown in the figure) in the considered experiment. Overall, these results are comparable with those obtained using nonlinear mapping technique [12].

The described above experiments allow us to make a conclusion that the proposed nonlinear methods can be successfully applied to hyperspectral image analysis when the spectral angle mapper measure (5) is used to measure the dissimilarity between pixels. The fact that SAED method outperformed SAPM methods allows us to make the assumption that SAM measures can be easily approximated by Euclidean distances for hyperspectral remote sensing scenes.

4   Conclusion

In this paper, we proposed three nonlinear mapping methods based on the spectral angle preserving principle, namely the spectral angle preserving mapping in Cartesian and hyperspherical coordinates, and the nonlinear mapping based on the approximation of spectral angles by Euclidean distances. For all the proposed methods we introduced spectral angle mapping errors and derived corresponding numerical optimization techniques.

We conducted the experimental study of the proposed methods, which showed that the proposed methods significantly decrease corresponding spectral angle mapping errors. Besides that, the proposed methods provided better classification accuracy than the PCA technique when the nearest neighbor classifier with SAM measure was used. In addition the experiments showed that SAM measures can be approximated by Euclidean distances for hyperspectral remote sensing scenes. Thus, we suppose that the proposed methods can be successfully applied to hyperspectral data analysis and visualization.

In the near future, we plan to apply the similar approach to study other dissimilarity measures used in hyperspectral image analysis.

Acknowledgments. The reported study was funded by RFBR according to the research project no. $15 - 07 - 01164-a$, $16 - 37 - 00202-mol\_a$.

References

1. Kim, D.H., Finkel, L.H.: Hyperspectral image processing using locally linear embedding. In: First International IEEE EMBS Conference on Neural Engineering, pp. 316–319 (2003)
2. Shen-En, Q., Guangyi, C.: A new nonlinear dimensionality reduction method with application to hyperspectral image analysis. In: IEEE International Geoscience and Remote Sensing Symposium, pp. 270–273 (2007)
3. Bachmann, C.M., Ainsworth, T.L., Fusina, R.A.: Exploiting manifold geometry in hyperspectral imagery. IEEE Trans. Geosci. Remote Sens. **43**(3), 441–454 (2005)
4. Journaux, L., Foucherot, I., Gouton, P.: Nonlinear reduction of multispectral images by curvilinear component analysis: application and optimization. In: International Conference on CSIMTA 2004 (2004)
5. Lennon, M., Mercier, G., Mouchot, M., Hubert-Moy, L.: Curvilinear component analysis for nonlinear dimensionality reduction of hyperspectral images. Proc. SPIE **4541**, 157–168 (2002)
6. Myasnikov, E.: Fast techniques for nonlinear mapping of hyperspectral data. Proc. SPIE **10341**, 103411D (2017)
7. Kruse, F.A., Boardman, J.W., Lefkoff, A.B., Heidebrecht, K.B., Shapiro, A.T., Barloon, P.J., Goetz, A.F.H.: The Spectral Image Processing System (SIPS) inter-active visualization and analysis of imaging spectrometer data. Remote Sens. Environ. **44**, 145–163 (1993)
8. Yan, L., Niu, X.: Spectral-angle-based Laplacian Eigenmaps for nonlinear dimensionality reduction of hyperspectral imagery. Photogramm. Eng. Remote Sens. **80**(9), 849–861 (2014)
9. Yan, L., Roy, D.P.: Improved time series land cover classification by missing-observation-adaptive nonlinear dimensionality reduction. Remote Sens. Environ. **158**, 478–491 (2015)
10. Kruskal, J.B.: Multidimensional scaling by optimizing goodness of fit to a non-metric hypothesis. Psychometrika **29**, 1–27 (1964)
11. Sammon, J.W.: A nonlinear mapping for data structure analysis. IEEE Trans. Comput. **18**(5), 401–409 (1969)
12. Myasnikov, E.: Evaluation of stochastic gradient descent methods for nonlinear mapping of hyperspectral data. In: Campilho, A., Karray, F. (eds.) ICIAR 2016. LNCS, vol. 9730, pp. 276–283. Springer, Cham (2016). doi:10.1007/978-3-319-41501-7_31
13. Blumenson, L.E.: A derivation of n-dimensional spherical coordinates. Am. Math. Mon. **67**(1), 63–66 (1960)
14. Hyperspectral Remote Sensing Scenes. http://www.ehu.eus/ccwintco/index.php?title=Hyperspectral_Remote_Sensing_Scenes

Object Triggered Egocentric Video Summarization

Samriddhi Jain$^{(\boxtimes)}$, Renu M. Rameshan, and Aditya Nigam

Indian Institute of Technology Mandi, Mandi, Himachal Pradesh, India
samriddhi_jain@students.iitmandi.ac.in, {renumr,aditya}@iitmandi.ac.in

Abstract. Egocentric videos are usually of long duration and contains lot of redundancy which makes summarization an essential task for such videos. In this work we are targeting object triggered egocentric video summarization which aims at extracting all the occurrences of an object in a given video, in near real time. We propose a modular pipeline which first aims at limiting the redundant information and then uses a Convolutional Neural Network and LSTM based approach for object detection. Following this we represent the video as a dictionary which captures the semantic information in the video. Matching a query object reduces to doing an And-Or Tree traversal followed by deepmatching algorithm for fine grained matching. The frames containing the object, which would have been missed at the pruning stage are retrieved by running a tracker on the frames selected by the pipeline mentioned. The modular pipeline allows replacing any module with its more efficient version. Performance tests ran on the overall pipeline for egocentric datasets, EDUB dataset and personal recorded videos, give an average recall of 0.76.

Keywords: Object based video summarization · Image retrieval · Object recognition · Egocentric vision

1 Introduction

There is a growing demand and access to new wearable cameras, leading to major breakthroughs in technologies related to action cameras like GoPro [1], Google Glasses. These egocentric cameras are widely used in daily life logging, capturing holiday memories and much more. They have also found applications in larger public activities like crowd surveillance [3], suspect identification and other security issues. This media generates an interesting first-person view and hence yields lots of essential information. It can help in understanding the social interactions, the behaviour and attention dynamics of the wearer. In [9], first person view is used to identify important objects around the wearer. Similar work has been done in [4], where gaze is used to identify the speaker and his/her characteristics. Researchers have also tried to classify parts of videos into some important activities, but so far they are limited only to a few activities. Also similar work has been done for storyline creation [23], in which the authors have proposed

© Springer International Publishing AG 2017
M. Felsberg et al. (Eds.): CAIP 2017, Part II, LNCS 10425, pp. 428–439, 2017.
DOI: 10.1007/978-3-319-64698-5_36

their approach over small videos and by using probabilistic models again considering only a limited number of events. Video analysis related previous work has been done in [24], where authors focussed on activity and gaze based video summarization. Another kind of summarization is object based summarization, in which the objective is to retrieve all the instances of a particular query object. Researchers have been working on object detection [6,16] and localisation in images since last decade. There are models based on CNN-LSTM [16,22] architectures which have given good performance on standard datasets and in competitions like ImageNet Challenges [19], but none of them have been ported to be used on egocentric videos. Work in [21], aims at efficient object retrieval from videos, but their approach is feature based and not by detecting the semantics of objects. Further in [12], egocentric image retrieval is targeted with the help of a features extracted from a Convolutional Neural Network, but the dataset used is not complete for comparison.

1.1 Motivation and Problem Statement

In this research work, we are presenting our work on object triggered video summarization, where all the occurrences of a query object are extracted automatically. This kind of work finds various applications in day-to-day life. Recently the research community is getting intrigued by the applications of object detection research in real life scenarios. It has become of importance lately by virtue of its applications, as stated above. For example, suppose a woman on a vacation has lost her purse and is not able to recall where she kept it the last. But she has life-logged her activities for the last 2–3 days and would like to go through the video thinking that might help. Now going through the huge volumes of this kind of data can be really cumbersome, tiring and time taking. However an automated platform which can just do all this quickly for her is the solution she needs. Similarly, assuming after the vacation, she wants to show a friend all the jewellery she saw in the market. Again it is extremely tough for her to sit back and summarize all the instances where jewellery appeared. Similarly, looking through hours of security footages for some suspecting object or noticing when a particular car arrives have also become quite cumbersome tasks and needs to be automated. Motivated by these problems, in this work we present a robust approach for object triggered egocentric video summarization where one can match and extract all the instances whenever a query object is present in the video as shown in Fig. 1, and in real time.

The problem statement of our work closely resembles the one in [12], where a content based image retrieval system is built using a Convolutional Neural Network. However our approach varies to a great extent as we first target limiting the computation and then use computation intensive costly algorithms on the pruned data. The problem addressed is similar to [21], but extended to egocentric vision by additionally identifying the semantics.

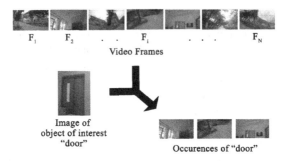

Fig. 1. Problem statement: Let $V = \{F_1, F_2...F_N\}$ be the video, where $F_i's$ are the frames and given the query image of a door; extract all the $F_i's$ containing a door. The user should be able to choose whether to extract any kind of door or specifically a particular door.

1.2   Challenges

While life logging, the egocentric videos raise the following challenges,

1. Large size of videos, a 3 h standard video shot at 30FPS will lead to 324,000 frames. Searching every frame is quite cumbersome and time consuming. As the recorded data increases, more computation and effort is needed to process it.
2. Most of the information in adjacent frames is redundant and irrelevant. Hence trivial methods of exhaustive object matching takes too much of time and repetitive computation.
3. The target objects in the video can have different orientation and position and the search algorithm should be able to handle these, in addition to scale, rotation and illumination variations.
4. While recording the egocentric video, constant head movement may repeatedly remove the target object from the frame, making it hard for object tracking.

In next section we propose an algorithm to handle these challenges upto some extent. We have created a pipeline which can resolve the issues mentioned in near real time. Section 2 details the proposed approach, Sect. 3 presents experimental results followed by conclusion in Sect. 4.

1.3   Contributions

In this paper, we propose an object centric model of retrieval from egocentric videos which ensures that costly operations are done only on limited number of frames, thereby ensuring near real time performance. The main contributions are,

1. Frame pruning by sampling only significant frames
2. Dictionary based two step enriched object matching approach which aims at extracting particular set of frames from the video

3. Usage of a simple tracking based method to extract frames possibly missed due to sampling
4. A modular pipeline where all stages are kept as independent modules and can be easily replaced with better algorithms in future.

2   Proposed Approach

Our objective is to summarize a video by extracting all the occurrences of a given query object, in a computationally efficient way. Let $V = \{F_1, F_2...F_N\}$ be the video, where $F_i's$ are the frames. Given an object template image $T(x, y)$ of size $P \times Q$, we need to generate a subset $U \subset V$ such that every element of U contains $T(x, y)$. Figure 1 gives a pictorial representation of the problem.

To make our queries robust we propose to learn the video structure beforehand and then run the matching techniques selectively during query time. By learning video structure we mean memorizing different objects which appeared in different frames. Learning the structure beforehand is essential because it is common for all the queries and memorizing the semantic information helps in directly querying in real time. In this section we propose our novel approach by which we first choose minimum size subset V_O from V to represent the whole video, and then use modifications of state of the art techniques for object matching and frame extraction. Figure 2 shows a summary of the full pipeline which has been explained in the following subsections. The first stage-frame pruning-aims at extracting the subset V_O, having minimum information loss. In next stage a dictionary structure is learnt over various objects present, which is then used in the final stage query extraction for object matching and retrieval.

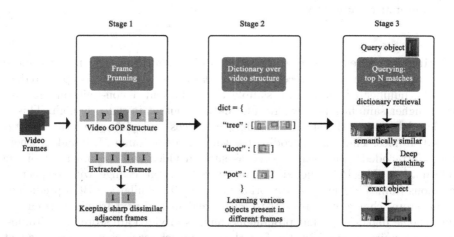

Fig. 2. Multi-stage pipeline: proposed multistage scheme consisting of three stages, (a) frame pruning, (b) dictionary generation and finally (c) querying the desired object. Each of these stages is described in subsequent sections.

2.1   Frame Pruning

As mentioned in the introduction, temporal redundancy is high in egocentric videos. For real time performance, we need a subset of V, which represents the video with minimal information loss. Figure 3 gives the steps in frame pruning, following which we could reduce temporal redundancy.

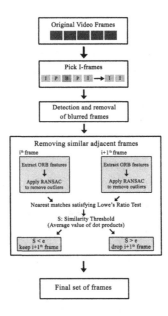

Fig. 3. Steps followed for frame pruning, the objective is to get minimum size subset of representative frames of V

Picking I-Frames. There exists video coding algorithms which can achieve large compression in the recorded video data size. As a first step in reducing the number of frames, we sample the frames in a non-uniform manner, with higher sampling rate where information content in video is high. This is achieved by using H.264 [20] encoding in videos and picking all the I(Intra-coded)-frames. The H.264 encoding has three kinds of frames, I, P and B, where the I(Intracoded) frames are stored as such in video and contain most of the information. The P(Predicted) frames store only the change with respect to previous frames in the form of motion vectors [20]. Similarly a B(Bi-predicted) frame stores change with respect to both previous and next frame. This encoding scheme makes sure that prominent changes in video are stored as I-frames. Non-uniformity comes from the fact that, wherever scene changes are frequent corresponding to more information, more I-frames are stored. Hence picking only the I-frames captures all such events and frames with redundant information are automatically dropped.

Removing Blurred Frames. A standard gradient based approach is used to remove the blurred frames as they interfere with algorithms for feature matching. We have used a Laplacian kernel M, given in Eq. (1) [14], and convolved it with the image I to get I' as in Eq. (2). Variance of I', denoted by $blurVar$ Eq. (3) is used to differentiate a blurred image from a sharp image. A high variance in convolved image indicates rich features as well as strong change in gradients, whereas a small variance implies a rather monotonic image. For example the images shown in Fig. 4a, the image is clearly blurred and gives $blurVar$ value as 13.42, where as the image in Fig. 4b which is sharp, gives the value 1024.36. Hence we keep a frame if its $blurVar$ is greater than some threshold e_{blur}, which is empirically evaluated to be between 100.0 to 150.0.

$$M = \begin{bmatrix} 0 & -1 & 0 \\ -1 & 4 & -1 \\ 0 & -1 & 0 \end{bmatrix} \tag{1}$$

$$I' = I \otimes M \tag{2}$$

$$blurVar = Var(I') \tag{3}$$

(a) Example of blurred frame, $blurVar = 13.42$

(b) Example of sharp frame, $blurVar = 1024.36$

Fig. 4. Handling blurred frames using Laplacian kernel M (1)

Further Reduction of Frames with ORB Feature Matching. After removing blurred frames, we are left with a subset of V, indicated by $V_I = \{F_{i_1}, F_{i_2}..., F_{i_n}\}$, where i_n is an index set containing the indices of sharp I-frames. While recording the egocentric videos a slight change in position of camera can result in huge change in object orientations in the video. Hence adjacent frames containing similar objects are also flagged as scene changes.

On the video representation as V_I, ORB [18] (Oriented FAST and rotated BRIEF) feature matching approach is used to further reduce the number of frames. Feature matching technique was selected since it can find matching objects in the presence of scale variation, translation and affine transformation. ORB features are fast, robust to extract and takes care of all previously mentioned challenges. All F_{i_k} frames satisfying a similarity criteria in Eq. (4) are picked and a new subset, $V_O = \{F'_1, F'_2..., F'_p\}$ is formed.

$$sim(F_{i_k}, F_{i_{k+1}}) < e \tag{4}$$

In order to evaluate the similarity score between two images, first ORB features are extracted from the two images and then RANSAC [5] algorithm is applied for removing the outliers with Hamming distance measure to find the nearest possible matches of features between the two images. The matching features are constrained to satisfy the Lowe's ratio test [10]. This test takes the ratio of distance from top two nearest neighbours and considers a good match when the ratio is only below 0.8. The motivation behind such a strategy is that a good match is more likely to lie near the query match rather than the second closest matching neighbour. Hence this test ensures that only good matches are selected while remaining are pruned. The average dot product of the matching features is taken as the similarity score, thresholding which helps in removing similar adjacent frames and further limiting the computation for the remaining pipeline. The threshold value, e has been empirically chosen as 0.3.

The above steps lead to a very condensed and dense video representation, which is then passed to the next stage of creating the dictionary structure.

2.2   Dictionary Structure

Over a condensed representation of any video, we have proposed to learn different objects present in every frame and create a dictionary/B+ tree that can be later used for querying individual objects. This method of forming dictionary can be done in parallel on different machines and can be combined later as shown in Fig. 5, which reduces the overall execution time. The combining can simply be done by appending the values for each key to form a bigger dictionary.

The major bottleneck is that we have to perform object detection and localisation in all the frames and for this we follow a state of the art object detection model densecap [8] and have taken top M prominent objects. From the obtained captions we have extracted the prominent named entities and weigh each with the caption score. A B+ tree based dictionary is learnt for each object frame combination as shown in Fig. 6. During test time only certain branch of the tree has to be traversed leading to much faster computation. We have also used

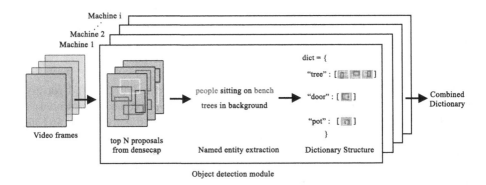

Fig. 5. Pipeline for creating the combined dictionary structure

word2vec [11] representation for grouping similar objects and reducing variations because of plurals. This gives our approach an NLP interface which can also be used in real time. It may be noted that in place of densecap, YOLO [15] can also be used as an object detection module.

```
dictionary = {   "object1" : {
                    {frame1, bounding_box_location1,
                     confidence_score1}
                    {frame2, bounding_box_location2,
                     confidence_score2}
                    ...}
                "object2" : {
                    {frame1, bounding_box_location1,
                     confidence_score1}
                    {frame2, bounding_box_location2,
                     confidence_score2}
                    ...}
                ....}
```

Fig. 6. Example schema of the dictionary structure

2.3 Query Matching

For each query, we formulate the frame extraction as an And-Or tree [13] as shown in Fig. 7. We identify the potential objects present in the query image and then extract the frames that semantically match at least one caption objects and finally union all in a set. Overall two kinds of matching are proposed,

1. **Semantic level matching**, where we extract all the objects semantically similar to query, that is, if the query object is a door, retrieve all the doors, wooden, glass or any kind. For this the semantic knowledge has already been captured in our dictionary and the And-Or tree and we can directly return the extracted set of frames.
2. **Exact object retrieval**, for extracting the same object as in the query image, we narrow down the search by considering only the resulting frames extracted from the And-Or tree. We calculate their similarity score with respect to the query image using the deepmatching algorithm [17] and select the ones above a certain threshold. Deepmatching algorithm is a two phase matching algorithm which takes two images and tries to match their corresponding patches (group of 4×4 pixels). In the first, bottom-up phase, it calculates correlation maps by convolving patches of two images and uses max-pooling and sub-sampling to aggregate features, hence generating a pyramid of correlation maps. Next the top down stage finds the local maximas at each correlation map and using those estimates the motion and position of atomic patches. Here we are using the number of matching features as our

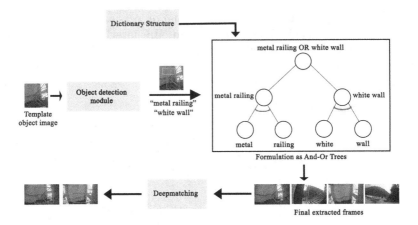

Fig. 7. Query time processing: Two stage image retrieval for a query image containing metal railing and white wall. Step 1 uses the dictionary structure to extract all the frames which contain either "metal railing" or "white wall", and union them in a set, S. This gives all the semantically similar images containing a metal railing or a white wall. Next a feature matching technique, deepmatching is used to extract frames from S which exactly match the query image.

similarity score. The correlation map pyramid in deepmatching algorithm helps in tracking non-rigid transformations and hence can efficiently detect the occluded and rotated objects. Other than these two, we also support textual query as discussed in previous section.

Tracking. The last two modules of the pipeline are applied on V_O, which is a pruned subset of original video V. Hence the set of extracted frames S, after the query may also be only a subset of actual desired output. It represents certain occurrences of the object, but to get all instances where the object appeared we are using a standard tracker STRUCK [7] on all elements of S in forward and backward directions.

3 Results

The first phase of our pipeline prunes the redundant information. The results for this are presented in Table 1, in which the sharp decrease in number of frames can be observed at the I-frames. Some videos used in this comparison are recorded by ourselves while others are egocentric videos taken from YouTube. After the ORB feature matching and removing similar frames we are left with only a small fraction of frames which provide a concise representation of the video and are passed on to the object detection module. Table 2 shows the performance measurements for different queries. These queries are run on the two videos, the first three queries are on a video which originally had 9849 frames and after the

Table 1. Reduction in number of frames at each step

	Video 1	Video 2	Video 3
Duration of video	19 m 29 s	5 m 28.63 s	10 m 3.25 s
Recorded at	30 fps	30 fps	30 fps
Original number of frames	35040	9849	18079
I-frames	2336	66	182
After removal of blurred scene	1536	60	126
Scene change detection (using ORB features)	360	35	42

pruning stage they were reduced to a set of 35 frames. The first query extracts all the frames where a person is present. The second one corresponds to the frames which have land and territories, which is a combination of queries for hills, mountains and fields. The third one is for frames which have occurrence of roads. All these queries are at a semantic level. In ground truth a particular frame is marked as positive if the query object is easy to recognise by a human eye. After extracting the set of frames, we have used the tracker STRUCK [7] to get the performance comparison over the original frames.

We also tested our full pipeline on EDUB 2015 dataset [2] (last two queries of Table 2). This dataset has images captured by a wearable camera for 8 different days, in total 4912 images covering 21 different classes. The first stage of frame pruning gave us 1940 images on which we ran our stage 2 for object detection.

Table 2. Performance measurements of various queries on a video. The first three queries are on video 2 which had 9849 original frames and 35 frames after pruning, the last two queries are on EDUB dataset [2] for the object bottle and bicycle.

	Video 2			EDUB dataset	
	Query1	Query2	Query3	Query4	Query5
Ground truth (number of frames containing object)					
Original frames	4259	7623	1166	202	10
Final frames after pruning	23	25	2	125	10
Performance measurements with respect to final set of frames					
Total frames returned by query module	22	21	4	134	21
True positives	20	21	2	52	9
Precision	0.909	1.0	0.5	0.389	0.428
Recall	0.869	0.84	1.0	0.416	0.9
Performance measurements with respect to original frames					
True positives	3715	6043	698	129	9
Recall	0.872	0.793	0.599	0.638	0.9

Although our results deviated with respect to final set of frames but after using tracker we were able to recover other frames too. We also observed that the ground truth of this dataset is coarse, limited and doesn't cover objects which our object detection module was able to capture correctly. Secondly the images are of low quality and blurred, which results in incorrect object detections. Hence the results are not upto the expectations and better results are expected upon improvisation in the ground truth and object detection models.

For extracting the exact object, we tested deepmatching algorithm on some object images. There are in total 245 frames and we got precision of 0.99 and recall of 0.98 for an object which is present in 180 frames. The last parts of pipelines shown in Figs. 7 and 2 give examples of semantic similar and exact objects extracted from video with respect to the query object of metal railing and door respectively. With the dictionary learnt beforehand, a single query takes around 10 s on CPU (Intel(R) Xeon(R) (E5-2650@2.00 GHz, 32 cores and 48 GB RAM)) and can be much faster on a GPU.

4 Conclusion

The objective of this research work was to summarize all the instances where an object appeared in the image in a computationally effective way. As shown in the results in Section 3, we are able to reduce the temporal redundancy to a large extent and even with the pruned set of frames we are able to retrieve almost all the frames containing the object. The object detection module results in some incorrect detections but our modular pipeline allows replacing it with better and more efficient models.

References

1. goPro Camera. https://gopro.com/
2. Bolaños, M., Radeva, P.: Ego-object discovery. CoRR abs/1504.01639 (2015). http://arxiv.org/abs/1504.01639
3. Corso, J.J., Alahi, A., Grauman, K., Hager, G.D., Morency, L.P., Sawhney, H., Sheikh, Y.: Video analysis for body-worn cameras in law enforcement. arXiv preprint (2016). arXiv:1604.03130
4. Fathi, A., Li, Y., Rehg, J.M.: Learning to recognize daily actions using gaze. In: Fitzgibbon, A., Lazebnik, S., Perona, P., Sato, Y., Schmid, C. (eds.) ECCV 2012. LNCS, vol. 7572, pp. 314–327. Springer, Heidelberg (2012). doi:10.1007/978-3-642-33718-5_23
5. Fischler, M.A., Bolles, R.C.: Random sample consensus: a paradigm for model fitting with applications to image analysis and automated cartography. Commun. ACM **24**(6), 381–395 (1981). http://doi.acm.org/10.1145/358669.358692
6. Girshick, R.B.: Fast R-CNN. CoRR abs/1504.08083 (2015). http://arxiv.org/abs/1504.08083

7. Hare, S., Golodetz, S., Saffari, A., Vineet, V., Cheng, M.M., Hicks, S.L., Torr, P.H.: Struck: structured output tracking with kernels. IEEE Trans. Pattern Anal. Mach. Intell. **38**(10), 2096–2109 (2016)

8. Johnson, J., Karpathy, A., Fei-Fei, L.: Densecap: fully convolutional localization networks for dense captioning. In: Proceedings of the IEEE Conference on Computer Vision and Pattern Recognition (2016)

9. Lee, Y.J., Ghosh, J., Grauman, K.: Discovering important people and objects for egocentric video summarization. In: CVPR, vol. 2, p. 7 (2012)

10. Lowe, D.G.: Distinctive image features from scale-invariant keypoints. Int. J. Comput. Vision **60**(2), 91–110 (2004)

11. Mikolov, T., Chen, K., Corrado, G., Dean, J.: Efficient estimation of word representations in vector space. arXiv preprint (2013). arXiv:1301.3781

12. Nebot, A., Binefa, X., de Mántaras, R.L.: Artificial intelligence research and development. In: Proceedings of the 19th International Conference of the Catalan Association for Artificial Intelligence, vol. 288, Barcelona, Catalonia, Spain. IOS Press, 19–21 October 2016

13. Nilsson, N.: Artificial Intelligence: A New Synthesis. Morgan Kaufmann Series in Arti. Morgan Kaufmann Publishers, Burlington (1998). https://books.google.co.in/books?id=LIXBRwkibdEC

14. Pech-Pacheco, J.L., Cristóbal, G., Chamorro-Martinez, J., Fernández-Valdivia, J.: Diatom autofocusing in brightfield microscopy: a comparative study. In: 15th International Conference on Pattern Recognition Proceedings, vol. 3, pp. 314–317. IEEE (2000)

15. Redmon, J., Farhadi, A.: YOLO9000: better, faster, stronger. CoRR abs/1612.08242 (2016). http://arxiv.org/abs/1612.08242

16. Ren, S., He, K., Girshick, R.B., Sun, J.: Faster R-CNN: towards real-time object detection with region proposal networks. CoRR abs/1506.01497 (2015). http://arxiv.org/abs/1506.01497

17. Revaud, J., Weinzaepfel, P., Harchaoui, Z., Schmid, C.: Deepmatching: hierarchical deformable dense matching. Int. J. Comput. Vision **120**, 1–24 (2015)

18. Rublee, E., Rabaud, V., Konolige, K., Bradski, G.: ORB: an efficient alternative to sift or surf. In: 2011 International Conference on Computer Vision, pp. 2564–2571. IEEE (2011)

19. Russakovsky, O., Deng, J., Su, H., Krause, J., Satheesh, S., Ma, S., Huang, Z., Karpathy, A., Khosla, A., Bernstein, M., et al.: Imagenet large scale visual recognition challenge. Int. J. Comput. Vision **115**(3), 211–252 (2015)

20. Schwarz, H., Marpe, D., Wiegand, T.: Overview of the scalable video coding extension of the h. 264/avc standard. IEEE Trans. Circuits Syst. Video Technol. **17**(9), 1103–1120 (2007)

21. Sivic, J., Schaffalitzky, F., Zisserman, A.: Efficient object retrieval from videos. In: 2004 12th European Signal Processing Conference, pp. 1737–1740, September 2004

22. Vinyals, O., Toshev, A., Bengio, S., Erhan, D.: Show and tell: lessons learned from the 2015 MSCOCO image captioning challenge. IEEE Trans. Pattern Anal. Mach. Intell. (2016)

23. Xiong, B., Kim, G., Sigal, L.: Storyline representation of egocentric videos with an applications to story-based search. In: Proceedings of the 2015 IEEE International Conference on Computer Vision (ICCV), ICCV 2015, pp. 4525–4533 (2015). doi:10.1109/ICCV.2015.514

24. Xu, J., Mukherjee, L., Li, Y., Warner, J., Rehg, J.M., Singh, V.: Gaze-enabled egocentric video summarization via constrained submodular maximization. In: Proceedings CVPR (2015)

3D Motion Consistency Analysis for Segmentation in 2D Video Projection

Wei Zhao$^{(\boxtimes)}$, Nico Roos, and Ralf Peeters

Department of Data Science and Knowledge Engineering, Maastricht University,
6200 MD Maastricht, The Netherlands
wei.zhao@maastrichtuniversity.nl

Abstract. Motion segmentation for 2D videos is usually based on tracked 2D point motions, obtained for a sequence of frames. However, the 3D real world motion consistency is easily lost in the process, due to projection from 3D space to the 2D image plane. Several approaches have been proposed in the literature to recover 3D motion consistency from 2D point motions. To further improve on this, we here propose a new criterion and associated technique, which can be used to determine whether a group of points show 2D motions consistent with joint 3D motion. It is also applicable for estimating the 3D motion information content. We demonstrate that the proposed criterion can be applied to improve segmentation results in two ways: finding the misclassified points in a group, and assigning unclassified points to the correct group. Experiments with synthetic data and different noise levels, and with real data taken from a benchmark, give insight in the performance of the algorithm under various conditions.

1 Introduction

Motion provides an important clue for the analysis of video sequences. It can be used for either detecting and segmenting the moving objects present in the scene, or recovering the 3D structure of a scene [2,4,5,18].

When an image is taken by a camera, it maps the 3D world onto a 2D image plane by a projective transformation. The motions of the scene objects as well as the camera, jointly cause the changes of corresponding pixels in the image. We can detect these 2D motions by estimating the displacements of pixels between frames, or by tracking salient feature points from video sequences [4,11,13,20]. Motion segmentation aims at grouping together the points (or pixels) that have the same motion in the video sequence. The key issue of motion segmentation is the definition of "same motion", which can be a 3D motion in the three-dimensional world, or simply a 2D motion of image pixels [24]. Motion segmentation is difficult because the detected motions of points (or pixels) are combined with displacements caused by camera motion and parallax caused by 3D structures [23].

Many motion segmentation approaches group together pixels undergoing the same 2D motion between successive images in the sequence [1,2,15,19,21,26].

© Springer International Publishing AG 2017
M. Felsberg et al. (Eds.): CAIP 2017, Part II, LNCS 10425, pp. 440–452, 2017.
DOI: 10.1007/978-3-319-64698-5_37

However, 3D geometric properties are typically affected by the transformation, such as shapes, angles and distances [9]. 3D motion consistency of points in the world coordinate frame is therefore not assured to be preserved in the 2D image coordinate frame. It implies that different parts of one and the same object, e.g. the three visible sides of a cube, can show different 2D motion patterns in the image plane. As a result, 2D motion based methods will fail to properly segment out the object. Another class of motion segmentation methods, tries to capture the 3D motion consistency with the help of constraints derived from geometric or physical models, such as rigidity of an object, 2D homography, an epipolar constraint, or a trilinear constraint [5, 7, 10, 24]. These constraints, with some success, allow to group points (or pixels) moving with the same 3D motion, based on their 2D motion information at the projected image.

In this paper, we propose a new criterion for measuring the 3D rigid motion consistency of a group of points, based on their 2D motions. This criterion can be used with singular value decomposition (SVD), to measure the 'quality' of segmented groups of points in a way that will be made more precise later. It is also applicable for recovering the parameters of 3D rigid motion giving the 2D motion information of a collection of points from the same object. This is used to detect misclassified points and to assign unclassified points to the correct group.

2   Related Work

A key problem of motion segmentation is to determine whether a set of points all have the same motion. Normally the motion of points (or pixels) is estimated by detecting their 2D positions at each frame from a video sequence. The 2D motion of each point (pixel) in image plane is a projection of a 3D motion in the scene. The "same motion" can be defined based on either their 2D motion consistency, or on their 3D motion consistency.

For segmentation of 2D motions in the image space, one straightforward way is to define the "same motion" with a parametric motion model. Parametric approaches use a 2D affine transformation to describe joint 2D motion [2, 21, 27]. The affine motion model neglects perspectivity effects, and is largely limited to approximate the rigid motion of planar surfaces far away from the camera. Non-parametric models, such as Gaussian processes, are more flexible and suitable for curved surfaces [22]. However, these methods usually segment 3D objects into multiple parts, because of discontinuities in projected 2D motions, caused by perspective effects, depth discontinuities, occlusions, etc. [24].

3D motion segmentation searches for multiple-view geometric constraints to measure the 3D motion consistency. The two-view-based approaches model the motion by a fundamental matrix, based on the epipolar geometry [12, 19]. Other approaches are based on the three-view geometry, and encapsulate the trilinear relations of corresponding points in three images [16, 25]. These methods handle 3D motion consistency by preserving the 3D relations of points. Motion segmentation based on such geometric constraints tends to suffer from a "chicken and egg" problem, as it requires prior knowledge of the number of objects [5].

A variety of solutions are proposed to avoid estimating the motion model explicitly, thus solving this "chicken and egg" problem. The factorization method introduced by Tomasi and Kanade [14], factorizes the trajectory matrix of points tracked in a video sequence into a motion matrix and a shape matrix. The rigidity of objects ensures the uniqueness of the shape matrix, and the feature trajectories belonging to an object are linearly dependent. Then the "same motion" is defined as belonging to a low-dimensional subspace, and trajectories lying in the same subspace are regarded as belonging to the same object. Many developments of the factorization methods are made by the following researchers [3,5,7,14,24]. The factorization methods can group together points moving with a consistent "behavior" over a long period of time, because they use the full temporal trajectory of every tracked point [24]. However, current factorization methods often fail to segment motions between only two frames [8].

In this paper, we investigate an efficient way of measuring 3D motion consistency using the 2D image motion between just two frames. The proposed criterion can be used for measuring the quality of segmented groups. Moreover, it is able to estimate the 3D structure in an efficient way.

3   Background

3.1   3D Rigid Body Motion

When a rigid object is moving in 3D space, all the points on the object will follow the same motion model. It can be modeled as a combination of 3D rotation and translation. Suppose a point moves from position $p = [X, Y, Z]^\mathsf{T}$ to $p' = [X', Y', Z']^\mathsf{T}$, then

$$p' = Rp + t \tag{1}$$

where R is a 3D rotation matrix and $t = [t_1, t_2, t_3]^\mathsf{T}$ is a translation vector carrying the displacements in the directions of the three axes. The matrix R can be parameterized as: $R = R_z(\varphi_z)R_y(\varphi_y)R_x(\varphi_x)$. where $\varphi_z \in (-\pi, \pi]$, $\varphi_y \in [-\frac{\pi}{2}, \frac{\pi}{2}]$, and $\varphi_x \in [-\frac{\pi}{2}, \frac{\pi}{2}]$ are yaw, pitch, and roll angles respectively. For convenience and conciseness we shall write $s_x = \sin\varphi_x, s_y = \sin\varphi_y, s_z = \sin\varphi_z$ and $c_x = \cos\varphi_x, c_y = \cos\varphi_y, c_z = \cos\varphi_z$. So:

$$R = \begin{bmatrix} c_z & -s_z & 0 \\ s_z & c_z & 0 \\ 0 & 0 & 1 \end{bmatrix} \begin{bmatrix} c_y & 0 & s_y \\ 0 & 1 & 0 \\ -s_y & 0 & c_y \end{bmatrix} \begin{bmatrix} 1 & 0 & 0 \\ 0 & c_x & -s_x \\ 0 & s_x & c_x \end{bmatrix} \tag{2}$$

We shall write r_{ij} to denote the entry in row i and column j of R.

3.2   Projections

The physical camera projects the 3D world onto a 2D image plane through some projection mechanism. Accurate knowledge of this projection mechanism may in principle be used to provide 3D information for understanding the images. However, in practice the lens system in a real camera is too complex to perform 3D reconstruction for. Instead, approximate camera models are developed for different applications, starting from corresponding simplifying assumptions.

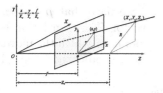

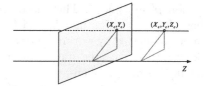

Fig. 1. Perspective camera model   **Fig. 2.** Orthographic camera model

General Perspective Projection. General perspective projection is an ide-
alized mathematical model for a real camera, which is widely used in computer
vision applications. It assumes that the camera is sufficiently small compared to
the viewed scenes and objects.

Figure 1 shows the simplest central-projection camera: the pinhole camera
model. The XYZ coordinate frame is centered at the camera, with the Z-axis
being the principal axis of the camera. The projected image plane coincides with
the focus plane, and employs the xy coordinate frame. The origin of this image
frame, o, is the projection of the camera center O on the image plane; their
distance is indicated by f. A point $\boldsymbol{X_c} = [X_c, Y_c, Z_c]^\mathsf{T}$ in the camera frame, is
mapped to the point $\boldsymbol{x} = [x, y]^\mathsf{T}$ in the image frame by

$$\boldsymbol{x} = \frac{f}{Z_c} P \boldsymbol{X_c} \qquad \text{where } P = \begin{bmatrix} 1 & 0 & 0 \\ 0 & 1 & 0 \end{bmatrix} \tag{3}$$

Orthographic Projection Model. If the camera is sufficiently far away from
the viewed scene, one may assume an infinite focal length. Then the points in the
camera frame are mapped to the image frame by parallel projection, as illustrated
in Fig. 2. For this orthographic projection model, the coordinate mapping takes
the form:

$$\boldsymbol{x} = P \boldsymbol{X_c} \tag{4}$$

4   3D Motion Consistency

To analyze 3D motion consistency, we address the situation in which we have a
given set of *matched pairs of feature points* from two image frames. We aim to find
a 3D rigid body motion consistent with all those matched pairs. Combining such
a 3D motion with the camera projection mechanism, we can set up an equation
relating the coordinates for each matched pair. With sufficiently many points
from the same object, an overdetermined system of equations will be obtained.
Due to the rigid body motion assumption, this system will have certain structural
properties. By using matrix factorization techniques we then can analyze how
to recover a 3D rigid body motion in the best possible way.

4.1   Theorems

Consider a point on an object undergoing a rigid body motion. Suppose it moves,
in the camera coordinate system, from some position $\boldsymbol{p}$ at time t to another

position p' at time t'. Then according to Eq. (1) we have that $p' = Rp + t$, where R is a rotation matrix and t a translation vector. The translation vector can be eliminated by working relative to a selected point for which the movement is known (e.g., a center of mass or any other point on the object): if it also holds that $p'_0 = Rp_0 + t$, then

$$p' - p'_0 = R(p - p_0). \tag{5}$$

If the scene is far away from the camera, and focal length is small compared to the distance of the object to the camera, then for every two points p and p' at distances Z_c and Z'_c, we can assume $\frac{f}{Z_c} \approx \frac{f}{Z'_c}$. Hence, we can ignore the effect of the scaling factor $\frac{f}{Z_c}$ and the camera projection can be approximated by an orthographic projection. Thus the point p at position $[x, y, z]^\mathsf{T}$ in the camera frame is mapped (up to a fixed factor) to position $[x, y]^\mathsf{T}$ in the image frame. The following theorems apply, subject to this orthographic projection assumption. The general situation is discussed in Sect. 4.3.

Theorem 1. *A set of $m + 1$ matched pairs of 2D points $(x_i, y_i)^\mathsf{T}$ and $(x'_i, y'_i)^\mathsf{T}$ (with $i = 0, \ldots, m$) can consistently be interpreted as the 2D coordinates of orthographic projections onto the image plane of $m + 1$ pairs of 3D points in camera space which are related by a single 3D rigid body motion, if and only if the $m \times 4$ data matrix*

$$M = \begin{bmatrix} \tilde{x}_1 & \tilde{y}_1 & \tilde{x}'_1 & \tilde{y}'_1 \\ \vdots & \vdots & \vdots & \vdots \\ \tilde{x}_m & \tilde{y}_m & \tilde{x}'_m & \tilde{y}'_m \end{bmatrix} \tag{6}$$

where $\tilde{x}_i = x_i - x_0$, $\tilde{y}_i = y_i - y_0$, $\tilde{x}'_i = x'_i - x'_0$ and $\tilde{y}'_i = y_i - y_0$, has a nontrivial null space containing a vector v of which the four entries satisfy

$$v_1^2 + v_2^2 = v_3^2 + v_4^2. \tag{7}$$

Typically, we will be interested in situations with sufficiently many matched pairs of data points (i.e., $m \geq 4$), for which non-rigid motions would otherwise produce the trivial null space $\{0\}$. For rigid body motion, the non-trivial null space of M will normally be of dimension 1, unless the rigid body motion is of a special type. The following theorem addresses the nature of the family of rigid body motions consistent with such data.

Theorem 2. *If the condition under Theorem 1 is satisfied, then there exists a family of rigid body motions, consistent with the data, having at least one real degree of freedom for the translation (corresponding to an arbitrary translation in the z-direction) and at least one real degree of freedom for the 3D rotation.*

Proof. From 2 and 5, we have

$$\begin{bmatrix} \tilde{x}'_1 \ldots \tilde{x}'_m \\ \tilde{y}'_1 \ldots \tilde{y}'_m \end{bmatrix} = \begin{bmatrix} c_z & -s_z \\ s_z & c_z \end{bmatrix} \begin{bmatrix} c_y & s_y s_x & s_y c_x \\ 0 & c_x & -s_x \end{bmatrix} \begin{bmatrix} \tilde{x}_1 \ldots \tilde{x}_m \\ \tilde{y}_1 \ldots \tilde{y}_m \\ \tilde{z}_1 \ldots \tilde{z}_m \end{bmatrix} \tag{8}$$

Every row $\tilde{x}_i'$ is a linear combination of $\tilde{x}_i$, implying that M has a rank of at most 3. We can rewrite the equation in terms of M:

$$M \begin{bmatrix} c_y & 0 \\ s_y s_x & c_x \\ -c_z & s_z \\ -s_z & -c_z \end{bmatrix} = \begin{bmatrix} \tilde{z}_1 \\ \vdots \\ \tilde{z}_m \end{bmatrix} \begin{bmatrix} -s_y c_x & s_x \end{bmatrix} \tag{9}$$

Multiplying the result by $\begin{bmatrix} c_x & s_x \\ -s_y s_x & s_y c_x \end{bmatrix}$:

$$M \begin{bmatrix} c_y c_x & c_y s_x \\ 0 & s_y \\ -c_z c_x - s_z s_y s_x & -c_z s_x + s_z s_y c_x \\ -s_z c_x + c_z s_y s_x & -s_z s_x - c_z s_y c_x \end{bmatrix} = \begin{bmatrix} \tilde{z}_1 \\ \vdots \\ \tilde{z}_m \end{bmatrix} \begin{bmatrix} -s_y & 0 \end{bmatrix} \tag{10}$$

If $s_y = 0$, we have a nontrivial $v = [1, 0, -c_z, -s_z]^{\mathsf{T}}$ such that $Mv = 0$; and if $s_y \neq 0$, we have a nontrivial vector $v = [c_y s_x, s_y, -c_z s_x + s_z s_y c_x, -s_z s_x - c_z s_y c_x]^{\mathsf{T}}$ such that $Mv = 0$. In both cases v satisfies $v_1^2 + v_2^2 = v_3^2 + v_4^2$. This proves one implication of Theorem 1.

Conversely, if a non-zero vector $v = [v_1, v_2, v_3, v_4]^{\mathsf{T}}$ is given in the kernel of M which happens to satisfy $v_1^2 + v_2^2 = v_3^2 + v_4^2$, we can proceed by the following two cases with respect to the value of v_2:

Case 1. Assume that $v_2 \neq 0$. Then let φ_y have an arbitrary nonzero value in the interval $\left(-\arctan\left|\frac{v_2}{v_1}\right|, \arctan\left|\frac{v_2}{v_1}\right|\right)$. Next, compute $\varphi_x = \arcsin\left(\frac{v_1}{v_2}\tan(\varphi_y)\right)$. Let $\lambda = \frac{v_2}{\sin \varphi_y}$ be a scaling factor, which is nonzero. Then $v_2 = \lambda s_y$ and $v_1 = \lambda c_y s_x$. Consequently λ can be computed from $v_1^2 + v_2^2 = \lambda^2(1 - c_y^2 c_x^2)$.

Note that $\begin{bmatrix} v_3 \\ v_4 \end{bmatrix} = \lambda \begin{bmatrix} -c_z & s_z \\ -s_z & -c_z \end{bmatrix} \begin{bmatrix} s_x \\ s_y c_x \end{bmatrix}$ should hold, which can be rewritten

in terms of c_z and s_z: $\begin{bmatrix} -v_3 & -v_4 \\ -v_4 & v_3 \end{bmatrix} \begin{bmatrix} c_z \\ s_z \end{bmatrix} = \lambda \begin{bmatrix} s_x \\ s_y c_x \end{bmatrix}$. The values of s_z and c_z are obtained, which uniquely specify $\varphi_z \in (-\pi, \pi]$.

Case 2. Assumes that $v_2 = 0$. Then let $\varphi_y = 0$ and note that $v_1 \neq 0$. Now choose φ_x to have an arbitrary nonzero value in the interval $\left(-\frac{\pi}{2}, \frac{\pi}{2}\right)$. Then set $\lambda = v_1$, φ_z is determined through $\begin{bmatrix} v_3 \\ v_4 \end{bmatrix} = \lambda \begin{bmatrix} -c_z \\ -s_z \end{bmatrix}$. It follows that $v_1^2 + v_2^2 = v_3^2 + v_4^2 = \lambda^2$.

In either of the two Cases 1 and 2, a nonzero scaling factor λ and suitable values for φ_z, φ_y and φ_x are obtained which make that the vector v is of the form $v = \lambda \begin{bmatrix} c_y s_x \\ s_y \\ -c_z s_x + s_z s_y s_x \\ -s_z s_x - c_z s_y c_x \end{bmatrix}$ or simplified form $v = \lambda \begin{bmatrix} 1 \\ 0 \\ -c_z \\ -s_z \end{bmatrix}$ when $s_y = 0$.

In *Case 1* (where $s_y \neq 0$), this allows one to construct a corresponding vector $(\tilde{z}_1, \ldots, \tilde{z}_m)^{\mathsf{T}}$ to satisfy the required identity. Because $\begin{pmatrix} c_x & s_x \\ -s_y s_x & s_y c_x \end{pmatrix}$ is invertible, $(\tilde{z}_1', \ldots, \tilde{z}_m')^{\mathsf{T}}$ can be obtained by reconsidering the omitted third row of R. Clearly, translations in the z-direction cannot be observed at all, while coordinate values in all directions can only be obtained relative to an arbitrarily chosen origin. For z_0 and z_0' one can introduce arbitrary values, which shows that the entry t_3 of translation vector $\boldsymbol{t}$ is completely free.

In *Case 2* (where $\varphi_y = 0$), both columns in the matrix following M in Eq. 10 are collinear; both columns are of the form $k(1, 0, -c_z, -s_z)^{\mathsf{T}}$ (because $c_y = 1$, and k can be either c_x or s_x).

With φ_x from the indicated range (which ensures that c_x and s_x are both nonzero), we have that both columns are collinear, and that the relationship in Eq. 10 is properly satisfied. However the matrix $\begin{pmatrix} c_x & s_x \\ -s_y s_x & s_y c_x \end{pmatrix}$ is no longer invertible, so to rewind our steps, we should reconsider Eq. 9. With $s_x \neq 0$ it follows that:

$$\begin{bmatrix} \tilde{z}_1 \\ \vdots \\ \tilde{z}_m \end{bmatrix} = \frac{1}{s_x} M \begin{bmatrix} 0 \\ c_x \\ s_z \\ -c_z \end{bmatrix} \tag{11}$$

Then we can proceed as in Case 1 to construct a rotation and translation which is consistent with the given observed data. This proves the converse implication of Theorem 1.

It also proves Theorem 2 upon noting that in both *Cases 1* and *2* a real degree of freedom for R (for the angles φ_y and φ_x, respectively) and for the coordinate t_3 was encountered. In special cases, i.e., when the rank of M is less than 3, more degrees of freedom may occur.

4.2   Applicability of Theoretical Results

Consistency with 3D Rigid Body Motion. In computer vision applications, motion based image segmentation is an important and fundamental topic. The aim is to partition visual elements (pixels or feature points) into groups, based on their motion features. Segmentation algorithms are used in tasks like object detection and tracking, where objects are represented by groups of points (or pixels). For videos from a monocular camera, the key challenge of motion segmentation is to segment the points w.r.t. their 3D motions, while only 2D projection-coordinates of points are available.

Theorem 1 can be used to determine whether the movements of a group of 2D points (matched point from consecutive images) are consistent with a 3D rigid body motion. Giving $m + 1$ pairs of points, we can decompose the $m \times 4$ data matrix M using the SVD:

$$M = UDV^{\mathsf{T}} \tag{12}$$

in which U is an $m \times m$ orthogonal matrix, V is a 4×4 orthogonal matrix, and $D = \mathrm{diag}\{d_1, d_2, d_3, d_4\}$ is an $m \times 4$ diagonal matrix with entries $d_1 \geq d_2 \geq d_3 \geq d_4 \geq 0$ on its main diagonal. Theorem 1 establishes that at least $d_4 = 0$ should hold if the movement of 2D points is consistent with a 3D rigid body motion. However, when working with real data, deviations may occur for various reasons, such as inaccuracies in feature extraction and motion detection. Moreover, the orthographic projection hypothesis - which disregards the perspective - is an approximation.

The value of d_4 can be taken as a measure for the (lack of) quality of 3D rigid body motion consistency, for the group of points being analyzed. According to Theorem 1, in case of a rigid 3D body motion, every vector v in the kernel satisfies $v_1^2 + v_2^2 = v_3^2 + v_4^2$ if $d_3 > 0$. This property can be also used as a quality measure for the rigid body motion consistency. Note that a vector v is obtained as the last column of matrix V if $d_4 = 0$ and $d_3 > 0$.

Reconstruction of 3D Rigid Body Motion. Theorems 1 and 2 also enable us to estimate the parameters of a 3D rigid body motion for a given set of matched pairs. Starting from data matrix M (Eq. 6) with a 1-dimensional null space, using Eq. 7, there will be one real degree of freedom when computing the 3D rotations φ_z, φ_y, φ_x. There is also one degree of freedom (the translation in the z direction) in determining t. However, the values $\tilde{z}_1, \ldots, \tilde{z}_m$ completely depend on the degree of freedom for the 3D rotation.

We may determine the value of φ_y or φ_x by minimizing a criterion function, such as the sum of squares of values $\tilde{z}_1, \ldots, \tilde{z}_m$. The idea is that the norm of the vector of changes in the (unobserved) z-direction, consistent with the computed rotation and translation, is minimized so that no unnecessarily large deviations are included in the rigid body motion.

4.3 Error Analysis

The proposed theorems are based on the orthographic projection, which is an approximation of the perspective projection. In this subsection, we analyse the errors of orthographic projection w.r.t. to the perspective projection.

Suppose a 3D point is moving from $(X_c, Y_c, Z_c)^\mathsf{T}$ at time t to $(X_c', Y_c', Z_c')^\mathsf{T}$ at time t', and two images are captured at the two time points. The coordinates of a projected point at time t and t' under the perspective projections are $(x_p, y_p)^\mathsf{T}$ and $(x_p', y_p')^\mathsf{T}$ respectively. The coordinates of the same projected point under orthographic projection are $(x, y)^\mathsf{T}$ and $(x', y')^\mathsf{T}$. According to the Eqs. 4 and 3: $[x, y]^\mathsf{T} = [X_c, Y_c]^\mathsf{T}$, $[x_p, y_p]^\mathsf{T} = \frac{f}{Z_c}[X_c, Y_c]^\mathsf{T}$, $[x', y']^\mathsf{T} = [X_c', Y_c']^\mathsf{T}$ and $[x_p', y_p']^\mathsf{T} = \frac{f}{Z_c'}[X_c', Y_c']^\mathsf{T}$. The perspective projection scales the $[X_c, Y_c]^\mathsf{T}$ with a factor $\frac{f}{Z_c}$. We can compensate for the scaling of $[x_p, y_p]^\mathsf{T}$ by multiplying $[x_p, y_p]^\mathsf{T}$ with $\mu = \frac{Z_c}{f}$. So, $[x, y]^\mathsf{T} = \mu[x_p, y_p]^\mathsf{T}$. By applying the same scaling to $[x_p', y_p']^\mathsf{T}$, we can compute the error caused by orthographic projection:

$$\begin{bmatrix} x' \\ y' \end{bmatrix} - \mu \begin{bmatrix} x_p' \\ y_p' \end{bmatrix} = (1 - \frac{Z_c}{Z_c'}) \begin{bmatrix} X_c' \\ Y_c' \end{bmatrix} \tag{13}$$

If the changes in z direction caused by translation and rotation are small, then $\frac{Z_c}{Z_c'} \approx 1$, and the error is approximately 0.

5   Experiments

In this section, we evaluate the applicability of the theorems on synthetic data in Subsect. 5.1, and subsequently on real video data in Subsect. 5.2. The theorems can also be used to estimate the 3D rigid motion of an object with one degree of freedom. We evaluate this aspect in Subsect. 5.3.

5.1   Improving Motion Segmentation Using Synthetic Data

There are two possible ways to apply our results in motion segmentation,

1. Giving the result of a segmentation, an object is represented as a group of points. Usually there are miss-classified points in each group, which make the result in an low precision. We can use the criterion in this paper to find out the miss-classified point in a class of points.
2. Given a group of points that are belonging to an object, and a set of new points without assignments, we can identify whether the new points belong to the object. Failing to identify these points result is a low recall.

We generated a 3D synthetic scene containing a cube, which follows a 3D transformation. Randomly chosen points on the surface of the cube are tracked. We also randomly generate some noise points that have arbitrary 3D motions. The motion of each point is represented by its initial position and the new position after transformation. Figure 3 illustrates the 3D motion flows of the points on the cube surface in camera space, while Fig. 4 shows the orthographic projection of these motion vectors on the image that is parallel to the xy plane.

For the first experiment, we choose 100 points from the cube object and n noise points. We aim at allocating these points into two subgroups: the "objects" and "noises" using the criterion in this paper. The accuracy is defined as the percentage of points that are successfully classified, which is illustrated in Fig. 5 w.r.t. different noise ratios (i.e. the percentage of noise points in the mixture set). We compare our method with a state-of-art method—sparse subspace clustering (SSC) [6], whose result is illustrate as the blue line in Fig. 5. The result shows that our method is more stable than SSC.

For the second experiment, m ($m \in [4, 100]$) points on the cube are chosen to represent the object, which is used to determine the classification of 200 other

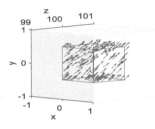

Fig. 3. 3D motion flows

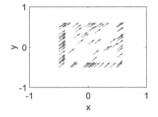

Fig. 4. 2D motion fields

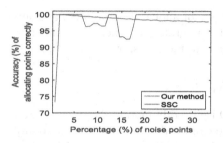
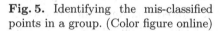
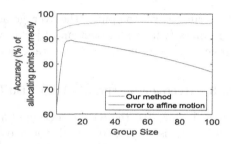

Fig. 5. Identifying the mis-classified points in a group. (Color figure online)

Fig. 6. Determining whether a new point belonging to an object.

points (half from the object and half from the noises) based on the criterion in our paper. We computed the error using $error = \left\| d_4 + \sqrt{v_1^2 + v_2^2 - v_3^2 - v_4^2} \right\|$. We assigned a point to the group if its error is lower than a threshold value (which is 0.005 in this experiment). We compared it with another error estimation method, which is computed by the error to the affine motion model that is estimated using the m points represent the object [26]. The performance is evaluated by the accuracy of correctly classifying the undetermined points, as shown in Fig. 6.

5.2 Improving Motion Segmentation Using Video Data

In this experiment, we used the real video sequences from the Hopkins155 benchmark data set [17], for which the feature points on the objects' surfaces and their motions are provided. We chose 25 video sequences from the category named "checkerboard". Each video contains 29 frames, which records a scene with 3 objects following distinct 3D motions (rotation and translation). There are 75 objects in total. In each experiment we chose one object and used the motion vectors between the frame pair $\{f_1, f_i\}$ ($i \in [2, 29]$). We computed the average accuracy of finding the misclassified points from a group over all objects w.r.t. different frame pairs and noise ratio, which is illustrated in Fig. 7. For the second

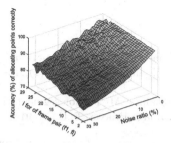
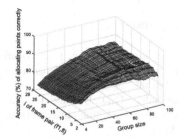

Fig. 7. The accuracy of finding misclassified points as function of the noise ratio and the distance between two frames.

Fig. 8. The accuracy of allocating a point correctly to an object w.r.t. the group size of the object and the distance between two frames.

experiment, we computed the average accuracy of allocating a point to a group of points (which represent an object), w.r.t. the group size (i.e. the number of points in the group) and the distance between frames, as shown in Fig. 8.

5.3 Recovering the 3D Rigid Body Motion

Exp. 5 addressed the reconstruction of a 3D rigid body motion. We investigated whether it is possible to handle the one degree of freedom for the 3D rotation by minimizing the sum of square of $\tilde{z}_1, \ldots, \tilde{z}_m$. Our initial experiments with points on the surfaces of a cube in the synthetic scene showed that for randomly chosen rotations smaller than $\pi/4$ rad, we can recover the rotation angles φ_x, φ_y and φ_z with average accuracies of 74.3%, 74.3%, 94.6% respectively.

6 Conclusion

This paper presented two theorems specifying properties of a 2D projection of a 3D rigid body movement. The theorems state that the data matrix of 2D projection of points on a 3D rigid body making a 3D movement, has a non-trivial kernel with a specific structure. The theorems also show that we can reconstruct the original 3D body movement with one degree of freedom for the translation in z-direction and one degree of freedom for the 3D rotation.

We used the theorems to measure the 3D rigid motion consistency of a group of 2D projection points. It can achieve above 95% accuracy in identifying misclassified points when the rate of the misclassified points is below 10%, and remains around 90% when the noise rate increases to 20%. We also used the theorems to determining whether new points belong to a known object. If we known more than 50 points belonging to the object, new points can be classified with an accuracy around 90%. These results suggest that the theorems can be used to improve the segmentation accuracy of existing motion segmentation algorithms.

Recovering the 3D rotation angle of a moving object has also been evaluated. The initial results are promising but further research is required.

References

1. Altunbasak, Y., Eren, P.E., Tekalp, A.M.: Region-based parametric motion segmentation using color information. Graph. Model. Image Process. **60**(1), 13–23 (1998)
2. Borshukov, G.D., Bozdagi, G., Altunbasak, Y., Tekalp, A.M.: Motion segmentation by multistage affine classification. IEEE Trans. Image Process. **6**(11), 1591–1594 (1997)
3. Boult, T.E., Brown, L.G.: Factorization-based segmentation of motions. In: 1991 Proceedings of the IEEE Workshop on Visual Motion, pp. 179–186. IEEE (1991)
4. Bovik, A.C.: Handbook of Image and Video Processing. Academic Press, London (2010)

5. Costeira, J., Kanade, T.: A multi-body factorization method for motion analysis. In: 1995 Proceedings of the Fifth International Conference on Computer Vision, pp. 1071–1076. IEEE (1995)

6. Elhamifar, E., Vidal, R.: Sparse subspace clustering. In: 2009 IEEE Conference on Computer Vision and Pattern Recognition, CVPR 2009, pp. 2790–2797. IEEE (2009)

7. Gruber, A., Weiss, Y.: Multibody factorization with uncertainty and missing data using the EM algorithm. In: 2004 Proceedings of the 2004 IEEE Computer Society Conference on Computer Vision and Pattern Recognition, CVPR 2004, vol. 1, p. I. IEEE (2004)

8. Gruber, A., Weiss, Y.: Incorporating non-motion cues into 3D motion segmentation. In: Leonardis, A., Bischof, H., Pinz, A. (eds.) ECCV 2006. LNCS, vol. 3953, pp. 84–97. Springer, Heidelberg (2006). doi:10.1007/11744078_7

9. Hartley, R.I., Zisserman, A.: Multiple View Geometry in Computer Vision, 2nd edn. Cambridge University Press, New York (2004). ISBN: 0521540518

10. Hartley, R., Zisserman, A.: Multiple View Geometry in Computer Vision. Cambridge University Press, Cambridge (2003)

11. Horn, B.K., Schunck, B.G.: Determining optical flow. In: 1981 Technical Symposium East, pp. 319–331. International Society for Optics and Photonics (1981)

12. Jian, Y.D., Chen, C.S.: Two-view motion segmentation with model selection and outlier removal by Ransac-enhanced Dirichlet process mixture models. Int. J. Comput. Vision **88**(3), 489–501 (2010)

13. Jodoin, P.M., Pierard, S., Wang, Y., Van Droogenbroeck, M.: Overview and benchmarking of motion detection methods (2014)

14. Tomasi, C., Kanade, T.: Shape and motion from image streams under orthography: a factorization method. Int. J. Comput. Vision **9**(2), 137–154 (1992)

15. Torr, P.H., Szeliski, R., Anandan, P.: An integrated Bayesian approach to layer extraction from image sequences. IEEE Trans. Pattern Anal. Mach. Intell. **23**(3), 297–303 (2001)

16. Torr, P.H.S., Zisserman, A.: Concerning Bayesian motion segmentation, model averaging, matching and the trifocal tensor. In: Burkhardt, H., Neumann, B. (eds.) ECCV 1998. LNCS, vol. 1406, pp. 511–527. Springer, Heidelberg (1998). doi:10. 1007/BFb0055687

17. Tron, R., Vidal, R.: A benchmark for the comparison of 3-D motion segmentation algorithms. In: 2007 IEEE Conference on Computer Vision and Pattern Recognition, CVPR 2007, pp. 1–8. IEEE (2007)

18. Ullman, S.: The interpretation of structure from motion. Proc. R. Soc. Lond. B Biol. Sci. **203**(1153), 405–426 (1979)

19. Vidal, R., Soatto, S., Ma, Y., Sastry, S.: Segmentation of dynamic scenes from the multibody fundamental matrix. Urbana **51**(61801), 1–2

20. Wang, H., Kläser, A., Schmid, C., Liu, C.L.: Dense trajectories and motion boundary descriptors for action recognition. Int. J. Comput. Vision **103**(1), 60–79 (2013)

21. Wang, J.Y., Adelson, E.H.: Layered representation for motion analysis. In: 1993 IEEE Proceedings of the Computer Society Conference on Computer Vision and Pattern Recognition, CVPR 1993, pp. 361–366. IEEE (1993)

22. Weiss, Y.: Smoothness in layers: motion segmentation using nonparametric mixture estimation. In: 1997 IEEE Proceedings of the Computer Society Conference on Computer Vision and Pattern Recognition, pp. 520–526. IEEE (1997)

23. Yuan, C.: Motion segmentation and dense reconstruction of scenes containing moving objects observed by a moving camera. ProQuest (2007)

24. Zelnik-Manor, L., Machline, M., Irani, M.: Multi-body factorization with uncertainty: revisiting motion consistency. Int. J. Comput. Vision **68**(1), 27–41 (2006)
25. Zhang, J., Shi, F., Liu, Y.: Motion segmentation by multibody trifocal tensor using line correspondence. In: 2006 Proceedings of the 18th International Conference on Pattern Recognition, ICPR 2006, vol. 1, pp. 599–602. IEEE (2006)
26. Zhao, W., Roos, N.: Motion based segmentation for robot vision using adapted em algorithm. In: Proceedings of the 11th International Conference on Computer Vision Theory and Applications, VISIGRAp 2016, pp. 649–656 (2016)
27. Zhao, W., Roos, N.: An EM based approach for motion segmentation of video sequence. In: Proceedings of the 24th International Conference in Central Europe on Computer Graphics, Visualization and Computer Vision, WSCG 2016, pp. 61–69 (2016)

Image Restoration II

Space-Variant Gabor Decomposition
for Filtering 3D Medical Images

Darian Onchis[1,2]([⊠]), Codruta Istin[3], and Pedro Real[4]

[1] University of Vienna, Vienna, Austria
darian.onchis@e-uvt.ro
[2] West University of Timisoara, Timisoara, Romania
[3] Politehnica University of Timisoara, Timisoara, Romania
[4] University of Seville, Seville, Spain

Abstract. This is an experimental paper in which we introduce the possibility to analyze and to synthesize 3D medical images by using multivariate Gabor frames with Gaussian windows. Our purpose is to apply a space-variant filter-like operation in the space-frequency domain to correct medical images corrupted by different types of acquisitions errors. The Gabor frames are constructed with Gaussian windows sampled on non-separable lattices for a better packing of the space-frequency plane. An implementable solution for 3D-Gabor frames with non-separable lattice is given and numerical tests on simulated data are presented.

1 Introduction

The study of noise removal in medical images was approached in different ways by numerous authors. From the first studies involving only convolutional filters [12], to the rank algorithms [13,14] and most recently to methods to average pixels depending on their neighborhood statistics [10,11], the search was both for maximizing the filter capabilities and to propose fast algorithms. While very successful for various signal analysis applications in medicine or telecommunications, time-frequency analysis and especially the Gabor frames expansions for image processing in 2D or 3D was of limited interest due to the large indexing problems i.e. a 3D image is analysed with a 6D lattice, and to the problems related to time consuming implementation.

In a fairly recent paper [15], the first author introduced a procedure to efficiently perform an nD Gabor frames decomposition. Based on that result, we propose in this paper a method to decompose, to filter and to reconstruct 3D medical images that overcomes both problems. We are using a tensor product decompositions [4] to reduce the decomposition to only 2D lattice case and also fast algorithms to implement this case. The result is an experimental framework for fast Gabor frames construction used to analyze 3D-data images sampled on quincunx-type lattices. The research is in the simulation stage and we present in here only tests done on the MATLAB MRI data set.

Therefore the main contribution of this conference paper is to introduce a completely invertible 3D Gabor transform applied to filter 3D medical images,

© Springer International Publishing AG 2017
M. Felsberg et al. (Eds.): CAIP 2017, Part II, LNCS 10425, pp. 455–461, 2017.
DOI: 10.1007/978-3-319-64698-5_38

different from existing examples in literature that present only 3D Gabor filters [7,9].

This paper is structured as follows: In the second section we recall the necessary results from Gabor analysis, while in the third section we present the 3D Gabor frames construction. The application to 3D medical imaging is given in the fourth section. In the last section, we present timings for the 3D Gabor construction and the conclusions are drawn.

2   Theoretical Preliminaries

Frames $(g_i)_{i \in I}$ generalize the idea of a basis in a Hilbert space $\boldsymbol{H}$ and they are formally defined as:

Definition 1. *A family* $(g_i)_{i \in I}$ *in a Hilbert space* $\boldsymbol{H}$ *is called a frame if there exist constants* $A, B > 0$ *such that for all* $g \in \boldsymbol{H}$

$$A\|g\|^2 \leq \sum_{i \in I} |\langle g, g_i \rangle|^2 \leq B\|g\|^2 \tag{1}$$

Every element $f \in \boldsymbol{H}$ has an expansion of the form:

$$f = SS^{-1}f = \sum_{i \in I} \langle S^{-1}f, g_i \rangle g_i = \sum_{i \in I} \langle f, S^{-1}g_i \rangle g_i$$

where S denotes the **invertible frame operator** [3]: $Sf = \sum_{i \in I} \langle f, g_i \rangle g_i$. The family $(\gamma_i)_{i \in I} = (S^{-1}g_i)_{i \in I}$ is again a frame with frame bounds B^{-1} and A^{-1} and is called a **canonical dual** frame. The main tool for time-frequency analysis is the **Short-Time Fourier Transform** in short STFT, defined for functions in $\boldsymbol{L}^2(\mathbb{R})$ as

$$V_g f(\lambda) = V_g f(a, b) = \langle f, M_b T_a g \rangle = \langle f, \pi(\lambda)g \rangle \tag{2}$$

where $T_a f(t) = f(t - a)$ is the translation (time shift) and $M_b f(t) = e^{2\pi i b \cdot t} f(t)$ is the modulation (frequency shift), for $\lambda = (a, b) in \mathbb{R}^2$. The operators $M_b T_a$ are called **time-frequency shifts**. Their composition is denoted by $\pi(\lambda)$. In order to obtain Gabor frames, the STFT is sampled over a time-frequency lattice. In the standard 1D case the regular lattice is of the form $\Lambda = a\mathbb{Z} \times b\mathbb{Z}$ with the condition $ab < 1$. Therefore a **Gabor frame** is defined as:

$$\mathcal{G}(g, a, b) := \{g_{k,l} = M_{bl}T_{ak}g, \quad k, l \in \mathbb{Z}\}, \tag{3}$$

i.e. the elements of the Gabor frame are translated and modulated versions of one atom g.

Theorem 1 (Dual Gabor Frames [5]**).** *If $\mathcal{G}(g, a, b)$ is a frame for $L^2(\mathbb{R})$, then the canonical dual frame takes the form $\mathcal{G}(\gamma, a, b)$ for $\gamma = S^{-1}g$. where the Gabor frame operator S is defined as:*

$$S := \sum_{k, l \in \mathbb{Z}} \langle f, M_{bl}T_{ak}g \rangle M_{bl}T_{ak}g \tag{4}$$

We introduce also the result of Bourouihiya [2], which extends the classical results of Lyubarski and Seip to higher dimensions [6] for $g_0(x) = 2^{-1/4}e^{-\pi x^2}$ with $x \in \mathbb{R}$:

Lemma 1. *Let $g = g_0 \otimes \cdots \otimes g_0$ (n factors) and $\Lambda_{ab} = (a_1\mathbb{Z} \times \cdots \times a_n\mathbb{Z}) \times (b_1\mathbb{Z} \times \cdots \times b_n\mathbb{Z})$. Then $\mathcal{G}(g, \Lambda_{ab})$ is a frame if and only if $a_jb_j < 1$ for $1 \le j \le n$.*

Based on this result, another extension to non-separable lattices of the form $N\Lambda_{a,b}$ is possible, which besides a multi-variate sampling in each dimension, give us a better packing of the time-frequency plane by using a (non-separable) hexagonal lattice that match with the circular contour lines of the Gaussian.

The representation of the non-separable lattice is based on the rectangular lattice via a shear operation. Therefore, we can give the following lemma:

Lemma 2. *Given a window $g = g_0 \otimes \cdots \otimes g_0$ (n factors) and a lattice of the form $N\Lambda_{a,b}$, where Λ is a rectangular lattice and N is a shear lattice, the system $\mathcal{G}(g, \Lambda_{ab})$ is a frame if and only if $a_jb_j < 1$ for $1 \le j \le n$.*

Proof. The shear matrix N is a symplectic matrix hence the determinant $det(N) = 1$. Therefore the volumes $vol(N\Lambda_{a,b}) = vol(\Lambda_{a,b})$ are equal and the Lemma 1 extends to non-separable lattices.

In applications sampled data of finite lengths are analysed; the process of sampling and periodization are employed [16]. In this way also the number of shifts in time and frequency becomes finite. The redundancy of a discrete system, not necessarily a Gabor system, is defined as the fraction of the number of used discrete function over the length of the domain, $\frac{\#shift}{L}$.

3   3D Gabor Frame Construction

In this section we present how to construct numerically a 3D Gabor transform for 3D image filtering. We will exploit the possibilities given by the Lemma 2 for choosing a non-separable lattice of quincunx-type. This situation can be easily expanded to more dimensions or can be reduced to less dimensions following the same procedure. We write n for data cube length in one direction and the modulation and translation operator are defined on $\mathbb{Z}_n$. For further discretization details [8] is a comprehensive source.

For the 3D case, we will use a multi-variate generalized Gaussian window, obtained as a tensor product of 1D windows in the form:

$$G_3 = g_1 \otimes g_2 \otimes g_3$$

We consider our 3D Gaussian window of size $n_1 \times n_2 \times n_3$ as a complex-valued function on the additive Abelian group $\mathcal{G} = n_1 \times n_2 \times n_3$. The joint position-frequency space is

$$\mathcal{G} \times \widehat{\mathcal{G}} = n_1 \times n_2 \times n_3 \times n_1 \widehat{\times n_2 \times n_3}.$$

Now, let's pay attention to the lattice. We would like to use a non-separable lattice obtained as in the hypothesis of the Lemma 2 by applying a shear matrix to the matrix generating the regular standard lattice. There are other symplectic matrices like the rotation matrix using the fractional Fourier transform [1] that can be used to transform a rectangular lattice into a quincunx-like lattice.

For our 3D case using a non-separable lattice, we will consider the following quincunx-like lattice:

$$\Lambda_3 = \begin{pmatrix} I_{3\times3} & Q_3 \\ O_3 & I_{3\times3} \end{pmatrix} A_6 \cdot \mathbb{Z}^6$$

where

$$Q_3 = \begin{pmatrix} a_1/2b_1 & 0 & 0 \\ 0 & a_2/2b_2 & 0 \\ 0 & 0 & a_3/2b_3 \end{pmatrix}$$

and

$$A_6 = diag(a_1, a_2, a_3, b_1, b_2, b_3)$$

This 6D non-separable lattice can be reduced to the case of 2D non-separability in time and frequency by the expansion in the canonical basis. Moreover, we can write the following tensor product:

$$\begin{pmatrix} 1 & a_1/2b_1 \\ 0 & 1 \end{pmatrix} \otimes \begin{pmatrix} 1 & 0 & 0 \\ 0 & 0 & 0 \\ 0 & 0 & 0 \end{pmatrix} + \begin{pmatrix} 1 & a_2/2b_2 \\ 0 & 1 \end{pmatrix} \otimes \begin{pmatrix} 0 & 0 & 0 \\ 0 & 1 & 0 \\ 0 & 0 & 0 \end{pmatrix} + \begin{pmatrix} 1 & a_3/2b_3 \\ 0 & 1 \end{pmatrix} \otimes \begin{pmatrix} 0 & 0 & 0 \\ 0 & 0 & 0 \\ 0 & 0 & 1 \end{pmatrix}$$

In this conditions it is feasible to perform a Gabor analysis followed by a Gabor synthesis at level of the constituting vectors, and therefore reducing the case of applying 3D matrices to the faster case of applying the transform over each dimension. Therefore, we can compute the coefficients of the expansion (e.g. the 3D Gabor filter) by using any fast algorithms developed for the 1D Gabor transform [8]. For a complete Gabor transform, we use the Theorem 1 and we invert the frame operator in order to obtain the dual frame. Using the computed dual frame, we can recover the 3D data cube.

4   Applications to 3D Medical Image Processing

We have tested our implementation for the MRI data set that comes with MAT-LAB2017. Loading mri.mat adds two variables to the workspace: D (128-by-128-by-1-by-27, class uint 8) and a grayscale colormap, map (89-by-3, class double). D comprises 27 128-by-128 horizontal slices from an MRI data scan of a human cranium. Values in D range from 0 through 88, so the colormap is needed to generate a figure with a useful visual range.

The procedure takes as inputs D = 3D medical image and the G = the Gaussian analysing atoms. The parameters a = time shift and b = frequency shift are defined depending of the application and under the liniar independence contraints. The output is DG = data cube recovered after Gabor analysis. The complete algorithm is summarized below:

Algorithm 1. 3D Gabor transform filtering with quincunx lattice

---

Input: G - Gaussian atom, D- 3D medical image
Output: DG - 3D Gabor filtered medical image
1: Generate Gabor matrix G for a quincux sampling lattice;
2: Compute the Gabor coefficients GC over each dimension
3: Remove the low energy Gabor coefficients
4: Recover DG over each dimension with the use of the dual atom

---

In the Figs. 1 and 2, we present the results of the applying the Gabor filtering.

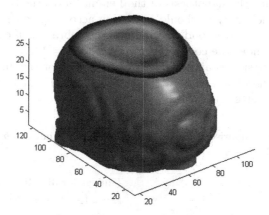

Fig. 1. Unfiltered 3D image

We observed that due to the reconstruction after the removal of the low energy coefficients the image becomes much smoother. The next step is to corroborate this results with the needs of medical practitioners or radiologists.

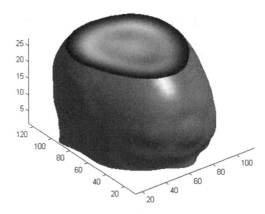

Fig. 2. Gabor filtered 3D image

5 Conclusions

In this paper, we described the construction of Gabor frames beyond the standard Gabor processing on regular lattices for signal and image processing. In this way, one has the freedom to choose a non-separable lattice and to sample the nD-Gaussian window in a multi-variate way, while still being assured that the result will be a Gabor frame. This added flexibility will be useful for application like 3D space-variant filtering or decomposition in the 3D Gabor domain for different features identifications (e.g. 3D plane waves).

The matrix we used for generating the lattice in our example, verifies the conditions under which the hypothesis of the Lemma 2 are true (i.e. shear matrix). The lattice parameters a_j, b_j should be chosen according to the time-frequency concentration of the corresponding windows g_j to obtain well concentrated systems. Therefore, under the conditions $a_j b_j < 1$ for $j = 1, 2, 3$, we obtain Gabor frames for decompositions and analysis. The timings for the implementation in $Matlab^{TM}$ on a notebook with Intel core i5 at 2.3 GHz and with 8.00 GB of RAM memory are given in Table 1.

Table 1. Timings comparison. The case 1. With redundancy 2, $ab = \frac{1}{2}$ the case 2., with redundancy 4, $ab = \frac{1}{4}$.

		Timings for 3D MRI data	
		1. n = 128	2. n = 128
Quincunx lattice	Coefficients	12.438 s	72.103 s
	Synthesis	13.134 s	78.363 s

Acknowledgments. The first author gratefully acknowledge the support of the Austrian Science Fund (FWF): project number P27516.

References

1. Bastiaans, M.J., van Leest, A.J.: From the rectangular to the quincunx Gabor lattice via fractional Fourier transformation. IEEE Signal Process. Lett. **5**(8), 203–205 (1998)
2. Bourouihiya, A.: The tensor product of frames. Sampl. Theory Signal Image Process. **7**(1), 65–76 (2008)
3. Christensen, O.: An Introduction to Frames and Riesz Bases. Applied and Numerical Harmonic Analysis. Birkhäuser, Boston (2003)
4. Christensen, O., Feichtinger, H., Paukner, S.: Gabor Analysis for Imaging. Handbook of Mathematical Methods in Imaging. Springer, Berlin (2010)
5. Gröchenig, K.: Foundations of Time-Frequency Analysis. Birkhäuser, Boston (2001)
6. Lyubarskii, Y.I.: Frames in the Bargmann space of entire functions. In: Entire and Subharmonic Functions, pp. 167–180. American Mathematical Society, Providence (1992)
7. Mikula, K., Sgallari, F.: Semi-implicit finite volume scheme for image processing in 3D cylindrical geometry. J. Comput. Appl. Math. **161**(1), 119–132 (2003)
8. Qiu, S., Feichtinger, H.G.: Discrete Gabor structures and optimal representation. IEEE Trans. Signal Process. **43**(10), 2258–2268 (1995)
9. Wang, Y., Chua, C.-S.: Face recognition from 2D and 3D images using 3D Gabor filters. Image Vis. Comput. **23**(11), 1018–1028 (2005)
10. Buades, A., Morel, J.M.: A Non-local algorithm for image denoising. In: IEEE Computer Society Conference on Computer Vision and Pattern Recognition, vol. 2, pp. 60–65, 20–26 June 2005
11. Dabov, K., Foi, A., Katkovnik, V., Egiazarian, K.: Image denoising by sparse 3D transform-domain collaborative filtering. IEEE Trans. Image Process. **16**(8), 2080–2095 (2007)
12. Pratt, W.K.: Digital Image Processing: PIKS Scientific inside, 4th edn. Wiley, Los Altos (2007)
13. Yaroslavsky, L.P., Kim, V.: Rank algorithms for picture processing. Comput. Vis. Graph. Image Process. **35**, 234–258 (1986)
14. Storozhilova, M., Lukin, A., Yurin, D., Sinitsyn, V.: 2.5D extension of neighborhood filters for noise reduction in 3D medical CT images. In: Gavrilova, M.L., Tan, C.J.K., Konushin, A. (eds.) Transactions on Computational Science XIX. LNCS, vol. 7870, pp. 1–16. Springer, Heidelberg (2013). doi:10.1007/978-3-642-39759-2_1
15. Onchis, D.M.: Optimized frames and multi-dimensional challenges in time-frequency analysis. Adv. Comput. Math. **40**(3), 703–709 (2014)
16. Søndergaard, P.L.: Gabor frames by sampling and periodization. Adv. Comput. Math. **27**(4), 355–373 (2007)

Learning Based Single Image Super Resolution Using Discrete Wavelet Transform

Selen Ayas[(⊠)] and Murat Ekinci

Computer Engineering Department, Karadeniz Technical University,
61080 Trabzon, Turkey
{selenguven,ekinci}@ktu.edu.tr

Abstract. Sparse representation has attracted considerable attention in image restoration field recently. In this paper, we study the implementation of sparse representation on single-image super resolution problem. In recent research, first and second-order derivatives are always used as features for patches to be trained as dictionaries. In this paper, we proposed a novel single image super resolution algorithm based on sparse representation with considering the effect of significant features. Therefore, the super resolution problem is approached from the viewpoint of preservation of high frequency details using discrete wavelet transform. The dictionaries are constructed from the distinctive features using K-SVD dictionary training algorithm. The proposed algorithm was tested on 'Set14' dataset. The proposed algorithm recovers the edges better as well as improving the computational efficiency. The quantitative, visual results and experimental time comparisons show the superiority and competitiveness of the proposed method over the simplest techniques and state-of-art SR algorithm.

Keywords: Super resolution · Sparse representation · Discrete wavelet transform

1 Introduction

In real world, the resolution of images is limited by the image acquisition device. However, high-resolution (HR) images are desired for many imaging applications such as satellite and aerial imaging, ultrasound imaging, medical image processing. The issue can be solved by hardware-based approaches, but these approaches are usually expensive. Therefore, algorithmic-based solutions are preferred instead of hardware-based approaches. Super resolution (SR) is a process of acquiring a HR image from a low-resolution (LR) input image. The goal of the process which can be achieved using algorithm-based solutions is to provide more detail information in an image [1].

Single image SR approaches can be classified generally into three major categories: interpolation-based methods [2–4], reconstruction-based methods [5–7] and learning-based methods [8–12]. Interpolation-based methods have been studied comprehensively. These methods work in similar way: pixels of HR image are

© Springer International Publishing AG 2017
M. Felsberg et al. (Eds.): CAIP 2017, Part II, LNCS 10425, pp. 462–472, 2017.
DOI: 10.1007/978-3-319-64698-5_39

assigned by computing weighted average of adjacencies pixels of the observed pixel in LR image. However, the super-resolved image cannot satisfy the expected image quality because of limited significant information of LR image. Besides, simple interpolation techniques cannot recover the high frequency information. Reconstruction-based approach is an ill-posed problem because of insufficient number of LR image. The generic image prior information in LR images is utilized to solve the inverse problem and reconstruct HR image. However, the insufficient number of LR image, as in the case of a single LR image, causes the poor-reconstructed HR image. Learning-based approaches have received considerable attention in recent years because of overcoming these difficulties. These methods learn the relationship between HR and their LR counterpart as a priori to reconstruct HR output [8–10,13–15].

Several learning-based approaches using trained learning model on training image set as priors have been proposed recently. Freeman et al. [13] used Markov Random Field trained by Bayesian belief propagation to infer HR image from LR input image. A Bayesian approach to single image SR problem was proposed by [14] where Primal Sketch priors were used to enhance the quality of HR image. Chang et al. [15] got inspired from locally linearly embedding which is a manifold learning method. The local geometry of LR image patches was generated and HR image patches were reconstructed by theirs neighbours. Among learning-based methods, sparse representation have been attracted attention in past years. A sparse coding based SR was firstly proposed in Yang et al. [8,9]. In [8,9], the sparse coding was proposed to learn coupled dictionaries of LR and HR images, therefore LR-HR image patches had same sparse representation. Sparse coefficients investigated by LR image patch over the LR dictionary was used for reconstructing the HR patch. Zeyde et al. [10] introduced K-SVD based dictionary training and orthogonal matching pursuit (OMP) based sparse coding approach to improve the algorithms in [8,9]. Also, the algorithms were simplified by reconstructing HR dictionary through pseudo-inverse expression resulting in reduction of the coupled dictionary training time.

Sparse representation over learned dictionaries in the wavelet domain is a relatively novel representation method. Nazzal et al. [16] proposed SR approach based on dictionary learning and sparse coding in wavelet domain using three subbands except for approximation subband in dictionary learning. The proposed algorithm was tested on two different data set one of which was Kodak and the other one was created using some benchmark images four of which were same as the dataset created by Zeyde et al. [10], although Zeyde et al. [10] used 14 images in their dataset, the other 4 benchmark images were taken from different dataset. Wu et al. [17] proposed an algorithm to learn four pairs of principal component analysis (PCA) dictionaries in order to describe the relation between the approximation subband and the left three subbands. They also applied nonlocal self-similarity (NLM) and iterative back projection (IBP) algorithms to HR image. The proposed algorithm was faster than the solution proposed by Yang et al. [8,9], however it was slower than the algorithm proposed by Zeyde et al. [10]. The aforementioned proposed algorithms which were indicated as superior to the baseline algorithm both visually and quantitatively were based on Zeyde's [10] solution.

In this paper, a single image SR algorithm using sparse representation based on discrete wavelet transform (DWT) is proposed in order to acquire HR image from LR input image. The dictionary learning and sparse coding are performed in two level wavelet domain. The features extracted from three subbands except for approximation subband in each level are learned in dictionary learning phase. However, approximation subband has a significant influence on dictionary learning as an intermediate process. Therefore, the difference image obtained by using approximation subbands is utilized to construct LR dictionaries, and indirectly HR dictionaries. Similarly, the difference image obtained by subtracting the LR test image and its interpolated first level approximation subband is used as an intermediate process to correct the interpolated high-frequency components in reconstruction phase. Since the constructed dictionaries and interpolated high-frequency components are handled to acquire super-resolved image, the intermediate process provides sharper HR image. The proposed technique has been compared with bicubic interpolation which is a standard interpolation technique, wavelet interpolation and ScSR algorithm [10]. In all wavelet steps of the proposed single image SR algorithm, Daubechies (db.4) wavelet transform has been used. The proposed algorithm overperforms available state-of-art method in single image SR problem. The visual and quantitative results are given in the experimental results section.

The rest of this paper is organized as follows. Section 2 represents the related work of our algorithms. Section 3 introduces the proposed wavelet based single image SR approach. Section 4 discusses the qualitative and quantitative results of the proposed method with the conventional interpolation and state-of-art SR algorithms. Conclusions are given in the final section.

2   Single Image Super Resolution Preliminaries

2.1   2D Discrete Wavelet Transform

The DWT has received considerable attention because of its time-scale localization characteristic. Due to the mentioned property, local properties in time and space are extracted from an image. The image is decomposed into four frequency subbands in each decomposition level. The next decomposition level is applied to sub image which has low frequency in horizontal and vertical level. The preprocessed images decomposed up to the first and second level using the Daubechies (db.4) wavelet for training and reconstruction phase in the proposed algorithm, respectively. Preserving the edges is required in SR problem. Therefore, DWT has been employed in order to reveal the high frequency details of images.

2.2   Sparse Representation

Given an input LR image, z_l, generated from HR image, y_h, by degradation factors, the objective of the single image SR problem is to reproduce a HR image, $\hat{y}$, with more details. An image degradation model is given as:

$$y_l = Qz_l = Ly_h + \hat{v} \tag{1}$$

where L, Q and $\hat{v}$ represents the degradation factors, interpolation operator and the additive i.i.d white Gaussian noise, respectively. The patches are extracted from y_l which is a super-resolved image of z_l by a simple interpolation algorithm. Let p_h be a HR image patch extracted from the image y_h, and it can be represented sparsely by γ over the dictionary D_h as:

$$p_h = D_h \gamma \tag{2}$$

The LR image patch, p_l, extracted from the image y_l is defined in (3) using relation between the HR and LR images given in (1).

$$p_l = L p_h + \hat{v} \tag{3}$$

Multiplying both sides of (2) by degradation factor L and then substituting this expression into (3) yields (4).

$$\| p_l - L D_h \gamma \|_2 \leq \epsilon \tag{4}$$

where ϵ is related to $\hat{v}$. A LR image patch, p_l, can be represented by the same sparse vector, γ, over the LR dictionary $D_l = L D_h$. LR dictionary, D_l, is obtained using LR image patches, p_l. HR dictionary, D_h, is acquired using the HR image patches, p_h, and sparse representation, γ, over the LR dictionary, D_l, in the training phase. For a given LR patch, p_l, its sparse representation, γ, is estimated and then HR patch, p_h, is acquired using the same representation and the HR dictionary, D_h, in the reconstruction phase. These observations are the keystones of the patch-based sparse representation algorithms for SR.

Algorithm 1. Wavelet domain based dictionary learning phase

Input: Training images $Y_i, i = 1, ..., n$
Output: Trained low and high-resolution dictionaries
1 **for** $i \leftarrow 1$ to n **do**
2   Convert the RGB image, Y_i, to YCbCr color space and extract Y component
3   Transform the training image Y_i by 1-level DWT
4   Transform the LL subband image by 1-level DWT
5   Apply 1-level IDWT to the subband images acquired in Line 4, separately
6   Subtract the interpolated LL subband image in Line 5 from LL subband in Line 3 to obtain difference image
7   Add the difference image to the high frequency subband images in Line 5
8 **end**
9 Extract features from subband images in Line 7 and concatenate them
10 Train low resolution dictionaries on features using K-SVD algorithm
11 Extract features from subband images in Line 3
12 Train high resolution dictionaries on features using sparse representation acquired in LR dictionary training

---

3   DWT-Based Single Image Super Resolution Approach

3.1   Wavelet Subbands Dictionary Learning

Dictionary learning intends to find dictionaries, D, yielding sparse representations for the training images. These learned dictionaries outperform predetermined dictionaries generated by some functions or transforms. The wavelet subbands dictionary learning phase is demonstrated in Algorithm 1.

The inputs and outputs of the dictionary learning phase are training HR images and trained low and high resolution dictionaries, respectively. The proposed dictionary learning algorithm is only applied to luminance channel, therefore all training images are converted to YCbCr color space. Then, the DWT has been employed in order to preserve the high frequency components of the images. 1-level DWT is applied to the images, and the images are separated into different subband images, i.e. LL, LH, HL and HH. The high frequency subbands are used to construct the HR dictionaries as described in Line 12.

Each of these training HR images is blurred and down scaled by a factor of s to construct LR images. However, down scaling process by a factor of s is provided by the characteristic of the wavelet transform which reduces the size of an image by half in each dimension. Therefore, down scale factor, s, is chosen as 2. In addition, the high frequency subbands are supposed to be blurred images.

The LR image obtained by low pass filtering of the HR image is used as the input for the second 1-level DWT as described in Line 4. Then, 1-level inverse discrete wavelet transform (IDWT)-based interpolation is applied to the high frequency subband images separately by setting the other three subband images to zero. The difference image obtained by subtracting interpolated LL subband from LL subband is utilized to remove the low frequencies. Then, the obtained difference image is added to the high frequency subband images. These interpolated subband images are used to construct the LR dictionaries.

To train LR dictionaries, local overlapping patches of size $\sqrt{n} * \sqrt{n}$ are extracted from the interpolated subband images, separately. These patches are then concatenated into one vector for each band, $p_l^i, i = lh, hl, hh$, of length nR where R indicates the total patch size and i denotes the LH, HL and HH subband images. In order to save computations in dictionary learning phase, Principal Component Analysis (PCA) is applied as a dimensionality reduction method to extracted patches for each subband images, p_l^i. The K-SVD dictionary training algorithm is applied to the LR patches and LR dictionaries are constructed according to (5).

$$D_l^i, \{\gamma^i\} = \arg \min_{D_l^i, \{\gamma^i\}} \| p_l^i - D_l^i \gamma^i \|^2 \ s.t. \ \| \gamma^i \|_0 \leq L \tag{5}$$

where L denotes the maximum sparsity level. This expression also produces sparse representation coefficients vector, γ^i, which belongs to the LR training patches, p_l^i.

According to (6), the corresponding HR dictionaries, D_h^i, are constructed for the HR patches using the same sparse representation coefficients vectors, γ^i, produced in LR dictionaries construction.

Algorithm 2. The high resolution image reconstruction phase

Input: Test image Y and trained low and high-resolution dictionaries
Output: HR output image

1 Convert the RGB image, Y, to YCbCr color space and extract Y component
2 Transform the test image Y by 1-level DWT
3 Apply 1-level IDWT to the subband images acquired in Line 2, separately
4 Subtract the interpolated LL subband image in Line 3 from Y to obtain difference image
5 Add the difference image to the high frequency subband images in Line 3
6 Compute grid and extract features from subband images in Line 5
7 Encode band features using OMP algorithm and LR dictionaries
8 Reconstruct HR patches using sparse representation and HR dictionaries
9 Combine all patches into one band and apply 1-level IDWT using acquired subbands and LR test image

---

$$D_h^i = p_h^i \gamma^{i^t} (\gamma^i \gamma^{i^t})^{-1} \tag{6}$$

where t and -1 denote transpose operator and pseudo-inverse, respectively.

The corresponding dictionaries accomplish wavelet subbands dictionary learning phase of the SR algorithm.

3.2 High Resolution Image Reconstruction

Image reconstruction intends to increase the resolution of an image. The reconstruction phase receives a LR test image and the dictionaries, which is constructed in dictionary learning phase, as inputs to produce HR super-resolved image. The HR image reconstruction phase is demonstrated in Algorithm 2.

The LR test image which is used as the input for the proposed DWT-based single image SR algorithm is obtained by low pass filtering of the HR image in the wavelet domain. The image preprocessing is same as in dictionary learning phase. The LR test image is decomposed into wavelet subband images and then the 1-level IDWT is applied to the each subband separately in order to obtain interpolated subband images.

In SR problem, preserving edges and small details are the crucial points. Therefore, an intermediate stage is used to correct the high frequency interpolated subband images. A difference image obtained by subtracting the interpolated low frequency subband image from the LR input image is used as an intermediate process. This improvement is achieved by adding the difference image to the high frequency subband images.

The next step is the patch extraction where local overlapping patches are extracted from the interpolated subband images. These patches are multiplied by the projection operator for dimensionality reduction. The OMP algorithm is applied to the patches in order to find the sparse representation vectors by utilizing the LR dictionaries. Sparse representation vectors are calculated as in (7).

$$\{\gamma^i\} = \arg\min_{\{\gamma^i\}} \| p_l^i - D_l^i \gamma^i \|^2 \; s.t. \; \| \gamma^i \|_0 \leq L \tag{7}$$

The representation vectors are multiplied by the HR dictionaries, D_h^i, and the HR subband patches are reconstructed using (8).

$$p_h^i = D_h^i \gamma^i \tag{8}$$

Then, the HR subband patches are merged by averaging in the overlap area to create the subband images. The HR output image is reconstructed by performing IDWT on the subband images and the LR input image as an approximate subband.

4   Experimental Results

In this section, the performance of the proposed algorithm is presented and compared to bicubic interpolation, wavelet interpolation and the ScSR method which is proposed by [10] both quantitatively and visually.

A high-resolution image set consists of 91 color images used by [8–10] was collected for the training phase. The LR image to be super-resolved by a factor of s was assumed to be generated from a high-resolution image. Therefore, HR 'Set14' dataset was used for the high resolution image reconstruction phase. The patch size and the dictionary size are the two parameters chosen empirically in patch-based sparse representation approach. The image patch size is set to 3×3 with 2 pixel overlapping scheme and the dictionary size is selected as 1024 for our proposed method empirically. All experiments were performed in MATLAB on a computer of Intel Core i7-4500U PC @2.4 GHz with 4 GB of RAM. The performances of the four methods were measured using the peak signal-to-noise ratio (PSNR), structural similarity index measure (SSIM) [18] and elapsed time. The single image scale-up based on sparse representation approach was firstly proposed by Yang et al. [8,9] and then Zeyde et al. [10] developed the solution with some modifications. The quantitative and visual results in [10] show that the proposed algorithm performs much better than Yang et al. [8,9] algorithm. Therefore, in this study, the proposed method was compared with three other methods, i.e. the bicubic interpolation, the wavelet interpolation and the ScSR method [10] with a magnification factor of 2.

The PSNR, SSIM and reconstruction time results of the bicubic interpolation, wavelet interpolation, ScSR and the proposed algorithm are given in Table 1. It is worth of note that the SSIM values were calculated at the appropriate scale according to [18]. Furthermore, the super-resolved images of 'Barbara' were compared at Fig. 1 where the original HR image is on the left and followed by the results obtained by wavelet interpolation, bicubic interpolation, ScSR and the proposed algorithm. In order to make better comparison, a small representative windows were enlarged by different factors and shown in the right sides of each image.

Table 1. The PSNR (dB), SSIM and reconstruction time (s) results obtained through the experiments.

Image	Wavelet Int.			Bicubic Int.			ScSR			Our method		
Baboon [500 × 480]	22.92	0.9827	0.0819	24.86	0.9551	0.0028	25.46	0.9837	6.2058	**25.46**	**0.9913**	4.0747
Barbara [720 × 576]	25.60	0.9854	0.1584	28.00	0.9632	0.0040	**28.66**	0.9850	11.8441	28.60	**0.9909**	6.6241
Bridge [512 × 512]	23.85	0.9875	0.1026	26.58	0.9737	0.0032	**27.54**	0.9914	6.7050	27.37	**0.9954**	4.1213
Coastguard [352 × 288]	25.33	0.7097	0.0362	29.12	0.7893	0.0028	**30.41**	0.8405	2.4516	30.13	**0.8447**	1.5724
Comic [250 × 261]	22.76	0.7315	0.0309	26.01	0.8491	0.0020	**27.64**	**0.8984**	2.2624	27.28	0.8915	1.3728
Face [276 × 276]	30.37	0.7987	0.0295	34.83	0.8623	0.0019	**35.57**	0.8820	1.8847	35.49	**0.8838**	1.1452
Flowers [500 × 362]	26.77	0.7938	0.0555	30.35	0.8983	0.0028	**32.25**	**0.9274**	4.3929	31.66	0.9229	2.7925
Foreman [352 × 288]	27.39	0.8559	0.0320	34.14	0.9518	0.0020	**36.38**	**0.9666**	2.4700	35.47	0.9630	1.5228
Lenna [512 × 512]	30.70	0.9938	0.0875	34.70	0.9904	0.0031	**36.22**	0.9965	6.4561	35.97	**0.9981**	4.0878
Man [512 × 512]	26.58	0.9894	0.0898	29.25	0.9813	0.0031	**30.49**	0.9935	6.6162	30.11	**0.9964**	4.0609
Monarch [768 × 512]	28.93	0.9958	0.1292	32.94	0.9949	0.0040	**35.72**	0.9985	10.1148	34.55	**0.9991**	6.1833
Pepper [512 × 512]	30.87	0.9943	0.0978	34.97	0.9929	0.0037	**36.57**	0.9971	6.6112	35.89	**0.9982**	4.0450
PPT3 [529 × 656]	24.36	0.9887	0.1146	26.87	0.9908	0.0037	**29.24**	0.9982	7.4794	28.35	**0.9987**	5.1141
Zebra [586 × 391]	26.38	0.9921	0.0667	30.62	0.9870	0.0028	**33.21**	0.9968	5.7738	32.16	**0.9983**	3.5724
Average	26.63	0.9142	0.0795	30.23	0.9414	0.0030	32.53	0.9611	5.8046	31.32	0.9623	3.5921

According to the given results in Table 1, the proposed algorithm performs quantitatively much better PSNR and SSIM improvement than wavelet and bicubic interpolation. ScSR algorithm only outperforms in PSNR value; however, the proposed algorithm is considerably better SSIM and reconstruction time improvements than ScSR on most of the images. Besides, in [18], it is proved that SSIM gives a much better indication of image quality. The proposed algorithm provides an average SSIM improvement of 0.0481, 0.0209 and 0.0012 in comparison with wavelet interpolation, bicubic interpolation and ScSR, respectively.

The performance of the proposed algorithm was visually evaluated by comparing with three other interpolation and SR algorithms as shown in Fig. 1. 'Barbara' image was super-resolved with magnification factor of 2. The quantitative SSIM value for the image is in accord with visual observation qualitatively. In Fig. 1, the vertical lines in headscarf gradually became stronger and the details in the hand also became clear in the super-resolved image. It is clear that the proposed method is considerably better than both of the interpolation methods and ScSR algorithm.

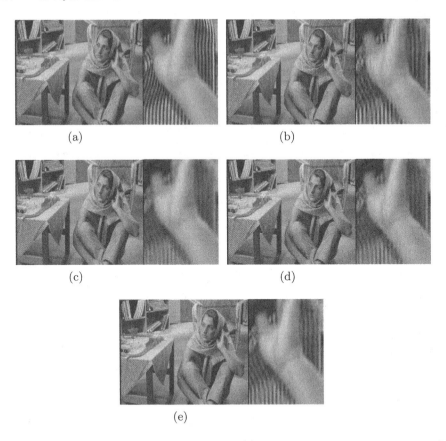

Fig. 1. Super-resolved image for 'Barbara'. (a) original high-resolution image; (b) wavelet interpolation; (c) bicubic interpolation; (d) ScSR method; (e) the proposed method. The enlarged representative windows on the right side of the each image.

The point worth mentioning is that the constructed LR and HR dictionaries recover the high frequency details with a great success rate in comparison to ScSR. The effectiveness of the proposed method can be discovered from both the visual result given in Fig. 1 and the quantitative values given in Table 1. In addition, the time complexity supports the effectiveness of the proposed algorithm.

5   Conclusion

A novel single image SR algorithm based on sparse representation using discrete wavelet transform was presented in this study. The main contribution of the study is that a DWT-based feature extraction algorithm providing better detection of high frequency details was performed in dictionary learning and sparse coding process. In addition, low frequency details were used as an intermediate process to improve information in dictionary learning and to correct the high-frequency components in image reconstruction. The experimental results on 'Set14' benchmark images show the superiority and competitiveness of the proposed algorithm in preserving edge details and reducing the computational time. Under the proposed SR approach, the usage of different magnification factors can be taken into account as a future investigation.

References

1. Nasrollahi, K., Moeslund, T.B.: Super-resolution: a comprehensive survey. Mach. Vision. Appl. **25**, 1423–1468 (2014)
2. Dai, S., Han, M., Xu, W., Wu, Y., Gong, Y.: Soft edge smoothness prior for alpha channel super resolution. In: IEEE Conference on Computer Vision and Pattern Recognition, pp. 1–8. IEEE Press, New York (2007)
3. Keys, R.G.: Cubic convolution interpolation for digital image processing. IEEE. T. Acoust. Speech. **29**, 1153–1160 (1981)
4. Li, X., Orchard, M.T.: New edge-directed interpolation. IEEE. T. Image. Process. **10**, 1521–1527 (2001)
5. Villena, S., Vega, M., Babacan, S.D., Molina, R., Katsaggelos, A.K.: Bayesian combination of sparse and non-sparse priors in image super resolution. Digit. Signal. Process. **23**, 530–541 (2013)
6. Villena, S., Vega, M., Molina, R., Katsaggelos, A.K.: Bayesian super-resolution image reconstruction using an L1 prior. In: 6th International Symposium on Image and Signal Processing and Analysis, pp. 152–157. IEEE Press, New York (2009)
7. Babacan, S.D., Molina, R., Katsaggelos, A.K.: Variational Bayesian super resolution. IEEE. T. Image. Process. **20**, 984–999 (2011)
8. Yang, J., Wright, J., Huang, T., Ma, Y.: Image super resolution as sparse representation of raw image patches. In: IEEE Conference on Computer Vision and Pattern Recognition, pp. 1–8. IEEE Press, New York (2008)
9. Yang, J., Wright, J., Huang, T., Ma, Y.: Image super-resolution via sparse representation. IEEE. T. Image. Process. **19**, 2861–2873 (2010)
10. Zeyde, R., Elad, M., Protter, M.: On single image scale-up using sparse-representations. In: Boissonnat, J.-D., Chenin, P., Cohen, A., Gout, C., Lyche, T., Mazure, M.-L., Schumaker, L. (eds.) Curves and Surfaces 2010. LNCS, vol. 6920, pp. 711–730. Springer, Heidelberg (2012). doi:10.1007/978-3-642-27413-8_47
11. Zhu, Z., Guo, F., Yu, H., Chen, C.: Fast single image super-resolution via self-example learning and sparse representation. IEEE. T. Multimedia **16**, 2178–2190 (2014)
12. Wang, S., Zhang, L., Liang, Y., Pan, Q.: Semi-coupled dictionary learning with applications to image super-resolution and photo-sketch synthesis. In: IEEE Conference on Computer Vision and Pattern Recognition, pp. 2216–2223. IEEE Press, New York (2012)
13. Freeman, W.T., Pasztor, E.C., Carmichael, O.T.: Learning low-level vision. Int. J. Comput. Vision. **40**, 25–47 (2000)
14. Sun, J., Zheng, N.N., Tao, H., Shum, H.Y.: Image hallucination with primal sketch priors. In: IEEE Computer Society Conference on Computer Vision and Pattern Recognition, pp. II–729. IEEE Press, New York (2003)
15. Chang, H., Yeung, D.Y., Xiong, Y.: Super-resolution through neighbor embedding. In: IEEE Computer Society Conference on Computer Vision and Pattern Recognition, p. I. IEEE Press, New York (2004)

16. Nazzal, M., Ozkaramanli, H.: Wavelet domain dictionary learning based single image superresolution. Signal. Image. Video. **9**, 1491–1501 (2015)
17. Wu, X., Fan, J., Xu, J., Wang, Y.: Wavelet domain multidictionary learning for single image super resolution. J. Electr. Comput. Eng. (2015). 37
18. Wang, Z., Bovik, A.C., Sheikh, H.R., Simoncelli, E.P.: Image quality assessment: from error visibility to structural similarity. IEEE. T. Image. Process. **13**, 600–612 (2004)

Directional Total Variation Based Image Deconvolution with Unknown Boundaries

Ezgi Demircan-Tureyen[1](✉) and Mustafa E. Kamasak[2]

[1] Department of Computer Engineering, Faculty of Engineering,
Istanbul Kultur University, 34156 Istanbul, Turkey
`e.demircan@iku.edu.tr`
[2] Department of Computer Engineering, Faculty of Computer and Informatics,
Istanbul Technical University, 34390 Istanbul, Turkey
`kamasak@itu.edu.tr`

Abstract. Like many other imaging inverse problems, image deconvolution suffers from ill-posedness and needs for an adequate regularization. Total variation (TV) is an effective regularizer; hence, frequently used in such problems. Various anisotropic alternatives to isotropic TV have also been proposed to capture different characteristics in the image. Directional total variation (DTV) is such an instance, which is convex, has the ability to capture the smooth boundaries as conventional TV does, and also handles the directional dominance by enforcing piecewice constancy through a direction. In this paper, we solve the deconvolution problem under DTV regularization, by using simple forward-backward splitting machinery. Besides, there are two bottlenecks of the deconvolution problem, that need to be addressed; one is the computational load revealed due to matrix inversions, second is the unknown boundary conditions (BCs). We tackle with the former one by switching to the frequency domain using fast Fourier transform (FFT), and the latter one by iteratively estimating a boundary zone to surrounder the blurred image by plugging a recently proposed framework into our algorithm. The proposed approach is evaluated in terms of the reconstruction quality and the speed. The results are compared to a very recent TV-based deconvolution algorithm, which uses a "partial" alternating direction method of multipliers (ADMM) as the optimization tool, by also plugging the same framework to cope with the unknown BCs.

Keywords: Deconvolution · Deblurring · Inpainting · Convex optimization · Directional total variation · Image reconstruction · Primal-dual algorithms

1 Introduction

Image deconvolution is an inverse task that aims to recover the underlying sharp image from its blurred observation, where the blur is caused by the convolution of the image with the shift invariant point spread function (PSF) of an imaging system. When that PSF is not known, the problem goes by the name of blind

© Springer International Publishing AG 2017
M. Felsberg et al. (Eds.): CAIP 2017, Part II, LNCS 10425, pp. 473–484, 2017.
DOI: 10.1007/978-3-319-64698-5_40

deconvolution (see [7], for a comprehensive review). However, the problem that is considered here is not blind, therefore the PSF is also used in the inversion process. Let the model be

$$g = h * \bar{f} + \eta \tag{1}$$

where $*$ is the convolution operator, h denotes a $d \times d$ convolution kernel and $\bar{f} \in \mathbb{R}^{\bar{m}\bar{n}}$ denotes the sharp image (say $F \in \mathbb{R}^{m \times n}$) that is surrounded by some synthetic pixels (to be able to include the boundary pixels into convolution) and stacked into column vector. The relations between the dimensions can be expressed as $\bar{m} = m + d - 1$ and $\bar{n} = n + d - 1$. The observed image (say $G \in \mathbb{R}^{m \times n}$) is denoted by $g \in \mathbb{R}^{mn}$, once again in vector form. Once for all, $\eta \in \mathbb{R}^{\bar{m}\bar{n}}$ stands for the additive noise.

Equation (1) can also be expressed as the multiplication of a linear sensing operator $H \in \mathbb{R}^{mn \times \bar{m}\bar{n}}$, and the padded image $\bar{f}$, as follows:

$$g = H\bar{f} + \eta \tag{2}$$

When the problem is modelled as in Eq. (2), solving it is so hard that cannot be carried out in a sufficient amount of time. This computational drawback is caused due to H being a block-Toeplitz-Toeplitz-block (BTTB) matrix, whose diagonalization (i.e. finding a diagonal basis in a transform domain), therefore inversion, cannot easily be achieved. Fortunately, one can modify this linear convolution model to be circular, such that H is block-circulant-circulant-block (BCCB) matrix whose diagonalization can easily be achieved by fast Fourier transform (FFT). The primitive approach is cropping the central region of F, then getting its size back to the original size $m \times n$, by periodically extending it in both directions. This approach is able to get over the mentioned drawback, but yields so-called ringing artifacts. Another approach suggests to extend F (periodically in both directions again) to get $\bar{f}$, but this time G should also be extended with a boundary zone of width $(d - 1)/2$. This boundary zone is mostly approximated by imposing synthetic BCs as in [14], however it also causes artifacts on the reconstructed image. The work [22] suggests estimating the unknown boundary zone, instead of imposing it. They propose a framework which alternately recovers the image and estimates the padded boundary zone through the current image. We also will make use of this framework, in order to cope with so-called unknown BCs.

The model given in Eq. (2) brings forth an ill-posed problem that requires an adequate regularization (Tikhonov-like [24], entropy based, wavelet based, total variation (TV) based, etc.). TV-based regularization is a popular variational technique that is used to solve image restoration problems. The first imaging model which utilizes TV as a regularization criteria was introduced by Rudin, Osher and Fatemi in [21]. The model was applied to stabilize denoising problem and gained attention for its edge-preserving nature. It was later generalized to be used for deconvolution [20]. Thereafter, many TV-based approaches have been proposed to tackle with deconvolution problem, as many other imaging inverse problems. Those approaches may vary from each other in terms of

(i) the deconvolution model that is considered (blind [9,15] or not [1,2]),
(ii) the way that they handle the BCs (the primitive approach mentioned above
[17], using the primitive approach but preprocessing the borders of the observation [19], surrounding the blurred image with sythetic boundary zone during
the inversion process [14], surrounding the blurred image with a boundary zone
that is estimated during the inversion process [22]), (iii) the definition of TV
regularizer (isotropic [9,20] or anisotropic – AnTV [11], and the anisotropy of
TV can further be diversified to its non-convex alternatives, L^p where $p \in (0,1)$
[10], weighted $L^1 - \alpha L^2$ [16], etc.), (v) the optimization tool that is used to solve
the problem (e.g., alternating direction algorithms [23,25] such as ADMM [2],
majorization–minimization approaches [6]), adaptivity of the parameters (e.g.,
a Bayesian approach to handle regularization parameter [18]).

In this paper, we contribute by solving the deconvolution problem under
directional total variation (DTV) regularization, which is quite useful when the
image to be recovered has a dominant direction inside. DTV was first introduced
in [3] as a convex alternative to AnTV. DTV's utility in noise removal was also
shown in the same paper. Later in [13], DTV is exploited to solve the reconstruction from sparse samples problem, which is often considered as a special case
of the inpainting problem. Even thought the settings of the inpainting and the
deconvolution are similar, there are two core issues in the latter one: handling
BCs and tackle with the computational load of matrix inversion/multiplication.
Here, we will treat the BCs as unknown, thus estimate them by plugging the
framework proposed in [22] to our algorithm. For the computational efficiency,
the algorithm will switch to frequency domain using FFT, when needed. We
make use of forward-backward splitting scheme (see [12]), as optimization tool,
in order to solve the DTV regularized problem. We also exploit Beck-Teboulle
fast gradient projection algorithm [4,5], thus the inversion process is accelerated.

The paper is organized as follows. In the next section, we review DTV and
forward-backward splitting method. Section 3 formulates the problem of DTV-
based image deconvolution and describes our algorithm by pluging the framework [22] to handle unknown boundaries. In Sect. 4, we present some experimental results before concluding the paper.

2   Background

Let padded observation be denoted by $\bar{g} \in \mathbb{R}^{\bar{m}\bar{n}}$. From the variational point of
view, the deconvolution problem van be formulated as a minimization problem
of the form:

$$\ddot{f} = \operatorname*{argmin}_{\bar{f}} \frac{1}{2}\left\| \bar{g} - H\bar{f} \right\|_2^2 + \lambda \Phi(\bar{f}) \tag{3}$$

by referring to the model given in Eq. (2). Here, $\ddot{f}$ denotes the optimal solution. On the other side, while the first term in the objective function encourages
the fidelity to the observed image, the second one promotes some prior properties of the exact solution, such as sparsity. Therefore, $\Phi(\cdot)$ corresponds to a

functional that is referred to as regularizer, and the regularization parameter λ is responsible to balance the contributions of those two terms to the objective function.

In TV-based regularization, $\Phi(\cdot)$ corresponds to isotropic or anisotropic TV functional, which we denote as $\|\cdot\|_{TV}$. The prior knowledge for TV regularization is having a piecewise constant (PWC) image, that is sparse in the gradient domain. Due to L^2 norm, isotropic (i.e. rotation invariant) TV measure can be viewed as a support function of the set B_2 (unit ball), and defined as follows:

$$\|\bar{f}\|_{TV} = \sum_{i,j} \|\Delta\bar{f}(i,j)\|_2 = \sum_{i,j} \sup_{t \in B_2} \langle \Delta\bar{f}(i,j), t \rangle \tag{4}$$

where $\Delta : \mathbb{R}^{\bar{m}\times\bar{n}} \to \mathbb{R}^{\bar{m}\times\bar{n}} \times \mathbb{R}^{\bar{m}\times\bar{n}}$ denotes the forward difference operator with two components as $\Delta\bar{f}(i,j) = \begin{pmatrix} \Delta_x\bar{f}(i,j) \\ \Delta_y\bar{f}(i,j) \end{pmatrix}$, where Δ_x and Δ_y standing for horizontal and vertical differences, respectively.

In the paper [3], it is suggested to use a support function of the set $E_{\alpha,\theta}$, which denotes an ellipse that is oriented along θ and has major axis length of $\alpha > 1$, while minor axis is 1. An ellipse can be formulated in terms of a rescaled and rotated unit ball as $E_{\alpha,\theta} = R_\theta \Lambda_\alpha B_2$, where $R_\theta = \begin{bmatrix} cos\theta & -sin\theta \\ sin\theta & cos\theta \end{bmatrix}$ and $\Lambda_\alpha = \begin{bmatrix} \alpha & 0 \\ 0 & 1 \end{bmatrix}$. Therefore, DTV measure can also defined as a support function of the form:

$$\|\bar{f}\|_{DTV} = \sum_{i,j} \sup_{t \in E_{\alpha,\theta}} \langle \Delta\bar{f}(i,j), t \rangle = \sum_{i,j} \sup_{t \in B_2} \langle \bar{f}(i,j), \Delta^{\mathsf{T}} R_\theta \Lambda_\alpha t \rangle \tag{5}$$

Here, $\Delta^{\mathsf{T}} : \mathbb{R}^{\bar{m}\times\bar{n}} \times \mathbb{R}^{\bar{m}\times\bar{n}} \to \mathbb{R}^{\bar{m}\times\bar{n}}$ is the adjoint of Δ operator, and known as $div(\cdot)$ operator. For more details on DTV, the reader is referred to [3].

In a DTV regularized minimization problem of the form given in Eq. (3); the second term of the objective function is not differentiable, even though the first term is differentiable. Therefore, the smooth optimization techniques cannot suffice to solve such problems. Fortunately, DTV preserves the convexity of TV functional. Hence, one can apply to the proximal splitting methods (see [12] for a review) which are typically used to solve the problems of the form $\min_{\bar{f}} J_1(\bar{f}) + J_2(\bar{f})$, where both J_1 and J_2 (therefore their summation) are convex.

Forward-backward splitting is a simple proximal splitting scheme, which suggests to iterate

$$\begin{aligned} \text{forward:} \quad & z = \bar{f} - \gamma\nabla J_1(\bar{f}) \\ \text{backward:} \quad & \ddot{f} = \operatorname*{argmin}_f \frac{1}{2\gamma}\|z - \bar{f}\|_2^2 + J_2(\bar{f}) \end{aligned} \tag{6}$$

where ∇ denotes the gradient of the differentiable function J_1, and thus "forward" step is corresponding to one-step gradient descent with step size γ. The minimization problem revealed in the "backward" step of Eq. (6) is the definition of the proximity operator of the non-differentiable function J_2 at z, abbreviated as $prox_{\gamma J_2}(z)$. This scheme will be exploited to solve the DTV regularized deconvolution problem at hand.

3   Proposed Method

As previously said, we will make use of the framework introduced in [22]. It basically suggests to estimate the boundary zone, that is padded around the observed image, explicitly. For that purpose, as in [22], we decompose and reorder $\bar{g}$, by using a permutation matrix $\mathcal{P}$, such that $\mathcal{P}\bar{g} = \begin{bmatrix} \bar{g}_K \\ \bar{g}_U \end{bmatrix}$, where K is the set of inner pixels corresponding to the observed (known) ones, and B is the set of padded (unknown) boundary pixels to be estimated. By the subscripts, we denote the domain of the function, e.g. $dom(\bar{g}_K) = K$. Therefore, the considered deconvolution problem of the form $\bar{g} = H\bar{f} + \eta$ (recall that the observation is surrounded with a boundary zone) requires to estimate not only $\bar{f}$, but also $\bar{g}_U$. In this case, our goal is minimizing the objective function, that is expressed as a combination of the data fidelity term and the DTV regularizer, with respect to both of $\bar{f}$ and $\bar{g}_U$, as follows:

$$\begin{bmatrix} \ddot{f} \\ \bar{g}_U \end{bmatrix} \in \underset{\bar{f},\bar{g}_U}{\operatorname{argmin}} \frac{1}{2} \left\| \begin{bmatrix} \bar{g}_K \\ \bar{g}_U \end{bmatrix} - \mathcal{P}H\bar{f} \right\|_2^2 + \lambda \|\bar{f}\|_{DTV} \tag{7}$$

By pursuing the framework [22], the minimizers of Eq. (7) can separately be estimated by alternating sequences of the following minimization problems:

$$(i) \quad \bar{f}^{(k+1)} \in \underset{\bar{f}}{\operatorname{argmin}} \frac{1}{2} \left\| \begin{bmatrix} \bar{g}_K \\ \bar{g}_U^{(k)} \end{bmatrix} - \mathcal{P}H\bar{f} \right\|_2^2 + \lambda \|\bar{f}\|_{DTV}$$

$$(ii) \quad \bar{g}_U^{(k+1)} \in \underset{\bar{g}_U}{\operatorname{argmin}} \left\| \bar{g}_U - (\mathcal{P}H\bar{f}^{(k+1)})_U \right\|_2^2 \tag{8}$$

where superscript k denotes the iteration number. The first equation of Eq. (8) estimates the image $\bar{f}^{(k+1)}$ given $\bar{g}$, currently surrounded by $\bar{g}_U^{(k)}$, and the kernel h, while the second equation is estimating $\bar{g}_U^{(k+1)}$ given $\bar{f}^{(k+1)}$ and the kernel.

Forward-backward splitting is a suitable tool for the first equation of Eq. (8), since both terms still demonstrates convexity, while only the first one is differentiable (with a τ-Lipschitz continuous gradient). When the "forward" step of Eq. (6) is realized on the data fidelity term of Eq. (8) - (i), one obtains

$$z^{(k)} = \bar{f}^{(k)} - \gamma H^\mathsf{T}\mathcal{P}^\mathsf{T}(\mathcal{P}H\bar{f}^{(k)} - \mathcal{P}\bar{g}^{(k)}) = \bar{f}^{(k)} - \gamma H^\mathsf{T}(H\bar{f}^{(k)} - \bar{g}^{(k)}) \tag{9}$$

Here the products like $H\bar{f}^{(k)}$ can efficiently be computed by means of FFT. Let $\mathcal{F}(\cdot)$ denote mapping from spatial to frequency domain using FFT, and $\mathcal{F}^{-1}(\cdot)$ is the inverse of it, the sequence of equations given in Eq. (10) is computed as

$$z^{(k)} = \bar{f}^{(k)} - \gamma\mathcal{F}^{-1}(\widehat{H}^*\mathcal{F}(\mathcal{F}^{-1}(\widehat{H}\widehat{f}^{(k)}) - \bar{g}^{(k)})) \tag{10}$$

where $\widehat{H} = \mathcal{F}(H)$ and $\widehat{f} = \mathcal{F}(\bar{f})$ are also used to denote the Fourier transforms of H and $\bar{f}$, for the sake of clarity. The superscript $*$ denotes complex conjugation. All products involving the operators H and H^{T} are performed as element-wise multiplications in the frequency domain. Also, recall that this multiplication results in circular convolution.

To continue from the "backward" step, let us first define DTV as

$$\|\bar{f}\|_{DTV} = \sup_{u(i,j)\in B_2} \langle \bar{f}, \Delta^{\mathsf{T}} R_\theta \Lambda_\alpha u \rangle \tag{11}$$

where u is a vector field consisting of the components $u(i,j) = [u_1(i,j), u_2(i,j)]^{\mathsf{T}}$. By plugging Eq. (11) into DTV's proximity operator at $z^{(k)}$, one obtains

$$\bar{f}^{(k+1)} \in \underset{\bar{f}}{\operatorname{argmin}} \frac{1}{2\gamma}\|\bar{f} - z^{(k)}\|_2^2 + \lambda \sup_{u(i,j)\in B_2} \langle \bar{f}, \Delta^{\mathsf{T}} R_\theta \Lambda_\alpha u \rangle \tag{12}$$

which can be expressed as a saddle point problem of the form:

$$\max_{u(i,j)\in B_2} \min_{\bar{f}\in\mathbb{R}^{\bar{m}\bar{n}}} \|\bar{f} - z^{(k)}\|_2^2 + 2\gamma\lambda\langle \bar{f}, \Delta^{\mathsf{T}} \Pi^{\mathsf{T}} u \rangle \tag{13}$$

where $\Pi^{\mathsf{T}} = R_\theta \Lambda_\alpha$ is used for the sake of simplicity.

The optimum solution of the inner minimization problem of Eq. (13) is equal to the solution of the following problem, which is derived by removing the constants that will not affect the minimizer.

$$\ddot{f} = \underset{\bar{f}}{\operatorname{argmin}}\|\bar{f} - (z^{(k)} - \lambda\gamma\Delta^{\mathsf{T}} \Pi^{\mathsf{T}} u)\|_2^2 \tag{14}$$

Here, the minimizer $\ddot{f} = z^{(k)} - \lambda\gamma\Delta^{\mathsf{T}} \Pi^{\mathsf{T}} u$. By plugging $\ddot{f}$ into Eq. (13), and converting it from maximization to the following minimization problem:

$$\ddot{u} \in \underset{u(i,j)\in B_2}{\operatorname{argmin}} \|z^{(k)} - \lambda\gamma\Delta^{\mathsf{T}} \Pi^{\mathsf{T}} u\|_2^2 \tag{15}$$

which gives us the dual problem of the primal Eq. (13), and allows us to use Chambolle's gradient projection technique (see [8]) which generates a decoupled sequence of gradient descent and projection onto B_2 operations, as follows:

$$u^{(j+1)} = P_{B_2}(u^{(j)} - \beta\nabla J(u^{(j)})) \tag{16}$$

Here, $\nabla J(u) = -2\lambda\gamma\Pi\Delta(z^{(k)} - \lambda\gamma\Delta^{\mathsf{T}} \Pi^{\mathsf{T}} u)$, where $J(\cdot)$ stands for the objective function to be minimized in Eq. (15). Also, $P_{B_2}(\cdot)$ is responsible for the projection[1] and β denotes the step size. β was found as $1/16\lambda^2\gamma^2(\alpha^2+1)$ in our previous work [13], by using the benefit of $\nabla J(u)$ being a τ-Lipschitz gradient. This expression is directly plugged in the proposed algorithm, as well.

Up to this point, we discoursed on the solution of equation (i) in Eq. (8), whose goal was to recover an intermediate estimation of $\bar{f}^{(k)}$. Once a $\bar{f}^{(k)}$ is

---

[1] $P_{B_2}(u) = u(i,j)/\max\{1, \|u(i,j)\|\}$ for all pairs of (i,j).

Algorithm 1. Proposed DTV Deconvolution using Framework [22]

1: $\epsilon \in (0, min\{1, 1/\tau\}), \quad \gamma \in [\epsilon, 2/\tau - \epsilon], \quad \lambda \in [\epsilon, 1]$ [12]

2: $\bar{f}^{(k=0)} := \bar{g}, \quad t^{(k=1)} := 1$

3: $\bar{v} \leftarrow \mathcal{F}^{-1}((\widehat{H}^*\widehat{H})^{-1}\widehat{H}^*\widehat{g})$

4: $\bar{g}_U^{(k=0)} \leftarrow (\mathcal{P}\bar{v})_U$

5: **while** $Eq.\,(7)$ is not $\epsilon - converged$ **do**

6: $\quad z^{(k)} = \bar{f}^{(k)} - \gamma\mathcal{F}^{-1}(\widehat{H}^*\mathcal{F}(\mathcal{F}^{-1}(\widehat{H}\widehat{f}^{(k)}) - \bar{g}^{(k)}))$

7: $\quad u^{(j=1)} := [0, 0], \quad s^{(j=1)} := 1$

8: $\quad$ **while** $Eq.\,(12)$ is not $\epsilon - converged$ **do**

9: $\quad\quad u^{(j)} \leftarrow P_{B_2}\Big(u^{(j)} + (2\lambda\gamma\beta)^{-1}\big(\Pi\Delta(z^{(k)} - \lambda\gamma\Delta^T\Pi^T u^{(j)})\big)\Big)$

10: $\quad\quad (u^{(j+1)}, s^{(j+1)}) \leftarrow$ FISTA $(s^{(j)}, u^{(j)}, u^{(j-1)})$

11: $\quad$ **end while**

12: $\quad \bar{f}^{(k)} \leftarrow z^{(k)} - \lambda\gamma div(\Pi^T u^{(j)})$

13: $\quad \bar{g}_U^{(k)} \leftarrow (\mathcal{F}^{-1}(\widehat{\mathcal{P}}\widehat{H}\widehat{f}^{(k)}))_U$

14: $\quad (t^{(k+1)}, \bar{f}^{(k+1)}) \leftarrow$ FISTA $(t^{(k)}, \bar{f}^{(k)}, \bar{f}^{(k-1)})$

15: **end while**

---

found, an intermediate boundary region $\bar{g}_U^{(k)}$ is estimated as the minimizer of equation (ii) in Eq. (8), which is a quadratic function. After each "backward" step, $\bar{g}_U^{(k)}$ can easily be computed as

$$\bar{g}_U^{(k)} = (\mathcal{P}H\bar{f}^{(k)})_U \tag{17}$$

which can be implemented in a computationally efficient way using FFT.

The pseudocode of the overall algorithm is given with the title of Algorithm 1. The equation used in line (3) defines the FFT-based least squares solution of the deconvolution model given in Eq. (2). Its result is then used to initialize $\bar{g}_U$ in line (4). The algorithm is also calling a subroutine named FISTA that stands for fast iterative shrinkage-thresholding algorithm proposed in [5], and is used to accelerate the rate of convergence. We attached the pseudocode of FISTA to Appendix A

On Estimating θ:

In case of not having a prior knowledge on θ, it is possible to capture the dominant direction by computing DTV measures of the blurred image for a range of θ values (while keeping $\alpha > 1$ constant), and picking the one which yields the smallest DTV. In the experiments that will be presented in the next section, we estimated θ from the observed image, as discussed here.

4   Experimental Evaluation

In this section, we report the experiments which are carried out on two images consisting of directional dominance inside. Those images are given in Figs. 1(a) and 2(a), and have the sizes of 609×609 and 553×553, respectively. We compare the empirical performance of the proposed method, that will be referred to as DTV-proposed (or DTV-pr. in short), with the TV-based deconvolution algorithm recently proposed in [22] together with the framework that we exploited. We will be referring to that algorithm as TV-AD during the experiments. We also include the results taken by using Algorithm 1 to perform isotropic TV-based regularization ($\alpha = 1$, TV-pr. will be used to denote) to be able to observe how close the results we get are to what TV-AD gets, and fairly compare the computational efficiencies of both frameworks.

We use structural similarity index (SSIM) [26] and peak signal-to-noise ration (PSNR) metrics to measure the reconstruction qualities. For each algorithm, we searched for the best average λ which is found as $3e - 5$ for TV-AD, $5e - 5$ for TV regularization with proposed layout, and $2e - 5$ for the proposed DTV-based algorithm. For the DTV, α is hand tuned to give the best PSNR. On the other

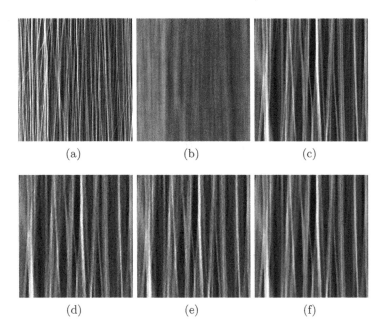

(a)   (b)   (c)

(d)   (e)   (f)

Fig. 1. Deblurring of (a) the original aligned nanofibers SEM image from (b) the observed image that is convolved using *motion* blur of length 30 and polluted by BSNR of 30 dB. (c) The detail image cropped from the original image is provided for the comparison. In the last row, the detailed cut views of the resulting (d) TV-AD based, (e) TV-based, and (f) DTV-based ($\alpha = 15$) deconvolution algorithms are given. The [SSIM, PSNR] values are (d) [0.79, 22.54], (e) [0.77, 21.89], (f) [0.82, 23.15]; and the run times are (d) 110.37 s, (e) 62.21 s, (f) 151.27 s, respectively.

hand, we selected θ by searching over $[-\pi/2, \pi/2]$ to find the one which produces the minimum DTV-norm given in Eq. (5) and applied on the observed image. The implementations are done in MATLAB, except the TV-AD method which has already been made available by the authors. The experiments are run on a 3.40 GHz Core 8 computer with 4 GB of memory.

In Figs. 1 and 2, the images were blurred by using linear *motion* kernel of length 30 and *Gaussian* kernel of size 13, respectively. They are also polluted by BSNR (Blurred Signal-to-Noise Ratio) of 30 dB and 32 dB. Also, in Table 1, the average performances of the algorithms, for various sizes of *motion blur* kernels applied on Fig. 1, and *Gaussian blur* kernels applied on Fig. 2 are tested for three different noise levels, including none.

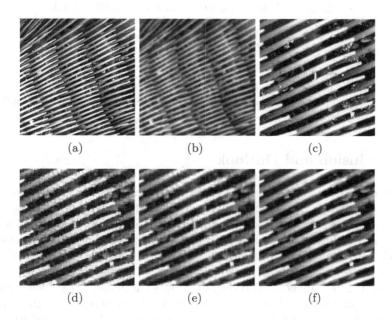

(a)   (b)   (c)

(d)   (e)   (f)

Fig. 2. Deblurring of (a) the original pollen series SEM image from (b) the observed image that is convolved using *Gaussian* blur of size 13 and polluted by BSNR of 32 dB. (c) The detail image cropped from the original image is provided for the comparison. In the last row, the detailed cut views of the resulting (d) TV-AD based, (e) TV-based, and (f) DTV-based ($\alpha = 9$) deconvolution algorithms are given. The [SSIM, PSNR] values are (d) [0.68, 19.45], (e) [0.72, 20.74], (f) [0.76, 21.61]; and the run times are (d) 46.99 s, (e) 36.06 s, (f) 65.32 s, respectively.

Given the experiments, when the observed image was noisy, TV-AD and TV-pr. produced similar results, but TV-pr. converged faster than TV-AD. On the other side, the images restored using DTV-pr. are distinguishably better than the ones obtained using TV-based regularizations, when the noise is the

Table 1. Comparison of TV-AD, TV-proposed, and DTV-proposed algorithms in terms of the reconstruction qualities in PSNR, and run times in seconds. Each entry is computed by averaging each run of the experiments performed by using various length $\in \{18, 21, 24, 27, 30\}$ *motion blur* on Fig. 1(a), and various size $\in \{9 \times 9, 11 \times 11, 13 \times 13, 15 \times 15, 17 \times 17\}$ *Gaussian blur* on Fig. 2(a), respectively corresponding to the first and last three rows.

BSNR	Average PSNR			Average run time		
	TV-AD	TV-pr.	DTV-pr.	TV-AD	TV-pr.	DTV-pr.
∞	25.76	24.31	25.28	91.54	65.07	101.06
35	24.81	23.93	25.05	89.12	63.91	121.58
30	22.84	22.33	24.58	90.12	46.79	136.77
∞	23.23	22.91	22.89	47.59	51.60	54.46
35	20.96	21.14	22.11	46.36	36.88	52.88
30	17.73	19.79	21.19	47.06	37.07	64.15

case, once again. However, when the blurred images are noise free, TV-AD and DTV-pr. are both competitive.

5   Conclusion and Outlook

We have devised an algorithm to solve deconvolution with unknown boundaries problem under DTV regularization. We have used forward-backward splitting scheme as the optimization tool and applied to FISTA as in [4] to speed up the algorithm. The idea behind the recently proposed framework [22] is exploited to tackle with the unknown boundary conditions, and FFT is used for efficient inversion and products. Experiments show that, if an appropriate α is selected and if the blurred image is also degraded by some noise, DTV yields images with better quality than the ones restored by TV-based schemes, at the expense of computational load which increases proportional to the α. As future work, we plan to apply the proposed approach to hybrid problems, such as inpainting while deblurring, and also extend it for blind deconvolution. Additionally, our ongoing work involves the extension of our approach, such that it can also be used for the images without directional dominance.

Acknowledgements. This work was supported by The Scientific and Technological Research Council of Turkey (TUBITAK) under 115E285.

Appendix A

Algorithm 2. FISTA [5]

1: **function:** $\left(f^{(next)}, t^{(next)}\right) = \text{FISTA}\left(t, f^{(cur)}, f^{(prev)}\right)$

2:    $t^{(next)} \leftarrow \left(1 + \sqrt{1 + 4t^2}\right)/2$

3:    $f^{(next)} \leftarrow f^{(cur)} + \left(\frac{t-1}{t^{(next)}}\right)\left(f^{(cur)} - f^{(prev)}\right)$

4: **end**

References

1. Afonso, M.V., Bioucas-Dias, J.M., Figueiredo, M.A.: Fast image recovery using variable splitting and constrained optimization. IEEE Trans. Image Process. **19**(9), 2345–2356 (2010)

2. Almeida, M.S., Figueiredo, M.: Deconvolving images with unknown boundaries using the alternating direction method of multipliers. IEEE Trans. Image Process. **22**(8), 3074–3086 (2013)

3. Bayram, I., Kamasak, M.E.: Directional total variation. IEEE Signal Process. Lett. **19**(12), 781–784 (2012)

4. Beck, A., Teboulle, M.: Fast gradient-based algorithms for constrained total variation image denoising and deblurring problems. IEEE Trans. Image Process. **18**(11), 2419–2434 (2009)

5. Beck, A., Teboulle, M.: A fast iterative shrinkage-thresholding algorithm for linear inverse problems. SIAM J. Imaging Sci. **2**(1), 183–202 (2009)

6. Bioucas-Dias, J.M., Figueiredo, M.A., Oliveira, J.P.: Total variation-based image deconvolution: a majorization-minimization approach. In: 2006 IEEE International Conference on Acoustics Speech and Signal Processing Proceedings, vol. 2, p. II. IEEE (2006)

7. Campisi, P., Egiazarian, K.: Blind Image Deconvolution: Theory and Applications. CRC Press, Boca Raton (2016)

8. Chambolle, A.: An algorithm for total variation minimization and applications. J. Math. Imaging Vis. **20**(1–2), 89–97 (2004)

9. Chan, T.F., Wong, C.K.: Total variation blind deconvolution. IEEE Trans. Image Process. **7**(3), 370–375 (1998)

10. Chartrand, R.: Exact reconstruction of sparse signals via nonconvex minimization. IEEE Signal Process. Lett. **14**(10), 707–710 (2007)

11. Choksi, R., van Gennip, Y., Oberman, A.: Anisotropic total variation regularized ℓ 1-approximation and denoising/deblurring of d bar codes. arXiv preprint arXiv:1007.1035 (2010)

12. Combettes, P.L., Pesquet, J.C.: Proximal splitting methods in signal processing. In: Bauschke, H., Burachik, R., Combettes, P., Elser, V., Luke, D., Wolkowicz, H. (eds.) Fixed-Point Algorithms for Inverse Problems in Science and Engineering, vol. 49, pp. 185–212. Springer, New York (2011)

13. Demircan-Tureyen, E., Kamasak, M.E.: Directional total variation based reconstruction from highly sparse subset of pixels (2017, submitted)
14. Fan, Y.W.D., Nagy, J.G.: Synthetic boundary conditions for image deblurring. Linear Algebra Appl. **434**(11), 2244–2268 (2011)
15. He, L., Marquina, A., Osher, S.J.: Blind deconvolution using tv regularization and bregman iteration. Int. J. Imaging Syst. Technol. **15**(1), 74–83 (2005)
16. Lou, Y., Zeng, T., Osher, S., Xin, J.: A weighted difference of anisotropic and isotropic total variation model for image processing. SIAM J. Imaging Sci. **8**(3), 1798–1823 (2015)
17. Ng, M.K., Chan, R.H., Tang, W.C.: A fast algorithm for deblurring models with neumann boundary conditions. SIAM J. Sci. Comput. **21**(3), 851–866 (1999)
18. Oliveira, J.P., Bioucas-Dias, J.M., Figueiredo, M.A.: Adaptive total variation image deblurring: a majorization-minimization approach. Signal Process. **89**(9), 1683–1693 (2009)
19. Reeves, S.J.: Fast image restoration without boundary artifacts. IEEE Trans. Image Process. **14**(10), 1448–1453 (2005)
20. Rudin, L.I., Osher, S.: Total variation based image restoration with free local constraints. In: IEEE International Conference on Image Processing, Proceedings ICIP 1994, vol. 1, pp. 31–35. IEEE (1994)
21. Rudin, L.I., Osher, S., Fatemi, E.: Nonlinear total variation based noise removal algorithms. Physica D Nonlinear Phenom. **60**(1), 259–268 (1992)
22. Simões, M., Almeida, L.B., Bioucas-Dias, J., Chanussot, J.: A framework for fast image deconvolution with incomplete observations. IEEE Trans. Image Process. **25**(11), 5266–5280 (2016)
23. Tao, M., Yang, J., He, B.: Alternating direction algorithms for total variation deconvolution in image reconstruction. TR0918, Department of Mathematics, Nanjing University (2009)
24. Tikhonov, A.N., Arsenin, V.I., John, F.: Solutions of Ill-Posed Problems, vol. 14. Winston, Washington DC (1977)
25. Wang, Y., Yang, J., Yin, W., Zhang, Y.: A new alternating minimization algorithm for total variation image reconstruction. SIAM J. Imaging Sci. **1**(3), 248–272 (2008)
26. Wang, Z., Bovik, A.C., Sheikh, H.R., Simoncelli, E.P.: Image quality assessment: from error visibility to structural similarity. IEEE Trans. Image Process. **13**(4), 600–612 (2004)

Author Index

Printed in the United States
By Bookmasters